The Fokker-Planck Equation

T0234601

Springer
Berlin
Heidelberg
New York
Barcelona
Budapest
Hong Kong
London
Milan
Paris
Singapore
Tokyo

Springer Series in Synergetics

Editor: Hermann Haken

An ever increasing number of scientific disciplines deal with complex systems. These are systems that are composed of many parts which interact with one another in a more or less complicated manner. One of the most striking features of many such systems is their ability to spontaneously form spatial or temporal structures. A great variety of these structures are found, in both the inanimate and the living world. In the inanimate world of physics and chemistry, examples include the growth of crystals, coherent oscillations of laser light, and the spiral structures formed in fluids and chemical reactions. In biology we encounter the growth of plants and animals (morphogenesis) and the evolution of species. In medicine we observe, for instance, the electromagnetic activity of the brain with its pronounced spatio-temporal structures. Psychology deals with characteristic features of human behavior ranging from simple pattern recognition tasks to complex patterns of social behavior. Examples from sociology include the formation of public opinion and cooperation or competition between social groups.

In recent decades, it has become increasingly evident that all these seemingly quite different kinds of structure formation have a number of important features in common. The task of studying analogies as well as differences between structure formation in these different fields has proved to be an ambitious but highly rewarding endeavor. The Springer Series in Synergetics provides a forum for interdisciplinary research and discussions on this fascinating new scientific challenge. It deals with both experimental and theoretical aspects. The scientific community and the interested layman are becoming ever more conscious of concepts such as self-organization, instabilities, deterministic chaos, nonlinearity, dynamical systems, stochastic processes, and complexity. All of these concepts are facets of a field that tackles complex systems, namely synergetics. Students, research workers, university teachers, and interested laymen can find the details and latest developments in the Springer Series in Synergetics, which publishes textbooks, monographs and, occasionally, proceedings. As witnessed by the previously published volumes, this series has always been at the forefront of modern research in the above mentioned fields. It includes textbooks on all aspects of this rapidly growing field, books which provide a sound basis for the study of complex systems.

A selection of volumes in the Springer Series in Synergetics:

1 **Synergetics** An Introduction
3rd Edition By H. Haken

8 **Stochastic Nonlinear Systems**
Editors: L. Arnold, R. Lefever

13 **Handbook of Stochastic Methods**
2nd Edition By C. W. Gardiner

15 **Noise-Induced Transitions**
Theory and Applications in Physics, Chemistry, and Biology
By W. Horsthemke, R. Lefever

18 **The Fokker-Planck Equation**
2nd Edition By H. Risken

19 **Chemical Oscillations, Waves, and Turbulence** By Y. Kuramoto

20 **Advanced Synergetics** 2nd Edition
By H. Haken

21 **Stochastic Phenomena and Chaotic Behaviour in Complex Systems**
Editor: P. Schuster

24 **Chaos and Statistical Methods**
Editor: Y. Kuramoto

32 **Dimensions and Entropies in Chaotic Systems** Quantification of Complex Behavior 2nd Corr. Printing
Editor: G. Mayer-Kress

40 **Information and Self-Organization**
A Macroscopic Approach to Complex Systems By H. Haken

51 **Foundations of Synergetics I** Distributed Active Systems By A. S. Mikhailov

52 **Foundations of Synergetics II**
Complex Pattern 2nd Edition
By A. S. Mikhailov, A. Yu. Loskutov

54 **Quantum Signatures of Chaos**
By F. Haake

56 **Quantum Noise** By C. W. Gardiner

57 **Nonlinear Nonequilibrium Thermodynamics I** Linear and Nonlinear Fluctuation-Dissipation Theorems
By R. Stratonovich

59 **Nonlinear Nonequilibrium Thermodynamics II** Advanced Theory
By R. Stratonovich

60 **Limits of Predictability**
Editor: Yu. A. Kravtsov

68 **Intelligent Measurements, Prediction and Control** By I. Grabec, W. Sachse

69 **Predictability of Complex Dynamical Systems** By Yu. A. Kravtsov, J. B. Kadtke

H. Risken

The Fokker-Planck Equation

Methods of Solution and Applications

Second Edition

With 95 Figures

Springer

Professor Dr. Hannes Risken †
Abteilung für Theoretische Physik, Universität Ulm, Oberer Eselsberg,
D-89081 Ulm, Germany

Series Editor:
Professor Dr. Dr. h.c.mult. Hermann Haken
Institut für Theoretische Physik und Synergetik der Universität Stuttgart
D-70550 Stuttgart, Germany
and
Center for Complex Systems, Florida Atlantic University
Boca Raton, FL 33431, USA

2nd Edition 1989
3rd Printing 1996

ISSN 0172-7389
ISBN 3-540-61530-X Study Edition
Springer-Verlag Berlin Heidelberg New York

ISBN 3-540-50498-2 Second Edition
Springer-Verlag Berlin Heidelberg New York

Springer-Verlag Berlin Heidelberg New York
a member of BertelsmannSpringer Science+Business Media GmbH

Offset Printing: Mercedes-Druck, Berlin. Bookbinding: Stürtz, Würzburg
SPIN 10874401 55/3111-5 4 3 - Printed on acid-free paper

Foreword

One of the central problems synergetics is concerned with consists in the study of macroscopic qualitative changes of systems belonging to various disciplines such as physics, chemistry, or electrical engineering. When such transitions from one state to another take place, fluctuations, i.e., random processes, may play an important role.

Over the past decades it has turned out that the Fokker-Planck equation provides a powerful tool with which the effects of fluctuations close to transition points can be adequately treated and that the approaches based on the Fokker-Planck equation are superior to other approaches, e.g., based on Langevin equations. Quite generally, the Fokker-Planck equation plays an important role in problems which involve noise, e.g., in electrical circuits.

For these reasons I am sure that this book will find a broad audience. It provides the reader with a sound basis for the study of the Fokker-Planck equation and gives an excellent survey of the methods of its solution. The author of this book, Hannes Risken, has made substantial contributions to the development and application of such methods, e.g., to laser physics, diffusion in periodic potentials, and other problems. Therefore this book is written by an experienced practitioner, who has had in mind explicit applications to important problems in the natural sciences and electrical engineering.

This book may be seen in the context of the book by C. W. Gardiner, "Handbook of Stochastic Methods", in this series which gives a broad and detailed overview of stochastic processes, and of the book by W. Horsthemke and R. Lefever, "Noise-induced Transitions", which treats a problem of particular current interest, namely, multiplicative noise.

Readers who are interested in learning more about the connection between the Fokker-Planck equation and other approaches within the frame of synergetics are referred to my introductory text "Synergetics. An Introduction".

H. Haken

Preface to the Second Edition

In this second edition various misprints of the first edition have been corrected; otherwise no changes have been made. Furthermore a supplement has been added to the material of the first edition. In this supplement a short review of some new developments with various recent references is given. It is my hope, that with the inclusion of the supplement this paperback edition will keep the book up to date in this fast growing field.

I wish to thank my colleague Dr. H.D. Vollmer and my coworkers Dipl. Phys. Th. Leiber and Dipl. Phys. K. Vogel for their help in preparing the supplement. Furthermore I wish to thank all of them and Dr. P. Jung, Miss B. Oder and Dr. K. Voigtländer for pointing out to me various misprints in the first edition. I am also grateful to Mrs. I. Gruhler and Mrs. B. Lossa for their secretarial assistance in preparing this second edition.

Ulm, December 1988 *H. Risken*

Preface to the First Edition

Fluctuations are a very common feature in a large number of fields. Nearly every system is subjected to complicated external or internal influences that are not fully known and that are often termed noise or fluctuations. The Fokker-Planck equation deals with those fluctuations of systems which stem from many tiny disturbances, each of which changes the variables of the system in an unpredictable but small way. The Fokker-Planck equation was first applied to the Brownian motion problem. Here the system is a small but macroscopic particle, immersed in fluid. The molecules of the fluid kick around the particle in an unpredictable way so the position of the particle fluctuates. Because of these fluctuations we do not know its position exactly, but instead we have a certain probability to find the particle in a given region. With the Fokker-Planck equation such a probability density can be determined. This equation is now used in a number of different fields in natural science, for instance in solid-state physics, quantum optics, chemical physics, theoretical biology and circuit theory.

This book deals with the derivation of the Fokker-Planck equation, the methods of solving it, and some of its applications. Whereas for some cases (e.g., linear problems, stationary problems with only one variable) the Fokker-Planck equation can be solved analytically, it is in general very difficult to obtain a solution. Various methods for solving the Fokker-Planck equation such as the simulation method, eigenfunction expansion, numerical integration, the variational method and the matrix continued-fraction method will be discussed. The last method especially, which turns out to be very effective in dealing with simple Fokker-Planck equations having two variables, is treated in detail. As far as I know it has not yet been presented in review or book form.

In the last part of the book the methods for solving the Fokker-Planck equation are applied to the statistics of a simple laser model and to Brownian motion in potentials, especially in periodic potentials. By investigating the statistical properties of laser light, I first became acquainted with the Fokker-Planck equation and I soon learned to appreciate it as a powerful tool for treating the photon statistics of lasers and the statistics of other nonlinear systems far from thermal equilibrium.

The main emphasis in the applications is made to the problem of Brownian motion in periodic potentials. This problem occurs, for instance, in solid-state physics (Josephson tunneling junction, superionic conductor), chemical physics (infrared absorption by rotating dipoles) and electrical circuit theory (phase-locked loops). Whereas the Fokker-Planck equation for this problem was solved many years ago for the overdamped case (large friction), solutions for arbitrary friction have been obtained only recently. It will be shown that the solution of the corresponding Fokker-Planck equation for Brownian motion in periodic potentials (as well as for other potentials) can be expressed in terms of matrix continued fractions which are very suitable for computer evaluation.

The present book is based on seminar and lecture notes, prepared and presented at the University of Ulm. Hopefully this book will be useful for graduate students in physics, chemical physics and electrical engineering to get acquainted with the Fokker-Planck equation and the methods of solving it, and that some parts of it will also be profitable to the research worker in these fields.

I wish to thank Prof. H. Haken for inviting me to write this monograph for the *Springer Series in Synergetics*. As a co-worker of Prof. Haken for nearly ten years, I had the privilege to work with him and his group in a very stimulating and creative atmosphere. I also want to express my gratitude to Dr. H. Lotsch and his staff of the Springer-Verlag for their co-operation. Next I wish to thank my co-worker and colleague Dr. H. D. Vollmer. Most of my research on the Fokker-Planck equation was done in close collaboration with him. With only few exceptions he has also provided me with the numerical results presented in this book. Furthermore, he has made many suggestions for improving the manuscript. The help of Dipl. Phys. P. Jung, Dr. M. Mörsch, and Dipl. Phys. K. Voigtländer for preparing the figures and for reading the proofs is also greatly appreciated. Last but not least I wish to thank Mrs. I. Gruhler and Mrs. H. Wenning for skilfully and patiently typing and correcting the manuscript.

Ulm, February 1984 *H. Risken*

Contents

1. Introduction

A Fokker-Planck equation was first used by *Fokker* [1.1] and *Planck* [1.2] to describe the Brownian motion of particles. To become familiar with this equation we first discuss the Brownian motion of particles in its simplest form.

1.1 Brownian Motion

1.1.1 Deterministic Differential Equation

If a small particle of mass m is immersed in a fluid a friction force will act on the particle. The simplest expression for such a friction or damping force is given by Stokes' law

$$F_c = -\alpha v\,. \tag{1.1}$$

Therefore the equation of motion for the particle in the absence of additional forces reads

$$m\dot{v} + \alpha v = 0 \tag{1.2}$$

or

$$\dot{v} + \gamma v = 0\,; \qquad \gamma = \alpha/m = 1/\tau\,. \tag{1.3}$$

Thus an initial velocity $v(0)$ decreases to zero with the relaxation time $\tau = 1/\gamma$ according to

$$v(t) = v(0)\,e^{-t/\tau} = v(0)\,e^{-\gamma t}\,. \tag{1.4}$$

The physics behind the friction is that the molecules of the fluid collide with the particle. The momentum of the particle is transferred to the molecules of the fluid and the velocity of the particle therefore decreases to zero. The differential equation (1.3) is a deterministic equation, i.e., the velocity $v(t)$ at time t is completely determined by its initial value according to (1.4).

1.1.2 Stochastic Differential Equation

The deterministic equation (1.2) is valid only if the mass of the particle is large so that its velocity due to thermal fluctuations is negligible. From the equipartition law, the mean energy of the particle is (in one dimension)

$$\tfrac{1}{2}m\langle v^2\rangle = \tfrac{1}{2}kT\,, \tag{1.5}$$

where k is Boltzmann's constant and T is the temperature. For smaller mass m the thermal velocity $v_{\text{th}} = \sqrt{\langle v^2\rangle} = \sqrt{kT/m}$ may be observable and therefore the velocity of a "small" particle cannot be described exactly by (1.3) with the solution (1.4). If the mass of the small particle is still large compared to the mass of the molecules, one expects (1.2) to be valid approximately. Equation (1.2) must, however, be modified so that it leads to the correct thermal energy (1.5). The modification consists in adding a fluctuating force $F_{\text{f}}(t)$ on the right-hand side of (1.2), i.e., the total force of the molecules acting on the small particle is decomposed into a continuous damping force $F_{\text{c}}(t)$ and a fluctuating force $F_{\text{f}}(t)$ according to [1.3].

$$F(t) = F_{\text{c}}(t) + F_{\text{f}}(t) = -\alpha v(t) + F_{\text{f}}(t)\,. \tag{1.6}$$

This force $F_{\text{f}}(t)$ is a stochastic or random force, the properties of which are given only in the average.

We now want to discuss why a stochastic force occurs. If we were to treat the problem exactly, we should have to solve the coupled equations of motion for all the molecules of the fluid and for the small particle, and no stochastic force would occur. Because of the large number of molecules in the fluid (the number is of the order 10^{23}), however, we cannot generally solve these coupled equations. Furthermore, since we do not know the initial values of all the molecules of the fluid, we cannot calculate the exact motion of the small particle immersed in the fluid. If we were to use another system (particle and fluid) identical to the first except for the initial values of the fluid, a different motion of the small particle results. As usually done in thermodynamics, we consider an ensemble of such systems (Gibbs ensemble). The force $F_{\text{f}}(t)$ then varies from system to system and the only thing we can do is to consider averages of this force for the ensemble.

Inserting (1.6) into (1.2) and dividing by the mass we get the equation of motion

$$\dot{v} + \gamma v = \Gamma(t)\,. \tag{1.7}$$

Here we have introduced the fluctuating force per unit mass

$$\Gamma(t) = F_{\text{f}}(t)/m\,, \tag{1.8}$$

which is called the Langevin force. Equation (1.7) is called a stochastic differential equation because it contains the stochastic force $\Gamma(t)$.

To proceed further one has to know some properties of this Langevin force $\Gamma(t)$. First we assume that its average over the ensemble should be zero

$$\langle \Gamma(t) \rangle = 0 , \tag{1.9}$$

because the equation of motion of the average velocity $\langle v(t) \rangle$ should be given by (1.2). If we multiply two Langevin forces at different times we assume that the average value is zero for time differences $t' - t$ which are larger than the duration time τ_0 of a collision, i.e.,

$$\langle \Gamma(t) \Gamma(t') \rangle = 0 \quad \text{for} \quad |t - t'| \geqq \tau_0 . \tag{1.10}$$

This assumption seems to be reasonable, because the collisions of different molecules of the fluid with the small particle are approximately independent. Usually, the duration time τ_0 of a collision is much smaller than the relaxation time $\tau = 1/\gamma$ of the velocity of the small particle. We may therefore take the limit $\tau_0 \to 0$ as a reasonable approximation, giving

$$\langle \Gamma(t) \Gamma(t') \rangle = q \delta(t - t') . \tag{1.11}$$

The δ function appears because otherwise the average energy of the small particle cannot be finite as it should be according to the equipartition law (1.5). This will be discussed in detail in Sect. 3.1, where it is furthermore shown that the noise strength q of the Langevin force is then given by

$$q = 2\gamma kT/m . \tag{1.12}$$

To determine higher correlations like $\langle v(t_1) v(t_2) \dots v(t_n) \rangle$ higher correlations of $\Gamma(t)$ must be known. One usually assumes that the $\Gamma(t)$ have a Gaussian distribution with δ correlation (Chap. 3). By integrating the Langevin equation (1.7) and by using (1.9, 11, 12) we can calculate the diffusion constant (Chap. 3). As is well known, this diffusion constant was first obtained by *Einstein* [1.4], who initiated the term theory of Brownian motion.

A noise force with the δ correlation (1.11) is called white noise, because the spectral distribution (Sect. 2.4.3) which is given by the Fourier transform of (1.11) is then independent of the frequency ω. If the stochastic forces $\Gamma(t)$ are not δ correlated, i.e., if the spectral density depends on the frequency, one uses the term colored noise.

1.1.3 Equation of Motion for the Distribution Function

Because in (1.7) $\Gamma(t)$ varies from system to system in the ensemble, i.e., it is a stochastic quantity, the velocity will also vary from system to system, i.e., it will become a stochastic quantity, too. We therefore may ask for the probability to find the velocity in the interval $(v, v + \mathrm{d}v)$, or in other words we may ask for the number of systems of the ensemble whose velocities are in the interval $(v, v+dv)$ divided by the total number of systems in the ensemble. Because v is a continuous variable we may ask for the probability density $W(v)$, also often called probability distribution in the physical literature. The probability density times the length of the interval dv is then the probability of finding the particle in the interval $(v, v+dv)$. This distribution function depends on time t and the initial distribution. The equation of motion for the distribution function $W(v,t)$ (Chap. 4) is given by

$$\frac{\partial W}{\partial t} = \gamma \frac{\partial (v W)}{\partial v} + \gamma \frac{kT}{m} \frac{\partial^2 W}{\partial v^2} . \tag{1.13}$$

Equation (1.13) is one of the simplest Fokker-Planck equations. By solving (1.13) starting with $W(v,0)$ for $t=0$ and subject to the appropriate boundary conditions, one obtains the distribution function $W(v,t)$ for all later times. Once we have found $W(v,t)$, any averaged value of the velocity can be calculated by integration [$h(v)$ arbitrary function of v]

$$\langle h(v(t)) \rangle = \int_{-\infty}^{\infty} h(v)\, W(v,t)\, dv . \tag{1.14}$$

As shown in Sect. 4.7.2, averaged values for multi-time functions may also, for certain processes, be evaluated by use of appropriate solutions of (1.13).

1.2 Fokker-Planck Equation

In this introductory chapter it is mainly discussed how a Fokker-Planck equation and some special forms of it look, how they arise and where and how one may use the Fokker-Planck equation. Many review articles and books on the Fokker-Planck equation exist [1.5–15].

1.2.1 Fokker-Planck Equation for One Variable

In Sect. 1.1 we found an equation of motion for the distribution function $W(v,t)$ for one-dimensional Brownian motion. As mentioned, it is a special Fokker-Planck equation. The general Fokker-Planck equation for one variable x has the form

$$\frac{\partial W}{\partial t} = \left[-\frac{\partial}{\partial x} D^{(1)}(x) + \frac{\partial^2}{\partial x^2} D^{(2)}(x) \right] W . \tag{1.15}$$

In (1.15) $D^{(2)}(x)>0$ is called the diffusion coefficient and $D^{(1)}(x)$ the drift coefficient. The drift and diffusion coefficients may also depend on time. Equation (1.13) is seen to be a special Fokker-Planck equation where the drift coefficient is linear and the diffusion coefficient is constant. Equation (1.15) is an equation of motion for the distribution function $W(x,t)$. Mathematically, it is a linear second-order partial differential equation of parabolic type. Roughly speaking, it is a diffusion equation with an additional first-order derivative with respect to x. In the mathematical literature, (1.15) is also called a forward Kolmogorov equation.

1.2.2 Fokker-Planck Equation for N Variables

A generalization of (1.15) to the N variables $x_1 \dots x_N$ has the form

$$\frac{\partial W}{\partial t} = \left[-\sum_{i=1}^{N} \frac{\partial}{\partial x_i} D_i^{(1)}(\{x\}) + \sum_{i,j=1}^{N} \frac{\partial^2}{\partial x_i \partial x_j} D_{ij}^{(2)}(\{x\}) \right] W . \tag{1.16}$$

The drift vector $D_i^{(1)}$ and the diffusion tensor $D_{ij}^{(2)}$ generally depend on the N variables $x_1, \dots, x_N = \{x\}$. The Fokker-Planck equation (1.16) is an equation for the distribution function $W(\{x\},t)$ of N macroscopic variables $\{x\}$. (Here x_i may be variables of different kinds for instance position and velocity.)

1.2.3 How Does a Fokker-Planck Equation Arise?

As discussed already for the Brownian motion case, the complete solution of a macroscopic system would consist in solving all the microscopic equations of the system. Because we cannot generally do this we use instead a stochastic description, i.e., we describe the system by macroscopic variables which fluctuate in a stochastic way. The Fokker-Planck equation is just an equation of motion for the distribution function of fluctuating macroscopic variables. For a deterministic treatment we neglect the fluctuations of the macroscopic variables. For the Fokker-Planck equation (1.16) this would mean that we neglect the diffusion term.

Equation (1.16) is then equivalent to the system of differential equations $(i = 1, \dots, N)$

$$\mathrm{d}x_i/\mathrm{d}t = \dot{x}_i = D_i^{(1)}(x_1, \dots, x_N) = D_i^{(1)}(\{x\}) \tag{1.17}$$

for the N macrovariables $\{x\}$. Table 1.1 gives a schematic representation of the following three stages of treating a system. A rigorous derivation of stochastic treatment should start with microscopic description. The deterministic treatment should then follow from the stochastic treatment by neglecting the fluctuations, as indicated by the big arrows. The drift and diffusion coefficients $D_i^{(1)}$ and $D_{ij}^{(2)}$ especially should be derived rigorously from the microscopic equations. Such a rigorous derivation may be very complicated or even impossible. In this case, one

Table 1.1. Three stages of treating a system

microscopic treatment		stochastic treatment		deterministic treatment
equation of motion for all microscopic variables	⇒	equation of motion for the distribution function of macro-scopic variables (or stochastic differen-tial equation)	⇒ ←	system of differential equations for macroscopic variables

⇒ rigorous derivation, ← heuristic derivation

may start with the deterministic equation and use heuristic arguments to obtain the stochastic description, as indicated by the small arrow in Table 1.1. In the heuristic treatment one usually adds some Langevin forces to the deterministic equation (1.17) and thus obtains a stochastic differential equation which is equivalent (for properly chosen Langevin forces) to a Fokker-Planck equation. The noise strength may then be determined by some other arguments, for example by use of the equipartition theorem. We thus obtain the Fokker-Planck equation (1.13) for Brownian motion of a particle as discussed in Sect. 1.1.

The Fokker-Planck equation is of course not the only equation of motion for the distribution function. Other equations like the Boltzmann equation and master equation are discussed shortly below. The Fokker-Planck equation is one of the simplest equations for continuous macroscopic variables. It usually appears for variables describing a macroscopic but small subsystem, like the position and velocity for the Brownian motion of a small particle, a current in an electrical circuit, the electrical field in a laser. If the subsystem is larger the fluctuations may then usually be neglected and thus one has a deterministic equation. In those cases, however, where the deterministic equations are not stable, a stochastic description is then necessary even for large systems.

1.2.4 Purpose of the Fokker-Planck Equation

By solving the Fokker-Planck equation one obtains distribution functions from which any averages of macroscopic variables are obtained by integration. Since the application of the Fokker-Planck equation is not restricted to systems near thermal equilibrium, we may as well apply it to systems far from thermal equilibrium, for instance, the laser. As shown in Chap. 12, the statistics of laser light may very well be treated by a Fokker-Planck equation. An ion in a superionic conductor under the influence of an additional strong external field would also be a system far from thermal equilibrium, a simple model of which will be treated by a Fokker-Planck equation in Chap. 11. The Fokker-Planck equation not only describes stationary properties, but also the dynamics of systems, if the appropriate time-dependent solution is used.

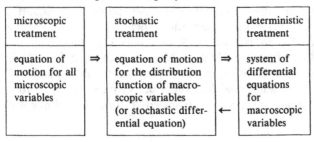

1.2.5 Solutions of the Fokker-Planck Equation

In this book we are mainly concerned with the methods for solving the Fokker-Planck equation and with its applications. Analytic solutions of the Fokker-Planck equation can be found in the following cases.

1) Linear drift vector and constant diffusion tensor. In this case, one obtains Gaussian distributions for the stationary as well as for the instationary solutions (see Sects. 3.2, 5.3, 6.5).
2) Detailed balance condition. If the drift vector and the diffusion matrix obey certain potential conditions (Sect. 6.4), the stationary solution is obtained by quadratures.
3) For a Fokker-Planck equation with one variable one also obtains the stationary solution by quadratures even if detailed balance is not valid, i.e., probability current not zero (Sect. 5.2).

In other special cases one may also find analytic solutions. Generally, however, it is difficult to obtain solutions of the Fokker-Planck equation especially if no separation of variables is possible or if the number of variables is large.

Various other methods of solution to be discussed in detail are: simulation methods (Sect. 3.6); transformation of a Fokker-Planck equation to a Schrödinger equation (Sects. 5.4, 6.3); numerical integration methods (Sects. 5.9.2, 6.6.4); analytic solutions for certain model potentials (Sect. 5.7) for a one-variable Fokker-Planck equation; matrix continued-fraction solutions for a two-variable Fokker-Planck equation (Sect. 6.6.6); and instationary solutions for time-varying small external fields (linear response, Chap. 7).

1.2.6 Kramers and Smoluchowski Equations

The *Klein-Kramers* or *Kramers* equation [1.16, 17] and the *Smoluchowski* [1.18] equation are special forms of the Fokker-Planck equation. The Kramers equation is an equation of motion for distribution functions in position and velocity space describing the Brownian motion of particles in an external field. In the one-dimensional case it has the form $[W = W(x, v, t)]$

$$\frac{\partial W}{\partial t} = \left[-\frac{\partial}{\partial x} v + \frac{\partial}{\partial v}\left(\gamma v - \frac{F(x)}{m}\right) + \frac{\gamma k T}{m} \frac{\partial^2}{\partial v^2} \right] W . \tag{1.18}$$

Here γ is the friction constant, m ist the mass of the particle, T is the temperature of the fluid, k Boltzmann's constant, and $F(x) = -mf'(x)$ is the external force where $mf(x)$ is the potential. Without any external force and x dependence (1.18) reduces to (1.13). The stochastic differential equation corresponding to (1.18) is

$$\left.\begin{aligned} &\dot{x} = v \\ &\dot{v} = -\gamma v + F(x)/m + \Gamma(t) \\ &\langle \Gamma(t')\Gamma(t)\rangle = 2\gamma(kT/m)\,\delta(t-t') . \end{aligned}\right\} \tag{1.19}$$

In the absence of the force $F(x)$ the last two equations of (1.19) reduce to (1.7, 11, 12). The first two equations of (1.19) may be written as the equation of motion

$$m\ddot{x}+m\gamma\dot{x}=F(x)+m\Gamma(t)\,. \tag{1.20}$$

Methods of solving (1.18) are treated in Chap. 10. The solutions and applications of (1.18) for a periodic potential are discussed in Chap. 11. Here we want to mention only that in the stationary state (and for suitable boundary conditions) the solution of (1.18) is given by the Boltzmann distribution (N is the normalization constant)

$$\begin{aligned} W_{\mathrm{st}}(x,v) &= N\exp[-E/(kT)] \\ E &= mv^2/2+mf(x)\,, \end{aligned} \tag{1.21}$$

as may be easily checked by insertion.

For large friction constants γ we may neglect the second derivative with respect to time in (1.20). We then obtain the stochastic differential equation

$$\dot{x}=F(x)/(m\gamma)+\Gamma(t)/\gamma \tag{1.22}$$

and the corresponding Fokker-Planck equation for the distribution function in position $[W=W(x,t)]$

$$\frac{\partial W}{\partial t}=\frac{1}{m\gamma}\left[-\frac{\partial}{\partial x}F(x)+kT\frac{\partial^2}{\partial x^2}\right]W\,. \tag{1.23}$$

This equation is called the Smoluchowski equation. The derivation of (1.23) from Kramers equation (1.18) and the higher corrections to (1.23) (inverse friction expansion) are discussed in Sect. 10.4. The Smoluchowski equation itself is treated in Chap. 5.

1.2.7 Generalizations of the Fokker-Planck Equation

Several generalizations of the Fokker-Planck equation (1.15) are in use. (For simplicity we discuss the case of one variable only.) First we consider an equation which does not stop after the second derivative with respect to x, but also contains higher derivatives. The general expansion with an infinite number of terms, i.e.,

$$\frac{\partial W}{\partial t}=\sum_{\nu=1}^{\infty}\left(-\frac{\partial}{\partial x}\right)^{\nu}D^{(\nu)}(x)\,W\,, \tag{1.24}$$

is called the *Kramers-Moyal* expansion [1.17, 19]. If x obeys a Langevin equation with Gaussian δ-correlated noise, it is shown in Sects. 3.3.2 and 4.1 that all coefficients $D^{(\nu)}$ with $\nu\geqq 3$ vanish and (1.24) then reduces to the Fokker-Planck equation (1.15). If x is a discrete variable the coefficients $D^{(\nu)}$ do not vanish generally, see Sect. 4.5 for an example. In this case we may truncate expansion (1.24) after the second term and again obtain as an approximation the Fokker-Planck equation (1.15). One may ask whether the truncation after some finite term, i.e.,

$$\frac{\partial W}{\partial t} = \sum_{\nu=1}^{N} \left(- \frac{\partial}{\partial x} \right)^{\nu} D^{(\nu)}(x)\, W; \qquad \infty > N > 2 \tag{1.25}$$

is useful. We shall see in Sect. 4.3 that for nonvanishing $D^{(N)}$ the transition probability [i.e, the solution of (1.25) with the initial distribution $\delta(x-x')$ at $t=0$] must have negative values for sufficiently small times. Because the probability density must always be positive, one therefore may think that (1.25) is of no use at all. However, a simple example in Sect. 4.6 shows that (1.25) may nevertheless be suitable to calculate the probability density quite accurately.

Another possibility to generalize (1.15) consists in taking memory effects into account. Then

$$\frac{\partial W(x,t)}{\partial t} = \int_{-\infty}^{t} \left[- \frac{\partial}{\partial x} D^{(1)}(x, t-\tau) + \frac{\partial^2}{\partial x^2} D^{(2)}(x, t-\tau) \right] W(x,\tau)\, \mathrm{d}\tau \,, \tag{1.26}$$

which we call, in association with the generalized master equation (Sect. 1.4), a generalized Fokker-Planck equation. Whereas for the Fokker-Planck equation (1.15) the distribution function is completely determined by the distribution function at $t = t_0$ (Markov process), the process described by (1.26) is determined by all earlier distributions (non-Markovian process). However, if the memory coefficients $D^{(1)}$ and $D^{(2)}$ decrease very rapidly in time, we recover the Fokker-Planck equation (1.15).

A more general equation would read

$$\partial W(x,t)/\partial t = \int_{-\infty}^{t} K(x, t-\tau)\, W(x,\tau)\, \mathrm{d}\tau \,, \tag{1.27}$$

where the memory kernel may either contain differential operators (finite or infinite order) or it may be an integral operator with respect to x or some other linear operator. An equation of the form (1.27), for instance, occurs if one tries to eliminate the velocity variable in the Kramers equation (Sect. 10.3.1).

1.3 Boltzmann Equation

The first equation of motion, which was derived for the distribution function of a dilute gas in position and velocity space, is the Boltzmann equation [1.20]. Here $f(\boldsymbol{x}, \boldsymbol{v}, t)\,\mathrm{d}^3x\,\mathrm{d}^3v$ is the number of gas molecules in the volume element $\mathrm{d}^3x\,\mathrm{d}^3v$ of the position and velocity space, also called μ space. For particles moving in an external force $\boldsymbol{F}(\boldsymbol{x})$ this equation takes the form

$$\left(\frac{\partial}{\partial t}+\boldsymbol{v}\cdot\nabla_r+\frac{\boldsymbol{F}(\boldsymbol{x})}{m}\nabla_v\right)f(\boldsymbol{x},\boldsymbol{v},t)=\left(\frac{\partial f}{\partial t}\right)_{\text{coll}}, \tag{1.28}$$

$$\left(\frac{\partial f}{\partial t}\right)_{\text{coll}}=\int d^3v_1\int d\Omega|\boldsymbol{v}-\boldsymbol{v}_1|\sigma(\boldsymbol{v},\boldsymbol{v}_1|\boldsymbol{v}',\boldsymbol{v}_1')$$
$$\times[f(\boldsymbol{x},\boldsymbol{v}',t)f(\boldsymbol{x},\boldsymbol{v}_1',t)-f(\boldsymbol{x},\boldsymbol{v},t)f(\boldsymbol{x},\boldsymbol{v}_1,t)]\,. \tag{1.29}$$

In (1.28) ∇_r and ∇_v denote the gradient with respect to position and velocity. In the collision operator (1.29) $\sigma(\boldsymbol{v},\boldsymbol{v}_1\,|\,\boldsymbol{v}',\boldsymbol{v}_1')$ is the differential scattering cross section of two colliding gas molecules with velocity $\boldsymbol{v}$ and $\boldsymbol{v}_1$ before and $\boldsymbol{v}'$ and $\boldsymbol{v}_1'$ after the collision. The space angle between $\boldsymbol{v}-\boldsymbol{v}_1$ and $\boldsymbol{v}'-\boldsymbol{v}_1'$ is denoted by Ω. Furthermore, certain symmetries for the differential cross section and the conservation laws for energy and momentum for each collision are assumed to be valid. A very high obstacle for finding a general solution of (1.28) is the nonlinearity occurring in the collision operator (1.29). For certain processes one may try to linearize the collision operator. A system in which one particle is very large compared to the others gives a linear equation for the distribution function of this particle. One may actually show that this equation reduces to the Fokker-Planck equation (1.13) or to its three-dimensional form [1.21, 22].

The complicated nonlinear collision operator (1.29) may also be approximated by some linear model operator. In the BGK model [1.23] one assumes that after each collision the velocity distribution becomes the Maxwell distribution. The BGK collision operator has the form ($1/\gamma$ is the relaxation time)

$$\left(\frac{\partial f}{\partial t}\right)_{\text{coll}}=\gamma\left[\sqrt{\frac{m}{2\pi kT}}^{\,3}\mathrm{e}^{-mv^2/(2kT)}\int f(\boldsymbol{x},\boldsymbol{v},t)\,\mathrm{d}^3v-f(\boldsymbol{x},\boldsymbol{v},t)\right]. \tag{1.30}$$

Considering only the one-dimensional case, (1.28) then reduces for the BGK model to

$$\left(\frac{\partial}{\partial t}+v\frac{\partial}{\partial x}+\frac{F}{m}\frac{\partial}{\partial v}\right)f=\boldsymbol{L}_{\text{coll}}f, \tag{1.31}$$

where the linear operator $\boldsymbol{L}_{\text{coll}}$ is given by

$$\boldsymbol{L}_{\text{coll}}=\boldsymbol{L}_{\text{coll}}^{\text{BGK}}=\gamma\left(\sqrt{\frac{m}{2\pi kT}}\,\mathrm{e}^{-mv^2/(2kT)}\int\ldots\mathrm{d}v-1\right). \tag{1.32}$$

It should be noted that the Kramers equation (1.18) may be written in the same form as (1.31), where the collision operator reads

$$L_{\text{coll}} = L_{\text{coll}}^{\text{K}} = \gamma\left(\frac{\partial}{\partial v}v + \frac{kT}{m}\frac{\partial^2}{\partial v^2}\right). \tag{1.33}$$

In the absence of an external force $F(x)$ the Maxwell distribution $\exp[-mv^2/(2kT)]$ is the stationary solution of (1.31) for both collision operators. As shown in App. 2, we may treat (1.31, 32) by the same matrix continued-fraction method used in Chap. 10 for the Kramers equation.

1.4 Master Equation

Very general linear equations for the probability density are the master equation and the generalized master equation [1.24–29]. If the variable x takes only integer values, the master equation has the form [1.24]

$$\partial W_n/\partial t = \dot{W}_n = \sum_m [w(m\to n)W_m - w(n\to m)W_n]. \tag{1.34}$$

In (1.34) W_n is the probability to find the integer value n and $w(m\to n)$ is the transition rate from m to n which must be positive. For a continuous variable the sum must be replaced by an integration, i.e., we then have

$$\partial W(x,t)/\partial t = \dot{W}(x,t) = \int [w(x'\to x)W(x',t) - w(x\to x')W(x,t)]\,dx'. \tag{1.35}$$

The probability at a later time is completely determined by the probability at time $t = t_0$; i.e., the process described by (1.34) is a Markov process. The Fokker-Planck equation (1.15) is a special form of the continuous master equation (1.35). Here the transition probability $w(x'\to x)$ is given by

$$w(x'\to x) = \left[-\frac{\partial}{\partial x}D^{(1)}(x) + \frac{\partial^2}{\partial x^2}D^{(2)}(x)\right]\delta(x-x'). \tag{1.36}$$

Inserting this expression into (1.35) leads to (1.15). Notice that in the first term on the right-hand side in (1.35) one can use $\int\delta(x-x')W(x',t)\,dx' = W(x,t)$ and that the last term on the right-hand side vanishes because of the derivative $\partial/\partial x'$.

The generalized master equation has the form [1.25–27] (for reviews, see [1.28, 29])

$$\dot{W}_n = \int_{-\infty}^{t}\left[\sum_m w(n\to m, t-\tau)W_m(\tau) - \sum_m w(m\to n, t-\tau)W_n(\tau)\right]d\tau. \tag{1.37}$$

The change of the probability not only depends on the probability at time t but also on the previous history. The previous probabilities enter with the weight function $w(n \rightarrow m, t-\tau)$ which describes the memory and so is called a memory

$m \rightarrow n$

function. An equation similar to (1.37) and its continuous analog follows from the microscopic equations by eliminating the irrelevant variables with the *Nakajima-Zwanzig* [1.26, 27] projector formalism (see also [1.28, 29] and the recent review by *Grabert* [1.30] on projection operator techniques).

2. Probability Theory

In this chapter we recapitulate some of the basic ideas and conceptions of probability theory needed to unterstand the other chapters. Though there are many text books on probability theory [2.1 – 6], a selection of basic ideas and concepts of probability theory in a simplified form may be in order for the reader not very familiar with probability theory.

2.1 Random Variable and Probability Density

We assume that there is a certain prescription how to obtain a number ξ. This prescription may consist for instance in the following experiments:

(i) tossing a coin and writing 0 for head and 1 for tail,
(ii) casting a die and counting the number of spots,
(iii) measuring the length of a rod.

We call ξ a random variable if the number ξ cannot be predicted [for instance because of lack of initial conditions and (or) of some other unknown factors]. By repeating the experiment N times (N realizations) we obtain N numbers

$$\xi_1, \xi_2, \ldots, \xi_N . \tag{2.1}$$

These numbers ξ_N may take only integer [cases (i), (ii)] or continuous [case (iii)] values.

Instead of repeating the experiment with one system N times we may also think that we have an ensemble of N identical systems and make one experiment for every system.

Whereas the numbers $\xi_1, \xi_2 \ldots$ cannot be predicted, some averages for $N \to \infty$ may be predicted and should give the same value for identical systems. The simplest average value is the mean value

$$\langle \xi \rangle = \lim_{N \to \infty} \frac{1}{N} (\xi_1 + \xi_2 + \ldots \xi_N) . \tag{2.2}$$

A general average value is

$$\langle f(\xi)\rangle = \lim_{N\to\infty} \frac{1}{N}[f(\xi_1)+\ldots+f(\xi_N)]\,, \tag{2.3}$$

where $f(\xi)$ is some arbitrary function.

Probability Density

If we choose for the function in (2.3) the shifted step function

$$f(\xi) = \Theta(x-\xi) \tag{2.4}$$

$$\Theta(x) = \begin{cases} 1 & x>0 \\ 1/2 \quad \text{for} & x=0 \\ 0 & x<0 \end{cases} \tag{2.5}$$

we obtain

$$\begin{aligned} P(\xi<x)+(1/2)P(\xi=x) &= \langle\Theta(x-\xi)\rangle \\ &= \lim_{N\to\infty}[\Theta(x-\xi_1)+\ldots+\Theta(x-\xi_N)]/N = \lim_{N\to\infty} M/N\,. \end{aligned} \tag{2.6}$$

The definition (2.6) differs from the usual one by a different weight of the probability at $\xi=x$. This is done because of our definition of the step function (2.5). [If we would have used $\Theta(x)=1$ for $x\geqq 0$ and $\Theta(x)=0$ for $x<0$ then the left hand side should be replaced by $P(\xi\leqq x)$]. For continuous processes, where the probability to find the discrete value $\xi=x$ is usually zero, both definitions agree.

In (2.6) M is the number of experiments (realizations) where $\xi\leqq x$. Thus M/N is the relative frequency where the random variable is equal to or less than x. In the limit $N\to\infty$ this relative frequency [2.1] is called the probability $P(\xi\leqq x)$ that the random variable is equal to or less than x. It follows from (2.6) that $P(\xi\leqq x)$ must be a nondecreasing function in x with $P(\xi\leqq\infty)=1$. The probability density function $W_\xi(x)$ of the random variable ξ is the derivative of P with respect to x

$$\begin{aligned} W_\xi(x) &= \frac{\mathrm{d}}{\mathrm{d}x}P(\xi\leqq x) = \frac{\mathrm{d}}{\mathrm{d}x}\langle\Theta(x-\xi)\rangle \\ &= \left\langle\frac{\mathrm{d}}{\mathrm{d}x}\Theta(x-\xi)\right\rangle = \langle\delta(x-\xi)\rangle\,. \end{aligned} \tag{2.7}$$

In (2.7) we introduced the Dirac δ function as the derivative of the step function (2.5). The probability $\mathrm{d}P$ to find the continuous stochastic variable ξ in the interval $x\leqq\xi\leqq x+\mathrm{d}x$ is given by (assuming that P is differentiable)

$$P(\xi\leqq x+\mathrm{d}x)-P(\xi\leqq x) = \frac{\mathrm{d}}{\mathrm{d}x}P(\xi\leqq x)\,\mathrm{d}x = W_\xi(x)\,\mathrm{d}x\,.$$

The probability density (2.7) is usually a smooth function for continuous random variables. For discrete random variables, P jumps at the discrete values x_n, and $W_\xi(x)$ then consists of a sum of δ functions

$$W_\xi(x) = \sum_n p_n \delta(x - x_n) . \tag{2.8}$$

In (2.8) p_n is the probability to find the discrete value x_n. By allowing δ function singularities for the probability density, we may formally treat the discrete case by the same expressions as those for the continuous case.

In the mathematical literature P is called the distribution function, whereas in the physical literature the probability density $W_\xi(x)$ is often also called the distribution function. We shall always use the probability density $W_\xi(x)$ and not P in our further considerations. For $W_\xi(x)$ we shall reserve the term distribution function.

The statistical properties of the random variable ξ are completely determined by the probability density, because any expectation value can be obtained from $W_\xi(x)$ by integration. This is seen as follows: because

$$f(\xi) = \int f(x)\, \delta(x - \xi)\, dx , \tag{2.9}$$

we get by taking averages

$$\begin{aligned} \langle f(\xi) \rangle &= \langle \int f(x)\, \delta(x - \xi)\, dx \rangle \\ &= \int f(x)\, \langle \delta(x - \xi) \rangle\, dx \\ &= \int f(x)\, W_\xi(x)\, dx . \end{aligned} \tag{2.10}$$

The normalization requires $[f(x) = 1]$

$$\int W_\xi(x)\, dx = 1 . \tag{2.11}$$

Remark on the Notation

The stochastic variable was denoted by ξ, whereas the variable in the distribution function was denoted by x. For those readers not yet very familiar with probability theory it is advisable to use different symbols for the stochastic variable and for the corresponding variable in the distribution function. To avoid initial confusion, this is done in this chapter and in the first part of Chap. 3, but later on, to save letters, the same symbols are used for the stochastic variables and the corresponding variables in the distribution function.

Transformation of a Random Variable

If we use the random variable

$$\eta = g(\xi) \tag{2.12}$$

instead of the random variable ξ, the probability density $W_\eta(y)$ of the random variable η is, according to (2.7, 10), given by

$$\begin{aligned} W_\eta(y) &= \langle \delta(y-\eta)\rangle = \langle \delta(y-g(\xi))\rangle \\ &= \int \delta(y-g(x))\, W_\xi(x)\,\mathrm{d}x\,. \end{aligned} \tag{2.13}$$

The last integral is easily evaluated. If $g_n^{-1}(y)$ is the nth simple root of $g(x)-y=0$, then

$$W_\eta(y) = \sum_n W_\xi(g_n^{-1}(y))\,[\,|\mathrm{d}g(x)/\mathrm{d}x|\,]^{-1}|_{x=x_n=g_n^{-1}(y)}\,, \tag{2.14}$$

from a well-known expression for the δ function, see e.g. [2.7]. (Another possibility to obtain (2.14) is by transformation of the differentials.)

As an example, we calculate from the one-dimensional Maxwell distribution

$$W(v) = \sqrt{\frac{m}{2\pi kT}}\,\exp\left(-\frac{mv^2}{2kT}\right) \tag{2.15}$$

the probability density of the energy

$$E = \tfrac{1}{2}mv^2 = g(v)\,. \tag{2.16}$$

Here we have

$$\begin{aligned} &v_{\substack{1\\2}} = g_{\substack{1\\2}}^{-1}(E) = \pm\sqrt{2E/m} \\ &\left|\frac{\mathrm{d}g}{\mathrm{d}v}\right|_{v=g_{\substack{1\\2}}^{-1}(E)} = |mv_{\substack{1\\2}}| = \sqrt{2mE} \end{aligned} \tag{2.17}$$

$$\begin{aligned} W(E) &= \sqrt{\frac{m}{2\pi kT}}\,\exp\left(-\frac{E}{kT}\right)\frac{1}{\sqrt{2mE}} + \sqrt{\frac{m}{2\pi kT}}\,\exp\left(-\frac{E}{kT}\right)\frac{1}{\sqrt{2mE}} \\ &= \frac{1}{\sqrt{\pi kTE}}\,\exp\left(-\frac{E}{kT}\right). \end{aligned} \tag{2.18}$$

2.2 Characteristic Function and Cumulants

The characteristic function $C_\xi(u)$ is the average

$$C_\xi(u) = \langle \mathrm{e}^{iu\xi}\rangle = \int \mathrm{e}^{iux}\, W_\xi(x)\,\mathrm{d}x\,. \tag{2.19}$$

From this characteristic function we obtain the nth moment M_n by differentiation

$$M_n = \langle \xi^n \rangle = \frac{1}{\mathrm{i}^n} \left. \frac{\mathrm{d}^n C_\xi(u)}{\mathrm{d}u^n} \right|_{u=0} . \tag{2.20}$$

Hence, the Taylor expansion of the characteristic function is given by the moments according to

$$C_\xi(u) = 1 + \sum_{n=1}^{\infty} (\mathrm{i}u)^n M_n / n! . \tag{2.21}$$

If we know all the moments, we thus have the characteristic function. If x runs from minus infinity to plus infinity the characteristic function (2.19) is the Fourier transform of the probability density $W_\xi(x)$ and the probability density $W_\xi(x)$ is the inverse Fourier transform of the characteristic function $C_\xi(u)$

$$W_\xi(x) = (2\pi)^{-1} \int C_\xi(u)\, \mathrm{e}^{-\mathrm{i}ux} \mathrm{d}u . \tag{2.22}$$

Because the probability density $W_\xi(x)$ must be positive $[W_\xi(x) \geqq 0]$, $C_\xi(u)$ must be a positive definite function, i.e., $C_\xi(u)$ must for every $n \geqq 1$ fulfill the relation

$$\sum_{k=1}^{n} \sum_{j=1}^{n} C_\xi(u_k - u_j) a_k^* a_j \geqq 0 . \tag{2.23}$$

In (2.23) $u_1, \ldots, u_n$ is an arbitrary set of real numbers and $a_1, \ldots, a_n$ is an arbitrary set of complex numbers. One may show that every positive definite function with the property $C_\xi(0) = 1$ is a characteristic function, i.e., its Fourier transform is positive, see for instance [Ref. 2.2, Chap. VII].

Cumulants

The cumulants K_n or semi-invariants are defined by one of the relations

$$C_\xi(u) = 1 + \sum_{n=1}^{\infty} \frac{(\mathrm{i}u)^n}{n!} M_n = \exp\left(\sum_{n=1}^{\infty} \frac{(\mathrm{i}u)^n}{n!} K_n \right), \tag{2.24}$$

$$\ln C_\xi(u) = \ln\left(1 + \sum_{n=1}^{\infty} \frac{(\mathrm{i}u)^n}{n!} M_n \right) = \sum_{n=1}^{\infty} \frac{(\mathrm{i}u)^n}{n!} K_n . \tag{2.25}$$

It follows from these relations that the first n cumulants can be expressed by the first n moments and vice versa. These relations up to $n = 4$ read explicitly

$$K_1 = M_1$$
$$K_2 = M_2 - M_1^2 \tag{2.26}$$

$$\begin{aligned} K_3 &= M_3 - 3M_1M_2 + 2M_1^3 \\ K_4 &= M_4 - 3M_2^2 - 4M_1M_3 + 12M_1^2M_2 - 6M_1^4 \end{aligned} \tag{2.26}$$

$$\begin{aligned} M_1 &= K_1 \\ M_2 &= K_2 + K_1^2 \\ M_3 &= K_3 + 3K_2K_1 + K_1^3 \\ M_4 &= K_4 + 4K_3K_1 + 3K_2^2 + 6K_2K_1^2 + K_1^4 . \end{aligned} \tag{2.27}$$

General expressions for the connection between cumulants and moments may be found in [Ref. 2.4, p. 165]. A very convenient form in terms of determinants reads

$$K_n = (-1)^{n-1} \begin{vmatrix} M_1 & 1 & 0 & 0 & 0 & \dots \\ M_2 & M_1 & 1 & 0 & 0 & \dots \\ M_3 & M_2 & \binom{2}{1}M_1 & 1 & 0 & \dots \\ M_4 & M_3 & \binom{3}{1}M_2 & \binom{3}{2}M_1 & 1 & \dots \\ M_5 & M_4 & \binom{4}{1}M_3 & \binom{4}{2}M_2 & \binom{4}{3}M_1 & \dots \\ \dots & \dots & \dots & \dots & \dots & \dots \end{vmatrix}_n \tag{2.28}$$

$$M_n = \begin{vmatrix} K_1 & -1 & 0 & 0 & 0 & \dots \\ K_2 & K_1 & -1 & 0 & 0 & \dots \\ K_3 & \binom{2}{1}K_2 & K_1 & -1 & 0 & \dots \\ K_4 & \binom{3}{1}K_3 & \binom{3}{2}K_2 & K_1 & -1 & \dots \\ K_5 & \binom{4}{1}K_4 & \binom{4}{2}K_3 & \binom{4}{3}K_2 & K_1 & \dots \\ \dots & \dots & \dots & \dots & \dots & \dots \end{vmatrix}_n \tag{2.29}$$

where the determinants $||_n$ contain n rows and n columns and where $\binom{n}{m} = \frac{n!}{(n-m)!\,m!}$ are the binomial coefficients. These connections between the expansion coefficients of (2.24 and 25) are found in [2.8]. [To obtain (2.28) the ith row of the corresponding determinant in [2.8] has to be multiplied with $(i-1)!$ and the jth column with the exception of the first column has to be divided by $(j-2)!$. For the derivation of (2.29) the ith row of the corresponding determinant has to be multiplied with $(i-1)!$ and the jth column has to be divided by $(j-1)!$].

As seen from (2.26), the first cumulant is equal to the first moment and the second cumulant is equal to the variance or mean-squared deviation

$$K_2 = M_2 - M_1^2 = \langle(\xi - \langle\xi\rangle)^2\rangle \geqq 0 . \tag{2.30}$$

It is sometimes useful to consider probability densities where all cumulants with the exception of the first two vanish. (It obviously does not make sense to consider probability densities where all moments with the exception of the first few vanish. It is immediately seen, for instance, that for $M_1 \neq 0$, $M_2 = M_3 = \ldots = 0$ relation (2.30) is violated and hence no positive probability density is possible). We have

for $K_2 = K_3 = \ldots = 0$

$$C_\xi(u) = \exp(\mathrm{i}uK_1) ; \qquad W_\xi(x) = \delta(x - K_1) \tag{2.31}$$

and for $K_3 = K_4 = \ldots = 0 , \quad K_2 \neq 0$

$$\begin{aligned} C_\xi(u) &= \exp(\mathrm{i}uK_1 - \tfrac{1}{2}u^2K_2) \\ W_\xi(x) &= (2\pi)^{-1} \int_{-\infty}^{\infty} \exp(-\mathrm{i}ux + \mathrm{i}uK_1 - \tfrac{1}{2}u^2K_2)\mathrm{d}u \\ &= (2\pi K_2)^{-1/2} \exp[-\tfrac{1}{2}(x - K_1)^2/K_2] . \end{aligned} \tag{2.32}$$

In the first case the probability density is a sharp distribution at K_1, whereas in the last case it is a shifted Gaussian distribution with the mean value K_1 and variance K_2.

If the cumulants vanish at some higher order $n \geqq 3$, i.e., $K_n \neq 0$; $K_{n+1} = K_{n+2} = \ldots = 0$, the probability density cannot be positive everywhere [2.9]. See also the similar discussion in Sect. 4.3.

2.3 Generalization to Several Random Variables

Generally we may have r random variables $\xi_1, \xi_2, \ldots, \xi_r$. As an example we may consider r dice, where ξ_i is the random variable corresponding to the ith die. By making N experiments or realizations, for each random variable ξ_i we get N numbers $\xi_{i1}, \xi_{i2}, \ldots, \xi_{iN}$. As in the case of one random variable we may take averages of an arbitrary function $f(\xi_1, \ldots, \xi_r)$ according to

$$\langle f(\xi_1, \ldots, \xi_r)\rangle = \lim_{N\to\infty} \frac{1}{N} [f(\xi_{11}, \ldots, \xi_{r1}) + \ldots + f(\xi_{1N}, \ldots, \xi_{rN})] . \tag{2.33}$$

Similarly to (2.7) we introduce the r-dimensional distribution function

$$W_{\xi_1,\ldots,\xi_r}(x_1, \ldots, x_r) = W_r(x_1, \ldots, x_r) = \langle\delta(x_1 - \xi_1) \ldots \delta(x_r - \xi_r)\rangle . \tag{2.34}$$

The index $\xi_1, \ldots, \xi_r$ of the distribution function will be omitted, to be replaced by the number r of variables. With the help of (2.34) the averages (2.33) can be calculated by integration

$$\langle f(\xi_1, \ldots, \xi_r)\rangle = \int \ldots \int f(x_1, \ldots, x_r)\, W_r(x_1, \ldots, x_r)\, dx_1 \ldots dx_r . \tag{2.35}$$

It immediately follows from (2.34) that we obtain the probability density of the first $i < r$ random variables by integration over the other variables

$$W_i(x_1, \ldots, x_i) = \int \ldots \int W_r(x_1, \ldots, x_i, x_{i+1}, \ldots, x_r)\, dx_{i+1} \ldots dx_r . \tag{2.36}$$

Similar to the one-variable case we may construct characteristic functions $C_r(u_1, \ldots, u_r)$ by the averages

$$\begin{aligned} C_r(u_1, \ldots, u_r) &= \langle \exp(\mathrm{i}u_1\xi_1 + \ldots + \mathrm{i}u_r\xi_r)\rangle \\ &= \int \ldots \int \exp[\mathrm{i}(u_1x_1 + \ldots + u_rx_r)]\, W_r(x_1, \ldots, x_r)\, dx_1 \ldots dx_r \end{aligned} \tag{2.37}$$

from which any mixed moments can be obtained by differentiation

$$\begin{aligned} M_{n_1, \ldots, n_r} &= \langle \xi_1^{n_1} \ldots \xi_r^{n_r}\rangle \\ &= \left(\frac{\partial}{\partial \mathrm{i}u_1}\right)^{n_1} \ldots \left(\frac{\partial}{\partial \mathrm{i}u_r}\right)^{n_r} C_r(u_1, \ldots, u_r)\bigg|_{u_1 = \ldots = u_r = 0} . \end{aligned} \tag{2.38}$$

Thus the moments (2.38) are the expansion coefficients of the characteristic function

$$C_r(u_1, \ldots, u_r) = \sum_{n_1, \ldots, n_r} M_{n_1, \ldots, n_r} \frac{(\mathrm{i}u_1)^{n_1}}{n_1!} \ldots \frac{(\mathrm{i}u_r)^{n_r}}{n_r!} , \tag{2.39}$$

where we have to sum over all $n_1, \ldots, n_r$ and where $M_{0, \ldots, 0} = 1$. Because of (2.37) the probability density W_r is given by the inverse Fourier transform of the characteristic function

$$\begin{aligned} W_r(x_1, \ldots, x_r) = (2\pi)^{-r} \int \ldots \int &\exp[-\mathrm{i}(u_1x_1 + \ldots + u_rx_r)] \\ &\times C_r(u_1, \ldots, u_r)\, du_1 \ldots du_r . \end{aligned} \tag{2.40}$$

If instead of the characteristic function in (2.38) we use its logarithm, we get the cumulants or semi-invariants

$$K_{n_1, \ldots, n_r} = \left(\frac{\partial}{\partial \mathrm{i}u_1}\right)^{n_1} \ldots \left(\frac{\partial}{\partial \mathrm{i}u_r}\right)^{n_r} \ln C_r(u_1, \ldots, u_r)\bigg|_{u_1 = \ldots = u_r = 0} . \tag{2.41}$$

For a general connection between these cumulants and the moments (2.38), see [Ref. 2.4, p. 165].

The characteristic function may therefore be expressed by the cumulants

$$C_r(u_1,\ldots,u_r) = \exp\left(\sum_{n_1,\ldots,n_r} K_{n_1,\ldots,n_r} \frac{(\mathrm{i}u_1)^{n_1}}{n_1!}\cdots\frac{(\mathrm{i}u_r)^{n_r}}{n_r!}\right). \tag{2.42}$$

2.3.1 Conditional Probability Density

If we consider only those realizations of the r random variables $\xi_1,\ldots,\xi_r$ where the last $r-1$ random variables take the fixed values $\xi_2=x_2,\ldots,\xi_r=x_r$, we obtain a certain probability density for the first random variable. This probability density is called conditional probability density and will be written as $P(x_1|x_2,\ldots,x_r)$.

Obviously the probability $W_r(x_1,\ldots,x_r)\,\mathrm{d}x_1\ldots\mathrm{d}x_r$ that the random variables $\xi_i (i=1,\ldots,r)$ are in the interval $x_i \leqq \xi_i \leqq x_i+\mathrm{d}x_i$ is the probability $P(x_1|x_2,\ldots,x_r)\,\mathrm{d}x_1$ that the first variable is in the interval $x_1 \leqq \xi_1 \leqq x_1+\mathrm{d}x_1$ and that the other variables have the sharp values $\xi_i = x_i\,(i=2,\ldots,r)$ times the probability $W_{r-1}(x_2,\ldots,x_r)\,\mathrm{d}x_2\ldots\mathrm{d}x_r$ that the last $r-1$ variables are in the interval $x_i \leqq \xi_i \leqq x_i+\mathrm{d}x_i\,(i=2,\ldots,r)$, i.e., we have

$$W_r(x_1,\ldots,x_r) = P(x_1|x_2,\ldots,x_r)\,W_{r-1}(x_2,\ldots,x_r)\,. \tag{2.43}$$

Because W_{r-1} follows from W_r (2.36), we may express the conditional probability density by W_r

$$\begin{aligned} P(x_1|x_2,\ldots,x_r) &= \frac{W_r(x_1,\ldots,x_r)}{W_{r-1}(x_2,\ldots,x_r)} \\ &= \frac{W_r(x_1,\ldots,x_r)}{\int W_r(x_1,\ldots,x_r)\,\mathrm{d}x_1}\,. \end{aligned} \tag{2.44}$$

Especially for two random variables the conditional probability density in terms of W_2, which is called the joint probability density, reads

$$P(x_1|x_2) = W_2(x_1,x_2)/\int W_2(x_1,x_2)\,\mathrm{d}x_1\,. \tag{2.45}$$

2.3.2 Cross Correlation

It may happen that the conditional probability density (2.45) does not depend on the value x_2 of the other random variable ξ_2. In that case we say that the random variables ξ_1 and ξ_2 are uncorrelated. It then follows from (2.45)

$$\begin{aligned} W_2(x_1,x_2) &= P(x_1)\int W_2(x_1,x_2)\,\mathrm{d}x_1 \\ &= W_1^{(1)}(x_1)\cdot W_1^{(2)}(x_2)\,, \end{aligned} \tag{2.46}$$

i.e., in that case the probability density W_2 decomposes into a product of two probability densities W_1.

In the other extreme, if ξ_1 is a function of ξ_2, i.e., $\xi_1 = f(\xi_2)$, the random variable ξ_1 is completely determined by the random variable ξ_2 and the probability density has the sharp value

$$P(x_1|x_2) = \delta(x_1 - f(x_2)) \tag{2.47}$$

and the joint probability density reads

$$W_2(x_1,x_2) = \delta(x_1 - f(x_2))\, W_1(x_2)\,. \tag{2.48}$$

Between these two extreme cases there may be intermediate cases, where the two random variables are partially correlated. For uncorrelated random variables, where the joint probability factorizes (2.46), the cross-correlation coefficient

$$\kappa(\xi_1,\xi_2) = \langle \xi_1 \xi_2 \rangle - \langle \xi_1 \rangle \langle \xi_2 \rangle \tag{2.49}$$

is obviously zero. The following correlation coefficient

$$R = \frac{\langle \xi_1 \xi_2 \rangle - \langle \xi_1 \rangle \langle \xi_2 \rangle}{\sqrt{\langle \xi_1^2 \rangle - \langle \xi_1 \rangle^2}\sqrt{\langle \xi_2^2 \rangle - \langle \xi_2 \rangle^2}} \tag{2.50}$$

measures the degree of correlation. One may easily show [Ref. 2.3, Vol. I, p. 276] that $|R| \leqq 1$. R is equal to ± 1 for linearly dependent random variables, i.e.,

$$\xi_1 = \pm a \xi_2 + b\,; \qquad a > 0\,. \tag{2.51}$$

If W_2 factorizes, R is zero. The reverse, however, is not true, i.e, $R = 0$ does not imply that W_2 factorizes. One may even construct examples where (2.48) is valid but where, with a suitable $f(x)$, the correlation coefficient (2.50) vanishes [Ref. 2.3, Vol. I, p. 236]. The cumulant (2.41) for two variables with $n_1 = n_2 = 1$ is identical to the correlation function (2.49).

In the general case we may consider the correlation coefficient (2.41) with $n_1 = \ldots = n_r = 1$

$$\kappa(\xi_1, \ldots, \xi_r) = K_{1,\ldots,1} = \left.\frac{\partial^r \ln C_r(u_1, \ldots, u_r)}{\partial i u_1 \ldots \partial i u_r}\right|_{u_1 = \ldots = u_r = 0}\,. \tag{2.52}$$

This coefficient not only vanishes if all random variables are independent, i.e., the probability distribution factorizes according to

$$W_r(x_1, \ldots, x_r) = W_1^{(1)}(x_1) \ldots W_1^{(r)}(x_r)\,. \tag{2.53}$$

(The upper index in W_1 indicates different distribution functions.) It also vanishes if any one of the random variables, for instance the first, is independent from the other variables (the others need not be independent), i.e., it vanishes if the probability density factorizes according to

$$W_r(x_1,\dots,x_r) = W_1^{(1)}(x_1)\,W_{r-1}(x_2,\dots,x_r)\,. \tag{2.54}$$

This is easily seen as follows. Because the characteristic function C_r then also factorizes

$$C_r(u_1,\dots,u_r) = C_1^{(1)}(u_1)\,C_{r-1}(u_2,\dots,u_r)\,, \tag{2.54a}$$

(2.52) must therefore be zero for $r \geqq 2$.

If all the cumulants (2.41) for $n_1 \geqq 1$, $n_2 \geqq 1, \dots, n_r \geqq 1$ vanish, the characteristic function C_r and hence also the probability density W_r factorize.

2.3.3 Gaussian Distribution

Next we consider only those probability densities where all cumulants (2.41) except those with $n_1+n_2+\dots+n_r \leqq 2$ vanish. We then must have

$$C_r(u_1,\dots,u_r) = \exp\left(\sum_{j=1}^{r} a_j \mathrm{i}u_j + \frac{1}{2}\sum_{j,k=1}^{r} \sigma_{jk}\mathrm{i}u_j\mathrm{i}u_k\right). \tag{2.55}$$

It follows from (2.38) that the first two moments are given by

$$\langle \xi_j \rangle = a_j\,, \tag{2.56}$$

$$\langle \xi_j \xi_k \rangle = \sigma_{jk} + a_j a_k\,. \tag{2.57}$$

Equations (2.56, 57) imply that the variance ($j=k$) and the covariance ($j \neq k$) read

$$\langle (\xi_j - \langle \xi_j \rangle)(\xi_k - \langle \xi_k \rangle) \rangle = \sigma_{jk}\,. \tag{2.58}$$

The probability density is the inverse Fourier transform of the characteristic function (2.55), i.e.,

$$W_r(x_1,\dots,x_r) = (2\pi)^{-r}\int\dots\int \exp\left[\sum_{j=1}^{r}(a_j - x_j)\mathrm{i}u_j - \frac{1}{2}\sum_{j,k=1}^{r}\sigma_{jk}u_j u_k\right] \times du_1\dots du_r\,. \tag{2.59}$$

The matrix $\sigma_{jk} = \sigma_{kj}$ is assumed to be positive definite. Then the inverse matrix $(\sigma^{-1})_{jk} = (\sigma^{-1})_{kj}$ and its square root $(\sigma^{1/2})_{jk} = (\sigma^{1/2})_{kj}$, as well as its inverse square root $(\sigma^{-1/2})_{jk} = (\sigma^{-1/2})_{kj}$, exist. (The square root of σ may be uniquely defined in such a way that it has positive eigenvalues.) To calculate the integral we introduce as integration variables

$$\alpha_j = \sum_k [(\sigma^{1/2})_{jk} u_k + \mathrm{i}(\sigma^{-1/2})_{jk}(x_k - a_k)] \,. \tag{2.60}$$

We may then write the exponential in (2.59) as

$$[\,] = -\frac{1}{2}\sum_j \alpha_j\alpha_j - \frac{1}{2}\sum_{j,k}(\sigma^{-1})_{jk}(x_j - a_j)(x_k - a_k) \,. \tag{2.61}$$

Because of the Jacobian

$$\begin{aligned}\frac{\mathrm{d}u_1 \ldots \mathrm{d}u_r}{\mathrm{d}\alpha_1 \ldots \mathrm{d}\alpha_r} &= \left(\frac{\mathrm{d}\alpha_1 \ldots \mathrm{d}\alpha_r}{\mathrm{d}u_1 \ldots \mathrm{d}u_r}\right)^{-1} = [\mathrm{Det}\,(\sigma^{1/2})_{jk}]^{-1} \\ &= (\mathrm{Det}\,\sigma_{jk})^{-1/2} \,,\end{aligned}$$

and because of

$$\begin{aligned}\int \ldots \int \exp\left(-\frac{1}{2}\sum_{j=1}^{r}\alpha_j\alpha_j\right)\mathrm{d}\alpha_1 \ldots \mathrm{d}\alpha_r &= \left(\int_{-\infty}^{\infty} \mathrm{e}^{-\alpha^2/2}\,\mathrm{d}\alpha\right)^r \\ &= (2\pi)^{r/2} \,,\end{aligned}$$

we get as the final result for the probability density the general Gaussian distribution

$$\begin{aligned}W_r(x_1, \ldots, x_r) &= (2\pi)^{-r/2}(\mathrm{Det}\,\sigma_{jk})^{-1/2} \\ &\quad \times \exp\left[-\frac{1}{2}\sum_{j,k}(\sigma^{-1})_{jk}(x_j - a_j)(x_k - a_k)\right] .\end{aligned} \tag{2.62}$$

The characteristic function for the random variables

$$\eta_j = \xi_j - a_j \tag{2.63}$$

is given by (2.55) with $a_j = 0$. One may then derive from this characteristic function the moments

$$\begin{aligned}\langle \eta_{j_1}\eta_{j_2} \cdots \eta_{j_{2n+1}} \rangle &= 0 \\ \langle \eta_{j_1}\eta_{j_2} \cdots \eta_{j_{2n}} \rangle &= \sum_{P_d} \sigma_{k_1,k_2}\sigma_{k_3,k_4} \cdots \sigma_{k_{2n-1},k_{2n}} \,,\end{aligned} \tag{2.64}$$

where we have to sum over only those $(2n)!/(2^n n!)$ permutations $(j_1 \ldots j_{2n}) \Rightarrow (k_1, \ldots, k_{2n})$ which lead to different expressions for $\sigma_{k_1,k_2} \cdots \sigma_{k_{2n-1},k_{2n}}$. Interchanging the indices of each individual σ (2^n possibilities) as well as interchanging pairs of indices of different σ($n!$ possibilities) does not lead to different expressions. For $n = 2$ for instance, we have

$$\langle \eta_i\eta_j\eta_k\eta_l \rangle = \sigma_{ij}\sigma_{kl} + \sigma_{ik}\sigma_{jl} + \sigma_{il}\sigma_{jk} \,. \tag{2.65}$$

Linear Transformation of Variables

It follows from (2.62) that a linear transformation of the stochastic variables, i.e.,

$$\xi_i' = \sum_j \alpha_{ij}\xi_j + b_i ,$$

leads again to a Gaussian distribution for the probability density in the transformed variables.

2.4 Time-Dependent Random Variables

We now consider a random variable ξ which depends on the time t, i.e., $\xi = \xi(t)$. Here we assume that we have an ensemble of systems and that each system leads to a number ξ which depends on time. This number ξ of one system (one realization) may look like the curve in Fig. 2.1.

Though the outcome for one system cannot be precisely predicted, we assume that ensemble averages exist and that these averages can be calculated. For the fixed time $t = t_1$ we may therefore define a probability density by

$$W_1(x_1, t) = \langle \delta(x_1 - \xi(t_1)) \rangle . \tag{2.66}$$

The bracket $\langle\ \rangle$ indicates the ensemble average. The probability to find the random variable $\xi(t_1)$ in the interval $x_1 \leqq \xi(t_1) \leqq x_1 + \mathrm{d}x_1$ is then given by $W_1(x_1, t)\,\mathrm{d}x_1$. Next we define the probability that $\xi(t_1)$ is in the interval $x_1 \leqq \xi(t_1) \leqq x_1 + \mathrm{d}x_1$, $\xi(t_2)$ is in the interval $x_2 \leqq \xi(t_2) \leqq x_2 + \mathrm{d}x_2, \ldots$, and that $\xi(t_n)$ is in the interval $x_n \leqq \xi(t_n) \leqq x_n + \mathrm{d}x_n$. This probability may be written as

$$W_n(x_n, t_n; \ldots; x_1, t_1)\,\mathrm{d}x_1 \ldots \mathrm{d}x_n ,$$

where the probability density W_n is given by

$$W_n(x_n, t_n; \ldots; x_1, t_1) = \langle \delta(x_1 - \xi(t_1)) \ldots \delta(x_n - \xi(t_n)) \rangle . \tag{2.66a}$$

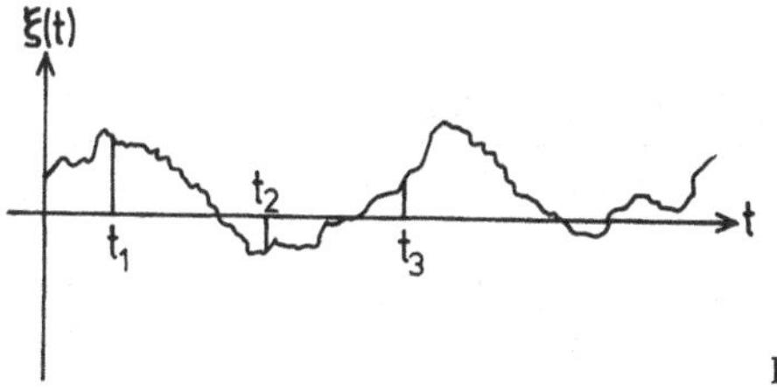

Fig. 2.1. A realization of the stochastic variable $\xi(t)$

If we know the infinite hierarchy of probability densities

$$\begin{aligned}&W_1(x_1,t_1)\\&W_2(x_2,t_2;x_1,t_1)\\&W_3(x_3,t_3;x_2,t_2;x_1,t_1)\\&\dots\end{aligned} \tag{2.67}$$

for every t_i in the interval $t_0 \leqq t_i \leqq t_0+T$, we know completely the time dependence of the process described by the random variable $\xi(t)$ in the interval $[t_0, t_0+T]$. With the probability density (2.66a) we obtain any averages by integration. The correlation function $\langle \xi(t)\xi(t')\rangle$, for instance, is given by

$$\langle \xi(t_2)\xi(t_1)\rangle = \iint x_2 x_1 W_2(x_2,t_2;x_1,t_1)\,dx_2 dx_1\,. \tag{2.68}$$

Because of (2.36) we obtain probability densities with lower number of variables from those with higher numbers by integration.

Stationary Processes

If the probability densities (2.67) are not changed be replacing t_i by t_i+T (T arbitrary) we call the process stationary. It then follows that W_1 does not depend on t and that W_2 can depend only on the time difference t_2-t_1.

2.4.1 Classification of Stochastic Processes

As in Sect. 2.3.1, we may define a conditional probability density as the probability density of the random variable ξ at time t_n under the condition that the random variable at the time $t_{n-1}<t_n$ has the sharp value x_{n-1}; at the time $t_{n-2}<t_{n-1}$ has the sharp value $x_{n-2},\dots$; and at the time $t_1<t_2$ has the sharp value x_1:

$$\begin{aligned}&P(x_n,t_n|x_{n-1},t_{n-1};\dots;x_1,t_1) = \langle \delta(x_n-\xi(t_n))\rangle\big|_{\xi(t_{n-1})=x_{n-1},\dots,\xi(t_1)=x_1}\\&t_n>t_{n-1}>\dots>t_1\,.\end{aligned} \tag{2.69}$$

In accordance with (2.44) we may express the conditional probability density by W_n in the following way:

$$\begin{aligned}P(x_n,t_n|x_{n-1},t_{n-1};\dots;x_1,t_1) &= \frac{W_n(x_n,t_n;\dots;x_1,t_1)}{W_{n-1}(x_{n-1},t_{n-1};\dots;x_1,t_1)}\\&= \frac{W_n(x_n,t_n;\dots;x_1,t_1)}{\int W_n(x_n,t_n;\dots;x_1,t_1)\,dx_n}\,.\end{aligned} \tag{2.70}$$

Following *Wang* and *Uhlenbeck* [1.7], we now classify the processes described by the random variable ξ as follows.

a) Purely Random Processes

We call the process a purely random process if the conditional probability density P_n ($n \geqq 2$ arbitrary) does not depend on the values $x_i = \xi(t_i)$ $(i < n)$ of the random variable at earlier times $t_i < t_n$

$$P(x_n, t_n | x_{n-1}, t_{n-1}; \dots; x_1, t_1) = P(x_n, t_n) . \tag{2.71}$$

It then follows that

$$W_n(x_n, t_n; \dots; x_1, t_1) = P(x_n, t_n)\, W_{n-1}(x_{n-1}, t_{n-1}; \dots; x_1, t_1) ,$$

or if we apply the same argument to W_{n-1} and so forth that the probability density W_n factorizes

$$W_n(x_n, t_n; \dots; x_1, t_1) = P(x_n, t_n) \dots P(x_1, t_1) . \tag{2.72}$$

Thus the complete information of the process is contained in $P(x_1, t_1) = W_1(x_1, t_1)$.

For the physical systems where the random variable $\xi(t)$ is a continuous function of time (Fig. 2.1), the random variable at two arbitrary close times t_n and $t_n - \varepsilon$, i.e., $\xi(t_n)$ and $\xi(t_n - \varepsilon)$, must have some correlation and the probability densities therefore cannot factorize. Thus a purely random process cannot describe physical systems where the random variable is a continuous function of time.

b) Markov Processes

For Markov processes, the conditional probability density depends only on the value $\xi(t_{n-1}) = x_{n-1}$ at the next earlier time but not on $\xi(t_{n-2}) = x_{n-2}$ and so on, i.e.,

$$P(x_n, t_n | x_{n-1}, t_{n-1}; \dots; x_1, t_1) = P(x_n, t_n | x_{n-1}, t_{n-1}) . \tag{2.73}$$

It then follows from (2.70) that

$$W_n(x_n, t_n; \dots; x_1, t_1) = P(x_n, t_n | x_{n-1}, t_{n-1})\, W_{n-1}(x_{n-1}, t_{n-1}; \dots; x_1, t_1) ,$$

or if we use the same argument for W_{n-1} and so on we may express W_n by a product of conditional probabilities and W_1

$$\begin{aligned} W_n(x_n, t_n; \dots; x_1, t_1) = {} & P(x_n, t_n | x_{n-1}, t_{n-1}) P(x_{n-1}, t_{n-1} | x_{n-2}, t_{n-2}) \\ & \dots P(x_2, t_2 | x_1, t_1)\, W_1(x_1, t_1) . \end{aligned} \tag{2.74}$$

Because of this relation, the conditional probabilities are also called transition probabilities.

For $n = 2$ (2.70) specializes to

$$P(x_2, t_2 | x_1, t_1) = \frac{W_2(x_2, t_2; x_1, t_1)}{W_1(x_1, t_1)} = \frac{W_2(x_2, t_2; x_1, t_1)}{\int W_2(x_2, t_2; x_1, t_1)\, dx_2} . \quad (2.75)$$

Thus for a Markov process the infinite hierarchy (2.67), i.e., the complete information about the process, is contained in $W_2(x_2, t_2; x_1, t_1)$.

We may interpret (2.71, 73) as follows: whereas for a purely random process there is no memory of values of the random variable at any preceding time, for a Markov process there is only a memory of the value of the random variable for the latest time, where we measured ξ. The time difference $t_2 - t_1$ of the conditional probability $P(x_2, t_2 | x_1, t_1)$ of a Markov process is arbitrary. If the time difference is large, the dependence of P on x_1 will be small (i.e., the memory of the value of the random variable is nearly lost). If, on the other hand, the time difference is infinitesimally small, the conditional probability will have the sharp value x_1, i.e.,

$$\lim_{t_2 \to t_1} P(x_2, t_2 | x_1, t_1) = \delta(x_1 - x_2) . \quad (2.76)$$

c) General Processes

Next one may consider processes where the conditional probability density depends only on the values of the random variable at the two latest times. In this case the complete information about the process is contained in W_3. Hence we may continue, i.e, we may have processes where the complete information is contained in W_4 and so on. Due to *Wang Uhlenbeck* [1.7], however, this further classification is not suitable to describe non-Markovian processes, i.e., processes where the complete information is not contained in W_2. For non-Markovian processes one may take into account besides $\xi(t) = \xi_1(t)$ more random variables $\xi_2(t), \ldots, \xi_r(t)$. By a proper choice of these additional variables one may then have a Markov process for the r random variables. (Several time-dependent variables are discussed in Sect. 2.5.) Another possibility is the following. As shown in Chap. 4, the equation of motion of the probability density for continuous Markov processes is the Fokker-Planck equation (1.15). For non-Markovian processes one may then use generalized Fokker-Planck equations (1.26, 27) which contain a memory function.

2.4.2 Chapman-Kolmogorov Equation

The probability density W_2 is obtained from W_3 by integrating over one coordinate. Thus we have

$$W_2(x_3, t_3; x_1, t_1) = \int W_3(x_3, t_3; x_2, t_2; x_1, t_1)\, dx_2 . \quad (2.77)$$

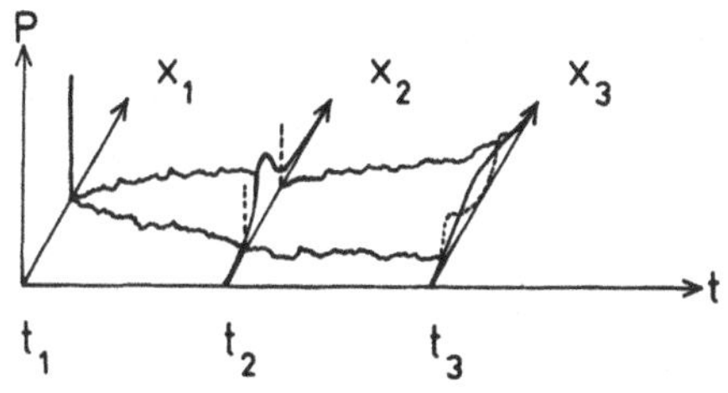

Fig. 2.2. Transition probabilities at times t_2 and t_3, with sharp initial value at time t_1. For two typical values the transition probabilities at t_3, which start at t_2 with a sharp value, are shown by a broken line. Two typical realizations of the stochastic variable $\xi(t)$ are also indicated

For a Markov process we may write (2.77) in the form [see (2.74)] (assuming $t_3 \geqq t_2 \geqq t_1$)

$$P(x_3, t_3 | x_1, t_1)\, W_1(x_1, t_1) = \int P(x_3, t_3 | x_2, t_2) P(x_2, t_2 | x_1, t_1)\, W_1(x_1, t_1)\, \mathrm{d}x_2 \, .$$

Because $W_1(x_1, t_1)$ is arbitrary, we obtain the Chapman-Kolmogorov equation

$$P(x_3, t_3 | x_1, t_1) = \int P(x_3, t_3 | x_2, t_2) P(x_2, t_2 | x_1, t_1)\, \mathrm{d}x_2 \, . \tag{2.78}$$

Equation (2.78) may be interpreted in the following way. The transition probability from x_1 at time t_1 to x_3 at time t_3 is the same as the transition probability from x_1 at time t_1 to x_2 at time t_2 times the transition probability from x_2 at time t_2 to x_3 at time t_3 for all possible x_2 (Fig. 2.2).

2.4.3 Wiener-Khintchine Theorem

Instead of the random variable $\xi(t)$ we may consider its Fourier transform

$$\tilde{\xi}(\omega) = \int_{-\infty}^{\infty} \mathrm{e}^{-\mathrm{i}\omega t} \xi(t)\, \mathrm{d}t \, , \tag{2.79}$$

which is also a random variable. For stationary processes the Wiener-Khintchine theorem expresses the correlation function $\langle \tilde{\xi}(\omega) \tilde{\xi}^*(\omega') \rangle$ of the Fourier transform of the random variable $\xi(t)$ by the correlation function $\langle \xi(t) \xi^*(t') \rangle$ of the random variable itself. (Here we take into account that the stochastic variable may have complex values.) Inserting (2.79) gives

$$\langle \tilde{\xi}(\omega) \tilde{\xi}^*(\omega') \rangle = \iint_{-\infty}^{\infty} \mathrm{e}^{-\mathrm{i}\omega t + \mathrm{i}\omega' t'} \langle \xi(t) \xi^*(t') \rangle\, \mathrm{d}t\, \mathrm{d}t' \, . \tag{2.80}$$

For stationary processes $\langle \xi(t) \xi^*(t') \rangle$ is only a function of the difference $t - t'$, i.e.,

$$\langle \xi(t) \xi^*(t') \rangle = \langle \xi(t - t') \xi^*(0) \rangle \, . \tag{2.81}$$

Introducing the new variables

$$\begin{aligned} \tau &= t - t' \\ t_0 &= (t' + t)/2 , \end{aligned} \tag{2.82}$$

we have

$$\langle \tilde{\xi}(\omega) \tilde{\xi}^*(\omega') \rangle = \int_{-\infty}^{\infty} e^{-i(\omega-\omega')t_0} dt_0 \int_{-\infty}^{\infty} e^{-i(\omega'+\omega)\tau/2} \langle \xi(\tau) \xi^*(0) \rangle d\tau . \tag{2.83}$$

The first integral is the δ function

$$2\pi\delta(\omega-\omega') = \int_{-\infty}^{\infty} e^{-i(\omega-\omega')t_0} dt_0 . \tag{2.84}$$

In the other integral we may therefore put $\omega = \omega'$ and thus finally obtain

$$\langle \tilde{\xi}(\omega) \tilde{\xi}^*(\omega') \rangle = \pi\delta(\omega-\omega') S(\omega) \tag{2.85}$$

$$S(\omega) = 2 \int_{-\infty}^{\infty} e^{-i\omega\tau} \langle \xi(\tau) \xi^*(0) \rangle d\tau . \tag{2.86}$$

Here $S(\omega)$ is called the spectral density. Thus the Wiener-Khintchine theorem states that the spectral density is the Fourier transform of the correlation function for stationary processes.

Remarks on the Spectral Density

The field strength $E(t)$ in an optical wave may be considered as a random variable. If such a wave goes through a prism, the prism separates the different Fourier components $\tilde{E}(\omega)$ of the incoming wave, and the intensity measured in a spectrometer is then given by $\tilde{E}(\omega)\tilde{E}^*(\omega)$. For stationary processes (infinitely long observation) this intensity is infinite [$\omega = \omega'$ in (2.85)]. For large but finite observation times T the intensity is given by the spectral density $S(\omega)$ (2.86) times the observation time T (method of *Rice* [2.10], see also Sect. 3.2.3).

2.5 Several Time-Dependent Random Variables

The generalization to r time-dependent random variables $\xi_1(t), \dots, \xi_r(t)$ is rather obvious and will now be discussed briefly. The probability density for the r variables at the n times $t_n, \dots, t_1$ is defined by

$$\begin{aligned} &W_n(x_1^{(n)}, \dots, x_r^{(n)}, t_n; \dots; x_1^{(1)}, \dots, x_r^{(1)}, t_1) \\ &\quad = \langle \delta(x_1^{(n)} - \xi_1(t_n)) \dots \delta(x_r^{(n)} - \xi_r(t_n)) \dots \delta(x_1^{(1)} - \xi_1(t_1)) \dots \delta(x_r^{(1)} - \xi_r(t_1)) \rangle . \end{aligned} \tag{2.87}$$

In (2.87) we wrote the index i of the times t_i as upper indices (in parenthesis to the variables $x_1^{(i)} \ldots x_r^{(i)}$).

For a Markov process the complete information is contained in the probability density for two times

$$W_2(x_1, \ldots, x_r, t; x_1', \ldots, x_r', t') .$$

The conditional probability

$$P(x_1, \ldots, x_r, t \,|\, x_1', \ldots, x_r', t') = \frac{W_2(x_1, \ldots, x_r, t; x_1', \ldots, x_r', t')}{\int \ldots \int W_2(x_1, \ldots, x_r, t; x_1', \ldots, x_r', t') \mathrm{d}x_1 \ldots \mathrm{d}x_r} \tag{2.88}$$

then obeys the Chapman-Kolmogorov equation ($t > t' > t''$)

$$P(x_1, \ldots, x_r, t \,|\, x_1'', \ldots, x_r'', t'') = \int \ldots \int P(x_1, \ldots, x_r, t \,|\, x_1', \ldots, x_r', t') \times P(x_1', \ldots, x_r', t' \,|\, x_1'', \ldots, x_r'', t'') \mathrm{d}x_1' \ldots \mathrm{d}x_r' . \tag{2.89}$$

If we eliminate one or more variables by integration, the remaining joint probability W_2 may no longer give us the complete information of process, i.e, for the reduced number of variables we may no longer have a Markov process. If for instance $W_2^{(2)}(x_1, x_2, t; x_1', x_2', t')$ is the two-times joint probability density describing the Markov process in the two random variables $\xi_1(t)$ and $\xi_2(t)$, the two-times joint probability density of the random variable $\xi_1(t)$ alone, i.e.,

$$W_2^{(1)}(x_1, t; x_1', t') = \iint W_2^{(2)}(x_1, x_2, t; x_1', x_2', t') \mathrm{d}x_2 \mathrm{d}x_2' , \tag{2.90}$$

will generally not contain the complete information for the random variable $\xi_1(t)$.

3. Langevin Equations

We first investigate the solution of the Langevin equation for Brownian motion. In Sect. 3.2 we treat a system of linear Langevin equations, followed in Sects. 3.3, 4 by general nonlinear Langevin equations.

3.1 Langevin Equation for Brownian Motion

We first look for solutions of the Langevin equation, see (1.7),

$$\dot{v} + \gamma v = \Gamma(t)\,, \tag{3.1}$$

where $\Gamma(t)$ is a Langevin force with zero mean and with a correlation function which is proportional to a δ function

$$\langle \Gamma(t) \rangle = 0\,, \qquad \langle \Gamma(t)\Gamma(t') \rangle = q\delta(t-t')\,. \tag{3.2}$$

The spectral density $S(\omega)$ of the Langevin force, which by the Wiener-Khintchine theorem (2.86) is the Fourier transform of the correlation function (3.2), is independent of the frequency ω:

$$S(\omega) = 2\int_{-\infty}^{\infty} \mathrm{e}^{-\mathrm{i}\omega\tau} q\delta(\tau)\,\mathrm{d}\tau = 2q\,. \tag{3.3}$$

Therefore the Langevin force (3.2) with a δ correlation is called white-noise force. In general, the spectral density would depend on ω and the noise is then termed colored noise. Because of the linearity of (3.1) it is sufficient to know the two-time correlation (3.2) of the Langevin force to calculate the two-time correlation function $\langle v(t_1)v(t_2)\rangle$ of the velocity. If one is interested in multitime correlation functions of the form $\langle v(t_1)v(t_2)v(t_3)\ldots\rangle$ [or if (3.1) were nonlinear], one has to know multitime correlation functions of the Langevin force. Here we assume that the random variables $\xi_i = \Gamma(t_i)$ are distributed according to the Gaussian distribution function (2.62) with zero mean. It then follows, see (2.64), that higher correlation functions are given by

$$\langle \Gamma(t_1)\Gamma(t_2)\dots\Gamma(t_{2n-1})\rangle = 0$$
$$\langle \Gamma(t_1)\Gamma(t_2)\dots\Gamma(t_{2n})\rangle = q^n\left[\sum_{P_d} \delta(t_{i_1}-t_{i_2})\,\delta(t_{i_3}-t_{i_4})\dots\delta(t_{i_{2n-1}}-t_{i_{2n}})\right], \tag{3.4}$$

where the sum has to be performed over those $(2n)!/(2^n n!)$ permutations which lead to different expressions for $\delta(t_{i_1}-t_{i_2})\dots\delta(t_{i_{2n-1}}-t_{i_{2n}})$ [see (2.64)]. In particular for $n=2$ we have

$$\langle \Gamma(t_1)\Gamma(t_2)\Gamma(t_3)\rangle = 0$$
$$\langle \Gamma(t_1)\Gamma(t_2)\Gamma(t_3)\Gamma(t_4)\rangle = q^2[\delta(t_1-t_2)\,\delta(t_3-t_4) + \delta(t_1-t_3)\,\delta(t_2-t_4) + \delta(t_1-t_4)\,\delta(t_2-t_3)]\,. \tag{3.5}$$

It should be noted that for singular correlation functions like (3.4, 5) the distribution function makes sense only, if the σ_{jk} are finite and if the limit $\sigma_{ij}\to 0$ is considered. We may use for instance the representation of the δ function

$$\delta_\varepsilon(t) = \begin{cases} \dfrac{1}{\varepsilon}, & -\dfrac{\varepsilon}{2} < t < \dfrac{\varepsilon}{2} \\ 0\,, & \text{elsewhere} \end{cases} \tag{3.6}$$

in (3.4, 5) and then take the limit $\varepsilon\to 0$.

We now want to solve (3.1) for the initial condition that at time $t=0$ the stochastic variable v has the sharp value v_0. For this initial condition the solution of (3.1) reads

$$v(t) = v_0 \mathrm{e}^{-\gamma t} + \int_0^t \mathrm{e}^{-\gamma(t-t')}\Gamma(t')\,\mathrm{d}t'\,. \tag{3.7}$$

By using (3.2) we obtain for the correlation function of the velocity

$$\langle v(t_1)v(t_2)\rangle = v_0^2 \mathrm{e}^{-\gamma(t_1+t_2)} + \int_0^{t_1}\int_0^{t_2} \mathrm{e}^{-\gamma(t_1+t_2-t_1'-t_2')}q\,\delta(t_1'-t_2')\,\mathrm{d}t_1'\,\mathrm{d}t_2'\,. \tag{3.8}$$

To calculate the double integral, we integrate over t_2' first. The integration over t_1' then runs only from 0 to t_2 or t_1 whatever is less (Fig. 3.1). We therefore have

$$\int_0^{t_1}\int_0^{t_2}\dots\mathrm{d}t_1'\,\mathrm{d}t_2' = q\int_0^{\min(t_1,t_2)} \mathrm{e}^{-\gamma(t_1+t_2-2t_1')}\mathrm{d}t_1'$$
$$= \frac{q}{2\gamma}(\mathrm{e}^{-\gamma|t_1-t_2|} - \mathrm{e}^{-\gamma(t_1+t_2)})\,.$$

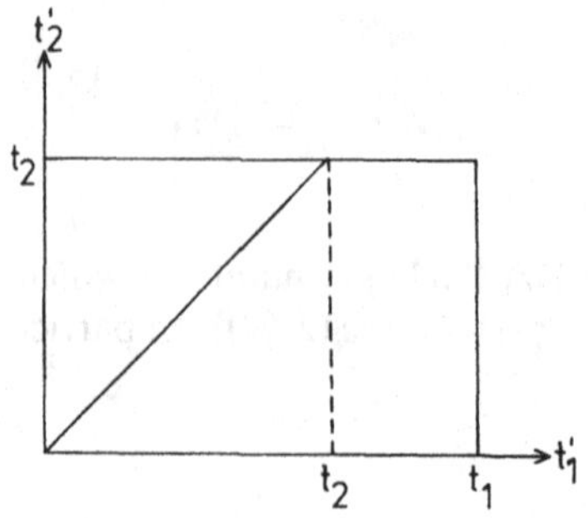

Fig. 3.1. Region of integration of the integral in (3.8) for $t_1 \geqq t_2$. The integrand is different from zero only on the line $t_2' = t_1'$

It is easily seen that (3.8) does not change by interchanging t_1 and t_2. Thus the final result for the velocity correlation function is given by

$$\langle v(t_1)\, v(t_2)\rangle = v_0^2 \mathrm{e}^{-\gamma(t_1+t_2)} + \frac{q}{2\gamma}(\mathrm{e}^{-\gamma|t_1-t_2|} - \mathrm{e}^{-\gamma(t_1+t_2)})\,. \tag{3.9}$$

For large t_1 and t_2, i.e., $\gamma t_1 \gg 1$, $\gamma t_2 \gg 1$, the velocity correlation function is independent of the initial velocity v_0 and is only a function of the time difference $t_1 - t_2$, i.e.,

$$\langle v(t_1)\, v(t_2)\rangle = \frac{q}{2\gamma}\mathrm{e}^{-\gamma|t_1-t_2|}\,. \tag{3.10}$$

In the stationary state the average energy of the Brownian particle is therefore given by

$$\langle E\rangle = \frac{1}{2}m\langle [v(t)]^2\rangle = \frac{1}{2}m\frac{q}{2\gamma}\,. \tag{3.11}$$

As mentioned in the introduction, the constant q is determined so that the average energy is given according to the equipartition law of classical statistical mechanics by

$$\langle E\rangle = \tfrac{1}{2}kT\,. \tag{3.12}$$

Hence we obtain for the constant q in (3.2)

$$q = 2\gamma kT/m\,. \tag{3.13}$$

3.1.1 Mean-Squared Displacement

For the Brownian motion of a particle it is difficult to measure the velocity correlation function (3.9). It is much easier to measure the mean-square value of its displacement. If we assume that the particle starts at time $t = 0$ at $x = x_0$ with the velocity $v = v_0$, the mean-square value of its displacement at time t is given by

$$\langle (x(t) - x_0)^2 \rangle = \left\langle \left[\int_0^t v(t_1)\,\mathrm{d}t_1 \right]^2 \right\rangle = \left\langle \int_0^t v(t_1)\,\mathrm{d}t_1 \int_0^t v(t_2)\,\mathrm{d}t_2 \right\rangle$$

$$= \int_0^t \int_0^t \langle v(t_1)\, v(t_2) \rangle \,\mathrm{d}t_1\,\mathrm{d}t_2 \,. \tag{3.14}$$

Here $\langle v(t_1)\, v(t_2) \rangle$ is the velocity correlation (3.9). Because

$$\int_0^t \int_0^t \mathrm{e}^{-\gamma(t_1 + t_2)} \mathrm{d}t_1\,\mathrm{d}t_2 = \left(\frac{1 - \mathrm{e}^{-\gamma t}}{\gamma} \right)^2 , \tag{3.15}$$

$$\int_0^t \int_0^t \mathrm{e}^{-\gamma|t_1 - t_2|} \mathrm{d}t_1\,\mathrm{d}t_2 = 2 \int_0^t \mathrm{d}t_1 \int_0^{t_1} \mathrm{e}^{-\gamma(t_1 - t_2)} \mathrm{d}t_2$$

$$= \frac{2}{\gamma} t - \frac{2}{\gamma^2} (1 - \mathrm{e}^{-\gamma t}) \,, \tag{3.16}$$

we obtain

$$\langle (x(t) - x_0)^2 \rangle = \left(v_0^2 - \frac{q}{2\gamma} \right) \frac{(1 - \mathrm{e}^{-\gamma t})^2}{\gamma^2} + \frac{q}{\gamma^2} t - \frac{q}{\gamma^3} (1 - \mathrm{e}^{-\gamma t}) \,. \tag{3.17}$$

Were we to start not with the sharp velocity v_0 but with an initial velocity distribution for the stationary state, the average square of the velocity is equal to $\langle v_0^2 \rangle = q/(2\gamma)$, see (3.11), and the first term on the right-hand side of (3.17) would vanish. For very large t ($\gamma t \gg 1$) the leading term in (3.17) for both cases is

$$\langle (x(t) - x_0)^2 \rangle = 2Dt \tag{3.18}$$

with

$$D = \frac{q}{2\gamma^2} = \frac{kT}{m\gamma} \,. \tag{3.19}$$

The last relation is the well-known Einstein result [1.4] for the diffusion constant D.

If one is interested only in this large time limit one may derive (3.18) in a shorter way. By neglecting the time derivative in (3.1) we obtain

$$\langle v(t_1)\, v(t_2) \rangle \approx \frac{1}{\gamma^2} \langle \Gamma(t_1)\, \Gamma(t_2) \rangle = \frac{q}{\gamma^2}\, \delta(t_1 - t_2) \,. \tag{3.20}$$

Insertion of this expression in (3.14) leads immediately to (3.18, 19), i.e.,

$$\langle (x(t)-x_0)^2\rangle \approx \frac{q}{\gamma^2}\int_0^t\int_0^t \delta(t_1-t_2)\,\mathrm{d}t_1\,\mathrm{d}t_2 = \frac{q}{\gamma^2}t\,.$$

If we think of the Langevin force as a successive number of peaked functions with nearly zero width and nearly infinite height, the velocity then consists of a successive number of peaked functions with width $\sim\gamma$ and height $\sim\gamma^{-1}$. Neglecting the time derivative in (3.1) is therefore equivalent to replacing the peaked functions with width γ by peaked functions with nearly zero width in the velocity. If we are interested only in the slow motion of (3.14) this replacement therefore leads to the same result.

3.1.2 Three-Dimensional Case

In three dimensions we have for each velocity component v_i an equation of type (3.1), i.e.,

$$\dot{v}_i = -\gamma v_i + \Gamma_i(t)\,; \qquad i=1,2,3\,. \tag{3.21}$$

The correlation of the Langevin forces between different components is zero for the isotropic case. Thus we have

$$\langle \Gamma_i(t)\rangle = 0\,; \qquad \langle \Gamma_i(t)\Gamma_j(t')\rangle = q\,\delta_{ij}\,\delta(t-t')\,. \tag{3.22}$$

Here again q is given by (3.13). Because the different components in (3.21) are not coupled we can immediately apply the result for the one-dimensional case and obtain for the mean (average) energy in the stationary state

$$E = \frac{1}{2}m\langle v^2\rangle = \frac{1}{2}m\sum_{i=1}^{3}\langle v_i^2\rangle = \frac{1}{2}m\cdot 3\frac{q}{2\gamma} = \frac{3}{2}kT\,. \tag{3.23}$$

The mean-squared displacement for large times reads

$$\langle (\boldsymbol{x}(t)-\boldsymbol{x}(0))^2\rangle = \sum_{i=1}^{3}\langle [x_i(t)-x_i(0)]^2\rangle \approx \frac{6kT}{m\gamma}t\,. \tag{3.24}$$

3.1.3 Calculation of the Stationary Velocity Distribution Function

To obtain the velocity distribution function we may first calculate all the moments $\langle v^{2n}\rangle$, from which we get the characteristic function (2.21), whose Fourier transform is the distribution function (2.22). As shown in Chap. 4, however, the distribution function can be calculated in a much simpler way with the help of the Fokker-Planck equation.

In the stationary state, i.e., for large times, (3.7) specializes to

$$v(t) = \int_0^\infty \mathrm{e}^{-\gamma\tau}\Gamma(t-\tau)\,\mathrm{d}\tau\,. \tag{3.25}$$

To derive (3.25) from (3.7) $t-t'=\tau$ was substituted, and then the range of integration was extended to infinity because of the factor $\exp(-\gamma t)$. By using (3.4) we obtain

$$\langle v(t)^{2n+1}\rangle = 0\,, \tag{3.26}$$

$$\begin{aligned}\langle v(t)^{2n}\rangle &= \int_0^\infty \dots \int_0^\infty \mathrm{e}^{-\gamma(\tau_1+\dots+\tau_{2n})}\langle\Gamma(t-\tau_1)\dots\Gamma(t-\tau_{2n})\rangle\mathrm{d}\tau_1\dots\mathrm{d}\tau_{2n}\\ &= \frac{(2n)!}{2^n n!}\left[\int_0^\infty\int_0^\infty \mathrm{e}^{-\gamma(\tau_1+\tau_2)}q\,\delta(\tau_1-\tau_2)\,\mathrm{d}\tau_1\,\mathrm{d}\tau_2\right]^n.\end{aligned} \tag{3.27}$$

The double integral is equal to $\langle v^2\rangle = q/(2\gamma)$ giving

$$\langle v(t)^{2n}\rangle = \frac{(2n)!}{2^n n!}\left(\frac{q}{2\gamma}\right)^n. \tag{3.28}$$

The characteristic function (2.21) becomes

$$\begin{aligned}C(u) &= 1 + \sum_{n=1}^\infty (\mathrm{i}u)^n\langle v(t)^n\rangle/n!\\ &= \sum_{n=0}^\infty \frac{(\mathrm{i}u)^{2n}\langle v(t)^{2n}\rangle}{(2n)!}\\ &= \sum_{n=0}^\infty \frac{(\mathrm{i}u)^{2n}}{2^n n!}\left(\frac{q}{2\gamma}\right)^n = \sum_{n=0}^\infty \frac{1}{n!}\left(-\frac{u^2 q}{4\gamma}\right)^n\\ &= \exp\left(-\frac{u^2 q}{4\gamma}\right)\end{aligned} \tag{3.29}$$

and the distribution function is therefore given by

$$\begin{aligned}W(v) = \langle\delta(v(t)-v)\rangle &= \frac{1}{2\pi}\int_{-\infty}^\infty C(u)\,\mathrm{e}^{-\mathrm{i}uv}\mathrm{d}u\\ &= \frac{1}{2\pi}\int_{-\infty}^\infty \exp\left(-\mathrm{i}uv - \frac{u^2 q}{4\gamma}\right)\mathrm{d}u \qquad (3.30)\\ &= \sqrt{\frac{\gamma}{\pi q}}\exp\left(-\frac{\gamma v^2}{q}\right) = \sqrt{\frac{m}{2\pi kT}}\exp\left(-\frac{mv^2}{2kT}\right),\end{aligned}$$

i.e., the stationary distribution is the Maxwell distribution. Here we have denoted the stochastic variable by $v(t)$ and the corresponding variable in the distribution function by v (see also notation remark, page 15).

Because (3.25) is a linear transformation of $\Gamma(t')$ into $v(t)$, it follows already from the remark concerning linear transformation of variables (Sect. 2.3.3) that $v(t)$ must be Gaussian distributed because $\Gamma(t)$ is Gaussian distributed.

3.2 Ornstein-Uhlenbeck Process

A Langevin equation of the type

$$\dot{\xi}_i + \sum_{j=1}^{N} \gamma_{ij}\xi_j = \Gamma_i(t)\,; \qquad i=1,\ldots,N \tag{3.31}$$

with δ-correlated Gaussian distributed Langevin forces

$$\langle \Gamma_i(t)\rangle = 0\,, \qquad \langle \Gamma_i(t)\Gamma_j(t')\rangle = q_{ij}\delta(t-t')\,, \qquad q_{ij} = q_{ji} \tag{3.32}$$

describes a process which is called an Ornstein-Uhlenbeck [1.5, 7] process. The essential feature is that the homogeneous equations are linear und that the coefficients q_{ij} describing the strength of the noise do not depend on the variables ξ_k. The Langevin equation (3.1) for Brownian motion is obviously the simplest form ($N=1$) of (3.31).

It should be noted that with a vanishing matrix γ_{ij} ($\gamma_{ij}=0$) the process described by (3.31, 32) is called a Wiener process.

We now want to find a solution of (3.31) with the initial condition describing a sharp value at $t=0$:

$$\xi_i(0) = x_i\,. \tag{3.33}$$

We first look for the homogeneous solution of (3.31) with the initial condition (3.33). This solution may be written in the form

$$\xi_i^{\mathrm{h}}(t) = G_{ij}(t)x_j\,, \tag{3.34}$$

where the Green's function has to satisfy the initial condition

$$G_{ij}(0) = \delta_{ij}\,. \tag{3.35}$$

From (3.34) onwards in this section Einstein's summation convention is used, i.e., the summation is performed over indices appearing twice in the equations without writing down the summation sign. Obviously Green's function must satisfy the system of differential equations

$$\dot{G}_{ij} + \gamma_{ik}G_{kj} = 0\,. \tag{3.36}$$

A formal solution of (3.36) reads in matrix notation

$$G(t) = \exp(-\gamma t) = I - \gamma t + \tfrac{1}{2}\gamma^2 t^2 \pm \ldots , \tag{3.37}$$

where the matrix elements of γ and G are given by γ_{ij} and G_{ij}, respectively.

For the inhomogeneous solution we make the ansatz (method of variation of the constant)

$$\xi_i^{\text{inh}}(t) = G_{ij}(t)\, c_j(t) , \tag{3.38}$$

which leads to

$$G_{ij}(t)\, \dot{c}_j(t) = \Gamma_i(t) . \tag{3.39}$$

Because the inverse of the matrix G (3.37) may be expressed by

$$G^{-1}(t) = G(-t) , \tag{3.40}$$

(the Green's function can be defined for positive as well as for negative times), and because

$$G(t)G^{-1}(t') = G(t)G(-t') = G(t-t') , \tag{3.41}$$

we finally obtain for the inhomogeneous solution with the initial value $\xi_i^{\text{inh}}(0) = 0$

$$\begin{aligned} \xi_i^{\text{inh}}(t) &= \int_0^t G_{ij}(t-t')\,\Gamma_j(t')\,dt' \\ &= \int_0^t G_{ij}(t')\,\Gamma_j(t-t')\,dt' . \end{aligned} \tag{3.42}$$

Therefore the general solution of (3.31) with the initial condition (3.33) is given by

$$\xi_i(t) = \xi_i^{\text{h}}(t) + \xi_i^{\text{inh}}(t) = G_{ij}(t)\, x_j + \int_0^t G_{ij}(t')\,\Gamma_j(t-t')\,dt' . \tag{3.43}$$

3.2.1 Calculation of Moments

Using (3.32) the following results are derived from (3.43) for the first moment and the variance

$$M_i(t) = \langle \xi_i(t) \rangle = G_{ij}(t)\, x_j , \tag{3.44}$$

$$\sigma_{ij}(t) = \sigma_{ji}(t) = \langle[\xi_i(t) - \langle\xi_i(t)\rangle][\xi_j(t) - \langle\xi_j(t)\rangle]\rangle$$

$$= \int_0^t\int_0^t G_{ik}(t_1')G_{js}(t_2')q_{ks}\delta(t_1' - t_2')\,dt_1'\,dt_2'$$

$$= \int_0^t G_{ik}(t')G_{js}(t')\,dt'\,q_{ks}\,. \tag{3.45}$$

We observe that σ_{ij} obeys the relation

$$\dot{\sigma}_{ij} = -\gamma_{ik}\sigma_{kj} - \gamma_{jk}\sigma_{ki} + q_{ij}\,. \tag{3.46}$$

This may be seen by differentiating (3.45) and by using (3.36)

$$\dot{\sigma}_{ij} = G_{ik}(t)G_{js}(t)q_{ks}\,,$$

$$\ddot{\sigma}_{ij} = \dot{G}_{ik}G_{js}q_{ks} + G_{ik}\dot{G}_{js}q_{ks} = -\gamma_{il}G_{lk}G_{js}q_{ks} - G_{ik}\gamma_{jl}G_{ls}q_{ks}\,,$$

i.e., we have

$$\ddot{\sigma}_{ij} = -\gamma_{il}\dot{\sigma}_{lj} - \gamma_{jl}\dot{\sigma}_{li}\,.$$

Integrating the last equation from zero to time t, we obtain (3.46) because $\dot{\sigma}_{ij}(0) = q_{ij}$ and $\sigma_{ij}(0) = 0$.

If the real parts of the eigenvalues of the matrix γ_{ij} are larger than zero the Green's function $G_{ij}(t)$ vanishes for *large times t*, giving

$$\sigma_{ij}(\infty) = \int_0^\infty G_{ik}(t)G_{js}(t)\,dt\,q_{ks}\,. \tag{3.47}$$

Using (3.37), we get from (3.44, 45) for *small times* $t \geqq 0$.

$$M_i(t) = \langle\xi_i(t)\rangle = x_i - \gamma_{ij}x_j t + \tfrac{1}{2}\gamma_{ik}\gamma_{kj}x_j t^2 \pm \dots\,, \tag{3.48}$$

$$\sigma_{ij}(t) = q_{ij}t - \tfrac{1}{2}(\gamma_{ik}q_{kj} + \gamma_{jk}q_{ik})t^2 \pm \dots\,. \tag{3.49}$$

Thus for small times the matrix γ_{ij} determines the motion of the first moment whereas the matrix q_{ij} determines the motion of the variance. For a one-dimensional diffusion process the variance $\langle[\xi(t) - \langle\xi(t)\rangle]^2\rangle$ is proportional to the time, where the proportionality constant is the diffusion coefficient. For a general Ornstein-Uhlenbeck process, in addition to the mean motion described by (3.44) there is a diffusion process which for small times is given by the matrix q_{ij}. For a *Wiener process* ($\gamma_{ij} = 0$), $G_{ij}(t) = \delta_{ij}$ and we obtain

$$\sigma_{ij}(t) = q_{ij}t \quad \text{for} \quad \gamma_{ij} = 0 \tag{3.50}$$

for all times $t \geqq 0$.

For later purposes it is mentioned that the higher central moments, defined by

$$\sigma_{i_1,\ldots,i_n}(t) = \langle [\xi_{i_1}(t) - \langle \xi_{i_1}(t)\rangle] \ldots [\xi_{i_n}(t) - \langle \xi_{i_n}(t)\rangle]\rangle , \tag{3.51}$$

vanish for all odd n (3.4), and that they are proportional to $t^{n/2}$ for small times for even n. For one variable, for instance for $n = 4$ for small times,

$$\begin{aligned}\sigma_{1,1,1,1}(t) &= q^2 \int_0^t\int_0^t\int_0^t\int_0^t [\delta(t_1-t_2)\,\delta(t_3-t_4) \\ &\quad + \delta(t_1-t_3)\,\delta(t_2-t_4) + \delta(t_1-t_4)\,\delta(t_2-t_3)]\,\mathrm{d}t_1\mathrm{d}t_2\mathrm{d}t_3\mathrm{d}t_4 \\ &= 3q^2t^2 .\end{aligned} \tag{3.52}$$

3.2.2 Correlation Function

One of the simplest two-time correlation functions is given by

$$K_{ij}(\tau, t) = \langle \xi_i(t+\tau)\,\xi_j(t)\rangle . \tag{3.53}$$

If we start with the initial value $\xi_i(t)$ at time t, the formal solution of (3.31) reads, compare (3.43),

$$\xi_i(t+\tau) = G_{is}(\tau)\,\xi_s(t) + \int_t^{t+\tau} G_{ij}(t-t')\,\Gamma_j(t')\,\mathrm{d}t' \;; \qquad \tau \geqq 0 . \tag{3.54}$$

If we insert this expression into (3.53) and take the average, the term containing the Langevin force drops out and we obtain for

$\tau \geqq 0$

$$\begin{aligned}K_{ij}(\tau, t) &= \langle \xi_i(t+\tau)\,\xi_j(t)\rangle \\ &= G_{is}(\tau)\,\langle \xi_s(t)\,\xi_j(t)\rangle = G_{is}(\tau)\,K_{sj}(0, t) .\end{aligned} \tag{3.55a}$$

Equation (3.55a) is called a regression theorem. It states that the two-time correlation function $\langle \xi_i(t+\tau)\,\xi_j(t)\rangle$ can be obtained from the one-time matrix $\langle \xi_i(t)\,\xi_j(t)\rangle$ by matrix multiplication with the Green's function of the noise-free equation (3.36). For

$\tau \leqq 0$

$$\begin{aligned}K_{ij}(\tau, t) &= \langle \xi_i(t-|\tau|)\,\xi_j(t-|\tau|+|\tau|)\rangle \\ &= G_{js}(|\tau|)\,\langle \xi_i(t-|\tau|)\,\xi_s(t-|\tau|)\rangle \\ &= G_{js}(|\tau|)\,K_{is}(0, t-|\tau|) .\end{aligned} \tag{3.55b}$$

Correlation Function for the Stationary State

If the eigenvalues of the evolution matrix γ_{ij} are all larger than zero, a stationary solution exists. One can then take the limit $t \to \infty$ and thus obtain the correlation function $K_{ij}(\tau) = K_{ij}(\tau, \infty)$ in the stationary state for

$\tau \geqq 0$

$$K_{ij}(\tau) = G_{is}(\tau) \langle \xi_s(\infty) \xi_j(\infty) \rangle = G_{is}(\tau) \sigma_{sj}(\infty) , \tag{3.56a}$$

and for

$\tau \leqq 0$

$$\begin{aligned} K_{ij}(\tau) = G_{js}(|\tau|) \langle \xi_i(\infty) \xi_s(\infty) \rangle &= G_{js}(|\tau|) \sigma_{is}(\infty) \\ &= G_{js}(|\tau|) \sigma_{si}(\infty) \\ &= K_{ji}(|\tau|) . \end{aligned} \tag{3.56b}$$

Thus for all τ

$$K_{ij}(\tau) = K_{ji}(-\tau) , \tag{3.57}$$

which generally holds for every stationary process. [For a stationary process, (3.53) does not depend on t. Replacing t by $t-\tau$ in (3.53) leads to (3.57)].

3.2.3 Solution by Fourier Transformation

Introducing the Fourier transform of the stochastic variables ξ_i and of the Langevin forces Γ_j

$$\begin{aligned} \tilde{\xi}_i(\omega) &= \int_{-\infty}^{\infty} \mathrm{e}^{-\mathrm{i}\omega t} \xi_i(t)\, \mathrm{d}t \\ \tilde{\Gamma}_j(\omega) &= \int_{-\infty}^{\infty} \mathrm{e}^{-\mathrm{i}\omega t} \Gamma_j(t)\, \mathrm{d}t \end{aligned} \tag{3.58}$$

we immediately obtain the Fourier transform of the inhomogeneous solution of (3.31). Because

$$\int_{-\infty}^{\infty} \mathrm{e}^{-\mathrm{i}\omega t} \dot{\xi}_j(t)\, \mathrm{d}t = \mathrm{e}^{-\mathrm{i}\omega t} \xi_j(t) \Big|_{-\infty}^{\infty} + \mathrm{i}\omega \int_{-\infty}^{\infty} \mathrm{e}^{-\mathrm{i}\omega t} \xi_j(t)\, \mathrm{d}t ,$$

then (neglecting the terms at $t = \pm\infty$)

$$(\mathrm{i}\omega \delta_{jk} + \gamma_{jk}) \tilde{\xi}_k(\omega) = \tilde{\Gamma}_j(\omega) ,$$

i.e.,

$$\tilde{\xi}_j(\omega) = (\gamma + \mathrm{i}\omega I)^{-1}_{jk} \tilde{\Gamma}_k(\omega) . \tag{3.59}$$

We now introduce the spectral density matrices of the stochastic variables and Langevin forces [cf. (2.85)]

$$\begin{aligned} \langle \tilde{\xi}_j(\omega)\,\tilde{\xi}_k^*(\omega')\rangle &= \pi S_{jk}^{(\xi)}(\omega)\,\delta(\omega-\omega') \\ \langle \tilde{\Gamma}_j(\omega)\,\tilde{\Gamma}_k^*(\omega')\rangle &= \pi S_{jk}^{(\Gamma)}(\omega)\,\delta(\omega-\omega')\,. \end{aligned} \tag{3.60}$$

In consequence of (3.59) both spectral densities are connected by

$$S_{jk}^{(\xi)}(\omega) = (\gamma + \mathrm{i}\,\omega I)_{jr}^{-1}(\gamma - \mathrm{i}\,\omega I)_{kl}^{-1} S_{rl}^{(\Gamma)}(\omega)\,. \tag{3.61}$$

The spectral density of the δ-correlated Langevin force (3.32) is, using (2.86),

$$S_{jk}^{(\Gamma)}(\omega) = 2q_{jk}\,. \tag{3.62}$$

Therefore we finally obtain for the spectral density matrix of the stochastic variables

$$S_{jk}^{(\xi)}(\omega) = 2(\gamma + \mathrm{i}\omega I)_{jr}^{-1}(\gamma - \mathrm{i}\,\omega I)_{kl}^{-1} q_{rl}\,. \tag{3.63}$$

This Fourier transformation method for obtaining the spectral density is sometimes called *Rice's* method [2.10].

From the Wiener-Khintchine theorem [see (2.86) for the case of one variable] the spectral density is the Fourier transform of the correlation matrix in the stationary state (3.56a, b)

$$\begin{aligned} S_{jk}(\omega) &= 2\int_{-\infty}^{\infty} K_{jk}(\tau)\,\mathrm{e}^{-\mathrm{i}\omega\tau}\mathrm{d}\tau \\ &= 2\int_0^{\infty} K_{jk}(\tau)\,\mathrm{e}^{-\mathrm{i}\omega\tau}\mathrm{d}\tau + 2\int_0^{\infty} K_{kj}(\tau)\,\mathrm{e}^{\mathrm{i}\omega\tau}\mathrm{d}\tau \\ &= 2(\gamma + \mathrm{i}\,\omega I)_{js}^{-1}\sigma_{sk}(\infty) + 2(\gamma - \mathrm{i}\,\omega I)_{ks}^{-1}\sigma_{sj}(\infty)\,. \end{aligned} \tag{3.64}$$

In deriving (3.64) we used the formal solution (3.37) and

$$\begin{aligned} \int_0^{\infty} G_{jk}(\tau)\,\mathrm{e}^{\pm\mathrm{i}\omega\tau}\mathrm{d}\tau &= \int_0^{\infty} [\mathrm{e}^{-(\gamma\mp\mathrm{i}\omega I)\tau}]_{jk}\mathrm{d}\tau \\ &= (\gamma \mp \mathrm{i}\,\omega I)_{jk}^{-1}\,. \end{aligned} \tag{3.65}$$

To see whether (3.64 and 63) are equivalent, on the right-hand side of (3.63) we introduce for q_{rl} the expression

$$q_{rl} = \gamma_{rs}\sigma_{sl}(\infty) + \gamma_{ls}\sigma_{sr}(\infty)\,, \tag{3.66}$$

which follows from (3.46) for the stationary state. Thus the right-hand side of (3.63) becomes

$$\text{rhs}\,(3.63) = 2(\boldsymbol{\gamma}+\mathrm{i}\,\omega\boldsymbol{I})^{-1}_{jr}\,\gamma_{rs}\sigma_{sl}(\infty)(\boldsymbol{\gamma}-\mathrm{i}\,\omega\boldsymbol{I})^{-1}_{kl}$$
$$+\,2(\boldsymbol{\gamma}+\mathrm{i}\,\omega\boldsymbol{I})^{-1}_{jr}(\boldsymbol{\gamma}-\mathrm{i}\,\omega\boldsymbol{I})^{-1}_{kl}\,\gamma_{ls}\sigma_{sr}(\infty)\,.$$

Because

$$(\boldsymbol{\gamma}\pm\mathrm{i}\,\omega\boldsymbol{I})^{-1}\boldsymbol{\gamma} = \boldsymbol{I}\mp\mathrm{i}\,\omega(\boldsymbol{\gamma}\pm\mathrm{i}\,\omega\boldsymbol{I})^{-1}$$

then

$$\text{rhs}\,(3.63) = 2\,\sigma_{jl}(\infty)(\boldsymbol{\gamma}-\mathrm{i}\,\omega\boldsymbol{I})^{-1}_{kl} + 2(\boldsymbol{\gamma}+\mathrm{i}\,\omega\boldsymbol{I})^{-1}_{jr}\,\sigma_{kr}(\infty)$$
$$-\,2\mathrm{i}\,\omega(\boldsymbol{\gamma}+\mathrm{i}\,\omega\boldsymbol{I})^{-1}_{js}\,\sigma_{sl}(\infty)(\boldsymbol{\gamma}-\mathrm{i}\,\omega\boldsymbol{I})^{-1}_{kl}$$
$$+\,2\mathrm{i}\,\omega(\boldsymbol{\gamma}+\mathrm{i}\,\omega\boldsymbol{I})^{-1}_{jr}\,\sigma_{sr}(\infty)(\boldsymbol{\gamma}-\mathrm{i}\,\omega\boldsymbol{I})^{-1}_{ks}\,.$$

By changing indices and using the symmetry relation (3.45), it is easily seen that the last two terms cancel and that the first two terms agree with the right-hand side of (3.64).

Finally I want to remark that the general solution (3.61) [as well as (3.43)] is valid for arbitrary spectral densities $S_{ij}^{(\Gamma)}(\omega)$, i.e., for arbitrary correlation matrices $\langle \Gamma_i(t)\Gamma_j(t')\rangle = f_{ij}(t-t')$.

3.3 Nonlinear Langevin Equation, One Variable

For one stochastic variable ξ, the general Langevin equation has the form

$$\dot{\xi} = h(\xi,t) + g(\xi,t)\,\Gamma(t)\,. \tag{3.67}$$

The Langevin force $\Gamma(t)$ is again assumed to be a Gaussian random variable with zero mean and δ correlation function (3.2). The constant q in (3.2) describing the noise strength may be absorbed into the function g. We choose the following normalization

$$\langle \Gamma(t)\rangle = 0\,; \qquad \langle \Gamma(t)\Gamma(t')\rangle = 2\delta(t-t')\,. \tag{3.68}$$

For constant g, (3.67) is called a Langevin equation with an additive noise force. For g depending on ξ one speaks of a Langevin equation with a multiplicative noise term. This distinction between additive and multiplicative noise may not be considered very significant because for the one variable equation (3.67), for time-independent h and g and for $g \neq 0$, the multiplicative noise always becomes an additive noise by a simple transformation of variables. Dividing (3.67) by g gives

$$\dot{\eta} = \frac{\dot{\xi}}{g(\xi)} = \frac{h(\xi)}{g(\xi)} + \Gamma(t) = h_1(\eta) + \Gamma(t)\,. \tag{3.69}$$

Thus for the variable

$$\eta \equiv f(\xi) = \int^{\xi} \frac{d\xi'}{g(\xi')}, \qquad \xi = f^{(-1)}(\eta) \tag{3.70}$$

we get an equation with an additive noise force where the function h_1 in (3.69) is given by

$$h_1(\eta) = h(f^{(-1)}(\eta))/g(f^{(-1)}(\eta)) . \tag{3.71}$$

For x-dependent g the following difficulty arises. Because the noise $\Gamma(t)$ has no correlation time (3.68), it is not yet clear which ξ value one has to use in the function g in (3.67). If, for instance, $\Gamma(t)$ is considered as a sum of peaked functions with no width, the stochastic variable $\xi(t)$ will jump at every time t_ν when such a peaked function occurs. The question then arises: Which ξ value must one use in g? One may use the value ξ just before t_ν or just after t_ν or some value between these two values. From a purely mathematical point one cannot answer this question, but one has to use some additional specification, for instance, the Itô or the Stratonovich definition, Sect. 3.3.3. Here we assume as it is usually done in physics that the $\delta(t)$ function is replaced by a function $\delta_\varepsilon(t)$ with a very small finite width ε, i.e., for instance by (3.6). In the final result one then has to take the limit $\varepsilon \to 0$. With this procedure we see that the average value of $\langle g(\xi, t)\Gamma(t)\rangle$ is no longer zero if g depends on ξ. The above average leads to the "spurious" or "noise-induced" drift.

3.3.1 Example

This spurious drift may be simply exemplified by the equation (a is a constant)

$$\dot{\xi} = a\xi\Gamma(t) . \tag{3.72}$$

It is a simplified form of the Kubo oscillator (App. 1). The formal solution reads

$$\xi(t) = x \exp\left[a\int_0^t \Gamma(t')dt'\right], \tag{3.73}$$

where $\xi(0) = x$ is assumed to be the sharp value of the stochastic variable at time $t = 0$. For this simple example the average of (3.73) can be calculated exactly. To perform the average, we need not assume that the noise force is δ-correlated. For the following derivation we need to assume only that the noise force, now denoted by $\varepsilon(t)$, is a stationary process with Gaussian distribution and zero mean value. By expanding the exponential into a power series we have

$$\left\langle \exp\left[a\int_0^t \varepsilon(t')\,dt'\right]\right\rangle = 1 + a\int_0^t \langle\varepsilon(t_1)\rangle dt_1 + \frac{1}{2!}a^2\int_0^t\int_0^t \langle\varepsilon(t_1)\,\varepsilon(t_2)\rangle dt_1\,dt_2$$

$$+\ldots+\frac{a^{2n}}{(2n)!}\int_0^t\ldots\int_0^t\langle\varepsilon(t_1)\ldots\varepsilon(t_{2n})\rangle dt_1\ldots dt_{2n}+\ldots\,.$$

The Gaussian property of $\varepsilon(t_i)$, $\langle\varepsilon(t_i)\rangle = 0$ and its stationarity requires (2.64)

$$\begin{aligned}&\langle\varepsilon(t_1)\,\varepsilon(t_2)\rangle = \Phi(t_1-t_2)\\ &\langle\varepsilon(t_1)\ldots\varepsilon(t_{2n-1})\rangle = 0\\ &\langle\varepsilon(t_1)\ldots\varepsilon(t_{2n})\rangle = \sum_{P_d}\Phi(t_{i_1}-t_{i_2})\,\Phi(t_{i_3}-t_{i_4})\ldots\Phi(t_{i_{2n-1}}-t_{i_{2n}})\,.\end{aligned} \tag{3.74}$$

Because there are $(2n)!/(2^n n!)$ different possibilities for the permutation of the $2n$ times t_i (the interchange of two times in the correlation function Φ and the permutation of the n correlation functions Φ do not lead to different results), then

$$\int_0^t\ldots\int_0^t\langle\varepsilon(t_1)\ldots\varepsilon(t_{2n})\rangle dt_1\ldots dt_{2n} = \frac{(2n)!}{2^n n!}\left[\int_0^t\int_0^t\Phi(t_1-t_2)\,dt_1\,dt_2\right]^n.$$

We are now able to perform the summation of the power series of the exponential, leading to

$$\left\langle \exp\left[a\int_0^t \varepsilon(t')\,dt'\right]\right\rangle = \exp\left[\frac{1}{2}a^2\int_0^t\int_0^t\langle\varepsilon(t_1)\,\varepsilon(t_2)\rangle dt_1\,dt_2\right]. \tag{3.75}$$

For non-Gaussian processes one obtains an expansion in the exponential on the right-hand side in terms of cumulants (similar to those in (2.37, 42) [3.1]. For the δ-correlated Langevin force $\Gamma(t)$ (3.68), the double integral in (3.75) reduces to

$$\int_0^t\int_0^t\langle\Gamma(t_1)\,\Gamma(t_2)\rangle dt_1\,dt_2 = 2t$$

and thus we finally arive at

$$\langle\xi(t)\rangle = x\exp(a^2 t)\,. \tag{3.76}$$

Obviously the average value $\langle\xi(t)\rangle$ obeys the differential equation

$$\langle\dot\xi(t)\rangle = a^2\langle\xi(t)\rangle \tag{3.76a}$$

with the initial condition $\langle\xi(0)\rangle = x$. Whereas for real a (positive or negative) the average value $\langle\xi(t)\rangle$ increases exponentially, it decreases in time for pure imaginary a. [For complex a it oscillates in time, increasing for $(\mathrm{Re}\{a\})^2 > (\mathrm{Im}\{a\})^2$ and decreasing for $(\mathrm{Re}\{a\})^2 < (\mathrm{Im}\{a\})^2$ in time].

For small times we thus get a drift from the stochastic force

$$\frac{\mathrm{d}}{\mathrm{d}t}\langle\xi(t)\rangle\big|_{t=0} = a^2 x\,, \tag{3.77}$$

called a spurious or noise-induced drift.

The average of every moment can also be calculated. We have

$$\langle[\xi(t)]^n\rangle = \left\langle x^n \exp\left[na\int_0^t \Gamma(t')\,\mathrm{d}t'\right]\right\rangle = x^n \exp[(na)^2 t]\,. \tag{3.78}$$

For later purposes we are interested in the nth moment centered at the initial value x $(n \geqq 1)$

$$\bar{M}_n(t) = \langle[\xi(t)-x]^n\rangle = x^n \sum_{\nu=0}^{n}\binom{n}{\nu}\exp[(\nu a)^2 t](-1)^{n-\nu}\,. \tag{3.79}$$

By expanding the exponential we get

$$\bar{M}_n(t) = x^n \sum_{m=1}^{\infty}\frac{(a^2 t)^m}{m!}\sum_{\nu=0}^{n}\binom{n}{\nu}\nu^{2m}(-1)^{n-\nu}\,, \tag{3.80}$$

where the sum over m starts at 1 because $\sum_{\nu=0}^{n}\binom{n}{\nu}(-1)^{n-\nu} = (1-1)^n = 0$. The last sum can be expressed by

$$\begin{aligned} S_{n,m} &= \sum_{\nu=0}^{n}\binom{n}{\nu}\nu^{2m}(-1)^{n-\nu} = \left(\frac{\mathrm{d}}{\mathrm{d}u}\right)^{2m}(\mathrm{e}^u-1)^n\bigg|_{u=0} \\ &= \left(\frac{\mathrm{d}}{\mathrm{d}u}\right)^{2m}\left(u+\frac{u^2}{2!}+\dots\right)^n\bigg|_{u=0}\,, \end{aligned} \tag{3.81}$$

which may be proved by using the binomial series. Since we are especially interested in the behavior of $\bar{M}_n(t)$ for small t, we therefore calculate the first non-vanishing element of (3.80). Using the last expression of (3.81) we find $(n \geqq 1)$

$$\begin{aligned} &S_{2n-1,m} = 0 \quad \text{for} \quad m < n\,, \quad S_{2n-1,n} = (n-\tfrac{1}{2})(2n)! \\ &S_{2n,m} \;\;\;= 0 \quad \text{for} \quad m < n\,, \quad S_{2n,n} = (2n)!\,. \end{aligned} \tag{3.82}$$

Hence we get the following expansion for small t $(n \geqq 1)$:

$$\bar{M}_{2n-1}(t) = x^{2n-1}\left(n-\frac{1}{2}\right)\frac{(2n)!}{n!}(a^2t)^n + \dots$$
$$\bar{M}_{2n}(t) = x^{2n}\frac{(2n)!}{n!}(a^2t)^n + \dots . \tag{3.83}$$

Defining Kramers-Moyal coefficients $D^{(n)}(x)$ as the first time derivative of the central moment divided by $n!$ (Sect. 3.3.2), i.e,

$$D^{(n)}(x) = \frac{1}{n!}\frac{\mathrm{d}}{\mathrm{d}t}\bar{M}_n(t)\bigg|_{t=0},$$

we have

$$D^{(1)}(x) = a^2x\,, \quad D^{(2)}(x) = a^2x^2\,, \quad D^{(n)}(x) = 0 \quad \text{for} \quad n \geqq 3\,. \tag{3.84}$$

3.3.2 Kramers-Moyal Expansion Coefficients

Usually a formal general solution of the stochastic differential equation (3.67) cannot be given. As shown in Chap. 4, we can set up a Fokker-Planck equation by which the probability density of the stochastic variable can be calculated. In this Fokker-Planck equation the following Kramers-Moyal expansion coefficients enter:

$$D^{(n)}(x,t) = \frac{1}{n!}\lim_{\tau\to 0}\frac{1}{\tau}\langle[\xi(t+\tau)-x]^n\rangle\bigg|_{\xi(t)=x}. \tag{3.85}$$

In (3.85) $\xi(t+\tau)$ $(\tau>0)$ is a solution of (3.67) which at time t has the sharp value $\xi(t)=x$. To derive these Kramers-Moyal expansion coefficients, we first write the differential equation (3.67) in the form of an integral equation

$$\xi(t+\tau)-x = \int_t^{t+\tau}[h(\xi(t'),t') + g(\xi(t'),t')\,\Gamma(t')]\,\mathrm{d}t' \tag{3.86}$$

and assume that h and g can be expanded according to ($\partial/\partial x$ performed on h and g is denoted by a prime, i.e., $(\partial/\partial\xi(t'))\,h(\xi(t'),t')\,|_{\xi(t')=x} \equiv (\partial/\partial x)\,h(x,t') \equiv h'(x,t')$ and similar for g)

$$h(\xi(t'),t') = h(x,t') + h'(x,t')(\xi(t')-x) + \dots$$
$$g(\xi(t'),t') = g(x,t') + g'(x,t')(\xi(t')-x) + \dots . \tag{3.87}$$

Inserting (3.87) in (3.86) leads to

$$\xi(t+\tau)-x = \int_t^{t+\tau} h(x,t')\mathrm{d}t' + \int_t^{t+\tau} h'(x,t')(\xi(t')-x)\mathrm{d}t' + \dots$$
$$+ \int_t^{t+\tau} g(x,t')\Gamma(t')\mathrm{d}t' + \int_t^{t+\tau} g'(x,t')(\xi(t')-x)\Gamma(t')\mathrm{d}t' + \dots . \tag{3.88}$$

For $\xi(t')-x$ in the integrand we iterate (3.88), producing

$$\begin{aligned}\xi(t+\tau)-x = &\int_t^{t+\tau} h(x,t')\mathrm{d}t' + \int_t^{t+\tau} h'(x,t')\int_t^{t'} h(x,t'')\mathrm{d}t''\mathrm{d}t' \\ &+ \int_t^{t+\tau} h'(x,t')\int_t^{t'} g(x,t'')\Gamma(t'')\mathrm{d}t''\mathrm{d}t' + \dots \\ &+ \int_t^{t+\tau} g(x,t')\Gamma(t')\mathrm{d}t' + \int_t^{t+\tau} g'(x,t')\int_t^{t'} h(x,t'')\Gamma(t')\mathrm{d}t''\mathrm{d}t' \\ &+ \int_t^{t+\tau} g'(x,t')\int_t^{t'} g(x,t'')\Gamma(t'')\Gamma(t')\mathrm{d}t''\mathrm{d}t' + \dots \,.\end{aligned} \tag{3.89}$$

By repeated iterations only Langevin forces and the known functions $g(x,t)$ and $h(x,t)$ and their derivatives appear on the right-hand side of (3.89). If we now take the average of (3.89) we have from (3.68)

$$\begin{aligned}\langle \xi(t+\tau)-x\rangle = &\int_t^{t+\tau} h(x,t')\mathrm{d}t' + \int_t^{t+\tau}\int_t^{t'} h'(x,t')h(x,t'')\mathrm{d}t''\mathrm{d}t' + \dots \\ &+ \int_t^{t+\tau} g'(x,t')\int_t^{t'} g(x,t'')2\delta(t''-t')\mathrm{d}t''\mathrm{d}t' + \dots \,.\end{aligned} \tag{3.90}$$

If we take for the δ function, for instance, the function (3.6) or any other representation $\delta_\varepsilon(t)$ symmetric around the origin, and finally take $\varepsilon\to 0$, we have

$$\begin{aligned}\int_t^{t'} g(x,t'')2\delta(t''-t')\mathrm{d}t'' &= g(x,t')\int_t^{t'} 2\delta(t''-t')\mathrm{d}t'' \\ &= g(x,t')\end{aligned} \tag{3.91}$$

and therefore get

$$\begin{aligned}\langle \xi(t+\tau)-x\rangle = &\int_t^{t+\tau} h(x,t')\mathrm{d}t' + \int_t^{t+\tau}\int_t^{t'} h'(x,t')h(x,t'')\mathrm{d}t''\mathrm{d}t' + \dots \\ &+ \int_t^{t+\tau} g'(x,t')g(x,t')\mathrm{d}t' + \dots \,.\end{aligned} \tag{3.92}$$

In the limit $\tau\to 0$ we thus arrive at

$$D^{(1)}(x,t) = h(x,t) + g'(x,t)\,g(x,t)\,. \tag{3.93}$$

The other integrals not written down in (3.92) do not contribute in the limit $\tau\to 0$. This is seen as follows: each Langevin force on the right-hand side of (3.89) is accompanied by an integral. The lowest terms are written down in (3.89). Higher terms vanish for the limit in (3.85). For instance, integrals of the form

$$\left\langle \int\limits_t^{t+\tau} \ldots \Gamma(t_1) \int\limits_t^{t_1} \ldots \Gamma(t_2) \int\limits_t^{t_2} \ldots \Gamma(t_3) \int\limits_t^{t_3} \ldots \Gamma(t_4) \mathrm{d}t_1 \mathrm{d}t_2 \mathrm{d}t_3 \mathrm{d}t_4 \right\rangle$$

can only give a contribution proportional to τ^2 which vanishes for the limit in (3.85). Integrals not containing the Langevin force are proportional to τ^n, where n is the number of integrals. For the limit in (3.85) thus only terms with only one such integral will survive. Using these arguments we obtain for the second coefficient

$$\begin{aligned} D^{(2)}(x,t) &= \frac{1}{2} \lim_{\tau \to 0} \frac{1}{\tau} \int\limits_t^{t+\tau} \int\limits_t^{t+\tau} g(x,t')\, g(x,t'')\, 2\delta(t'-t'')\mathrm{d}t'\, \mathrm{d}t'' \\ &= g^2(x,t) \,. \end{aligned} \tag{3.94}$$

By using these arguments for the higher coefficients $D^{(n)}$ we conclude that they all vanish for $n \geqq 3$. The final result is

$$\begin{aligned} D^{(1)}(x,t) &= h(x,t) + \frac{\partial g(x,t)}{\partial x} g(x,t) \\ D^{(2)}(x,t) &= g^2(x,t) \\ D^{(n)}(x,t) &= 0 \quad \text{for} \quad n \geqq 3 \,. \end{aligned} \tag{3.95}$$

In addition to the deterministic drift $h(x,t)$, $D^{(1)}$ contains a term which is called the spurious drift or the noise-induced drift

$$D^{(1)}_{\text{noise-ind}} = \frac{\partial g(x,t)}{\partial x} g(x,t) = \frac{1}{2} \frac{\partial}{\partial x} D^{(2)}(x,t) \,. \tag{3.96}$$

It stems from the fact that during a change of $\Gamma(t)$ also $\xi(t)$ changes and therefore $\langle g(\xi(t),t)\Gamma(t)\rangle$ is no longer zero. For example, we see that $D^{(1)}$ in (3.84) is just the noise-induced drift.

3.3.3 Itô's and Stratonovich's Definitions

As already mentioned, from a mathematical point of view the stochastic differential equation (3.67) with a Langevin force which is Gaussian distributed and having the δ correlation function (3.68) is not completely defined.

It is obviously impossible to plot a realization with a δ-correlated noise, i.e., one where $\Gamma(t_1)$ and $\Gamma(t_2)$ are completely independent for arbitrary small $|t_2 - t_1|$. To see something, the correlation function of $\Gamma(t)$ must have at least a width of the order of the line thickness in the drawing. In Fig. 3.2 a realization of the Langevin force with the finite correlation time τ_c is shown. As seen from the drawing, the integrated variable W is a much smoother function. It even exists if the correlation time of $\Gamma(t)$ is zero.

In the beginning of this section we assumed that the correlation function of the Langevin force has a very small but finite width, let us say of the order ε.

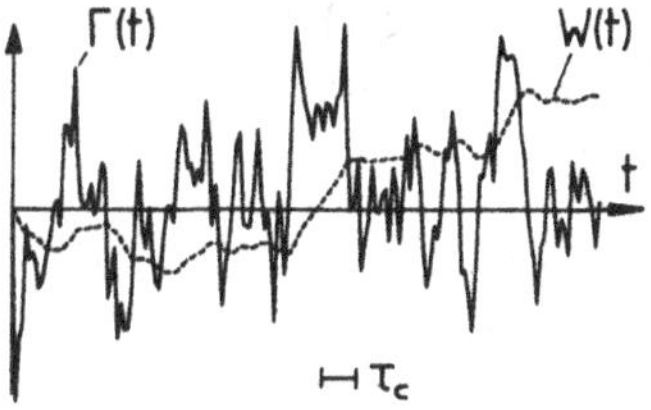

Fig. 3.2. A crude picture of a realization of the Langevin force $\Gamma(t)$ and of $W(t) = \int_0^t \Gamma(t')\,dt'$

This width must be small compared to the time variations incorporated in the system. For an Ornstein-Uhlenbeck process, for instance, it must be small compared to all the inverse eigenvalues of the evolution matrix γ_{ij}. If we have a finite correlation time ε, the spectral density of the Langevin force cannot be independent of the frequency any longer. The spectral density must vanish for frequencies larger than $1/\varepsilon$. For applications it seems reasonable to use a spectral density of the noise which is cut off at some large frequency, because otherwise the total power of the noise would be infinitely large. With this procedure in mind, the Langevin equation (3.67) leads to the drift and diffusion coefficients (3.95).

Mathematicians do not use this "physical procedure". (Similarly they do not use the physicists' method of dealing with the δ function.) They write down a modified form of (3.67) where the limit $\varepsilon \to 0$ can already be performed from the beginning, starting with the integral formulation (3.86) of (3.67) written as a Stieltjes integral

$$\xi(t+\tau) = x + \int_t^{t+\tau} h(\xi, t')\,dt' + \int_t^{t+\tau} g(\xi, t')\,dW(t')\,. \tag{3.97}$$

The Stieltjes integral must be used, because for $\varepsilon \to 0$ $\dot{W} = \Gamma(t)$ does not exist and therefore $dW = \dot{W}dt = \Gamma(t)\,dt$ cannot be used. In (3.97) $W(t)$ is a Wiener process. In our notation $\dot{W} = \Gamma(t)$, and the increment is

$$w(\tau) = W(t+\tau) - W(t) = \int_t^{t+\tau} \Gamma(t')\,dt'\,. \tag{3.98}$$

The distribution of $w(\tau)$ is Gaussian because $\Gamma(t)$ is Gaussian distributed. All correlation functions are obtained from (3.4), and after integration all δ functions in (3.4) then disappear. We have for instance ($q = 2$, $\tau \geqq 0$, $\tau_1 \geqq 0$, $\tau_2 \geqq 0$)

$$\begin{aligned} w(0) &= 0 \\ \langle w(\tau)\rangle &= 0 \\ \langle w(\tau_2)\, w(\tau_1)\rangle &= \begin{cases} 2\tau_2 & \text{for} \quad \tau_1 \geqq \tau_2\,, \\ 2\tau_1 & \text{for} \quad \tau_1 \leqq \tau_2\,. \end{cases} \end{aligned} \tag{3.99}$$

[Equation (3.99) agrees with (3.9) for $\gamma \to 0$, $q = 2$ and $v_0 = 0$]. Thus the Wiener process defined by (3.99) and its Gaussian properties form a well-defined process existing in the limit $\varepsilon \to 0$.

In (3.97) we may now eliminate the stochastic variable $\xi(t)$ by the same iteration procedure as in (3.86). One then obtains integrals of the form

$$A_{\substack{\mathrm{I}\\ \mathrm{S}}} = \begin{matrix}(\mathrm{I})\\(\mathrm{S})\end{matrix}\int_0^{\tau} \Phi(w(\tau'),\tau')\,\mathrm{d}w(\tau')\,, \tag{3.100}$$

which are not yet defined. The *Itô* (I) [3.2] and *Stratonovich* (S) [3.3] definitions are

$$A_{\mathrm{I}} = \lim_{\Delta\to 0}\sum_{i=0}^{N-1} \Phi(w(\tau_i),\tau_i)[w(\tau_{i+1})-w(\tau_i)]\,, \tag{3.101}$$

$$A_{\mathrm{S}} = \lim_{\Delta\to 0}\sum_{i=0}^{N-1} \Phi\left(\frac{w(\tau_i)+w(\tau_{i+1})}{2},\frac{\tau_i+\tau_{i+1}}{2}\right)[w(\tau_{i+1})-w(\tau_i)]\,, \tag{3.102}$$

where

$$\Delta = \max(\tau_{i+1}-\tau_i)\,;\qquad 0=\tau_0<\tau_1<\ldots<\tau_N=\tau\,.$$

In the Itô definition $\Phi(w(\tau_i),\tau_i)$ is independent of the increment $w(\tau_{i+1})-w(\tau_i)$, i.e., it depends only on the value of $w(\tau_i)$ at the last point τ_i, whereas in the Stratonovich definition both points τ_i and τ_{i+1} contribute in a symmetric way. If Φ does not depend on $w(t)$, (3.100) is an ordinary Stieltjes integral [for continuous $\Phi(\tau)$] and both definitions agree.

We now want to apply (3.101, 102) to (3.97). The first iteration of (3.97) leads to (putting $\xi = x$ in the integral and substituting $t' = t+\tau$)

$$\begin{aligned}\xi^{(1)}(t)-x &= \int_0^{\tau} h(x,t+\tau')\,\mathrm{d}\tau' + \int_0^{\tau} g(x,t+\tau')\,\mathrm{d}w(\tau') + \ldots\\ &= \tau h(x,t+\Theta_1\tau) + g(x,t+\Theta_2\tau)\,w(\tau) + \ldots\end{aligned} \tag{3.103}$$

with $0 \leq \Theta_i \leq 1$. Because of (3.99) the next iteration leads after averaging to [cf. (3.90)] using $\langle w(\tau)\rangle = 0$ and retaining only terms proportional to τ

$$\begin{aligned}\langle \xi^{(2)}(t)-x\rangle &= \tau h(x,t+\Theta_1\tau) + \left\langle \int_0^{\tau} g'(x,t+\tau')\,g(x,t+\Theta_2\tau')\,w(\tau')\,\mathrm{d}w(\tau')\right\rangle\\ &= \tau h(x,t+\Theta_1\tau)\\ &\quad + g'(x,t+\Theta_3\tau)\,g(x,t+\Theta_3\Theta_2\tau)\left\langle \int_0^{\tau} w(\tau')\,\mathrm{d}w(\tau')\right\rangle\,.\end{aligned} \tag{3.104}$$

The stochastic integral is different for Stratonovich (S) and Itô (I) definitions. We have (see [1.15, 2.6, 3.4, 5] and the next page for a derivation)

$$\left\langle (\mathrm{I}) \int_0^\tau w(\tau')\,\mathrm{d}w(\tau') \right\rangle = \left\langle \frac{w^2(\tau)}{2} - \tau \right\rangle = 0\,, \tag{3.105}$$

$$\left\langle (\mathrm{S}) \int_0^\tau w(\tau')\,\mathrm{d}w(\tau') \right\rangle = \left\langle \frac{w^2(\tau)}{2} \right\rangle = \tau\,. \tag{3.106}$$

We therefore obtain for the drift coefficients (3.85)

$$\begin{aligned} &\mathrm{I:} \quad D^{(1)} = h(x,t) \\ &\mathrm{S:} \quad D^{(1)} = h(x,t) + g'(x,t)\,g(x,t)\,, \end{aligned} \tag{3.107}$$

i.e., the spurious drift is missing in the Itô definition. For the diffusion coefficient it is sufficient to use (3.103). Hence no stochastic integral is necessary and we obtain in both cases

$$D^{(2)} = \frac{1}{2} \lim_{\tau \to 0} \frac{1}{\tau} g^2(x, t + \Theta_2 \tau) \langle w^2(\tau) \rangle + \ldots = g^2(x,t)\,. \tag{3.108}$$

Derivation of (3.105, 106)

If we use (3.99) we easily obtain (3.105, 106):

$$\begin{aligned} \left\langle (\mathrm{I}) \int_0^\tau w(\tau')\,\mathrm{d}w(\tau') \right\rangle &= \left\langle \sum_{i=0}^{N-1} w(\tau_i)[w(\tau_{i+1}) - w(\tau_i)] \right\rangle \\ &= \sum_{i=0}^{N-1} [\langle w(\tau_i)\, w(\tau_{i+1}) \rangle - \langle w(\tau_i)\, w(\tau_i) \rangle] \\ &= \sum_{i=0}^{N-1} (2\tau_i - 2\tau_i) = 0\,, \end{aligned}$$

$$\begin{aligned} \left\langle (\mathrm{S}) \int_0^\tau w(\tau')\,\mathrm{d}w(\tau') \right\rangle &= \frac{1}{2} \sum_{i=0}^{N-1} [w(\tau_i) + w(\tau_{i+1})]\,[w(\tau_{i+1}) - w(\tau_i)] \\ &= \frac{1}{2} \sum_{i=0}^{N-1} [\langle w(\tau_i)\, w(\tau_{i+1}) \rangle + \langle w(\tau_{i+1})\, w(\tau_{i+1}) \rangle \\ &\quad - \langle w(\tau_i)\, w(\tau_i) \rangle - \langle w(\tau_{i+1})\, w(\tau_i) \rangle] \\ &= \frac{1}{2} \sum_{i=0}^{N-1} [2\tau_i + 2\tau_{i+1} - 2\tau_i - 2\tau_i] \\ &= \sum_{i=0}^{N-1} (\tau_{i+1} - \tau_i) = \tau\,. \end{aligned}$$

The Stratonovich definition is consistent with (3.90, 91) and leads to the drift coefficient (3.95) which is consistent with the example (3.72), see (3.84). One advantage of the Itô definition is that in (3.97) or its differential form

$$d\xi = h(\xi, t)\,dt + g(\xi, t)\,dW \tag{3.109}$$

the drift coefficient $h(x,t) = D^{(1)}$ appears directly. The disadvantage is, however, that one must learn new rules for integration [e.g., (3.105)] and differentiation, i.e., the Itô calculus. Here, we will always use the Stratonovich definition.

3.4 Nonlinear Langevin Equations, Several Variables

For N variables $\{\xi\} = \xi_1, \xi_2, \ldots, \xi_N$ the general Langevin equations have the form $(i = 1, 2, \ldots, N)$

$$\dot{\xi}_i = h_i(\{\xi\}, t) + g_{ij}(\{\xi\}, t)\Gamma_j(t)\,. \tag{3.110}$$

In this section we use Einstein's summation convention. The $\Gamma_j(t)$ are again Gaussian random variables with zero mean and with correlation functions proportional to the δ function. We may normalize these Langevin forces $\Gamma_i(t)$ in such a way that the correlation functions for different indices i are zero and that the factor in front of the δ function is 2, i.e.,

$$\langle \Gamma_i(t) \rangle = 0\,, \qquad \langle \Gamma_i(t)\Gamma_j(t') \rangle = 2\delta_{ij}\delta(t-t')\,. \tag{3.111}$$

We now want to calculate the Kramers-Moyal coefficients. Similar to the one-variable case, all Kramers-Moyal cofficients higher than $n = 2$ are zero. We therefore need only the drift and the diffusion coefficients defined by

drift coefficient:

$$\begin{aligned} D_i(\{x\}, t) &\equiv D_i^{(1)}(\{x\}, t) \\ &= \lim_{\tau \to 0} \frac{1}{\tau} \langle \xi_i(t+\tau) - x_i \rangle |_{\xi_k(t) = x_k} \qquad k = 1, 2, \ldots, N\,, \end{aligned} \tag{3.112}$$

diffusion coefficient:

$$\begin{aligned} D_{ij}(\{x\}, t) &\equiv D_{ij}^{(2)}(\{x\}, t) \\ &= \frac{1}{2} \lim_{\tau \to 0} \frac{1}{\tau} \langle [\xi_i(t+\tau) - x_i][\xi_j(t+\tau) - x_j] \rangle |_{\xi_k(t) = x_k} \\ & \quad k = 1, 2, \ldots, N\,. \end{aligned} \tag{3.113}$$

In (3.112, 113) $\xi_i(t+\tau)$ $(\tau > 0)$ is a solution of (3.110) which at time t has the sharp value $\xi_k(t) = x_k$ for $k = 1, 2, \ldots, N$. We now proceed in the same way as in Sect. 3.3.2 first writing (3.110) as an integral equation

$$\xi_i(t+\tau)-x_i=\int_t^{t+\tau}[h_i(\{\xi(t')\},t')+g_{ij}(\{\xi(t')\},t')\Gamma_j(t')]\,dt' \;, \tag{3.114}$$

inserting the expansions

$$\begin{aligned} h_i(\{\xi(t')\},t') &= h_i(\{x\},t') + \left[\frac{\partial}{\partial x_k}h_i(\{x\},t')\right]\cdot(\xi_k(t')-x_k)+\dots \\ g_{ij}(\{\xi(t')\},t') &= g_{ij}(\{x\},t') + \left[\frac{\partial}{\partial x_k}g_{ij}(\{x\},t')\right]\cdot(\xi_k(t')-x_k)+\dots \end{aligned} \tag{3.115}$$

into (3.114). We then obtain

$$\begin{aligned} \xi_i(t+\tau)-x_i &= \int_t^{t+\tau} h_i(\{x\},t')\,dt' \\ &+\int_t^{t+\tau}\left[\frac{\partial}{\partial x_k}h_i(\{x\},t')\right]\cdot[\xi_k(t')-x_k]\,dt'+\dots \\ &+\int_t^{t+\tau} g_{ij}(\{x\},t')\Gamma_j(t')\,dt' \\ &+\int_t^{t+\tau}\left[\frac{\partial}{\partial x_k}g_{ij}(\{x\},t')\right]\Gamma_j(t')[\xi_k(t')-x_k]\,dt'+\dots . \end{aligned} \tag{3.116}$$

For $[\xi_k(t')-x_k]$ under the integrals we iterate (3.116). By taking the average and taking into account (3.111) we then get

$$\begin{aligned} D_i(\{x\},t) = h_i(\{x\},t) + \lim_{\tau\to 0}\frac{1}{\tau}\int_t^{t+\tau}\int_t^{t'}\left[\frac{\partial}{\partial x_k}g_{ij}(\{x\},t')\right]g_{kl}(\{x\},t'') \\ \times 2\delta_{jl}\delta(t'-t'')\,dt'\,dt'' \;. \end{aligned} \tag{3.117}$$

From

$$\int_t^{t'} g_{kj}(\{x\},t'')2\delta(t'-t'')\,dt'' = g_{kj}(\{x\},t')$$

[similar to (3.91) which is equivalent to the Stratonovich rule] we finally obtain for the drift coefficient

$$D_i(\{x\},t) = h_i(\{x\},t) + g_{kj}(\{x\},t)\frac{\partial}{\partial x_k}g_{ij}(\{x\},t) \;. \tag{3.118}$$

The last term is the noise-induced drift or spurious drift. Similar to the one-variable case, the diffusion coefficient is

$$D_{ij}(\{x\},t) = g_{ik}(\{x\},t)\,g_{jk}(\{x\},t) \;. \tag{3.119}$$

All higher Kramers-Moyal coefficients are zero

$$D^{(\nu)}_{i_1\ldots i_\nu}(\{x\},t) = \frac{1}{\nu!}\lim_{\tau\to 0}\frac{1}{\tau}\langle[\xi_{i_1}(t+\tau)-x_{i_1}]\ldots[\xi_{i_\nu}(t+\tau)-x_{i_\nu}]\rangle$$
$$= 0 \quad \text{for} \quad \nu \geqq 3 . \tag{3.120}$$

3.4.1 Determination of the Langevin Equation from Drift and Diffusion Coefficients

As shown in Chap. 4, the drift and diffusion coefficients determine the Fokker-Planck equation which describes the evolution of the probability density. The drift and diffusion coefficients D_i and D_{ij} are uniquely determined by the functions h_i and g_{ij} of the Langevin equation as given by (3.118, 119) in the Stratonovich sense. The question now arises whether the Langevin equations, i.e., h_i and g_{ij}, are uniquely determined by the drift and diffusion coefficients D_i and D_{ij}. For N variables we have N equations (3.118) and, because of the symmetry of $D_{ij} = D_{ji}$, $\frac{1}{2}N(N+1)$ equations (3.119) for matrices D_{ij} being positive definite. The number of unknown elements h_i is N and the number of unknown elements g_{ij} is N^2. The degree of freedom f is given by

$$f = \text{total number of unknown elements} - \text{total number of equations}$$
$$= N + N^2 - N - \tfrac{1}{2}N(N+1) = \tfrac{1}{2}N(N-1) .$$

Thus for $N = 1$ we have no choice ($f = 0$) and up to the $\pm$ sign of g, h and g are uniquely determined by $D^{(1)}$ and $D^{(2)}$ (3.95). For $N = 2, f = 1$ and we may impose one additional relation on the unknown elements. For $N = 3, f = 3$ and we may impose 3 additional relations on the unknown elements. For $N \geqq 3$ for instance, according to [3.6] it is possible to require that the deterministic drift term in the Langevin equation is zero. Thus without any additional relations the Langevin equations are not uniquely determined by drift and diffusion coefficients for $N \geqq 2$. For instance, the following two systems of Langevin equations

$$\begin{aligned} \dot{\xi}_1 &= 2\xi_2 + \Gamma_1 \\ \dot{\xi}_2 &= -2\xi_1 + \Gamma_2 \end{aligned} \tag{3.121}$$

and

$$\begin{aligned} \dot{\xi}_1 &= \cos(\xi_1^2 + \xi_2^2)\Gamma_1 + \sin(\xi_1^2 + \xi_2^2)\Gamma_2 \\ \dot{\xi}_2 &= -\sin(\xi_1^2 + \xi_2^2)\Gamma_1 + \cos(\xi_1^2 + \xi_2^2)\Gamma_2 \end{aligned} \tag{3.122}$$

with $\langle\Gamma_i(t)\Gamma_j(t')\rangle = 2\delta_{ij}\delta(t-t')$ in both cases lead to the same drift and diffusion coefficients

$$D_1 = 2x_2, \quad D_2 = -2x_1, \quad D_{ij} = \delta_{ij} \tag{3.123}$$

and are therefore physically equivalent. In system (3.122) the deterministic drift coefficient h_i is zero and the noise-induced drift is different from zero, whereas the reverse is true for system (3.121).

One way to obtain h_i and g_{ij} of the Langevin equations would be the following:
by an orthogonal transformation, which may depend on $\{x\}$ and t, we can diagonalize the positive definite matrix D_{ij}, then take the positive root of the eigenvalues and transform it back. Thus we get a well-defined root of the matrix D_{ij} and obtain one expression for g_{ij} and h_i in terms of drift and diffusion coefficients

$$\begin{aligned} g_{ij} &= (D^{1/2})_{ij} = (D^{1/2})_{ji} \\ h_i &= D_i - (D^{1/2})_{kj} \frac{\partial}{\partial x_k} (D^{1/2})_{ij} . \end{aligned} \tag{3.124}$$

General solutions can be obtained by multiplying the matrix $(D^{1/2})_{ij}$ with arbitrary orthogonal matrices $O_{ij}(\{x\}, t)$. For further investigations and especially for the case where the matrix D_{ij} is singular (i.e., Det $D_{ij} = 0$), see [3.6].

3.4.2 Transformation of Variables

The Langevin equations are very convenient to calculate the drift and diffusion coefficients if a variable transformation is performed. If we introduce new variables ξ_i' in (3.110) by

$$\xi_i' = \xi_i'(\{\xi\}, t) , \tag{3.125}$$

we obtain for the new variables the Langevin equations (the Langevin forces are not changed)

$$\begin{aligned} \dot{\xi}_i' &= \frac{\partial \xi_i'}{\partial t} + \frac{\partial \xi_i'}{\partial \xi_k} \dot{\xi}_k \\ &= \frac{\partial \xi_i'}{\partial t} + \frac{\partial \xi_i'}{\partial \xi_k} h_k + \frac{\partial \xi_i'}{\partial \xi_k} g_{kj} \Gamma_j \\ &= h_i'(\{\xi'\}, t) + g_{ij}'(\{\xi'\}, t) \Gamma_j . \end{aligned} \tag{3.126}$$

Thus the transformed quantities h' and g' are given by

$$h_i' = \frac{\partial \xi_i'}{\partial t} + \frac{\partial \xi_i'}{\partial \xi_k} h_k ; \qquad g_{ij}' = \frac{\partial \xi_i'}{\partial \xi_k} g_{kj} . \tag{3.127}$$

Hence, for the transformed drift and diffusion coefficients (writing x_i' instead of ξ_i' in the argument)

$$\begin{aligned}
D_i' &= h_i' + g_{lj}' \frac{\partial g_{ij}'}{\partial x_l'} \\
&= \frac{\partial x_i'}{\partial t} + \frac{\partial x_i'}{\partial x_k} h_k + \frac{\partial x_l'}{\partial x_r} g_{rj} \frac{\partial}{\partial x_l'} \left(\frac{\partial x_i'}{\partial x_k} g_{kj} \right) \\
&= \frac{\partial x_i'}{\partial t} + \frac{\partial x_i'}{\partial x_k} D_k - \frac{\partial x_i'}{\partial x_k} g_{rj} \frac{\partial g_{kj}}{\partial x_r} + g_{rj} \frac{\partial}{\partial x_r} \left(\frac{\partial x_i'}{\partial x_k} g_{kj} \right) \\
&= \frac{\partial x_i'}{\partial t} + \frac{\partial x_i'}{\partial x_k} D_k + \frac{\partial^2 x_i'}{\partial x_r \partial x_k} g_{rj} g_{kj} ,
\end{aligned}$$

i.e.,

$$D_i' = \frac{\partial x_i'}{\partial t} + \frac{\partial x_i'}{\partial x_k} D_k + \frac{\partial^2 x_i'}{\partial x_r \partial x_k} D_{rk} , \tag{3.128}$$

$$D_{ij}' = \frac{\partial x_i'}{\partial x_r} \frac{\partial x_j'}{\partial x_k} D_{rk} . \tag{3.129}$$

If the transformation (3.125) is nonlinear, then the old diffusion coefficients affect the new drift coefficients.

3.4.3 How to Obtain Drift and Diffusion Coefficients for Systems

If we know the Langevin equations for a system we immediately obtain the drift and diffusion coefficients by (3.118, 119). The problem, however, is how to get these equations. In the case of Brownian motion we may obtain the Langevin equations by heuristic arguments, as already discussed in the introduction for a particle not moving in an additional field of force. For the Brownian motion of a particle in a potential $mf(x)$, it seems reasonable to assume that the Langevin force is not affected by the potential. (The collision of the molecules with the particle is described by a Langevin force. If the motion of the molecules is not affected by the potential which acts on the particle, the Langevin force is unchanged.) We then obtain the Langevin equations

$$\begin{aligned}
\dot{v}(t) &= -\gamma v(t) - f'(x(t)) + \Gamma(t) \\
\dot{x}(t) &= v(t)
\end{aligned} \tag{3.130}$$

with $\Gamma(t)$ given by (3.2, 4). Thus the drift and diffusion coefficients now read (3.13, 118, 119)

$$\begin{aligned}
D_v &= -\gamma v - f'(x) ; \quad & D_{vv} &= \tfrac{1}{2} q = \gamma k T / m \\
D_x &= v \quad & ; \quad D_{xx} &= D_{xv} = 0 .
\end{aligned} \tag{3.131}$$

Instead of using different symbols we have distinguished the stochastic variables from the starting values at time t by adding the time argument. Sometimes the system of the two first-order equations (3.130) is written in one second-order equation

$$\ddot{x}(t) + \gamma \dot{x}(t) + f'(x(t)) = \Gamma(t) . \tag{3.132}$$

We also obtain a Langevin equation for those cases where a given external noise acts on a system without any further noise. For instance, if we think of an electrical circuit, we know the deterministic differential equations. These differential equations can always be cast into the form of first-order equations, i.e., we obtain the functions h_i. If we then couple the circuit to a noise generator, i.e., if we add a given noise, we get the Langevin equation (3.110), where g_{ij} is assumed to be known from the external noise. (For such a physical noise with a finite noise power we have to use the Stratonovich rule and not the Itô rule.)

In systems without external noise we speak of internal noise [3.7]. For instance, the fluctuations in a circuit due to the discreteness of the moving charges cannot be removed from the circuit. It is therefore termed internal noise. Because we cannot switch off the noise we cannot obtain h_i directly, although we may, however, measure the drift and the diffusion coefficients.

If we know some general equation like the master equation for the system, we may then derive from it the Fokker-Planck equation by some approximation techniques, for instance, by the $1/\Omega$ expansion of *van Kampen* [1.24] or the $1/N$ expansion of *Haken* and *Vollmer* [3.8]. One may also calculate directly from more general equations the drift and diffusion coefficients. These were calculated for the variables describing a laser system where the quantum fluctuations have to be considered as internal noise, via quantum-mechanical operator equations [3.9].

3.5 Markov Property

The process described by the Langevin equation (3.67) with δ-correlated Langevin force (3.68) is a Markov process, i.e., its conditional probability at time t_n depends only on the value $\xi(t_{n-1}) = x_{n-1}$ at the next earlier time (2.73). This follows from the fact that a first-order differential equation is uniquely determined by its initial value and that the δ-correlated Langevin force $\Gamma(t)$ at a former time $t < t_{n-1}$ cannot change the conditional probability at a later time $t > t_{n-1}$. The same is true for the system of Langevin equations (3.110) with Gaussian Langevin forces according to (3.111).

This Markovian property is destroyed if $\Gamma(t)$ is no longer δ-correlated. For instance, the process described by

$$\dot{\xi} = h(\xi) + \tilde{\Gamma}(t) \tag{3.133}$$

with a Gaussian distributed force $\tilde{\Gamma}(t)$, which has the correlation function

$$\langle \tilde{\Gamma}(t_1)\, \tilde{\Gamma}(t_2) \rangle = \frac{q}{2\gamma} \mathrm{e}^{-\gamma|t_1 - t_2|} , \tag{3.134}$$

is no longer a Markov process for finite γ (although $\tilde{\Gamma}$ itself is still a Markov process). However, if we introduce an additional variable η, then a Markov process results for the two-dimensional process described by

$$\begin{aligned} \dot{\xi} &= h(\xi) + \eta \\ \dot{\eta} &= -\gamma\eta + \Gamma(t) \end{aligned} \tag{3.135}$$

with δ-correlated noise

$$\langle \Gamma(t_1)\,\Gamma(t_2)\rangle = q\,\delta(t_1 - t_2)\,. \tag{3.136}$$

Because of (3.1, 2 and 10) it is easily seen that (3.135, 136) are equivalent to (3.133, 134). Thus by introducing new random variables, non-Markovian processes may be reduced to Markovian processes (see App. A1 for an application).

3.6 Solutions of the Langevin Equation by Computer Simulation

The one-variable Langevin equation (3.67) or the multivariable Langevin equation (3.110) may be solved by computer simulation, also called the molecular dynamics method. The main idea is to simulate the Langevin force on a computer, integrate the equation of motion with the simulated Langevin force and then take the average for a large number of realizations. The computer simulation of the Langevin force and the numerical integration may be performed simultaneously. To explain the main idea we treat the one-variable Langevin equation (3.67) first.

To integrate the Langevin equation starting at $t = 0$ with the value

$$\xi(0) = \xi_0$$

to the finite time $t = T$, we first divide the time interval T into N small finite steps of length τ

$$t_n = \tau n\,, \qquad \tau = T/N\,, \qquad n = 1, 2, \dots, N\,. \tag{3.137}$$

The stochastic variable at a later time t_{n+1}

$$\xi_{n+1} \equiv \xi(t_{n+1}) = \xi(\tau \cdot (n+1))$$

is calculated according to ($n = 0, 1, \dots, N-1$)

$$\xi_{n+1} = \xi_n + D^{(1)}(\xi_n, t_n)\,\tau + \sqrt{D^{(2)}(\xi_n, t_n)\,\tau}\; w_n\,. \tag{3.138}$$

Here, $w_0, w_1, \dots, w_{N-1}$ are independent Gaussian-distributed random variables with zero mean and with variance 2, i.e.,

$$\langle w_n \rangle = 0\,, \qquad \langle w_n w_{n'} \rangle = 2\,\delta_{nn'} \tag{3.139}$$

and $D^{(1)}$ and $D^{(2)}$ are given by (3.95). Obviously, for $\tau \to 0$ (3.138) leads to the correct Kramers-Moyal coefficients (3.85). [For the Itô definition (Sect. 3.3.3) we may use $D^{(1)} = h$ and $D^{(2)} = g^2$.]

Usually, a random number generator produces random numbers r_ν in the range $0 \leqq r_\nu < 1$. To obtain the random variables w_n we may use

$$w_n = \sqrt{24/M} \sum_{\nu=1}^{M} (r_\nu - 1/2)\,, \tag{3.140}$$

where M is a large number, e.g., 10. Obviously, the average of (3.140) is zero. From the central limit theorem, w_n are Gaussian distributed for large M. Furthermore, the variance is equal to 2

$$\begin{aligned}\langle w_n^2 \rangle &= (24/M) \sum_{\nu,\mu=1}^{M} \langle (r_\nu - 1/2)(r_\mu - 1/2) \rangle \\ &= \frac{24}{M} \sum_{\nu=1}^{M} \langle (r_\nu - 1/2)^2 \rangle = \frac{24}{M} M \int_{-1/2}^{1/2} x^2 \mathrm{d}x = 2\,.\end{aligned}$$

In this way we obtain a realization $\xi_N = \xi(T)$. The error of this procedure for one step is of the order τ^2 and for N steps of the order τ. As shown in [3.10], one may improve the method in such a way that the error for N steps is of the order τ^2 or even of some higher order [3.11]. By calculating a large number of such realizations (one must, of course, be careful that the random numbers r_ν are independent for each n and for each realization) one may perfom an average and thus obtain $\langle \xi(T) \rangle$ or any other expectation value $\langle f(\xi(T)) \rangle$. The initial value ξ_0 may have either a sharp value or it may already be a stochastic variable, the probability of which follows from the initial distribution.

For the Langevin equation with N_v variables we proceed in the same way. Denoting the square root of the diffusion matrix by g_{ij} (3.124), and the drift coefficient by D_i (3.118), we obtain the stochastic variables $\xi_{in} = \xi_i(t_n)$ at times $t_n = \tau n$ by ($n = 0, 1, \ldots, N-1$; $i = 1, 2, \ldots, N_v$)

$$\xi_{i(n+1)} = \xi_{in} + D_i(\{\xi_n\}, t_n)\,\tau + \sum_{j=1}^{N_v} g_{ij}(\{\xi_n\}, t_n) \sqrt{\tau}\, w_{jn}\,. \tag{3.141}$$

The starting value at $t = 0$ is denoted by

$$\xi_{i0} \equiv \xi_i(0)\,.$$

In (3.141) $w_{j0}, w_{j1}, \ldots, w_{j(N-1)}$ are $N \times N_v$ independent Gaussian variables with zero mean and with variance equal to 2, i.e.,

$$\langle w_{jn} w_{kn'} \rangle = 2\,\delta_{jk}\,\delta_{nn'}\,. \tag{3.142}$$

Each of the variables w_{jn} may be generated according to (3.140).

It is usually not very difficult to perform the iterations (3.138 or 141) and then take the average over various realizations. The accuracy of the method is, however, not very good: in particular it is very hard to improve the accuracy by an order of magnitude. (For linear processes, where the solution is Gaussian, the relative error would be proportional to $N_s^{-1/2}$, where N_s is the number of realizations.) For this reason, the method is most appropriate to get some estimate for $\langle \xi(T) \rangle$ ($\langle \xi_i(T) \rangle$), if the time interval T is not too large compared to the time constants implied in $D^{(1)}$ and $D^{(2)}$ (D_i and D_{ij}).

4. Fokker-Planck Equation

As shown in Sects. 3.1, 2 we can immediately obtain expectation values for processes described by the linear Langevin equations (3.1, 31). For nonlinear Langevin equations (3.67, 110) expectation values are much more difficult to obtain, so here we first try to derive an equation for the distribution function. As mentioned already in the introduction, a differential equation for the distribution function describing Brownian motion was first derived by *Fokker* [1.1] and *Planck* [1.2]: many review articles and books on the Fokker-Planck equation now exist [1.5 – 15].

Our derivation starts with an expansion of the distribution function, known as Kramers-Moyal expansion [1.17, 19]. In this equation, only the Kramers-Moyal coefficients (3.95, 118 – 120) will enter. As seen in Sect. 3.3, these Kramers-Moyal coefficients can also be calculated for the nonlinear Langevin equations. As it turned out, these coefficients vanish for $n \geqq 3$ for the Langevin equations (3.67, 110) with δ-correlated Gaussian-distributed Langevin forces, and only the drift and diffusion coefficients (3.107, 108, 118, 119) enter in the distribution function equation. Hence the Kramers-Moyal expansion with an infinite number of terms stops after the second term. This equation is then the Fokker-Planck equation or the forward Kolmogorov equation.

The problem of obtaining averages is thus reduced to the problem of solving this Fokker-Planck equation. For pedagogic reasons we first treat the one-variable case and then the more complicated case of N variables.

4.1 Kramers-Moyal Forward Expansion

It follows from the definition of the transition probability (2.69) that the probability density $W(x,t+\tau)$ at time $t+\tau$ and the probability density $W(x,t)$ at time t are connected by ($\tau \geqq 0$)

$$W(x,t+\tau) = \int P(x,t+\tau|x',t)\, W(x',t)\,\mathrm{d}x' \,. \tag{4.1}$$

To derive an expression for the differential $\partial W(x,t)/\partial t$, we must know the transition probability $P(x,t+\tau|x',t)$ for small τ. We first assume that we know all the moments ($n \geqq 1$)

$$M_n(x',t,\tau) = \langle[\xi(t+\tau)-\xi(t)]^n\rangle|_{\xi(t)=x'} = \int(x-x')^n P(x,t+\tau|x',t)\,dx\,, \quad (4.2)$$

where $|_{\xi(t)=x'}$ means that at time t the random variable has the sharp value x'. We now derive a general expansion of the transition probability in three different ways.

First Way

If all the moments are given, we can construct the characteristic function (x' is to be considered as a parameter) (2.19, 21)

$$\begin{aligned} C(u,x',t,\tau) &= \int_{-\infty}^{\infty} e^{iu(x-x')}P(x,t+\tau|x',t)\,dx \\ &= 1 + \sum_{n=1}^{\infty}(iu)^n M_n(x',t,\tau)/n!\,. \end{aligned} \quad (4.3)$$

Because the characteristic function is the Fourier transform of the probability density and vice versa (2.22) we can express the transition probability by the moments M_n

$$\begin{aligned} P(x,t+\tau|x',t) &= \frac{1}{2\pi}\int_{-\infty}^{\infty} e^{-iu(x-x')}C(u,x',t,\tau)\,du \\ &= \frac{1}{2\pi}\int_{-\infty}^{\infty} e^{-iu(x-x')}\left[1+\sum_{n=1}^{\infty}(iu)^n M_n(x',t,\tau)/n!\right]du\,. \end{aligned} \quad (4.4)$$

Because ($n \geqq 0$)

$$\frac{1}{2\pi}\int_{-\infty}^{\infty}(iu)^n e^{-iu(x-x')}du = \left(-\frac{\partial}{\partial x}\right)^n \delta(x-x') \quad (4.5)$$

and

$$\delta(x-x')f(x') = \delta(x-x')f(x)\,, \quad (4.6)$$

we have

$$P(x,t+\tau|x',t) = \left[1+\sum_{n=1}^{\infty}\frac{1}{n!}\left(-\frac{\partial}{\partial x}\right)^n M_n(x,t,\tau)\right]\delta(x-x')\,. \quad (4.7)$$

Second Way

Equation (4.7) may be derived without using the characteristic function in the following way [4.1]: starting from the identity

$$P(x,t+\tau|x',t) = \int\delta(y-x)\,P(y,t+\tau|x',t)\,dy \quad (4.8)$$

and using the formal Taylor series expansion of the δ function in the form

$$
\begin{aligned}
\delta(y-x) &= \delta(x'-x+y-x') \\
&= \sum_{n=0}^{\infty} \frac{(y-x')^n}{n!}\left(\frac{\partial}{\partial x'}\right)^n \delta(x'-x) \\
&= \sum_{n=0}^{\infty} \frac{(y-x')^n}{n!}\left(-\frac{\partial}{\partial x}\right)^n \delta(x'-x)\,,
\end{aligned} \tag{4.9}
$$

we get

$$
\begin{aligned}
P(x,t+\tau|x',t) &= \sum_{n=0}^{\infty} \frac{1}{n!}\left(-\frac{\partial}{\partial x}\right)^n \int (y-x')^n P(y,t+\tau|x',t)\,\mathrm{d}y\,\delta(x'-x) \\
&= \left[1+\sum_{n=1}^{\infty}\frac{1}{n!}\left(-\frac{\partial}{\partial x}\right)^n M_n(x',t,\tau)\right]\delta(x'-x) \\
&= \left[1+\sum_{n=1}^{\infty}\frac{1}{n!}\left(-\frac{\partial}{\partial x}\right)^n M_n(x,t,\tau)\right]\delta(x-x')\,.
\end{aligned} \tag{4.10}
$$

In deriving the second line of (4.10) we used (4.2) and for the last line $\delta(x-x') = \delta(x'-x)$ and (4.6).

Inserting (4.7) or (4.10) into (4.1) leads in both cases to

$$
\begin{aligned}
W(x,t+\tau) - W(x,t) &= \frac{\partial W(x,t)}{\partial t}\tau + O(\tau^2) \\
&= \sum_{n=1}^{\infty}\left(-\frac{\partial}{\partial x}\right)^n \int \delta(x-x')\,M_n(x,t,\tau)\,W(x',t)\,\mathrm{d}x'/n! \\
&= \sum_{n=1}^{\infty}\left(-\frac{\partial}{\partial x}\right)^n [M_n(x,t,\tau)/n!]\,W(x,t)\,.
\end{aligned} \tag{4.11}
$$

Third Way

The formal Taylor series expansion (4.9) is convenient for deriving (4.11). After multiplying (4.9) by a function of y and x' and then integrating the equation over y and x', we end with a Taylor series expansion of this function (only for this expansion can the Taylor series converge). Therefore (4.11) may be derived by avoiding any δ function and its derivatives and using only Taylor series expansion for the distribution function and the transition probability. This derivation of (4.11) runs as follows. Introducing $\Delta = x - x'$, the integrand in (4.1) may be expanded in a Taylor series according to

$$
\begin{aligned}
P(x,t+\tau|x',t)\,W(x',t) &= P(x-\Delta+\Delta,t+\tau|x-\Delta,t)\,W(x-\Delta,t) \\
&= \sum_{n=0}^{\infty}\frac{(-1)^n}{n!}\Delta^n\left(\frac{\partial}{\partial x}\right)^n P(x+\Delta,t+\tau|x,t)\,W(x,t)\,.
\end{aligned}
$$

Inserting this expression in (4.1) and integrating over Δ we directly obtain (4.11). (The negative sign of the differential $d\Delta = -dx'$ may be absorbed into the integration boundaries.)

We now assume that the moments M_n can be expanded into a Taylor series with respect to τ ($n \geqq 1$)

$$M_n(x,t,\tau)/n! = D^{(n)}(x,t)\tau + O(\tau^2) . \tag{4.12}$$

The term with τ^0 must vanish, because for $\tau = 0$ the transition probability P has the initial value

$$P(x,t|x',t) = \delta(x-x') , \tag{4.13}$$

which leads to vanishing moments (4.2). By taking into account only the linear terms in τ we thus have

$$\frac{\partial W(x,t)}{\partial t} = \sum_{n=1}^{\infty}\left(-\frac{\partial}{\partial x}\right)^n D^{(n)}(x,t)\,W(x,t) = L_{\mathrm{KM}} W , \tag{4.14}$$

where the differential symbol acts on $D^{(n)}(x,t)$ and $W(x,t)$. The Kramers-Moyal operator L_{KM} is defined by

$$L_{\mathrm{KM}}(x,t) = \sum_{n=1}^{\infty}(-\partial/\partial x)^n D^{(n)}(x,t) . \tag{4.15}$$

Equation (4.14) is the Kramers-Moyal expansion.

For non-Markovian processes, the conditional probability in (4.1) depends on the values of the stochastic variable $\xi(t')$ at all earlier times $t'<t$ (2.69). Hence also the moments (4.2) and their expansion coefficients $D^{(n)}$ which occur in (4.14) depend on these earlier times for non-Markovian processes. For Markov processes, $D^{(n)}$ do not depend on the values of $\xi(t')$ at these earlier times. With respect to time t, (4.14) is then a differential equation of first order and the distribution function $W(x,t)$ is uniquely determined by integration of (4.14) starting with the initial distribution $W(x,t_0)$ ($t>t_0$) and for appropriate boundary conditions. Therefore we assume that the process described by the probability density $W(x,t)$ is a Markov process.

The transition probability $P(x,t\,|\,x',t')$ is the distribution $W(x,t)$ for the special initial condition $W(x,t') = \delta(x-x')$. Thus the transition probability must also obey (4.14), i.e.,

$$\partial P(x,t|x',t')/\partial t = L_{\mathrm{KM}}(x,t)\,P(x,t|x',t') , \tag{4.16}$$

where the initial condition of P is given by (4.13) with t replaced by t'.

4.1.1 Formal Solution

A formal solution of (4.16) with the initial value (4.13) for time-independent L_{KM} reads

$$P(x,t|x',t') = \mathrm{e}^{L_{\mathrm{KM}}(x)(t-t')}\delta(x-x') . \tag{4.17}$$

For time-dependent Kramers-Moyal operators we have to take into account that L_{KM} does not need to commute with itself for different times. The general solution of (4.16) with the initial value (4.13) may be found by iteration of (4.16) (Dyson series [4.2])

$$\begin{aligned} P(x,t|x',t') &= \delta(x-x') + \int_{t'}^{t} L_{\mathrm{KM}}(x,t_1)\mathrm{d}t_1\delta(x-x') \\ &\quad + \int_{t'}^{t}\mathrm{d}t_1\int_{t'}^{t_1}\mathrm{d}t_2 L_{\mathrm{KM}}(x,t_1)L_{\mathrm{KM}}(x,t_2)\,\delta(x-x') + \dots \\ &= \left[1 + \sum_{n=1}^{\infty}\int_{t'}^{t}\mathrm{d}t_1\int_{t'}^{t_1}\mathrm{d}t_2\dots\int_{t'}^{t_{n-1}}\mathrm{d}t_n L_{\mathrm{KM}}(x,t_1)\dots L_{\mathrm{KM}}(x,t_n)\right] \\ &\quad \times\delta(x-x') . \end{aligned} \tag{4.18}$$

If we introduce the time-ordering operator $\overleftarrow{T}$ which interchanges the time-dependent operators in such a way that the operators with larger times stand to the left of operators with smaller times, (4.18) becomes [4.2]

$$\begin{aligned} P(x,t|x',t') &= \overleftarrow{T}\left[1 + \sum_{n=1}^{\infty}\frac{1}{n!}\int_{t'}^{t}\mathrm{d}t_1\int_{t'}^{t}\mathrm{d}t_2\dots\int_{t'}^{t}\mathrm{d}t_n L_{\mathrm{KM}}(x,t_1)\dots L_{\mathrm{KM}}(x,t_n)\right] \\ &\quad \times\delta(x-x') \\ &= \overleftarrow{T}\exp\left[\int_{t'}^{t}L_{\mathrm{KM}}(x,t'')\mathrm{d}t''\right]\delta(x-x') . \end{aligned} \tag{4.19}$$

For small time differences $\tau = t-t'$ (4.18) reduces to

$$P(x,t+\tau|x',t) = [1+L_{\mathrm{KM}}(x,t)\cdot\tau + O(\tau^2)]\,\delta(x-x') \tag{4.20}$$

in agreement with (4.7, 12, 15).

4.2 Kramers-Moyal Backward Expansion

In (4.15, 16) we derived an equation of motion for the transition probability $P(x,t|x',t')$. In (4.15, 16) differential operators with respect to x and t occur, i.e., with respect to the value of the stochastic variable $\xi(t)$ at the later time $t>t'$. Backward expansions are equations of motion for P where we differentiate with respect to x' and t', i.e., with respect to the value of the stochastic variable $\xi(t')$ at the earlier time $t'<t$. As shown at the end of this section, both equations lead to the same result for P and thus either one can be used for determining P.

For the derivation we follow closely the procedure of the second way in Sect. 4.1

Starting from the Chapman-Kolmogorov equation (2.78) in the form ($t \geqq t' + \tau \geqq t'$)

$$P(x,t|x',t') = \int P(x,t|x'',t'+\tau)\, P(x'',t'+\tau|x',t')\,\mathrm{d}x'' \tag{4.21}$$

we write as in (4.8)

$$P(x'',t'+\tau|x',t') = \int \delta(y-x'')\, P(y,t'+\tau|x',t')\,\mathrm{d}y\,. \tag{4.22}$$

Furthermore, we make a Taylor series expansion of the δ function in the form

$$\begin{aligned} \delta(y-x'') &= \delta(x'-x''+y-x') \\ &= \sum_{n=0}^{\infty} \frac{(y-x')^n}{n!} \left(\frac{\partial}{\partial x'}\right)^n \delta(x'-x'') \end{aligned} \tag{4.23}$$

and obtain

$$\begin{aligned} P(x'',t'+\tau|x',t') &= \sum_{n=0}^{\infty} \frac{1}{n!} \int (y-x')^n P(y,t'+\tau|x',t')\,\mathrm{d}y \left(\frac{\partial}{\partial x'}\right)^n \delta(x'-x'') \\ &= \left[1 + \sum_{n=1}^{\infty} \frac{1}{n!} M_n(x',t',\tau) \left(\frac{\partial}{\partial x'}\right)^n\right] \delta(x'-x'')\,. \end{aligned} \tag{4.24}$$

Inserting (4.24) in (4.21) yields

$$\begin{aligned} P(x,t|x',t') - P(x,t|x',t'+\tau) &= -\frac{\partial P(x,t|x',t')}{\partial t'}\tau + O(\tau^2) \\ &= \sum_{n=1}^{\infty} \frac{1}{n!} M_n(x',t',\tau) \left(\frac{\partial}{\partial x'}\right)^n P(x,t|x',t'+\tau) \\ &= \tau \sum_{n=1}^{\infty} D^{(n)}(x',t') \left(\frac{\partial}{\partial x'}\right)^n P(x,t|x',t') + O(\tau^2)\,. \end{aligned} \tag{4.25}$$

In deriving the last line we used (4.12). By taking into account only the linear terms in τ we get

$$\frac{\partial P(x,t|x',t')}{\partial t'} = -L^{+}_{\mathrm{KM}}(x',t')\, P(x,t|x',t') \tag{4.26}$$

with

$$L^{+}_{\mathrm{KM}}(x',t') = \sum_{n=1}^{\infty} D^{(n)}(x',t')(\partial/\partial x')^n\,. \tag{4.27}$$

As may be easily checked, (4.27) is the adjoint operator of (4.15). Equations (4.26, 27) form the desired Kramers-Moyal backward expansion.

4.2.1 Formal Solution

A formal solution of (4.26) with the initial value (4.13) reads for time-independent L_{KM}^{+}

$$P(x,t|x',t') = \mathrm{e}^{L_{\mathrm{KM}}^{+}(x')(t-t')}\delta(x-x') . \tag{4.28}$$

For a time-dependent operator we have the Dyson series

$$\begin{aligned} P(x,t|x',t') &= \left[1 + \sum_{n=1}^{\infty} \int_{t'}^{t} \mathrm{d}t_1 \int_{t_1}^{t} \mathrm{d}t_2 \ldots \int_{t_{n-1}}^{t} \mathrm{d}t_n L_{\mathrm{KM}}^{+}(x',t_1) \ldots L_{\mathrm{KM}}^{+}(x',t_n)\right] \\ &\quad \times \delta(x-x') \\ &= \vec{T}\left[1 + \sum_{n=1}^{\infty} \frac{1}{n!} \int_{t'}^{t} \mathrm{d}t_1 \int_{t'}^{t} \mathrm{d}t_2 \ldots \int_{t'}^{t} \mathrm{d}t_n L_{\mathrm{KM}}^{+}(x',t_1) \ldots L_{\mathrm{KM}}^{+}(x',t_n)\right] \\ &\quad \times \delta(x-x') \\ &= \vec{T} \exp\left[\int_{t'}^{t} L_{\mathrm{KM}}^{+}(x',t'')\,\mathrm{d}t''\right] \delta(x-x') . \end{aligned} \tag{4.29}$$

In (4.29) the time-ordering operator $\vec{T}$ arranges the operators $L_{\mathrm{KM}}^{+}(x',t)$ so that the time in the products of L_{KM}^{+} increases from left to right. For small time differences $\tau = t - t'$ (4.29) reduces to

$$P(x,t+\tau|x',t) = [1 + L_{\mathrm{KM}}^{+}(x',t)\,\tau + O(\tau^2)]\,\delta(x-x') . \tag{4.30}$$

4.2.2 Equivalence of the Solutions of the Forward and Backward Equations

To show the equivalence of (4.28, 29 and 30) with (4.17, 19 and 20), respectively, we first derive the relation

$$A(x)\,\delta(x-x') = A^{+}(x')\,\delta(x-x') . \tag{4.31}$$

Here $A(x)$ is a general real operator containing only differential operators with respect to x and functions depending only on x. For a derivation of (4.31) we first observe that $A(x)\,\varphi(x)$ can be written in two different ways:

$$\begin{aligned} A(x)\,\varphi(x) &= A(x) \int \delta(x-x')\,\varphi(x')\,\mathrm{d}x' \\ &= \int A(x)\,\delta(x-x')\,\varphi(x')\,\mathrm{d}x' \\ &= \int \varphi(x')\,A(x)\,\delta(x-x')\,\mathrm{d}x' , \end{aligned} \tag{4.32}$$

$$A(x)\,\varphi(x)=\int\delta(x-x')\,A(x')\,\varphi(x')\,dx'$$
$$=\int\varphi(x')\,A^{+}(x')\,\delta(x-x')\,dx'\,. \tag{4.33}$$

By subtracting both expressions we get

$$0=\int\varphi(x')\,[A(x)\,\delta(x-x')-A^{+}(x')\,\delta(x-x')]\,dx'\,. \tag{4.34}$$

Because $\varphi(x)$ is an arbitrary function the bracket in (4.34) must be zero.

The equivalence of (4.20, 30) follows immediately from (4.31) for $A(x)=L_{\mathrm{KM}}(x)$. Furthermore, one easily shows the equivalence of (4.28, 29) with (4.17, 19) by using (4.31) for

$$A(x)=e^{L_{\mathrm{KM}}(x)(t-t')}\,;\qquad A^{+}(x)=e^{L^{+}_{\mathrm{KM}}(x)(t-t')} \tag{4.35}$$

and for

$$A(x)=\overleftarrow{T}\exp\left[\int_{t'}^{t}L_{\mathrm{KM}}(x,t'')\,dt''\right]$$
$$A^{+}(x)=\vec{T}\exp\left[\int_{t'}^{t}L^{+}_{\mathrm{KM}}(x,t'')\,dt''\right]. \tag{4.36}$$

The last relation follows from the fact that the adjoint of a product reverses its order

$$(ABC\dots)^{+}=\dots C^{+}B^{+}A^{+}\,. \tag{4.37}$$

4.3 Pawula Theorem

For the solution of (4.14) it is important to know how many terms of expansion (4.15) must be taken into account. We first derive the theorem of *Pawula* [4.3], which states that for a positive transition probability P, the expansion (4.15) may stop either after the first term or after the second term, if it does not stop after the second term it must contain an infinite number of terms. If expansion (4.15) stops after the second term, (4.15, 16) are then called the Fokker-Planck or forward Kolmogorov equation, and (4.26, 27) is then called the backward Kolmogorov equation.

To derive the Pawula theorem we need the generalized Schwartz inequality

$$\left[\int f(x)\,g(x)\,P(x)\,dx\right]^2\leqq\int f^2(x)\,P(x)\,dx\int g^2(x)\,P(x)\,dx\,. \tag{4.38}$$

In (4.38) $P(x)$ is a nonnegative function and $f(x)$ and $g(x)$ are arbitrary functions. The inequality may be derived from

$$\iint [f(x)\,g(y) - f(y)\,g(x)]^2 P(x)\,P(y)\,\mathrm{d}x\,\mathrm{d}y \geqq 0\,,$$

which obviously holds for nonnegative P. We now apply (4.38) with $(n, m \geqq 0)$

$$f(x) = (x-x')^n; \quad g(x) = (x-x')^{n+m};$$
$$P(x) = P(x, t+\tau | x', t')$$

and thus obtain for the moments (4.2) the inequality

$$M_{2n+m}^2 \leqq M_{2n} \cdot M_{2n+2m}\,. \tag{4.39}$$

For $n=0$ we have $M_m^2 \leqq M_{2m}$. This relation is obviously fulfilled for $m=0$ $(M_0=1)$. For $m \geqq 1$ no restriction follows from this relation for the short time expansion coefficients $D^{(n)}$ of M_n (4.12). For $m=0$, $M_{2n}^2 \leqq M_{2n}^2$, which is obviously fulfilled for every n. Thus we need to consider (4.39) only for $n \geqq 1$ and $m \geqq 1$. By inserting (4.12) into (4.39), dividing the resulting inequality by τ^2 and taking the limit $\tau \to 0$ we then obtain the following inequality for the expansion coefficients $D^{(n)}$ $(n \geqq 1,\ m \geqq 1)$:

$$[(2n+m)!\,D^{(2n+m)}]^2 \leqq (2n)!\,(2n+2m)!\,D^{(2n)} D^{(2n+2m)}\,. \tag{4.40}$$

If $D^{(2n)}$ is zero, $D^{(2n+m)}$ must be zero, too, i.e.,

$$D^{(2n)} = 0 \Rightarrow D^{(2n+1)} = D^{(2n+2)} = \ldots = 0 \quad (n \geqq 1)\,. \tag{4.41}$$

Furthermore if $D^{(2n+2m)}$ is zero, $D^{(2n+m)}$ must be zero, too, i.e.,

$$D^{(2r)} = 0 \Rightarrow D^{(r+n)} = 0 \quad (n = 1, \ldots, r-1)\,, \qquad \text{i.e.,}$$
$$D^{(2r-1)} = \ldots = D^{(r+1)} = 0 \quad (r \geqq 2)\,. \tag{4.42}$$

From (4.41) and the repeated use of (4.42), one concludes that if any $D^{(2r)} = 0$ for $r \geqq 1$ all coefficients $D^{(n)}$ with $n \geqq 3$ must vanish, i.e.,

$$D^{(2r)} = 0 \Rightarrow D^{(3)} = D^{(4)} = \ldots = 0 \quad (r \geqq 1)\,. \tag{4.43}$$

The Pawula theorem immediately follows from the last statement. (In contrast to (4.43) for even coefficients a vanishing odd coefficient does not lead to restrictions.)

The Pawula theorem, however, does not say that expansions truncated at $n \geqq 3$ are of no use. As we shall discuss in Sect. 4.6 for a simple example, one may very well use Kramers-Moyal expansions truncated at $n \geqq 3$ for calculating distribution functions. Though the transition probability must then have negative values at least for sufficiently small times, these negative values may be very small. For the example discussed in Sect. 4.6, the distribution function obtained by the Kramers-Moyal expansion truncated at a proper $n \geqq 3$ is in better agreement with the exact distribution than the distribution function following from the Kramers-Moyal expansion truncated at $n = 2$.

4.4 Fokker-Planck Equation for One Variable

If the Kramers-Moyal expansion (4.14) stops after the second term we get the Fokker-Planck equation ($\partial/\partial t$ is denoted by a dot)

$$\dot{W}(x,t) = L_{\mathrm{FP}} W(x,t)\,, \tag{4.44}$$

$$L_{\mathrm{FP}} = -\frac{\partial}{\partial x} D^{(1)}(x,t) + \frac{\partial^2}{\partial x^2} D^{(2)}(x,t)\,. \tag{4.45}$$

For the nonlinear Langevin equation (3.67) with (3.68) the drift coefficient $D^{(1)}$ and the diffusion coefficient $D^{(2)}$ are given by (3.93, 94) in terms of the function occurring in (3.67). All higher Kramers-Moyal coefficients $D^{(n)}$ with $n \geqq 3$ are zero [see the last equation in (3.95)] and therefore (4.44) with L_{FP} given by (4.45) is the exact equation for the probability density $W(x,t)$. For another derivation, see App. A5.

Equations (4.44, 45) may be written in the form

$$\frac{\partial W}{\partial t} + \frac{\partial S}{\partial x} = 0\,, \tag{4.46}$$

$$S(x,t) = \left[D^{(1)}(x,t) - \frac{\partial}{\partial x} D^{(2)}(x,t)\right] W(x,t)\,. \tag{4.47}$$

Because (4.46) is a continuity equation for a probability distribution, S has to be interpreted as a probability current. If this probability current vanishes at the boundaries $x = x_{\min}$ and $x = x_{\max}$, (4.46) then guarantees that the normalization is preserved

$$\int_{x_{\min}}^{x_{\max}} W(x,t)\,\mathrm{d}x = \text{const}\,. \tag{4.48}$$

For natural boundary conditions ($x_{\min} = -\infty, x_{\max} = \infty$), $W(x,t)$ and the probability current (4.47) also vanish at $x = \pm\infty$.

For a stationary process the probability current must be constant. With natural boundary conditions, the probability current must be zero. To demonstrate the usefulness of the Fokker-Planck equation we calculate the stationary distribution function for the Brownian motion process described by the Langevin equation (3.1) with (3.2). Here we have

$$D^{(1)} = -\gamma v\,, \qquad D^{(2)} = q/2 = \gamma kT/m \tag{4.49}$$

and we immediately get from

$$S = \left(-\gamma v - \frac{\gamma kT}{m}\frac{\partial}{\partial v}\right) W = 0 \tag{4.50}$$

and from the normalization condition the Maxwell distribution (3.30), i.e.,

$$W(v) = \sqrt{\frac{m}{2\pi kT}} \exp\left(-\frac{mv^2}{2kT}\right). \tag{4.51}$$

4.4.1 Transition Probability Density for Small Times

We now derive an expression for the transition probability density for small τ in another form than (4.20) specialized for the Fokker-Planck operator, i.e.,

$$P(x, t+\tau|x', t) = [1 + L_{\mathrm{FP}}(x, t)\tau + O(\tau^2)]\,\delta(x-x') \tag{4.52}$$

with

$$L_{\mathrm{FP}}(x, t) = -\frac{\partial}{\partial x} D^{(1)}(x, t) + \frac{\partial^2}{\partial x^2} D^{(2)}(x, t). \tag{4.53}$$

Inserting (4.53) into (4.52) we get up to corrections of the order τ^2

$$\begin{aligned} P(x, t+\tau|x', t) &= \left[1 - \frac{\partial}{\partial x} D^{(1)}(x', t)\tau + \frac{\partial^2}{\partial x^2} D^{(2)}(x', t)\tau\right] \delta(x-x') \\ &= \exp\left[-\frac{\partial}{\partial x} D^{(1)}(x', t)\tau + \frac{\partial^2}{\partial x^2} D^{(2)}(x', t)\tau\right] \delta(x-x'). \end{aligned} \tag{4.54}$$

In deriving (4.54) we replaced x by x' (4.6) in the drift and diffusion coefficients. If we now introduce the representation of the δ function in terms of a Fourier integral, we obtain for small τ

$$\begin{aligned} P(x, t+\tau|x', t) &= \exp\left[-\frac{\partial}{\partial x} D^{(1)}(x', t)\tau + \frac{\partial^2}{\partial x^2} D^{(2)}(x', t)\tau\right] \frac{1}{2\pi} \int_{-\infty}^{\infty} \mathrm{e}^{\mathrm{i}u(x-x')} du \\ &= \frac{1}{2\pi} \int_{-\infty}^{\infty} \exp[-\mathrm{i}uD^{(1)}(x', t)\tau - u^2 D^{(2)}(x', t)\tau + \mathrm{i}u(x-x')]\,du \\ &= \frac{1}{2\sqrt{\pi D^{(2)}(x', t)\tau}} \exp\left(-\frac{[x-x'-D^{(1)}(x', t)\tau]^2}{4D^{(2)}(x', t)\tau}\right). \end{aligned} \tag{4.55}$$

For drift and diffusion coefficients independent of x and t, (4.55) is not only valid for small τ, but for arbitrary $\tau > 0$. [The last line in (4.54) is then the formal solution (4.17).] We now want to check that (4.55) leads to the correct moments

$$M_n(x', t, \tau) = \int (x-x')^n P(x, t+\tau|x', t)\,dx.$$

Using [4.4]

$$\int\limits_{-\infty}^{\infty} x^n \exp[-(x-\beta)^2]\,dx = (2i)^{-n}\sqrt{\pi}\,H_n(i\beta)\,, \tag{4.56}$$

where $H_n(x)$ are the Hermite polynomials ($H_0 = 1$, $H_1 = 2x$, $H_2 = 4x^2 - 2, \dots$) we obtain from (4.55)

$$\begin{aligned} M_n(x', t, \tau) &= \left[-i\sqrt{D^{(2)}(x',t)\tau}\right]^n \\ &\quad \times H_n\left\{\tfrac{1}{2} i D^{(1)}(x',t)\sqrt{\tau/D^{(2)}(x',t)}\right\}. \end{aligned} \tag{4.57}$$

For the expansion coefficients of M_n linear in τ we therefore have ($M_0 = 1$)

$$\lim_{\tau\to 0}\frac{1}{\tau} M_n(x', t, \tau)/n! = \begin{cases} D^{(1)}(x',t) & \\ D^{(2)}(x',t) & \text{for} \\ 0 & \end{cases} \begin{aligned} &n = 1 \\ &n = 2 \\ &n \geqq 3\,. \end{aligned}$$

Thus (4.55) [as well as (4.52, 53)] leads to the correct drift and diffusion coefficients, i.e., it leads to expectation values which are correct up to terms linear in τ.

The form (4.55) is not unique. A class of equivalent forms has been derived [4.5, 6]. One of these forms may be obtained as follows: by performing the differentiation for the drift and diffusion coefficient in (4.53) we get

$$\begin{aligned} L_{FP}(x,t) &= -\frac{\partial D^{(1)}(x,t)}{\partial x} + \frac{\partial^2 D^{(2)}(x,t)}{\partial x^2} \\ &\quad - \left[D^{(1)}(x,t) - 2\frac{\partial D^{(2)}(x,t)}{\partial x}\right]\frac{\partial}{\partial x} + D^{(2)}(x,t)\frac{\partial^2}{\partial x^2}\,. \end{aligned} \tag{4.53a}$$

If we insert this expression into (4.52) and replace $\partial/\partial x$ by $-\partial/\partial x'$, we can perform the same steps as before, leading for small time τ to

$$\begin{aligned} P(x, t+\tau | x', t) &= \frac{1}{2\sqrt{\pi D^{(2)}(x,t)\tau}} \exp\left(-\frac{\partial D^{(1)}(x,t)}{\partial x}\tau + \frac{\partial^2 D^{(2)}(x,t)}{\partial x^2}\tau \right. \\ &\quad \left. - \frac{\{x - x' - [D^{(1)}(x,t) - 2\partial D^{(2)}(x,t)/\partial x]\,\tau\}^2}{4D^{(2)}(x,t)\tau}\right). \end{aligned} \tag{4.55a}$$

Notice that here x instead of x' appears in the drift and diffusion coefficients.

4.4.2 Path Integral Solutions

The transition probabilities are needed for the path integral solutions [1.14, 4.5–12]. They are derived as follows: by repeatedly applying the Chapman-Kolmogorov equation (2.78) we can express the evolution of $W(x,t)$ from the initial distribution $W(x_0, t_0)$ in terms of the transition probability. Dividing the time difference $t - t_0$ in N small time intervals of length $\tau = (t - t_0)/N$, we have $(t_n = t_0 + n\tau)$

$$\begin{aligned} W(x,t) = \int dx_{N-1} \int dx_{N-2} \dots \int dx_0 \\ P(x, t | x_{N-1}, t_{N-1}) P(x_{N-1}, t_{N-1} | x_{N-2}, t_{N-2}) \dots \\ P(x_1, t_1 | x_0, t_0) W(x_0, t_0) . \end{aligned} \tag{4.58}$$

For $N \to \infty$ we may use for the transition probability the expression (4.55) for small τ, which then gives correct expectation values of $W(x, t)$ in the limit $N \to \infty$. (Every integral is correct up to the order $1/N^2$ and the product of the $N+1$ integrals is then correct up to the order $1/N$ [4.6].) Inserting (4.55) into (4.58) and taking the limit $N \to \infty$ we obtain with $x_N = x$, $[\tau = (t - t_0)/N]$

$$\begin{aligned} W(x,t) = \lim_{N \to \infty} \underbrace{\int \dots \int}_{N \text{ times}} \prod_{i=0}^{N-1} \{[4\pi D^{(2)}(x_i, t_i)\tau]^{-1/2} dx_i\} \\ \times \exp\left(- \sum_{i=0}^{N-1} \frac{[x_{i+1} - x_i - D^{(1)}(x_i, t_i)\tau]^2}{4D^{(2)}(x_i, t_i)\tau} \right) W(x_0, t_0) . \end{aligned} \tag{4.59}$$

If we use (4.55a) instead of (4.55) in (4.58), we obtain a slightly different expression.

Positivity of the Distribution Function

Because in (4.59) all the factors in front of $W(x_0, t_0)$ are positive, the distribution function must remain positive if we start with a positive distribution $W(x_0, t_0)$.

Generalized Onsager-Machlup Function

By writing

$$x_{i+1} - x_i = \dot{x}(t_i)\tau$$

we may put the negative term in the exponent in (4.59) for the limit $N \to \infty$ in the form

$$\sum_{i=0}^{N-1} \frac{[\dot{x}(t_i) - D^{(1)}(x_i, t_i)]^2}{4D^{(2)}(x_i, t_i)} \tau = \int_{t_0}^{t} \frac{[\dot{x}(t') - D^{(1)}(x(t'), t')]^2}{4D^{(2)}(x(t'), t')} dt' . \tag{4.60}$$

The function under the integral is called a generalized Onsager-Machlup function. (*Onsager* and *Machlup* [4.7] investigated such forms for a linear drift coefficient and a constant diffusion coefficient.) Expression (4.59), where the sum in the exponent is replaced by (4.60), seems at first glance to be quite evident. For small diffusion $D^{(2)}$, for instance, only the pathes near the deterministic solution of

$$\dot{x} = D^{(1)}(x,t),$$

contribute to W. It was pointed out in [1.14, 4.6], however, that this and similar other continuous forms are meaningless if the discretization process is not specified. Hence, only discrete forms such as (4.59) should be used.

4.5 Generation and Recombination Processes

To exemplify a process containing an infinite number of Kramers-Moyal coefficients $D^{(n)}$ we consider a process in which the stochastic variable $\xi(t)$ can take on only the discrete values $x_m = lm$ ($m = 1, \ldots, M$) and in which only transitions to nearest-neighbor states occur. If the transition rate from state x_m to state x_{m+1} (generation rate) is denoted by $G(x_m, t)$ and if the transition rate from state x_m to state x_{m-1} (recombination rate) is denoted by $R(x_m, t)$, the equation of motion for the probability $W(x_m, t)$ of state x_m is given by the following master equation [special case of (1.34) for nearest-neighbor transitions]

$$\begin{aligned}\dot{W}(x_m, t) &= G(x_{m-1}, t)\, W(x_{m-1}, t) - G(x_m, t)\, W(x_m, t) \\ &\quad + R(x_{m+1}, t)\, W(x_{m+1}, t) - R(x_m, t)\, W(x_m, t)\,. \end{aligned} \tag{4.61}$$

This equation may be easily read off Fig. 4.1. For $x_m = m$, $G(m) = \mu m$, $R(m) = \nu m$, (4.61) describes a birth and death process, whereas for $x_m = m$, $G(m) = \mu$, $R(m) = 0$, (4.61) describes a Poisson process. Exact solutions of (4.61) for various processes are given in Table 2.1 of [1.12]; for multidimensional generation and recombination processes, see [1.11c].

Because

$$f(x \pm l) = \exp(\pm l\partial/\partial x) f(x)$$

we may immediately write the master equation (4.61) in form of the Kramers-Moyal expansion (4.14). Denoting the variable x_m by x we have

$$\begin{aligned}\dot{W}(x,t) &= [\exp(-l\partial/\partial x) - 1]\, G(x,t)\, W(x,t) + [\exp(l\partial/\partial x) - 1]\, R(x,t)\, W(x,t) \\ &= \sum_{n=1}^{\infty} (-\partial/\partial x)^n D^{(n)}(x,t)\, W(x,t)\,, \end{aligned} \tag{4.62}$$

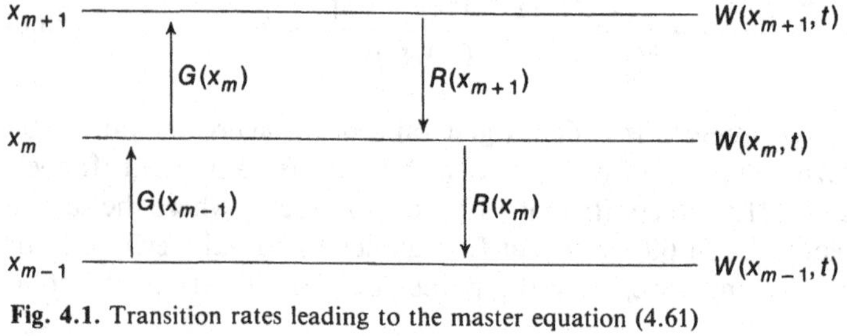

Fig. 4.1. Transition rates leading to the master equation (4.61)

where the Kramers-Moyal coefficients are given by

$$D^{(n)}(x,t)=(l^n/n!)[G(x,t)+(-1)^n R(x,t)]\,. \tag{4.63}$$

In particular, the drift and diffusion coefficients $D^{(1)}$ and $D^{(2)}$ read

$$\begin{aligned} D^{(1)} &= l(G-R) = l(\text{rate in} - \text{rate out}) \\ D^{(2)} &= (l^2/2)(G+R) = (l^2/2)(\text{rate in} + \text{rate out})\,. \end{aligned} \tag{4.63 a}$$

If the difference l between the discrete steps becomes smaller, higher Kramers-Moyal coefficients also become smaller and we may truncate expansion (4.62) at some finite value n. For an actual system we cannot change l. If, for instance, x describes electric charges, l will be the elementary charge e, which cannot be changed. We may of course increase the size of system. If we increase the size of the system by a factor L, i.e. $m=1,\dots,ML$, extensive quantities will also increase by this factor, i.e., $x=ml=Lx_{\text{nor}}$. If the rates G and R and the probability depend only on the intensive quantities $x_{\text{nor}}=x/L=(m/L)\,l$, then we get

$$\begin{aligned} \dot{W}(x_{\text{nor}},t) &= \sum_{n=1}^{\infty}(-\partial/\partial x_{\text{nor}})^n D^{(n)}(x_{\text{nor}},t)\, W(x_{\text{nor}},t) \\ D^{(n)}(x_{\text{nor}},t) &= (\alpha^n/n!)[G(x_{\text{nor}},t)+(-1)^n R(x_{\text{nor}},t)] \end{aligned}$$

with

$$\alpha^n=(l/L)^n\,. \tag{4.64}$$

Thus by increasing the size of the system the Kramers-Moyal coefficients also decrease more rapidly in n ($1/\Omega$ expansion by *van Kampen* [1.24]). Thus, if we truncate expansion (4.62) after the second term we obtain the Fokker-Planck equation (4.44, 45) with drift and diffusion coefficients given by (4.64). Other possibilities to truncate (4.62) are discussed in the following section for the Poisson process.

4.6 Application of Truncated Kramers-Moyal Expansions

A continuous stochastic variable obeying the Langevin equation (3.67) with δ-correlated Gaussian Langevin forces (3.68) leads to a Fokker-Planck equation, i.e., to the Kramers-Moyal expansion (4.14), which stops after the second term. We have seen in the last section that for a generation and recombination process, where the stochastic variable takes on only discrete values, the Kramers-Moyal expansion has an infinite number of terms. An equation with an infinite number of terms cannot be treated numerically and the question arises whether one can approximate the infinite Kramers-Moyal expansion by a Kramers-Moyal expansion truncated at a finite order. One may conclude from the Pawula theorem

(Sect. 4.3) that the Kramers-Moyal expansion can be truncated only after the first or second terms because the transition probability calculated from the Kramers-Moyal expansion truncated at some finite term of the order $N \geqq 3$ must have negative values at least for small enough times. However, an approximate distribution function does not need to be positive everywhere. As long as the negative values and the region where they occur are small this approximate distribution function may be very useful.

We now want to investigate the different approximations of expansion (4.14) for the simple example [4.13] of the Poisson process, for which the master equation (4.61) reduces to ($l = 1$, $x_m = m \geqq 0$, $G(m) = \mu$, $R(m) = 0$)

$$\dot{W}(m,t) = \mu W(m-1,t) - \mu W(m,t)\,. \tag{4.65}$$

The solution of (4.65) with the initial value

$$W(m,0) = \delta_{m,0} \tag{4.66}$$

is the Poisson distribution

$$W(m,t) = \tau^m \mathrm{e}^{-\tau}/m! \quad \text{with} \quad \tau = \mu t\,. \tag{4.67}$$

The cumulants K_n (2.21, 25) of this Poisson distribution are all equal ($K_n = \tau$ for $n \geqq 1$). If m is substituted by the continuous variable $x(-\infty < x < \infty)$ and $W(x-1,t)$ is expanded into a Taylor series we get the infinite Kramers-Moyal expansion

$$\dot{W}(x,t) = \sum_{n=1}^{\infty} \mu(-\partial/\partial x)^n W(x,t)/n!\,. \tag{4.68}$$

If we truncate the expansion (4.68) after the Nth term we have

$$\dot{W}_N(x,t) = \sum_{n=1}^{N} \mu(-\partial/\partial x)^n W_N(x,t)/n!\,. \tag{4.69}$$

In the continuous case we should use as initial condition

$$W(x,0) = \delta(x)\,. \tag{4.70}$$

In order to see how (4.69) approximates (4.67), we have to solve (4.69). By making a Fourier transform with respect to x it is easily seen that the solution of (4.69) with the initial condition (4.70) is given by

$$W_N(x,t) = \frac{1}{2\pi}\int_{-\infty}^{\infty} \exp\left[\mathrm{i}kx + \sum_{n=1}^{N}(-\mathrm{i}k)^n \mu t/n!\right]\mathrm{d}k\,. \tag{4.71}$$

By performing the integration we easily get for $N=1$ and $N=2$

$$W_1(x,t) = \delta(x-\tau)\,, \tag{4.72}$$

$$W_2(x,t) = (2\pi\tau)^{-1/2} e^{-(x-\tau)^2(2\tau)^{-1}}\,. \tag{4.73}$$

For higher N the integration cannot be done analytically. For a numerical integration we write (4.71) in the real form

$$W_N(x,t) = \frac{1}{\pi}\int_0^\infty \exp\left[\sum_{m=1}^{[N/2]}(-k^2)^n\tau/(2n)!\right] \times \cos\left[kx - k\tau\sum_{n=0}^{[(N-1)/2]}(-k^2)^n/(2n+1)!\right] dk\,. \tag{4.74}$$

Here $[a]$ is the integer part of the number a and the sum has to be omitted if the lower index is larger than the upper one. Due to the exponential function in (4.71), however, only the approximations for $N=1,2,3,6,7,10,11,\ldots$ exist.

To compare (4.74) with the exact result (4.67) it is convenient to treat n as a continuous variable in (4.67). We therefore use [$\Gamma(x)$ is the gamma function]

$$W(x,t) = \tau^x e^{-\tau}/\Gamma(x+1)\,, \tag{4.75}$$

which agrees with (4.67) for integer $x \geqq 0$. From the argument of positivity of the distribution function we conclude that (4.65) can be approximated only by truncation at $N=2$, i.e., by a Fokker-Planck equation or the exact solution of (4.65). Figure 4.2 shows the exact solution (4.75) together with (4.73) and higher-order

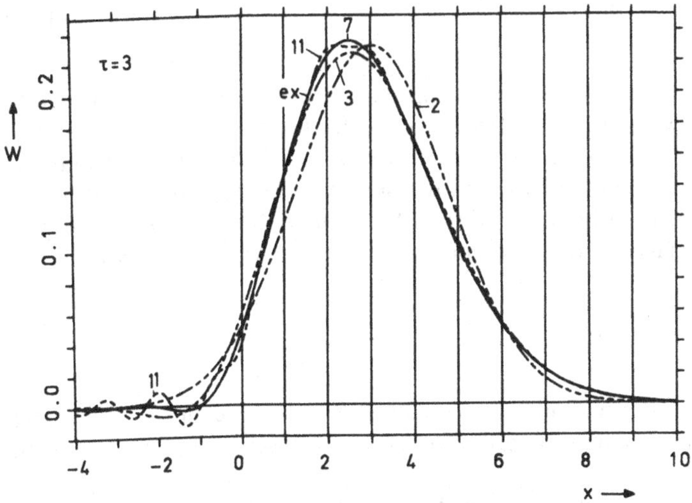

Fig. 4.2. Plot of the exact distribution (ex) and of the finite-order approximate distributions (4.71) for $N=2,3,7,11$ and $\tau=3$. The approximation for $N=7$ agrees with the exact distribution within the linewidth. For the Poisson process only the positive integer values of x have to be considered

Table 4.1. The exact normalization $M(0)$ and the exact first five moments $M(p)$, $p=1,\ldots,5$ and their successive approximations (4.76) for $N=2,3,6,7,10,11$ and $\tau=3$

Approx.	$M(0)$	$M(1)$	$M(2)$	$M(3)$	$M(4)$	$M(5)$
Exact	1.000	3.000	12.000	57.000	309.000	1866.000
2	0.980	3.025	11.963	54.063	269.835	1457.907
3	1.002	2.992	12.017	56.968	306.013	1817.669
6	1.000	3.000	12.000	57.004	309.088	1867.829
7	1.000	3.000	12.000	57.000	308.993	1865.893
10	1.001	2.976	11.522	47.337	115.036	−1994.323
11	0.994	3.006	12.004	56.976	308.914	1866.233

approximations W_N calculated numerically [4.13]. It may be seen that the main virtue of W_2 is to be positive everywhere. Some higher approximations are closer to the exact solution in the sense of least-squared deviation, as seen especially for $N=7$ where no difference is perceptible. Like the exact solution (4.75), the distribution W_7 is negative for some negative x values. For large x there are also very small negative values of W_7. As suggested from the numerical results even the approximation $N=3$, that is significantly better than that for $N=2$, seems to stay positive for all $x \geqq 0$ and therefore has properties similar to (4.75). As is seen, furthermore, terms of order higher than $N=7$ tend to have larger mean-squared deviations; so the approximation (4.71) seems to be a semiconvergent series, converging only for $\tau \to \infty$ in the strict sense (for smaller τ lower approximations seem to be better, i.e., $N=2$ for $\tau \approx 0.1$). Table 4.1 shows the moments if they are calculated either analytically (exact) or numerically by summing up the approximations (4.71) at the integer values $x=0,1,2,\ldots$

$$M_N(p) = \sum_{m=0}^{\infty} m^p W_N(m,t)\,. \tag{4.76}$$

It is seen that the first higher-order approximations lead to more accurate moments. This shows that (4.71) also converges to the exact distribution (4.75); also this convergence seems to be asymptotic. It was found in [4.13] that the even-numbered approximations to the distribution oscillate more than the odd-numbered ones. The negative value of the fifth moment for $N=10$ is a result of negative values for large x. If the moments are calculated by integration

$$\tilde{M}_N(p) = \int_{-\infty}^{\infty} x^p W_N(x,t)\,dx$$

it may be seen that the cumulants up to the order $p=N$ are identical to the exact ones and that higher cumulants vanish. Therefore the moments $\tilde{M}_N(p)$ agree with the exact ones up to the order $p=N$.

Thus for certain parameters in the Poisson process the absolute amount of negative values of the distribution function calculated by (4.69) for appropriate $N \geqq 3$ gets extremely small in the relevant region of variables, and the solution of the Fokker-Planck equation [i.e., (4.69) for $N=2$] deviates from the exact solution much more than the solution of (4.69) deviates for some suitable $N \geqq 3$ values. From this example we conclude that for approximate calculations of distribution functions, Kramers-Moyal expansion truncated at some suitable $N \geqq 3$ term may sometimes be used. Because the convergence seems to be asymptotic, its N value should not be chosen too large. (To estimate the appropriate N value without knowing the exact result will, however, be a difficult task.)

4.7 Fokker-Planck Equation for N Variables

For N stochastic variables

$$\{\xi\} = \xi_1, \xi_2, \ldots, \xi_N \tag{4.77}$$

we proceed similarly to the one-variable case. We start with the extension of (4.1) for N variables, i.e., with

$$W(\{x\}, t+\tau) = \int P(\{x\}, t+\tau \,|\, \{x'\}, t)\, W(\{x'\}, t)\, \mathrm{d}^N x' \,. \tag{4.78}$$

In (4.78) the volume element is denoted by

$$\mathrm{d}^N x' = \mathrm{d}x_1' \, \mathrm{d}x_2' \ldots \mathrm{d}x_N' \tag{4.79}$$

and N integrations have to be performed over the N variables (only one integration sign is written down). Denoting the δ function for the N variables by

$$\delta(\{x\}) = \delta(x_1)\,\delta(x_2)\ldots\delta(x_N)\,, \tag{4.80}$$

we may write

$$P(\{x\}, t+\tau \,|\, \{x'\}, t) = \int \delta(\{y\} - \{x\})\, P(\{y\}, t+\tau \,|\, \{x'\}, t)\, \mathrm{d}^N y \,. \tag{4.81}$$

It is now convenient to use the summation convention, i.e., we perform the summation over latin indices appearing twice in the expressions without writing down the summation signs. A Taylor series expansion at $\{y\} = \{x'\}$ of the δ function appearing in (4.81) then has the form

$$\begin{aligned}
\delta(\{y\} - \{x\}) &= \delta(\{x'\} - \{x\} + \{y\} - \{x'\}) \\
&= \sum_{\nu=0}^{\infty} \frac{1}{\nu!} (y_{j_1} - x'_{j_1})(y_{j_2} - x'_{j_2}) \ldots (y_{j_\nu} - x'_{j_\nu}) \frac{\partial^\nu}{\partial x'_{j_1} \partial x'_{j_2} \ldots \partial x'_{j_\nu}} \delta(\{x'\} - \{x\}) \\
&= \sum_{\nu=0}^{\infty} \frac{1}{\nu!} \frac{(-\partial)^\nu}{\partial x_{j_1} \partial x_{j_2} \ldots \partial x_{j_\nu}} (y_{j_1} - x'_{j_1})(y_{j_2} - x'_{j_2}) \ldots (y_{j_\nu} - x'_{j_\nu})\, \delta(\{x'\} - \{x\}) \,.
\end{aligned} \tag{4.82}$$

In deriving the last line we used

$$(\partial/\partial x_i')\,\delta(\{x'\}-\{x\}) = (-\partial/\partial x_i)\,\delta(\{x'\}-\{x\})\,.$$

The summation convention implies that we have to sum over the indices $j_1, j_2, \dots j_\nu$. Inserting (4.82) into (4.81) yields

$$P(\{x\},t+\tau|\{x'\},t)$$
$$= \left[1+\sum_{\nu=1}^{\infty}\frac{1}{\nu!}\frac{(-\partial)^\nu}{\partial x_{j_1}\partial x_{j_2}\dots\partial x_{j_\nu}}M^{(\nu)}_{j_1,j_2,\dots,j_\nu}(\{x\},t,\tau)\right]\delta(\{x\}-\{x'\})\,, \quad (4.83)$$

where the νth moment is defined by

$$M^{(\nu)}_{j_1,j_2,\dots j_\nu}(\{x'\},t,\tau) = \int (y_{j_1}-x'_{j_1})(y_{j_2}-x'_{j_2})\dots(y_{j_\nu}-x'_{j_\nu})$$
$$\times P(\{y\},t+\tau|\{x'\},t)\,d^N y\,. \quad (4.84)$$

In deriving (4.83) we used in accordance with the one-dimensional case (4.10) the extension of (4.6) to the N-variable δ function and $\delta(\{x\}-\{x'\}) = \delta(\{x'\}-\{x\})$. Expanding the moments for small τ (4.12)

$$M^{(\nu)}_{j_1,j_2,\dots,j_\nu}(\{x\},t,\tau)/\nu! = D^{(\nu)}_{j_1,j_2,\dots,j_\nu}(\{x\},t)\,\tau + O(\tau^2)\,, \quad (4.85)$$

we obtain the forward Kramers-Moyal expansion for N variables by inserting (4.83) into (4.78), dividing the resulting equation by τ and taking the limit $\tau\to 0$:

$$\partial W(\{x\},t)/\partial t = \sum_{\nu=1}^{\infty}\frac{(-\partial)^\nu}{\partial x_{j_1}\dots\partial x_{j_\nu}}D^{(\nu)}_{j_1,\dots,j_\nu}(\{x\},t)\,W(\{x\},t)\,. \quad (4.86)$$

The solution of (4.86) with the initial condition

$$W(\{x'\},t') = P(\{x\},t'|\{x'\},t') = \delta(\{x\}-\{x'\}) \quad (4.87)$$

is the transition probability P. Thus the forward equation for this probability density reads

$$\partial P(\{x\},t|\{x'\},t')/\partial t = L_{\mathrm{KM}}(\{x\},t)\,P(\{x\},t|\{x'\},t') \quad (4.88)$$

with

$$L_{\mathrm{KM}}(\{x\},t) = \sum_{\nu=1}^{\infty}\frac{(-\partial)^\nu}{\partial x_{j_1}\dots\partial x_{j_\nu}}D^{(\nu)}_{j_1,\dots,j_\nu}(\{x\},t)\,. \quad (4.89)$$

The corresponding backward equation takes the form

$$\partial P(\{x\},t|\{x'\},t')/\partial t' = -L^{+}_{\mathrm{KM}}(\{x'\},t')\,P(\{x\},t|\{x'\},t')\,, \quad (4.90)$$

$$L^+_{\rm KM}(\{x\},t)=\sum_{\nu=1}^{\infty}D^{(\nu)}_{j_1,\dots,j_\nu}(\{x\},t)\frac{\partial^\nu}{\partial x_{j_1}\dots\partial x_{j_\nu}}, \tag{4.91}$$

where the initial condition reads

$$P(\{x\},t\,|\,\{x'\},t)=\delta(\{x\}-\{x'\}). \tag{4.92}$$

The backward equation may easily be derived by extending the derivation in Sect. 4.2 to the N-variable case. Formal solutions of (4.88, 90) with initial conditions (4.87, 92) are given by (4.17 – 19, 28, 29), where one has to replace x and x' by $\{x\}$ and $\{x'\}$. The equivalence of the formal solutions of the forward and backward equations may be shown by using the N-variable version of (4.31), i.e.,

$$A(\{x\})\,\delta(\{x\}-\{x'\})=A^+(\{x'\})\,\delta(\{x\}-\{x'\}), \tag{4.93}$$

as was done for the one-variable case. In (4.93) $A(\{x\})$ is an operator containing functions and derivatives of the variables $x_1,\dots,x_N$.

For a process which is described by the Langevin equation (3.110) with δ-correlated Gaussian Langevin forces (3.111) all coefficients $D^{(\nu)}$ with $\nu\geqq 3$ vanish (3.120). The transition probability then satisfies the equations (summation convention, $t\geqq t'$).

Fokker-Planck or Forward Kolmogorov Equation

$$\partial P(\{x\},t\,|\,\{x'\},t')/\partial t=L_{\rm FP}(\{x\},t)\,P(\{x\},t\,|\,\{x'\},t'), \tag{4.94}$$

$$L_{\rm FP}(\{x\},t)=-\frac{\partial}{\partial x_i}D_i(\{x\},t)+\frac{\partial^2}{\partial x_i\partial x_j}D_{ij}(\{x\},t), \tag{4.95}$$

Backward Kolmogorov Equation

$$\partial P(\{x\},t\,|\,\{x'\},t')/\partial t'=-L^+_{\rm FP}(\{x'\},t')\,P(\{x\},t\,|\,\{x'\},t'), \tag{4.96}$$

$$L^+_{\rm FP}(\{x'\},t')=D_i(\{x'\},t')\frac{\partial}{\partial x'_i}+D_{ij}(\{x'\},t')\frac{\partial^2}{\partial x'_i\,\partial x'_j}. \tag{4.97}$$

The initial condition in both cases is

$$P(\{x\},t\,|\,\{x'\}t)=P(\{x\},t'\,|\,\{x'\},t')=\delta(\{x\}-\{x'\}). \tag{4.98}$$

If we multiply (4.94) by $W(\{x'\},t')$ and integrate over x' we obtain the Fokker-Planck equation for the probability density $W(\{x\},t)$, i.e.,

$$\partial W(\{x\},t)/\partial t=L_{\rm FP}(\{x\},t)\,W(\{x\},t). \tag{4.94a}$$

In (4.95, 97) we omitted the upper index 1 in the drift coefficient and the upper index 2 in the diffusion coefficient, because both coefficients are distinguished by the number of lower indices. The drift coefficient or drift vector, the diffusion coefficient or diffusion matrix are defined by [cf. (4.84, 85)]:

drift vector

$$D_i(\{x\},t) = \lim_{\tau\to 0}\frac{1}{\tau}\langle \xi_i(t+\tau)-\xi_i(t)\rangle\Big|_{\xi_k(t)=x_k}, \qquad (4.99)$$

diffusion matrix

$$\begin{aligned} D_{ij}(\{x\},t) &= D_{ji}(\{x\},t) \\ &= \frac{1}{2}\lim_{\tau\to 0}\frac{1}{\tau}\langle[\xi_i(t+\tau)-\xi_i(t)][\xi_j(t+\tau)-\xi_j(t)]\rangle\Big|_{\xi_k(t)=x_k}, \end{aligned} \qquad (4.100)$$

where $|_{\xi_k(t)=x_k}$ means that the stochastic variable ξ_k at time t has the sharp value x_k $(k=1,2,\dots,N)$.

As seen from the definition, the diffusion matrix is a symmetric matrix. Furthermore, it is semidefinite, which follows from (a_i is an arbitrary vector, $\tau > 0$)

$$\begin{aligned} 2D_{ij}a_ia_j &= \lim_{\tau\to 0}\frac{1}{\tau}\langle\{[\xi_i(t+\tau)-\xi_i(t)]\,a_i[\xi_j(t+\tau)-\xi_j(t)]\,a_j\}\rangle\Big|_{\xi_k(t)=x_k} \\ &= \lim_{\tau\to 0}\frac{1}{\tau}\langle\{[\xi_i(t+\tau)-\xi_i(t)]\,a_i\}^2\rangle\Big|_{\xi_k(t)=x_k} \geqq 0\,. \end{aligned} \qquad (4.101)$$

Sometimes we assume that D_{ij} is positive definite, i.e.,

$$D_{ij}a_ia_j > 0 \quad \text{for} \quad a_ia_i > 0\,. \qquad (4.102)$$

Then the inverse of the diffusion matrix will exist.

4.7.1 Probability Current

The Fokker-Planck equation (4.94a) with (4.95) may be written in the form of the continuity equation

$$\frac{\partial W}{\partial t} + \frac{\partial S_i}{\partial x_i} = 0\,, \qquad (4.103)$$

where the probability current S_i is defined by

$$S_i = D_iW - (\partial/\partial x_j)D_{ij}W\,. \qquad (4.104)$$

If the probability current vanishes at an $N-1$ dimensional surface F of the N-dimensional space, the continuity equation (4.103) ensures that the total probability remains constant inside this surface F. If it is normalized to 1 at time $t=t'$, the normalization will always be 1, i.e.,

$$\int_{V(F)} W(\{x\},t)\,\mathrm{d}^N x = 1\,. \tag{4.105}$$

In (4.105) $V(F)$ is the volume inside the surface F. For natural boundary conditions the probability W and the current S_i vanish at infinity and therefore the normalization condition reads

$$\int W(\{x\},t)\,\mathrm{d}^N x = 1\,. \tag{4.105 a}$$

(If we do not indicate any integration boundaries, an integration from $-\infty$ to $+\infty$ is understood.)

4.7.2 Joint Probability Distribution

As discussed in Sect. 2.4.1, the complete information of a Markov process is contained in the joint probability distribution $W_2(\{x\},t;\{x'\},t')$ which can be expressed by the transition probability density (4.98) and the distribution at time t',

$$W_2(\{x\},t;\{x'\},t') = P(\{x\},t\,|\,\{x'\},t')\,W(\{x'\},t')\,. \tag{4.106}$$

If the drift and diffusion coefficients do not depend on time, a stationary solution may exist. In this case, P can depend only on the time difference $t-t'$, and we may write for the joint probability distribution in the stationary state for

$t \geqq t'$

$$W_2(\{x\},t;\{x'\},t') = P(\{x\},t-t'\,|\,\{x'\},0)\,W_{\mathrm{st}}(\{x'\})\,, \tag{4.106 a}$$

$t \leqq t'$

$$W_2(\{x\},t;\{x'\},t') \equiv W_2(\{x'\},t';\{x\},t) = P(\{x'\},t'-t\,|\,\{x\},0)\,W_{\mathrm{st}}(\{x\})\,. \tag{4.106 b}$$

4.7.3 Transition Probability Density for Small Times

The extension of (4.52) for the N-variable case reads

$$P(\{x\},t+\tau\,|\,\{x'\},t) = [1 + L_{\mathrm{FP}}(\{x\},t)\cdot\tau + O(\tau^2)]\,\delta(\{x\}-\{x'\})\,. \tag{4.107}$$

If we insert the operator (4.95) here we may write up to terms of the order τ^2

$$P(\{x\},t+\tau|\{x'\},t)$$
$$\approx\left[1-\frac{\partial}{\partial x_i}D_i(\{x'\},t)\tau+\frac{\partial^2}{\partial x_i\partial x_j}D_{ij}(\{x'\},t)\tau\right]\delta(\{x\}-\{x'\})$$
$$\approx\exp\left[-\frac{\partial}{\partial x_i}D_i(\{x'\},t)\tau+\frac{\partial^2}{\partial x_i\partial x_j}D_{ij}(\{x'\},t)\tau\right]\delta(\{x\}-\{x'\})\,.$$

Here we replaced $\{x\}$ by $\{x'\}$ in the drift and diffusion coefficient. By inserting the δ function expression

$$\delta(\{x\}-\{x'\})=(2\pi)^{-N}\int\exp[\mathrm{i}u_j(x_j-x_j')]\,\mathrm{d}^Nu \tag{4.108}$$

we obtain the extension of (4.55) to N variables

$$P(\{x\},t+\tau|\{x'\},t)=(2\sqrt{\pi\tau})^{-N}[\mathrm{Det}\{D_{sl}(\{x'\},t)\}]^{-1/2}$$
$$\times\exp\left\{-\frac{1}{4\tau}[D^{-1}(\{x'\},t)]_{jk}[x_j-x_j'-D_j(\{x'\},t)\tau][x_k-x_k'-D_k(\{x'\},t)\tau]\right\}. \tag{4.109}$$

In (4.109) we assumed that the diffusion matrix is positive definite so that the inverse of the diffusion matrix exists and $\mathrm{Det}\{D_{sl}\}\neq 0$. It may be shown in a way similar to the one-variable case that the drift vector (4.99) and the diffusion matrix (4.100) are recovered from (4.109), whereas all higher Kramers-Moyal expansion coefficients vanish. Path integral solutions may be derived from the transition probability density for small τ, i.e., from (4.109), in the same way as in the one-variable case, Sect. 4.4.2. With this path integral solution it can again be shown that the solution of the Fokker-Planck equation stays positive, if it was initially positive.

4.8 Examples for Fokker-Planck Equations with Several Variables

We now list a few examples of Fokker-Planck equations with more than one variable.

4.8.1 Three-Dimensional Brownian Motion without Position Variable

The equation of motion for the velocity of a particle without any external force is the Langevin equation (3.21) with a Gaussian δ-correlated Langevin force, whose strength is given by (3.22, 13). Therefore we now have 3 variables and the drift and diffusion coefficients read

$$D_i=-\gamma v_i\,,\quad D_{ij}=\tfrac{1}{2}q\,\delta_{ij}=(\gamma kT/m)\,\delta_{ij}\,. \tag{4.110}$$

As is seen, the diffusion matrix is positive definite. The Fokker-Planck equation takes the form $[W = W(v_1, v_2, v_3, t)]$

$$\frac{\partial W}{\partial t} = \gamma\left(\frac{\partial}{\partial v_i} v_i + \frac{kT}{m} \frac{\partial^2}{\partial v_i \partial v_i}\right) W$$

$$= \gamma\left(\nabla_v v + \frac{kT}{m} \Delta_v\right) W. \tag{4.111}$$

In the last line of (4.111) we have introduced vector notation, the ∇ operator and the Laplace operator act with respect to the velocity coordinate. Equation (4.111) describes a special Ornstein-Uhlenbeck process. The general solution of this process will be given in Sect. 6.5.

4.8.2 One-Dimensional Brownian Motion in a Potential

The equations of motion for the velocity and position coordinate for the Brownian motion of a particle in the potential $mf(x)$ are given by (3.130) and the corresponding drift and diffusion coefficients by (3.131). In this case the diffusion matrix is singular. The corresponding Fokker-Planck equation

$$\frac{\partial W(x, v, t)}{\partial t} = \left\{-\frac{\partial}{\partial x} v + \frac{\partial}{\partial v}[\gamma v + f'(x)] + \frac{\gamma kT}{m} \frac{\partial^2}{\partial v^2}\right\} W(x, v, t) \tag{4.112}$$

is often called Kramers equation. In (4.112) $mf'(x) = -F(x)$ is the negative force. This equation is investigated further in Chap. 10.

4.8.3 Three-Dimensional Brownian Motion in an External Force

For three-dimensional Brownian motion in an external field of force $\boldsymbol{F}(\boldsymbol{x})$ there are 6 coordinates. The Fokker-Planck equation then reads $[W = W(x_1, x_2, x_3, v_1, v_2, v_3, t)]$

$$\frac{\partial W}{\partial t} = \left[-\frac{\partial}{\partial x_i} v_i + \frac{\partial}{\partial v_i}\left(\gamma v_i - \frac{F_i}{m}\right) + \frac{\gamma kT}{m} \frac{\partial^2}{\partial v_i \partial v_i}\right] W$$

$$= \left[-\nabla_x v + \nabla_v\left(\gamma v - \frac{F}{m}\right) + \frac{\gamma kT}{m} \Delta_v\right] W. \tag{4.113}$$

In the last expression we have used vector notation.

4.8.4 Brownian Motion of Two Interacting Particles in an External Potential

For Brownian motion of two particles with mass m_1, m_2 in one dimension, the Langevin equations are

$$\begin{aligned} \dot{x}_1 &= v_1; \quad \dot{v}_1 = -\gamma_1 v_1 - f_a'(x_1) - \frac{m_1+m_2}{m_1}\frac{\partial}{\partial x_1} f_w(x_1-x_2) + \Gamma_1, \\ \dot{x}_2 &= v_2; \quad \dot{v}_2 = -\gamma_2 v_2 - f_a'(x_2) - \frac{m_1+m_2}{m_2}\frac{\partial}{\partial x_2} f_w(x_1-x_2) + \Gamma_2. \end{aligned} \tag{4.114}$$

In (4.114) $x_1(x_2)$ and $v_1(v_2)$ are the position and velocity of the first (second) particle; $mf_a(x_1)$ is the external force for the two particles and $\frac{(m_1+m_2)}{2} f_w(x_1-x_2)$ is the interaction potential of the two particles. If we assume that the Langevin forces Γ_1 and Γ_2 acting on particles 1 and 2 are not correlated, then

$$\langle \Gamma_\nu(t)\Gamma_\mu(t')\rangle = 2\frac{\gamma_\nu kT}{m_\nu}\delta_{\nu\mu}\delta(t-t') \tag{4.115}$$

(no summation convention) and the Fokker-Planck equation for the distribution function $W = W(x_1, v_1; x_2, v_2; t)$ takes the form

$$\begin{aligned} \partial W/\partial t = \Bigg\{ &-\frac{\partial}{\partial x_1} v_1 + \frac{\partial}{\partial v_1}\left[f_a'(x_1) + \frac{m_1+m_2}{m_1}\frac{\partial f_w(x_1-x_2)}{\partial x_1} + \gamma_1 v_1 \right] \\ &+ \gamma_1 \frac{kT}{m_1}\frac{\partial^2}{\partial v_1^2} - \frac{\partial}{\partial x_2} v_2 + \frac{\partial}{\partial v_2} \\ &\times \left[f_a'(x_2) + \frac{m_1+m_2}{m_2}\frac{\partial f_w(x_1-x_2)}{\partial x_2} + \gamma_2 v_2 \right] + \gamma_2 \frac{kT}{m_2}\frac{\partial^2}{\partial v_2^2} \Bigg\} W. \end{aligned} \tag{4.116}$$

A numerical solution of this equation for an external cos-potential and some models for the interaction potential are given in [4.14].

4.9 Transformation of Variables

If instead of the N variables $\{x\} = x_1, \ldots, x_N$ we use other new N variables $\{x'\} = x_1', \ldots, x_N'$ which are given by the old variables in the form

$$x_i' = x_i'(\{x\}, t) = x_i'(x_1, \ldots, x_N, t); \quad i = 1, \ldots, N, \tag{4.117}$$

the Fokker-Planck equation (4.94a, 95) may be expressed in terms of the new variables. It is the purpose of this chapter to find the transformation of the old drift and diffusion coefficients to the new ones. Though these new drift and diffusion coefficients were already obtained in Sect. 3.4.2 by transforming the Langevin equation (3.110), we now want to derive this transformation by using only the Fokker-Planck equation (4.94a).

By going over from one variable system to another the probability in the volume element does not change, i.e.,

$$W \mathrm{d}^N x = W' \mathrm{d}^N x' \,. \tag{4.118}$$

Because the volume elements are transformed according to the Jacobian J

$$\mathrm{d}^N x/\mathrm{d}^N x' = J = |\mathrm{Det}\{\partial x_i/\partial x_j'\}| = 1/J' = 1/|\mathrm{Det}\{\partial x_i'/\partial x_j\}| \,, \tag{4.119}$$

the probability densities W and W' are connected by

$$W' = JW = W/J' \,. \tag{4.120}$$

To find the transformation of the Fokker-Planck equation we must first know the derivative of the Jacobian. Because

$$\frac{\partial x_i'}{\partial x_j}\frac{\partial x_j}{\partial x_k'} = \delta_{ik} \,, \tag{4.121}$$

the cofactor or minor A^{jk} of the element $a_{ji} = \partial x_i'/\partial x_j$ is given by

$$A^{jk} = J' \frac{\partial x_j}{\partial x_k'} \,. \tag{4.122}$$

Therefore we may express the derivative of J' with respect to the element $a_{jk} = \partial x_k'/\partial x_j$ by

$$\frac{\partial J'}{\partial a_{jk}} = A^{jk} = J' \frac{\partial x_j}{\partial x_k'} \,. \tag{4.123}$$

Using the chain rule we thus obtain

$$\begin{aligned}
-\frac{1}{J}\frac{\partial J}{\partial x_i} &= -\frac{\partial \ln J}{\partial x_i} = \frac{\partial \ln J'}{\partial x_i} = \frac{1}{J'}\frac{\partial J'}{\partial x_i} = \frac{1}{J'}\frac{\partial J'}{\partial a_{jk}}\frac{\partial a_{jk}}{\partial x_i} \\
&= \frac{\partial x_j}{\partial x_k'}\frac{\partial}{\partial x_i}\frac{\partial x_k'}{\partial x_j} = \frac{\partial x_j}{\partial x_k'}\frac{\partial}{\partial x_j}\frac{\partial x_k'}{\partial x_i} \\
&= \frac{\partial}{\partial x_k'}\frac{\partial x_k'}{\partial x_i} \,.
\end{aligned} \tag{4.124}$$

Similarly, we obtain for the time derivative of J

$$-\frac{1}{J}\left(\frac{\partial J}{\partial t}\right)_x=\frac{1}{J'}\left(\frac{\partial J'}{\partial t}\right)_x=\frac{1}{J'}\frac{\partial J'}{\partial a_{jk}}\left(\frac{\partial a_{jk}}{\partial t}\right)_x$$
$$=\frac{\partial x_j}{\partial x'_k}\left(\frac{\partial}{\partial t}\right)_x\frac{\partial x'_k}{\partial x_j}=\frac{\partial x_j}{\partial x'_k}\frac{\partial}{\partial x_j}\left(\frac{\partial x'_k}{\partial t}\right)_x$$
$$=\frac{\partial}{\partial x'_k}\left(\frac{\partial x'_k}{\partial t}\right)_x . \tag{4.125}$$

The index x indicates that the old variables are kept constant. This index is necessary if the transformation (4.117) depends on t. We obviously have

$$\left(\frac{\partial}{\partial t}\right)_x=\left(\frac{\partial}{\partial t}\right)_{x'}+\left(\frac{\partial x'_k}{\partial t}\right)_x\frac{\partial}{\partial x'_k} . \tag{4.126}$$

To express the derivative $\partial/\partial x_i$ in terms of the derivatives of the new variables, we, again, use the chain rule to get

$$\frac{\partial}{\partial x_i}=\frac{\partial x'_k}{\partial x_i}\frac{\partial}{\partial x'_k}=\frac{\partial}{\partial x'_k}\frac{\partial x'_k}{\partial x_i}-\left[\frac{\partial}{\partial x'_k}\frac{\partial x'_k}{\partial x_i}\right]=\frac{\partial}{\partial x'_k}\frac{\partial x'_k}{\partial x_i}+\frac{1}{J}\frac{\partial J}{\partial x_i},$$

where the bracket indicates that the operator does not act outside this bracket. Because

$$\frac{\partial}{\partial x_i}=\frac{1}{J}\frac{\partial}{\partial x_i}J-\frac{1}{J}\frac{\partial J}{\partial x_i},$$

we get the useful result

$$\frac{\partial}{\partial x_i}=\frac{1}{J}\frac{\partial}{\partial x'_k}\frac{\partial x'_k}{\partial x_i}J . \tag{4.127}$$

Applying (4.127) twice we obtain

$$\frac{\partial^2}{\partial x_i\partial x_j}=\frac{1}{J}\frac{\partial}{\partial x'_k}\frac{\partial x'_k}{\partial x_i}J\frac{1}{J}\frac{\partial}{\partial x'_r}\frac{\partial x'_r}{\partial x_j}J$$
$$=\frac{1}{J}\frac{\partial^2}{\partial x'_k\partial x'_r}\frac{\partial x'_k}{\partial x_i}\frac{\partial x'_r}{\partial x_j}J-\frac{1}{J}\frac{\partial}{\partial x'_k}\left[\frac{\partial}{\partial x'_r}\frac{\partial x'_k}{\partial x_i}\right]\frac{\partial x'_r}{\partial x_j}J$$
$$=\frac{1}{J}\frac{\partial^2}{\partial x'_k\partial x'_r}\frac{\partial x'_k}{\partial x_i}\frac{\partial x'_r}{\partial x_j}J-\frac{1}{J}\frac{\partial}{\partial x'_k}\frac{\partial^2 x'_k}{\partial x_i\partial x_j}J . \tag{4.128}$$

For the derivative with respect to t we have similarly

$$\left(\frac{\partial}{\partial t}\right)_x = \frac{1}{J}\left(\frac{\partial}{\partial t}\right)_x J - \frac{1}{J}\left(\frac{\partial J}{\partial t}\right)_x$$

$$= \frac{1}{J}\left(\frac{\partial}{\partial t}\right)_{x'} J + \frac{1}{J}\left(\frac{\partial x_k'}{\partial t}\right)_x \frac{\partial}{\partial x_k'} J - \frac{1}{J}\left(\frac{\partial J}{\partial t}\right)_x ,$$

$$\left(\frac{\partial x_k'}{\partial t}\right)_x \frac{\partial}{\partial x_k'} = \frac{\partial}{\partial x_k'}\left(\frac{\partial x_k'}{\partial t}\right)_x - \left[\frac{\partial}{\partial x_k'}\left(\frac{\partial x_k'}{\partial t}\right)_x\right]$$

$$= \frac{\partial}{\partial x_k'}\left(\frac{\partial x_k'}{\partial t}\right)_x + \frac{1}{J}\left(\frac{\partial J}{\partial t}\right)_x ,$$

i.e., we get

$$\left(\frac{\partial}{\partial t}\right)_x = \frac{1}{J}\left(\frac{\partial}{\partial t}\right)_{x'} J + \frac{1}{J}\frac{\partial}{\partial x_k'}\left(\frac{\partial x_k'}{\partial t}\right)_x J . \tag{4.129}$$

By inserting (4.127 – 129) into the Fokker-Planck equation (4.94a, 95) we easily obtain the Fokker-Planck equation for the new variables [$W' = JW$, (4.120)]

$$\left(\frac{\partial W'}{\partial t}\right)_{x'} = \left(-\frac{\partial}{\partial x_k'} D_k' + \frac{\partial^2}{\partial x_k' \partial x_r'} D_{kr}'\right) W' , \tag{4.130}$$

$$D_k' = \left(\frac{\partial x_k'}{\partial t}\right)_x + \frac{\partial x_k'}{\partial x_i} D_i + \frac{\partial^2 x_k'}{\partial x_i \partial x_j} D_{ij} , \tag{4.131}$$

$$D_{kr}' = \frac{\partial x_k'}{\partial x_i}\frac{\partial x_r'}{\partial x_j} D_{ij} . \tag{4.132}$$

These transformed drift and diffusion coefficients agree with (3.128, 129).

4.10 Covariant Form of the Fokker-Planck Equation

The transformation to new variables may be seen best by writing the Fokker-Planck equation in covariant form, i.e., in a form where only scalars, contravariant or covariant vectors and tensors and covariant derivatives occur. In this chapter we restrict ourselves to coordinate transformations, which are independent of time. If we go over to new coordinates x'^i which are functions of the old coordinates x^i and vice versa, i.e.,

$$x'^i = x'^i(x^1, \dots, x^N), \quad i = 1, \dots, N$$
$$x^i = x^i(x'^1, \dots, x'^N), \tag{4.133}$$

contravariant vectors A^i and covariant vectors A_i are transformed according to

$$A'^i = \frac{\partial x'^i}{\partial x^j} A^j, \quad A'_i = \frac{\partial x^j}{\partial x'^i} A_j. \tag{4.134}$$

The coordinate differential dx^i is a contravariant vector, i.e.,

$$dx'^i = \frac{\partial x'^i}{\partial x^j} dx^j \tag{4.135}$$

(therefore one usually puts the index of the coordinate in the upper place), whereas the gradient of a scalar is a covariant vector, i.e.,

$$\frac{\partial \varphi}{\partial x'^i} = \frac{\partial x^j}{\partial x'^i} \frac{\partial \varphi}{\partial x^j}. \tag{4.136}$$

A scalar is not changed by a coordinate transformation, i.e.,

$$\varphi' = \varphi. \tag{4.137}$$

For the transformation of a tensor of rank n with p contravariant and $q = n - p$ covariant indices we have to use $\partial x'/\partial x$ p times and $\partial x/\partial x'$ q times [4.15 – 17]. As seen from (4.132), the diffusion tensor $\bar{D}^{ij} \equiv D_{ij}$ $(n = p = 2)$ is a purely contravariant tensor

$$\bar{D}'^{kr} = \frac{\partial x'^k}{\partial x_i} \frac{\partial x'^r}{\partial x_j} \bar{D}^{ij} \tag{4.138}$$

and the indices should therefore be put in the upper place. Obviously, the probability density is not a scalar because it transforms according to (4.120) and furthermore the drift vector is not a contravariant vector because of the last term in (4.131) ($\partial x'_k/\partial t$ is zero because in this section we assume that transformation (4.133) does not depend on time). Thus we first have to find a scalar $\bar{W}$ which may be used instead of the probability density W and a contravariant vector $\bar{D}^i$ which may be used instead of the drift vector. Following *Graham* [4.18] (see also [4.19]), we introduce a scalar $\bar{W}$ defined by

$$\bar{W} = \sqrt{\text{Det}}\, W \quad \text{with} \quad \text{Det} = \text{Det}\{\bar{D}^{ij}\}. \tag{4.139}$$

This transformation can be done only if Det > 0, i.e., if the diffusion matrix is positive definite. Because of the transformation (4.138) it is easily seen that Det transforms according to

$$\begin{aligned}\mathrm{Det}' &= \mathrm{Det}\{\bar{D}'^{ij}\} = (\mathrm{Det}\,\partial x'^k/\partial x_i)^2\,\mathrm{Det}\{\bar{D}^{sj}\}\\ &= J^{-2}\,\mathrm{Det}\{\bar{D}^{sj}\} = J^{-2}\,\mathrm{Det}\,,\end{aligned} \tag{4.140}$$

where J is the Jacobian (4.119). We therefore have ($W' = JW$)

$$\bar{W}' = \sqrt{\mathrm{Det}'}\,W' = J^{-1}\sqrt{\mathrm{Det}}\,JW = \bar{W}\,, \tag{4.141}$$

which shows that $\bar{W}$ is indeed a scalar. Next we introduce the contravariant drift vector [4.18]

$$\bar{D}^i = D_i - \sqrt{\mathrm{Det}}\frac{\partial}{\partial x^j}\frac{\bar{D}^{ij}}{\sqrt{\mathrm{Det}}}\,, \tag{4.142}$$

which transforms in accordance with (4.134), as shown by the following equation:

$$\begin{aligned}\frac{\partial x'^k}{\partial x^i}\bar{D}^i &= \frac{\partial x'^k}{\partial x^i}D_i - \frac{\partial x'^k}{\partial x^i}\sqrt{\mathrm{Det}}\frac{\partial}{\partial x^j}\frac{\bar{D}^{ij}}{\sqrt{\mathrm{Det}}}\\ &= D'_k - \frac{\partial^2 x'^k}{\partial x^i\partial x^j}\bar{D}^{ij} - \frac{\partial x'^k}{\partial x^i}\frac{\sqrt{\mathrm{Det}}}{J}\frac{\partial}{\partial x'^r}\frac{\partial x'^r}{\partial x^j}J\frac{\bar{D}^{ij}}{\sqrt{\mathrm{Det}}}\\ &= D'_k - \frac{\partial^2 x'^k}{\partial x^i\partial x^j}\bar{D}^{ij} - \sqrt{\mathrm{Det}'}\frac{\partial}{\partial x'^r}\frac{\partial x'^k}{\partial x^i}\frac{\partial x'^r}{\partial x^j}\frac{\bar{D}^{ij}}{\sqrt{\mathrm{Det}'}}\\ &\quad + \sqrt{\mathrm{Det}'}\left(\frac{\partial}{\partial x'^r}\frac{\partial x'^k}{\partial x^i}\right)\frac{\partial x'^r}{\partial x^j}\frac{\bar{D}^{ij}}{\sqrt{\mathrm{Det}'}}\\ &= D'_k - \sqrt{\mathrm{Det}'}\frac{\partial}{\partial x'^r}\frac{\bar{D}'^{kr}}{\sqrt{\mathrm{Det}'}} = \bar{D}'^k\,.\end{aligned} \tag{4.143}$$

In deriving (4.143), in the second line we used (4.127, 131) and in the third line, (4.140), the chain rule, $(\partial x'^r/\partial x^j)\,\partial/\partial x'^r = \partial/\partial x^j$ and (4.138).

Instead of (4.104) we now use the contravariant probability current

$$\bar{S}^i = \bar{D}^i\bar{W} - \bar{D}^{ij}\frac{\partial\bar{W}}{\partial x^j}\,. \tag{4.144}$$

The contraction of a contravariant tensor of rank 2 with a covariant vector $A_i = \partial\bar{W}/\partial x^j$ [which appears in (4.144)] is a contravariant vector as may be seen by the transformation law

$$\begin{aligned}B'^i = \bar{D}'^{ij}A'_j &= \frac{\partial x'^i}{\partial x^k}\frac{\partial x'^j}{\partial x^r}\frac{\partial x^s}{\partial x'^j}\bar{D}^{kr}A_s = \frac{\partial x'^i}{\partial x^k}\delta_{rs}\bar{D}^{kr}A_s\\ &= \frac{\partial x'^i}{\partial x^k}B^k\,.\end{aligned}$$

Next we must find an expression for the divergence of a vector. We have already seen that the derivative of a scalar, i.e.,

$$\bar{W}_{;i} \equiv \frac{\partial \bar{W}}{\partial x^i}, \tag{4.145}$$

is a covariant vector. It is easy to show that

$$\bar{S}^i_{;i} \equiv \sqrt{\mathrm{Det}}\,\frac{\partial}{\partial x^i}\,\frac{\bar{S}^i}{\sqrt{\mathrm{Det}}} \tag{4.146}$$

is a scalar, which is the desired expression for the divergence of a vector. From (4.127, 140) we have

$$\begin{aligned} \bar{S}^i_{;i} &= \frac{\sqrt{\mathrm{Det}}}{J}\,\frac{\partial}{\partial x'^k}\,\frac{\partial x'^k}{\partial x^i}\,\frac{J}{\sqrt{\mathrm{Det}}}\,\bar{S}^i \\ &= \sqrt{\mathrm{Det}'}\,\frac{\partial}{\partial x'^k}\,\frac{\bar{S}'^k}{\sqrt{\mathrm{Det}'}} = \bar{S}'^k_{;k}\,. \end{aligned} \tag{4.147}$$

Thus the equation

$$\begin{aligned} \partial \bar{W}/\partial t &= -\bar{S}^i_{;i} = [-\bar{D}^i \bar{W} + \bar{D}^{ij}\bar{W}_{;j}]_{;i} \\ &= \sqrt{\mathrm{Det}}\,\frac{\partial}{\partial x^i}\,\frac{1}{\sqrt{\mathrm{Det}}}\left(-\bar{D}^i\bar{W} + \bar{D}^{ij}\frac{\partial \bar{W}}{\partial x^j}\right) \end{aligned} \tag{4.148}$$

has the correct covariant form, i.e., it has same the form for every coordinate transformation. Using (4.139, 142), it is easy to show that (4.148) is identical to the Fokker-Planck equation (4.94a), i.e.,

$$\begin{aligned} \sqrt{\mathrm{Det}}\,\frac{\partial W}{\partial t} = \sqrt{\mathrm{Det}}\,\frac{\partial}{\partial x^i}\,\frac{1}{\sqrt{\mathrm{Det}}}\Bigg\{ &-D_i\sqrt{\mathrm{Det}}\,W + \sqrt{\mathrm{Det}}\left[\frac{\partial}{\partial x^j}\,\frac{\bar{D}^{ij}}{\sqrt{\mathrm{Det}}}\right] \\ &\times \sqrt{\mathrm{Det}}\,W + D^{ij}\frac{\partial \sqrt{\mathrm{Det}}\,W}{\partial x^j}\Bigg\}, \end{aligned}$$

$$\begin{aligned} \frac{\partial W}{\partial t} &= \frac{\partial}{\partial x^i}\left\{-D_i W + \left[\frac{\partial}{\partial x^j}\,\frac{\bar{D}^{ij}}{\sqrt{\mathrm{Det}}}\right]\sqrt{\mathrm{Det}}\,W + \frac{1}{\sqrt{\mathrm{Det}}}\bar{D}^{ij}\frac{\partial\sqrt{\mathrm{Det}}\,W}{\partial x^j}\right\} \\ &= \frac{\partial}{\partial x^i}\left(-D_i W + \frac{\partial}{\partial x^j}D_{ij}W\right). \end{aligned}$$

The comparison with tensor analysis [4.15 – 17] shows that the diffusion matrix (4.100) may serve as the contravariant metric tensor [4.18], i.e.,

$$g^{ij} = \bar{D}^{ij} = D_{ij} . \tag{4.149}$$

The covariant metric tensor $g_{ij} = \bar{D}_{ij}$ is then the inverse of the diffusion matrix (4.100), i.e., $g_{ij} = (D^{-1})_{ij}$. The Christoffel symbols of first and second kinds and the Riemann curvature tensor are expressed by the metric tensor in the following way [4.15 – 17]:

$$[ij,k] = \frac{1}{2}\left(\frac{\partial g_{ik}}{\partial x^j} + \frac{\partial g_{jk}}{\partial x^i} - \frac{\partial g_{ij}}{\partial x^k}\right), \tag{4.150}$$

$$\left\{{i \atop jk}\right\} = g^{il}[jk,l] , \tag{4.151}$$

$$R^l_{ijk} = \frac{\partial}{\partial x^j}\left\{{l \atop ik}\right\} - \frac{\partial}{\partial x^k}\left\{{l \atop ij}\right\} + \left\{{l \atop mj}\right\}\left\{{m \atop ik}\right\} - \left\{{l \atop mk}\right\}\left\{{m \atop ij}\right\} . \tag{4.152}$$

If the Riemann curvature tensor vanishes the space is Euclidean. By using a proper coordinate transformation, the metric tensor and therefore also the diffusion tensor can then be reduced to the metric tensor of Euclidean cartesian coordinates [4.17], i.e., to

$$g^{ij} = g_{ij} = D^{ij} = \delta_{ij} . \tag{4.153}$$

If the Riemann curvature tensor does not vanish, it is impossible to find a transformation where (4.153) is valid globally (it may then be valid only locally, i.e., near some fixed point $\{x_0\}$). If we have only one variable, the Riemann curvature tensor always vanishes. Then we can find a transformation so that the diffusion coefficient $D^{(2)} > 0$ is normalized to unity, see also Sect. 5.1. For two variables with $D_{12} = D_{21} = 0$ for instance the Riemann curvature tensor vanishes only if

$$\frac{\partial}{\partial x_1}\sqrt{D_{11}D_{22}}\frac{\partial}{\partial x_1}\frac{1}{D_{22}} + \frac{\partial}{\partial x_2}\sqrt{D_{11}D_{22}}\frac{\partial}{\partial x_2}\frac{1}{D_{11}} = 0 \tag{4.154}$$

is fulfilled.

5. Fokker-Planck Equation for One Variable; Methods of Solution

We now want to discuss methods for solving the one-variable Fokker-Planck equation (4.44, 45) with time-independent drift and diffusion coefficients, assuming $D^{(2)}(x) > 0$

$$\partial W(x,t)/\partial t = L_{\mathrm{FP}} W(x,t) = -(\partial/\partial x)\, S(x,t), \tag{5.1}$$

$$L_{\mathrm{FP}}(x) = -\frac{\partial}{\partial x} D^{(1)}(x) + \frac{\partial^2}{\partial x^2} D^{(2)}(x)\,. \tag{5.2}$$

In (5.1) S is the probability current (4.47).

The stochastic Langevin equation (3.67), for instance, with Gaussian δ-correlated Langevin forces and time-independent h and g leads to (5.1, 2) with $D^{(2)}(x)$ and $D^{(1)}(x)$ given by (3.95).

The Smoluchowski equation (1.23) describing one-dimensional Brownian motion of a particle in the potential $f(x)$ in the high-friction limit is a special case of (5.1, 2), where the drift and diffusion coefficients are given by

$$D^{(1)} = (m\gamma)^{-1} F(x) = -(m\gamma)^{-1} \tilde{f}'(x)\,, \tag{5.3}$$

$$D^{(2)} = kT(m\gamma)^{-1}\,. \tag{5.4}$$

In (5.3) $F(x) = -\tilde{f}'(x)$ is the force due to the potential $\tilde{f}(x)$, m is the mass of the particle, γ is the friction constant, k is Boltzmann's constant and T is the temperature of the surrounding heat bath. The derivation of this Smoluchowski equation from the two-variable Fokker-Planck equation in position and velocity space (i.e., Kramers equation) is discussed in detail in Sect. 10.4.

5.1 Normalization

By a suitable transformation $x' \equiv y = y(x)$ the x-dependent diffusion coefficient can be transformed to an arbitrary constant $D > 0$. For the one-variable case this transformation according to (4.132) reads

$$D'^{(2)} \equiv D = \left(\frac{\mathrm{d}y}{\mathrm{d}x}\right)^2 D^{(2)}(x) . \tag{5.5}$$

Thus this transformation is given by

$$y = y(x) = \int_{x_0}^{x} \sqrt{D/D^{(2)}(\xi)}\,\mathrm{d}\xi . \tag{5.6}$$

The transformed drift coefficient then takes the form [see (4.131)]

$$\begin{aligned} D'^{(1)}(y) &= \frac{\mathrm{d}y}{\mathrm{d}x} D^{(1)}(x) + \frac{\mathrm{d}^2 y}{\mathrm{d}x^2} D^{(2)}(x) \\ &= \sqrt{\frac{D}{D^{(2)}(x)}} \left[D^{(1)}(x) - \frac{1}{2} \frac{\mathrm{d}D^{(2)}(x)}{\mathrm{d}x} \right] \end{aligned} \tag{5.7}$$

and the transformed Fokker-Planck equation reads ($D = \text{const}$)

$$\frac{\partial W'(y,t)}{\partial t} = \left[-\frac{\partial}{\partial y} D'^{(1)}(y) + D \frac{\partial^2}{\partial y^2} \right] W'(y,t) , \tag{5.8}$$

where W' is given by, cf. (4.119),

$$W' = J \cdot W = (\mathrm{d}y/\mathrm{d}x)^{-1} W = \sqrt{D^{(2)}(x)/D}\, W . \tag{5.9}$$

In (5.7 and 9) $x = x(y)$ has to be expressed by the y variable according to (5.6). Without loss of generality we may thus treat the equation with constant diffusion coefficient, i.e.,

$$\frac{\partial W}{\partial t} = \left[\frac{\partial}{\partial x} f'(x) + D \frac{\partial^2}{\partial x^2} \right] W = -\frac{\partial}{\partial x} S(x,t) , \tag{5.10}$$

where S is the probability current.

Here we have introduced the potential

$$f(x) = -\int^{x} D^{(1)}(x')\,\mathrm{d}x' . \tag{5.11}$$

Up to a constant the potential (5.11) agrees with the potential $\tilde{f}$ of the Smoluchowski equation.

Because D is arbitrary, we may use $D = 1$. This normalization is, however, not very convenient if the low-noise limit $D \to 0$ is considered and we therefore retain the constant D.

Transformation (5.6) can also be done in the Langevin equation, see (3.69, 70, 95).

5.2 Stationary Solution

For stationary solutions the probability current in (5.1) must be constant. Thus, if the probability current vanishes at some x the current must be zero for any x. Then for $S=0$

$$D^{(1)}(x)\,W_{st}(x)=\frac{D^{(1)}(x)}{D^{(2)}(x)}D^{(2)}(x)\,W_{st}(x)=\frac{\partial}{\partial x}D^{(2)}(x)\,W_{st}(x)\,. \tag{5.12}$$

We can immediately integrate (5.12), yielding

$$W_{st}(x)=\frac{N_0}{D^{(2)}(x)}\exp\left(\int^x\frac{D^{(1)}(x')}{D^{(2)}(x')}\mathrm{d}x'\right)=N\mathrm{e}^{-\Phi(x)}\,, \tag{5.13}$$

where N_0 is the integration constant, which has to be chosen such that W_{st} is normalized. In (5.13) we introduced the potential

$$\Phi(x)=\ln D^{(2)}(x)-\int^x\frac{D^{(1)}(x')}{D^{(2)}(x')}\mathrm{d}x'\,. \tag{5.14}$$

For the case of the Smoluchowski equation (5.3, 4) we may put $\Phi(x)=\tilde{f}(x)/(kT)$ and for (5.10) $\Phi(x)=f(x)/D$ because the potential $\Phi(x)$ is defined only up to an additive constant and therefore the $\ln D^{(2)}$ term may be omitted. Introducing this potential the probability current may be written in the form

$$S(x,t)=-D^{(2)}(x)\,\mathrm{e}^{-\Phi(x)}\frac{\partial}{\partial x}[\mathrm{e}^{\Phi(x)}W(x,t)]\,. \tag{5.15}$$

In the stationary state, where S is constant, we thus have for arbitrary S

$$W_{st}(x)=N\mathrm{e}^{-\Phi(x)}-S\mathrm{e}^{-\Phi(x)}\int^x\frac{\mathrm{e}^{\Phi(x')}}{D^{(2)}(x')}\mathrm{d}x'\,. \tag{5.16}$$

One of the integration constants in (5.16) is determined by the normalization

$$\int W_{st}(x)\,\mathrm{d}x=1\,, \tag{5.17}$$

the other constant must be determined from the boundary conditions, so the problem arises as to which boundary conditions must be used. (For a further discussion of boundary conditions, see Sect. 5.4.) For problems where x extends to $\pm\infty$, we require that the integral (5.17) exists. In that case, W and also S must vanish at $\pm\infty$ (natural boundary conditions) and therefore $S=0$ for every x. If the stochastic variable ξ cannot reach values smaller than x_{min}, we require that the probability current must be zero at x_{min}. In the stationary state, S then also vanishes for every x, i.e., (5.16) reduces to (5.13). There may, of course, also be other boundary conditions. If for instance x is an angle variable we usually require that the distribution function is periodic. In that case, S is determined by this periodicity condition. The current will then be zero only if $f(x)$ is also periodic. For a further discussion of boundary conditions, see Sect. 5.4.

An important question is whether every initial distribution finally decays to the stationary distribution. For some restrictions of the drift and diffusion coefficients and of the boundary conditions one can prove that any two solutions of the Fokker-Planck equation agree for large times. Thus if a stationary solution exists, every solution must finally decay to that solution. We postpone the derivation of the proof to Sect. 6.1, where a proof is given for the general N-variable case. We further show in Sect. 5.4 that all eigenvalues with the exception of the stationary eigenvalue $\lambda = 0$ are larger than zero, which also answers the above question positively.

5.3 Ornstein-Uhlenbeck Process

Nonstationary solutions of the Fokker-Planck equation (5.1, 2) are more difficult to obtain. A general expression for the nonstationary solution can be found only for special drift and diffusion coefficients.

Wiener Process

A process which is described by (5.1, 2) with vanishing drift coefficient ($D^{(1)} = 0$) and constant diffusion coefficient $D^{(2)}(x) = D$ is called a Wiener process. The equation for the transition probability $P = P(x, t \mid x', t')$ is then the diffusion equation

$$\frac{\partial P}{\partial t} = D \frac{\partial^2 P}{\partial x^2} \tag{5.18}$$

with the initial condition

$$P(x, t' \mid x', t') = \delta(x - x') \, . \tag{5.19}$$

The solution for $t > t'$ reads [5.1]

$$P(x, t \mid x', t') = \frac{1}{\sqrt{4\pi D(t-t')}} \exp\left(-\frac{(x-x')^2}{4D(t-t')}\right) . \tag{5.20}$$

The general solution for the probability density with the initial distribution $W(x', t')$ is then given by

$$W(x, t) = \int P(x, t \mid x', t') \, W(x', t') \, dx' \, . \tag{5.21}$$

Thus the transition probability serves as the Green's function of (5.18).

For the Ornstein-Uhlenbeck process the drift coefficient is linear and the diffusion coefficient is constant, i.e.,

Ornstein-Uhlenbeck Process

$$D^{(1)}(x) = -\gamma x; \quad D^{(2)}(x) = D = \text{const}. \tag{5.22}$$

With these coefficients the Fokker-Planck equation is the same as the Smoluchowski equation for a harmonically bound oscillator. In this case $\gamma = \omega_0^2/m$ is positive.

The equation for the transition probability now reads

$$\frac{\partial P}{\partial t} = \gamma \frac{\partial}{\partial x}(xP) + D\frac{\partial^2}{\partial x^2}P \tag{5.23}$$

with the initial condition (5.19). The solution of (5.23) is best found by making a Fourier transform with respect to x, i.e.,

$$P(x,t|x',t') = (2\pi)^{-1}\int e^{ikx}\tilde{P}(k,t|x',t')\,dk. \tag{5.24}$$

The equation for the Fourier transform is given by (replace $\partial/\partial x$ by ik and x by $i\partial/\partial k$)

$$\frac{\partial \tilde{P}}{\partial t} = -\gamma k \frac{\partial}{\partial k}\tilde{P} - Dk^2\tilde{P}, \tag{5.25}$$

which is simpler than (5.23) because only first-order derivatives with respect to k occur. Because of (5.19) the initial condition for the Fourier transform is

$$\tilde{P}(k,t'|x',t') = e^{-ikx'}. \tag{5.26}$$

The first-order equation (5.25) may be solved by the methods of characteristics [5.1]. The solution of (5.25) with the initial condition (5.26) reads ($t>t'$)

$$\tilde{P}(k,t|x',t') = \exp[-ikx'e^{-\gamma(t-t')} - Dk^2(1-e^{-2\gamma(t-t')})/(2\gamma)], \tag{5.27}$$

as may easily be checked by insertion. By performing the integral in (5.24) [cf. (2.32)] we finally get the Gaussian distribution ($t>t'$)

$$P(x,t|x',t') = \sqrt{\frac{\gamma}{2\pi D(1-e^{-2\gamma(t-t')})}}\exp\left[-\frac{\gamma(x-e^{-\gamma(t-t')}x')^2}{2D(1-e^{-2\gamma(t-t')})}\right]. \tag{5.28}$$

In the limit $\gamma \to 0$ we recover the result (5.20) for the Wiener process.

Equation (5.28) is valid for positive and negative γ. For positive γ and large time differences $\gamma(t-t') \gg 1$, (5.28) passes over to the stationary distribution

$$W_{st}(x) = \sqrt{\gamma/(2\pi D)}\, \exp[-\gamma x^2/(2D)] \tag{5.29}$$

in agreement with (5.13). For $\gamma \leqq 0$ no stationary solution exists.

The Ornstein-Uhlenbeck process may equally well be described by a linear Langevin equation of the type (3.1) with Gaussian Langevin forces. The stochastic variable and the Langevin force are then connected by the linear transformation (3.7). Because for a linear transformation of variables Gaussian distributions will remain Gaussian (see the remark at the end of Sect. 2.3.3), the transition probability must also be a Gaussian distribution.

Joint Probability Density

In the stationary state the joint probability density for the variables $\xi(t)$ and $\xi(t')$ may be expressed by P and W_{st}. For $t \geqq t'$

$$W_2(x,t;x',t') = P(x,t|x',t')\, W_{st}(x')\,, \tag{5.30}$$

and for $t \leqq t'$,

$$W_2(x,t;x',t') = P(x',t'|x,t)\, W_{st}(x)\,. \tag{5.31}$$

By inserting (5.28, 29) in (5.30, 31) we obtain in both cases

$$W_2(x,t;x',t') = \frac{\gamma}{2\pi D\sqrt{1-e^{-2\gamma|t-t'|}}} \exp\left(-\gamma \frac{x^2+x'^2-2xx'\,e^{-\gamma|t-t'|}}{2D(1-e^{-2\gamma|t-t'|})}\right). \tag{5.32}$$

For large time differences $\gamma|t-t'| \gg 1$, (5.32) decomposes into a product of two stationary distribution functions (5.29), meaning that the distributions for x and x' become independent.

5.4 Eigenfunction Expansion

In this chapter we are looking for nonstationary solutions of (5.1, 2). A separation ansatz for $W(x,t)$

$$W(x,t) = \varphi(x)\, e^{-\lambda t} \tag{5.33}$$

leads to

$$L_{FP}\varphi = -\lambda\varphi\,. \tag{5.34}$$

Here $\varphi(x)$ and λ are the eigenfunctions and eigenvalues of the Fokker-Planck operator with appropriate boundary conditions. Before we proceed it is necessary to talk about boundary conditions.

Boundary Conditions

If the potential $\Phi(x)$ (e.g., $f(x)$ for x-independent diffusion) jumps to an infinite high positive value, the particles cannot penetrate in the region $x > x_{max}$ and therefore the probability current S must vanish at that point. The infinite high potential then acts as a reflecting wall (Fig. 5.1 a). If the potential jumps to an infinite large negative value it follows from the continuity condition for the probability current that $e^{\Phi}W$ should vanish at this point (5.84). In this case we talk about an absorbing wall (Fig. 5.1 b). For the left boundary at $x = x_{min}$ similar considerations are valid. Because of the two possibilities at each side, there are four possibilities B1 ... B4, as shown in Table 5.1.

For finite x_{max} and x_{min}, $e^{\Phi}W = 0$ requires that W should be zero. If $\Phi(x)$ goes to plus infinity for $x \to \pm\infty$ (e.g., $\Phi = \alpha x^2$), we have a reflecting wall at $x_{max} \to +\infty$ and $x_{min} \to -\infty$ and the probability current should vanish there. It then follows from the Fokker-Planck equation that $\int_{-\infty}^{\infty} W(x,t)\,dx$ is constant (4.48), and that this constant is equal to 1, if it is initially equal to 1. This normalization requires that the distribution function W goes to zero for $x \to \pm\infty$. (As seen for the parabolic potential in Sect. 5.5.1 $e^{\Phi}W$, however, remains finite.) This boundary condition is called natural boundary condition. It may also happen that $\Phi(x)$ goes to minus infinity for $x \to +\infty$ or for $x \to -\infty$ or for both $x \to \pm\infty$. In this case we require in analogy to (5.84) that $e^{\Phi}W$ is zero at $x \to +\infty$ or at $x \to -\infty$ or at both $x \to \pm\infty$, respectively. As shown in Sect. 5.5.2 for the inverted parabolic potential, W and S are then finite but $e^{\Phi}W$ vanishes at $x \to \pm\infty$. Thus for the boundary condition in Table 5.1 x_{min}, x_{max} or both can reach $-\infty$, $+\infty$ or $\pm\infty$, respectively.

To obtain eigenvalues by numerical integration for potentials, where $\Phi(x)$ goes to plus infinity for $x \to \pm\infty$, we may require that $S = 0$ at $x = \pm A$ for some large A. Alternatively, we may require that $W = 0$ at $x = \pm A$. Though at finite A the eigenvalues for the $S = 0$ condition will be different from the eigenvalues of the $W = 0$ condition, both eigenvalues will coincide in the limit $A \to \infty$. (Obviously, we cannot require that both S and W are zero at $x = \pm A$ with finite A.)

Besides these boundary conditions we may have periodic boundary conditions with period L. If, for instance, x is an angle variable and if we do not distinguish whether a full rotation is made or not, the distribution function and therefore also the probability current must be periodic with period $L = 2\pi$. These periodic boundary conditions can be fulfilled only if the drift and diffusion coef-

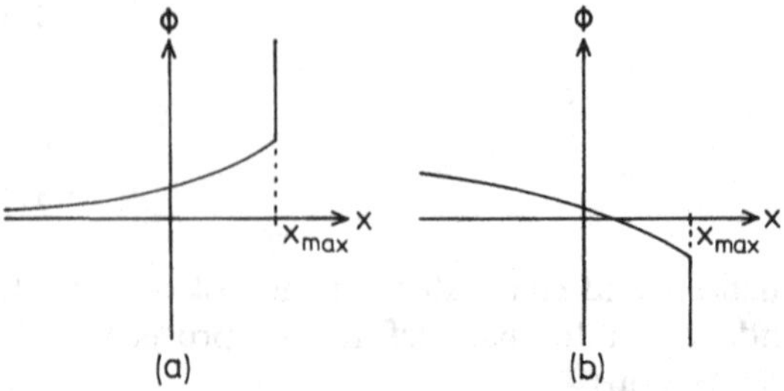

Fig. 5.1. Reflecting (**a**) and absorbing (**b**) wall

Table 5.1. Boundary conditions discussed in the text

x_{min} \ x_{max}	$S=0$ (reflecting wall)	$e^{\Phi}W=0$ (absorbing wall)
$S=0$ (reflecting wall)	B1	B2
$e^{\Phi}W=0$ (absorbing wall)	B3	B4

Natural boundary conditions:
$S=0$ for $x_{min}\to-\infty$, $x_{max}\to+\infty$
(i.e. B1 for $x_{min}\to-\infty$, $x_{max}\to+\infty$)

Periodic boundary conditions:
$W(x,t)=W(x+L,t)$, $S(x,t)=S(x+L,t)$

ficients are also periodic with the period L. An example of this boundary condition is given in Sect. 11.3.

A stationary solution can occur only for the boundary condition B1 (this includes natural boundary conditions) or for periodic boundary conditions. In the first case, the stationary solution of the Fokker-Planck equation is given by (5.13), for the other, by (5.16).

Transformation of the Fokker-Planck Operator

The Fokker-Planck operator (5.2), which may be written in the form, cf. (5.1, 15),

$$L_{FP}=\frac{\partial}{\partial x}D^{(2)}(x)\,e^{-\Phi(x)}\frac{\partial}{\partial x}e^{\Phi(x)}, \tag{5.35}$$

is obviously not a Hermitian operator. If the two functions W_1 and W_2 both satisfy the same boundary conditions listed in Table 5.1, we have

$$\begin{aligned}
&\int_{x_{min}}^{x_{max}} W_1 e^{\Phi} L_{FP} W_2\,dx\\
&\quad=\int_{x_{min}}^{x_{max}} W_1 e^{\Phi}\frac{\partial}{\partial x}D^{(2)}e^{-\Phi}\frac{\partial}{\partial x}e^{\Phi}W_2\,dx\\
&\quad=-\int_{x_{min}}^{x_{max}}\left[\frac{\partial}{\partial x}W_1e^{\Phi}\right]D^{(2)}e^{-\Phi}\left[\frac{\partial}{\partial x}e^{\Phi}W_2\right]dx\\
&\quad=\int_{x_{min}}^{x_{max}} W_2 e^{\Phi}\frac{\partial}{\partial x}D^{(2)}e^{-\Phi}\frac{\partial}{\partial x}W_1e^{\Phi}\,dx\\
&\quad=\int_{x_{min}}^{x_{max}} W_2 e^{\Phi} L_{FP} W_1\,dx\,.
\end{aligned} \tag{5.36}$$

In deriving the third and the fourth line we have used partial integration and

$$\underset{2}{W_1} e^{\Phi} D^{(2)} e^{-\Phi} \frac{\partial}{\partial x} e^{\Phi} \underset{1}{W_2} \Big|_{x_{\min}}^{x_{\max}} = - \underset{2}{W_1} e^{\Phi} \underset{1}{S_2} \Big|_{x_{\min}}^{x_{\max}} = 0 . \tag{5.37}$$

Hence for all boundary conditions of Table 5.1 the adjoint of the operator $e^{\Phi} L_{FP}$ is given by

$$(e^{\Phi} L_{FP})^+ \equiv L_{FP}^+ e^{\Phi} = e^{\Phi} L_{FP} , \tag{5.38}$$

i.e., $e^{\Phi} L_{FP}$ as well as

$$L = e^{-\Phi/2} e^{\Phi} L_{FP} e^{-\Phi/2} = e^{\Phi/2} L_{FP} e^{-\Phi/2} \tag{5.39}$$

is an Hermitian operator.

Orthogonality of Eigenfunctions

The eigenvalues may be discrete or continuous or both. In the following we use the notation for discrete eigenvalues denoted by an index n. If continuous eigenvalues occur, one should proceed in the same way as discussed in quantum mechanics [5.3], i.e., the Kronecker symbol δ_{nm} has to be replaced by the δ function and the occurring sums by integrations. If $\varphi_n(x)$ are the eigenfunctions of the Fokker-Planck operator L_{FP} with the eigenvalue λ_n (5.34), the functions

$$\psi_n(x) = e^{\Phi(x)/2} \varphi_n(x) \tag{5.40}$$

are eigenfunctions of L with the same eigenvalues λ_n

$$L \psi_n = -\lambda_n \psi_n . \tag{5.41}$$

Because L is an Hermitian operator, the eigenvalues are real and two eigenfunctions ψ_1 and ψ_2 with different eigenvalues $\lambda_1 \neq \lambda_2$ must be orthogonal. If we normalize the eigenfunctions we thus have the orthonormality relation

$$\int_{x_{\min}}^{x_{\max}} \psi_n \psi_m dx = \int_{x_{\min}}^{x_{\max}} e^{\Phi} \varphi_n \varphi_m dx = \delta_{nm} . \tag{5.42}$$

Positivity of Eigenvalues

By using the first and the third line of (5.36) with $W_1 = W_2 = \varphi_n(x)$ and (5.35) we get

$$\int_{x_{\min}}^{x_{\max}} \varphi_n e^{\Phi} L_{\mathrm{FP}} \varphi_n \mathrm{d}x = \int_{x_{\min}}^{x_{\max}} \psi_n L \psi_n \mathrm{d}x = -\lambda_n$$

$$= -\int_{x_{\min}}^{x_{\max}} \left(\frac{\partial}{\partial x} \psi_n e^{\Phi/2} \right)^2 D^{(2)} e^{-\Phi} \mathrm{d}x \leqq 0 . \tag{5.43}$$

The equals sign in (5.43) is valid only for the stationary solution

$$\psi_0(x) = \sqrt{N} e^{-\Phi(x)/2}; \quad \lambda_0 = 0 . \tag{5.44}$$

All other eigenvalues λ_n ($n \geqq 1$) must be larger than zero. For finite potentials $\Phi(x)$ a stationary solution cannot exist for the boundary conditions B2 – B4. Thus all eigenvalues are larger than zero for these boundary conditions. For (5.44) to exist under natural boundary conditions, $\Phi(x)$ must be positive and increase with increasing $|x|$ at least asymptotically.

Other eigenfunctions with $\lambda_n > 0$ can exist for asymptotically negative $\Phi(x)$, with appropriate boundary conditions, see the example in Sect. 5.5.2.

Completeness Relations

Eigenfunctions of Hermitian operators usually form a complete set [5.1, 2, 4]. The completeness relation for the eigenfunctions ψ_n or φ_n may be expressed by

$$\begin{aligned} \delta(x-x') &= \sum_n \psi_n(x)\, \psi_n(x') \\ &= e^{\Phi(x)/2 + \Phi(x')/2} \sum_n \varphi_n(x)\, \varphi_n(x') \\ &= e^{\Phi(x)} \sum_n \varphi_n(x)\, \varphi_n(x') \\ &= e^{\Phi(x')} \sum_n \varphi_n(x)\, \varphi_n(x') . \end{aligned} \tag{5.45}$$

Transition Probability Density

By using the last expression of (5.45) to represent the δ function and the formal solution (4.17) for the Fokker-Planck operator, we immediately obtain the expansion of the transition probability into eigenfunctions ($t \geqq t'$)

$$\begin{aligned} P(x,t\,|\,x',t') &= e^{L_{\mathrm{FP}}(x)(t-t')} \delta(x-x') \\ &= e^{\Phi(x')} \sum_n e^{L_{\mathrm{FP}}(x)(t-t')} \varphi_n(x)\, \varphi_n(x') \\ &= e^{\Phi(x')} \sum_n \varphi_n(x)\, \varphi_n(x')\, e^{-\lambda_n(t-t')} \\ &= e^{\Phi(x')/2 - \Phi(x)/2} \sum_n \psi_n(x)\, \psi_n(x')\, e^{-\lambda_n(t-t')} . \end{aligned} \tag{5.46}$$

Joint Probability Density

In the stationary state the joint probability density for the variables $\xi(t)$ and $\xi(t')$ may be obtained from (5.30, 31). If the stationary distribution $W_{st}(x) = [\psi_0(x)]^2$ exists, we have

$$W_2(x,t;x',t') = \psi_0(x)\,\psi_0(x')\sum_n \psi_n(x)\,\psi_n(x')\,\mathrm{e}^{-\lambda_n|t-t'|}\,. \tag{5.47}$$

The symmetry of W_2, i.e., $W_2(x,t;x',t') = W_2(x',t';x,t)$, is immediately seen.

Explicit Form of L

Because of (5.35) the transformed operator (5.39) takes the form

$$L = \mathrm{e}^{\Phi/2}\frac{\partial}{\partial x}\sqrt{D^{(2)}}\,\mathrm{e}^{-\Phi/2}\sqrt{D^{(2)}}\,\mathrm{e}^{-\Phi/2}\frac{\partial}{\partial x}\mathrm{e}^{\Phi/2} = -\hat{a}a\,, \tag{5.48}$$

where a and $\hat{a}$ are defined by

$$\begin{aligned} a &= \sqrt{D^{(2)}}\,\mathrm{e}^{-\Phi/2}\frac{\partial}{\partial x}\mathrm{e}^{\Phi/2} \\ &= \sqrt{D^{(2)}}\frac{\partial}{\partial x} + \frac{1}{2}\left(\frac{\mathrm{d}D^{(2)}}{\mathrm{d}x} - D^{(1)}\right)\Big/\sqrt{D^{(2)}}\,, \end{aligned} \tag{5.49}$$

$$\begin{aligned} \hat{a} &= -\mathrm{e}^{\Phi/2}\frac{\partial}{\partial x}\sqrt{D^{(2)}}\,\mathrm{e}^{-\Phi/2} \\ &= -\frac{\partial}{\partial x}\sqrt{D^{(2)}} + \frac{1}{2}\left(\frac{\mathrm{d}D^{(2)}}{\mathrm{d}x} - D^{(1)}\right)\Big/\sqrt{D^{(2)}}\,. \end{aligned} \tag{5.50}$$

The second lines follow by use of (5.14).

For natural boundary conditions a and $\hat{a}$ are the adjoints of each other, i.e., $\hat{a} = a^+$. It then also follows from (5.48) that all eigenvalues λ must be non-negative. By inserting the last expressions for a and $\hat{a}$ into (5.48) we get the operator of the Sturm-Liouville equation [5.1, 2, 4]

$$L = \frac{\partial}{\partial x}D^{(2)}\frac{\partial}{\partial x} - V\,, \tag{5.51}$$

$$V(x) = \frac{1}{4}\left(\frac{\mathrm{d}D^{(2)}}{\mathrm{d}x} - D^{(1)}\right)^2\Big/D^{(2)} + \frac{1}{2}\,\frac{\mathrm{d}D^{(1)}}{\mathrm{d}x} - \frac{1}{2}\,\frac{\mathrm{d}^2D^{(2)}}{\mathrm{d}x^2}\,. \tag{5.52}$$

The eigenvalues are usually arranged in increasing order

$$0 \leqq \lambda_0 < \lambda_1 < \lambda_2 < \dots . \tag{5.53}$$

The first eigenfunction ψ_0 has no zeros, the next eigenfunction ψ_1 has one zero and so on [5.2]. If a stationary solution exists the first eigenvalue λ_0 is zero; otherwise it is larger than zero. Whereas a degeneracy cannot occur for the boundary conditions B1 – B4, it can occur for periodic boundary conditions, see Sect. 11.3.2 for an example.

Transformation to a Schrödinger Equation

By using a proper transformation of the variable, the one-variable Fokker-Planck equation can always be transformed to (5.10) where the diffusion constant is x-independent. Then L has the same form as the negative single-particle Hamilton operator in quantum mechanics, i.e.,

$$L = D\frac{\partial^2}{\partial x^2} - V_S(x) \tag{5.54}$$

with the potential

$$V_S(x) = \tfrac{1}{4}[f'(x)]^2/D - \tfrac{1}{2}f''(x) . \tag{5.55}$$

For the potential $\Phi(x)$ we may now use

$$\Phi(x) = f(x)/D \tag{5.56}$$

because we can neglect the additive constant $\ln D$ (5.14). The form (5.55) guarantees that the eigenvalue of the stationary solution $\psi_0 = \sqrt{N}\exp[-f(x)/(2D)]$ is zero. The eigenvalue problem (5.41) is the same as the eigenvalue problem of the Schrödinger equation. The operators a and $\hat{a}$ simplify to

$$\begin{aligned} a &= \sqrt{D}\frac{\partial}{\partial x} + \frac{1}{2}\frac{f'(x)}{\sqrt{D}} = \sqrt{D}\exp\left(-\frac{1}{2}\frac{f(x)}{D}\right)\frac{\partial}{\partial x}\exp\left(\frac{1}{2}\frac{f(x)}{D}\right) \\ \hat{a} &= -\sqrt{D}\frac{\partial}{\partial x} + \frac{1}{2}\frac{f'(x)}{\sqrt{D}} = -\sqrt{D}\exp\left(\frac{1}{2}\frac{f(x)}{D}\right)\frac{\partial}{\partial x}\exp\left(-\frac{1}{2}\frac{f(x)}{D}\right) \end{aligned} \tag{5.57}$$

and their commutator is given by

$$a\hat{a} - \hat{a}a = f'' . \tag{5.58}$$

If the transformation (5.40) is applied to the probability density $W(x,t)$, the Fokker-Planck equation (5.10) is formally equivalent to the time-dependent

Schrödinger equation with imaginary times $t_{\text{Schröd}} = -\mathrm{i}\hbar t$ and with a mass given by $m_{\text{Schröd}} = \hbar^2/(2D)$. Transformation of an equation of the type (5.1, 2) to the Schrödinger form is also found in [5.2, 4].

5.5 Examples

We now want to discuss the eigenvalues, eigenfunctions, the potential $f(x)$ and the potential $V_S(x)$ of the corresponding Schrödinger equation for some examples. We first notice that every soluble example of the Schrödinger equation may serve as a soluble example of the normalized Fokker-Planck equation (5.10) [5.5–9]. From (5.44, 56) the potential $f(x)$ of the normalized Fokker-Planck equation is then expressed by the lowest eigenfunction $\psi_0(x)$ of the Schrödinger equation

$$\begin{aligned} f(x) &= -2D\ln\psi_0(x) + D\ln N \\ f'(x) &= -2D\psi_0'(x)/\psi_0(x)\,. \end{aligned} \tag{5.59}$$

Here we have assumed that the stationary solution exists, i.e., that the eigenvalue λ_0 is zero. As will be seen from the third example, simple forms of the potential $V_S(x)$ of the Schrödinger equation may lead to complicated forms of the potential $f(x)$ of the Fokker-Planck equation. This is also seen in [5.6, 7] where simple bistable models for $V_S(x)$ (potential box with a square barrier in the middle) lead to more complicated expressions for $f(x)$. In Sect. 5.7 it is shown that for simple expressions of $f(x)$ (i.e., a box with a rectangular barrier in the middle) eigenvalues and eigenfunctions can also be obtained.

5.5.1 Parabolic Potential

For the parabolic potential of the Fokker-Planck equation

$$f(x) = \tfrac{1}{2}\gamma x^2; \qquad \gamma > 0 \tag{5.60}$$

the potential (5.55) of the Schrödinger equation is also parabolic

$$V_S(x) = \gamma\left(\frac{\gamma}{4D}x^2 - \frac{1}{2}\right). \tag{5.61}$$

Introducing the boson operators similar to a and $\hat{a}$ in (5.57)

$$b = \frac{a}{\sqrt{\gamma}} = \frac{1}{\sqrt{2}}\left(\frac{\partial}{\partial\xi} + \xi\right); \qquad \xi = \sqrt{\frac{\gamma}{2D}}\,x$$

$$b^+ = \frac{\hat{a}}{\sqrt{\gamma}} = \frac{1}{\sqrt{2}}\left(-\frac{\partial}{\partial \xi} + \xi\right); \quad bb^+ - b^+b = 1\,, \tag{5.62}$$

the transformed Fokker-Planck operator (5.48) takes the form

$$L = -\gamma b^+ b\,. \tag{5.63}$$

Eigenvalues and normalized eigenfunctions are given by the well-known expressions [5.3]

$$\begin{aligned} &\lambda_n = \gamma n\,; \quad n = 0, 1, 2, \ldots \\ &\psi_0(x) = \sqrt[4]{\frac{\gamma}{2\pi D}}\, e^{-\xi^2/2} \\ &\psi_n(x) = \frac{(b^+)^n}{\sqrt{n!}}\,\psi_0(x) = \sqrt[4]{\frac{\gamma}{2\pi D}}\,\frac{1}{\sqrt{2^n n!}}\,\mathrm{H}_n(\xi)\, e^{-\xi^2/2}\,, \end{aligned} \tag{5.64}$$

where $\mathrm{H}_n(x)$ are the Hermite polynomials.

If we apply the following summation formula for the Hermite polynomials ([5.10]; $|\alpha| < \frac{1}{2}$)

$$\sum_{n=0}^{\infty} \frac{\alpha^n}{n!}\,\mathrm{H}_n(x)\mathrm{H}_n(y) = \frac{1}{\sqrt{1-4\alpha^2}}\exp\left[\frac{4\alpha}{1-4\alpha^2}(xy - \alpha x^2 - \alpha y^2)\right], \tag{5.65}$$

we recover from (5.46) the transition probability (5.28) and from (5.47) the joint distribution (5.32) for the Ornstein-Uhlenbeck process.

5.5.2 Inverted Parabolic Potential

For the inverted parabolic potential

$$\bar{f}(x) = -\tfrac{1}{2}\bar{\gamma}x^2\,; \quad \bar{\gamma} > 0 \tag{5.66}$$

no stationary solution exists. Nevertheless we can make the transformation (5.39, 40) with $\Phi(x) = \bar{f}(x)/D = -\frac{1}{2}\bar{\gamma}x^2/D$ and obtain the following potential $\bar{V}_S(x)$ of the Schrödinger equation:

$$\bar{V}_S(x) = \bar{\gamma}\left(\frac{\bar{\gamma}}{4D}x^2 - \frac{1}{2}\right) + \bar{\gamma}\,. \tag{5.67}$$

A comparison with (5.61) shows that the normalized eigenfunctions are the same as in Sect. 5.5.1 with γ replaced by $\bar{\gamma}$, i.e.,

$$\bar{\psi}_n(x) = \sqrt[4]{\frac{\bar{\gamma}}{2D\pi}} \frac{1}{\sqrt{2^n n!}} \mathrm{H}_n(\bar{\xi}) \mathrm{e}^{-\bar{\xi}^2/2}; \quad \bar{\xi} = \sqrt{\frac{\bar{\gamma}}{2D}} x \tag{5.68}$$

and that the eigenvalues are raised by $\bar{\gamma}$, i.e., they start with $\lambda_0 = \bar{\gamma}$

$$\bar{\lambda}_n = \bar{\gamma}(n+1); \quad n = 0, 1, 2, \ldots . \tag{5.69}$$

(By formally changing γ to $\bar{\gamma}$ in (5.62) and ξ in $\mathrm{i}\bar{\xi}$ it is seen that $-a/\sqrt{\bar{\gamma}}$ now becomes a creation operator b^+ and correspondingly that $-\hat{a}/\sqrt{\bar{\gamma}}$ now becomes an annihilation operator b, cf. Sect. 5.8.)

Using both (5.46, 65) we obtain

$$P(x,t|x',t') = \sqrt{\frac{\bar{\gamma}}{2\pi D(1-\mathrm{e}^{-2\bar{\gamma}(t-t')})}} \exp\left(-\frac{\bar{\gamma}(x\mathrm{e}^{-\bar{\gamma}(t-t')} - x')^2}{2D(1-\mathrm{e}^{-2\bar{\gamma}(t-t')})}\right) \mathrm{e}^{-\bar{\gamma}(t-t')}, \tag{5.70}$$

which is identical to (5.28) if γ in (5.28) is replaced by $-\bar{\gamma}$. Though no stationary solution exists for an inverted parabolic potential, eigenfunctions with the boundary condition B4 in Table 5.1. for $x_{\substack{max\\min}} \to \pm\infty$ do exist, they can be normalized according to (5.42) and they may be used to calculate the transition probability. (The probability current S for these eigenfunctions is finite for $x \to \pm\infty$.)

5.5.3 Infinite Square Well for the Schrödinger Potential

One of the simplest eigenvalue problems for the Schrödinger equation is the rectangular-well potential with infinitely high walls, Fig. 5.2. The lowest eigenfunction $\psi_0(x) = a^{-1/2}\cos[\pi x/(2a)]$ for $-a<x<a$ leads to the potential (5.59)

$$f(x) = -2D\ln\{\cos[\pi x/(2a)]\} \tag{5.71}$$

of the Fokker-Planck equation, plotted in Fig. 5.2. [In (5.71) we have normalized the potential by $f(0) = 0$, i.e., $N = a^{-1}$.] At $x = \pm a$ the potential $f(x)$ becomes singular. It may easily be checked that the probability density as well as the probability current are zero at $x = \pm a$. Higher eigenvalues and normalized eigenfunctions of the transformed Fokker-Planck operator (5.54) are

even solutions ($n = 1, 2, \ldots$)

$$\lambda_{2n} = D\pi^2 a^{-2}(n^2 + n), \tag{5.72}$$

$$\psi_{2n}(x) = a^{-1/2}\cos[(n+1/2)\pi x/a], \tag{5.73}$$

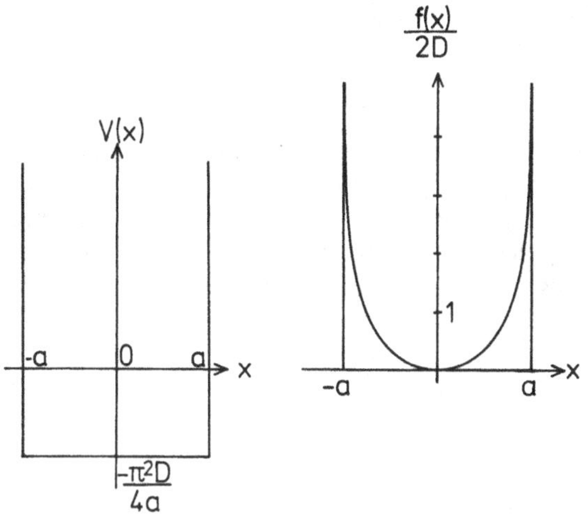

Fig. 5.2. Rectangular-well potential $V(x)$ and the corresponding potential $f(x)$ of the Fokker-Planck equation, (5.71)

odd solutions ($n = 1, 2, \ldots$)

$$\lambda_{2n-1} = D\pi^2 a^{-2}(n^2 - 1/4)\,, \tag{5.74}$$

$$\psi_{2n-1}(x) = a^{-1/2}\sin(n\pi x/a)\,. \tag{5.75}$$

The transition probability for the potential (5.71) is obtained by inserting (5.72 – 75) into (5.46).

5.5.4 V-shaped Potential for the Fokker-Planck Equation

If the potential of the Fokker-Planck equation is given by the V-shaped form

$$f(x) = D\kappa|x|; \qquad \kappa > 0\,, \tag{5.76}$$

the Schrödinger potential $V_S(x)$ consists, see (5.55), of an attractive δ potential

$$V_S(x) = D\kappa^2/4 - D\kappa\,\delta(x)\,. \tag{5.77}$$

Only the stationary eigenfunction

$$\psi_0(x) = \sqrt{\kappa/2}\,\mathrm{e}^{-\kappa|x|/2} \tag{5.78}$$

has the discrete eigenvalue $\lambda_0 = 0$. The other eigenvalues form a continuum ($k > 0$)

$$\lambda_k = D\kappa^2/4 + Dk^2; \tag{5.79}$$

their eigenfunctions normalized to the δ function are [5.11]

$$\psi_k^{\mathrm{s}}(x) = [(4k^2+\kappa^2)\pi]^{-1/2}(2k\cos kx - \kappa \sin k|x|)$$
$$\psi_k^{\mathrm{a}}(x) = \pi^{-1/2}\sin kx \,. \tag{5.80}$$

The symbols s and a indicate even or symmetric and odd or antisymmetric eigenfunctions.

The transition probability for the potential (5.76) is obtained from (5.46), where the sum in (5.46) must be replaced by an integration over k for the continuous eigenfunctions while the discrete eigenfunctions must be retained as a single term.

5.6 Jump Conditions

For the Schrödinger equation one often uses potential models where $V_S(x)$ jumps at certain points of x and is constant elsewhere. One may ask whether such models may also be used for the potential $f(x)$ of the Fokker-Planck equation. As is seen from (5.55), jumps of the potential $f(x)$ lead to higher singularities (first derivative of the δ function and square of the δ function) for the potential $V_S(x)$ than δ function singularities which are usually treated in quantum mechanics. We first derive the jump conditions for the unnormalized Fokker-Planck equation (5.1, 2) and then specialize the result to the normalized Fokker-Planck equation (5.10).

Finite Jump

We assume that a finite jump of the potential $\Phi(x)$ (5.14) occurs at $x=0$. A finite jump may occur either if the diffusion coefficient $D^{(2)}(x)$ has a finite jump or if the drift coefficient $D^{(1)}(x)$ has a δ function singularity. If we assume that the time derivative of the probability density is finite at the jump, it follows from the continuity equation (4.46) that the probability current (5.15) must be continuous ($\partial/\partial x$ is denoted by a prime)

$$S(+0,t) = -D^{(2)}(+0)[\Phi'(+0)\,W(+0,t) + W'(+0,t)]$$
$$= S(-0,t) = -D^{(2)}(-0)[\Phi'(-0)\,W(-0,t) + W'(-0,t)] \,. \tag{5.81}$$

Here $\lim\limits_{\varepsilon\to 0} f(\pm|\varepsilon|)$ was abbreviated by $f(\pm 0)$. (If the probability current would not be continuous at $x=0$ this would mean that at $x=0$ particles are added or removed, i.e. that we have a probability source or sink at $x=0$.) Furthermore it follows from (5.15) that we may write

$$\frac{\partial}{\partial x}[\mathrm{e}^{\Phi(x)}\,W(x,t)] = -S(x,t)\,\mathrm{e}^{\Phi(x)}/D^{(2)}(x) \,.$$

By formally integrating this expression we get

$$e^{\Phi(+0)}W(+0,t)-e^{\Phi(-0)}W(-0,t) = -\lim_{\varepsilon\to 0}\int_{-\varepsilon}^{\varepsilon} S(x,t)e^{\Phi(x)}/D^{(2)}(x)\,dx\,.$$

If only finite jumps in $\Phi(x)$ and $D^{(2)}(x)>0$ occur, the integral vanishes for $\varepsilon\to 0$, i.e., we have

$$e^{\Phi(+0)}W(+0,t) = e^{\Phi(-0)}W(-0,t)\,. \tag{5.82}$$

Equations (5.81, 82) are the jump conditions for the probability density.

Infinite Jumps

If the integral in (5.14) is finite for $x \leqq x_{max}$ but has an infinite positive value for $x > x_{max}$ no diffusion into the region $x > x_{max}$ can occur. Therefore the probability current (5.15) must be zero for $x = x_{max}$, i.e.,

$$\Phi'(x_{max})\,W(x_{max},t) = -W'(x_{max},t)\,. \tag{5.83}$$

If the integral in (5.14) is finite for $x \leqq x_{max}$ but has an infinite negative value for $x > x_{max}$ and if we assume that $W(x,t)$ is finite for $x > x_{max}$ it follows from the jump condition (5.82) that $\exp(\Phi)\,W$ must be zero for $x \to x_{max}$, i.e.

$$\exp[\Phi(x_{max}-0)]\,W(x_{max}-0,t) = 0\,.$$

For finite $\Phi(x_{max}-0)$ this reduces to the condition that the probability distribution itself must be zero for $x \to x_{max}$,

$$W(x_{max}-0,t) = 0\,. \tag{5.84}$$

Similar results are valid if the jump occurs at x_{min}.

Jump Conditions for the Eigenfunctions

For the normalized equation (5.10) the jump conditions for the eigenfunctions (5.40) of the operator (5.54) corresponding to (5.81 – 84) then take the form [5.12]

$$\exp\left(-\frac{f(+0)}{2D}\right)\left[\psi_n'(+0)+\frac{f'(+0)}{2D}\,\psi_n(+0)\right]$$

$$= \exp\left(-\frac{f(-0)}{2D}\right)\left[\psi_n'(-0)+\frac{f'(-0)}{2D}\,\psi_n(-0)\right] \tag{5.81 a}$$

$$\exp\left(\frac{f(+0)}{2D}\right)\psi_n(+0) = \exp\left(\frac{f(-0)}{2D}\right)\psi_n(-0)\,, \tag{5.82a}$$

$$2D\,\psi_n'(x_{\max}-0) = -f'(x_{\max}-0)\,\psi_n(x_{\max}-0)\,, \tag{5.83a}$$

$$\psi_n(x_{\max}-0) = 0\,. \tag{5.84a}$$

These jump conditions are valid for any potential with continuous values and derivatives between the jumps. However, if the potential is linear between the jumps, the transformed potential $V_S(x)$ is a constant and the solutions of the differential equation are immediately obtained between the jumps. Each jump condition then leads to one homogeneous equation, the whole set of those equations having only nontrivial solutions if the determinant is zero. This condition is in general a transcendental equation, which determines the eigenvalues and eigenfunctions. For simple potential wells the transcendental equation may be solved analytically, as in the following example.

5.7 A Bistable Model Potential

As an example we treat the following bistable rectangular potential well (Fig. 5.3)

$$\begin{aligned} f(x) &= f_0\,, && |x| \leqq L/2 \\ f(x) &= 0\,, && L/2 < |x| \leqq L \\ f(x) &= \infty\,, && x > L\,. \end{aligned} \tag{5.85}$$

It turns out that for this special bistable model all eigenvalues and eigenfunctions can be obtained analytically. (If the width of the barrier in the middle is not half the total width of the box a transcendental equation has to be solved.)

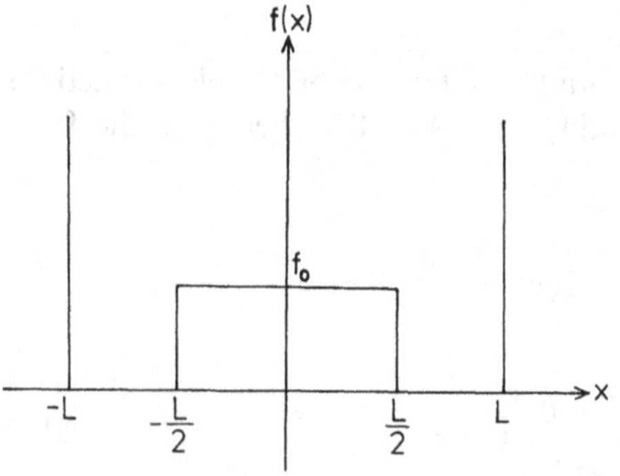

Fig. 5.3. Bistable potential model

With the help of the jump conditions (5.81 a – 83 a) we easily obtain the following eigenvalues and normalized eigenfunctions:

even eigenfunctions and their eigenvalues

$$\lambda_{4n} = (\pi^2 D/L^2)(2n)^2; \qquad (n = 0, 1, 2, \ldots) \tag{5.86}$$

$$\psi_0 = [L(1 + e^{-f_0/D})]^{-1/2} e^{-f(x)/(2D)}, \tag{5.87}$$

$$\psi_{4n} = \sqrt{2}[L(1 + e^{-f_0/D})]^{-1/2} e^{-f(x)/(2D)} \cos 2n\pi x/L; \qquad (n = 1, 2, 3, \ldots), \tag{5.88}$$

$$\lambda_{4n+2} = (\pi^2 D/L^2)(2n+1)^2; \qquad (n = 0, 1, 2, \ldots), \tag{5.89}$$

$$\psi_{4n+2} = \sqrt{2}[L(1 + e^{f_0/D})]^{-1/2} e^{f(x)/(2D)} \cos(2n+1)\pi x/L, \tag{5.90}$$

odd eigenfunctions and their eigenvalues

$$\lambda_{4n+1} = (\pi^2 D/L^2)(2n+\nu)^2; \qquad n = 0, 1, 2, \ldots, \tag{5.91}$$

$$\begin{aligned} \psi_{4n+1} &= L^{-1/2} \sin[(2n+\nu)x\pi/L]; \qquad 0 \leqq x < L/2 \\ \psi_{4n+1} &= L^{-1/2} \cos[(2n+\nu)(L-x)\pi/L]; \qquad L/2 < x < L \end{aligned} \tag{5.92}$$

$$\lambda_{4n-1} = (\pi^2 D/L^2)(2n-\nu)^2; \qquad n = 1, 2, 3, \ldots, \tag{5.93}$$

$$\begin{aligned} \psi_{4n-1} &= L^{-1/2} \sin[(2n-\nu)x\pi/L]; \qquad 0 \leqq x < L/2 \\ \psi_{4n-1} &= L^{-1/2} \cos[(2n-\nu)(L-x)\pi/L]; \qquad L/2 < x < L. \end{aligned} \tag{5.94}$$

Here ν is defined by

$$\nu = (2/\pi) \arctan\{\exp[-f_0/(2D)]\}; 0 < \nu < 1. \tag{5.95}$$

Some of the lowest eigenvalues and their eigenfunctions are shown in Figs. 5.4, 5. In particular, the lowest nonzero eigenvalue reads

$$\lambda_1 = (4D/L^2)[\arctan\{\exp[-f_0/(2D)]\}]^2, \tag{5.96}$$

which in the limit of large barrier heights is proportional to the Boltzmann factor, i.e.,

$$\lambda_1 = (4D/L^2) \exp(-f_0/D) \qquad \text{for} \qquad f_0/D \gg 1. \tag{5.96a}$$

Some other bistable models and a soluble metastable and a periodic potential model are given in [5.12]. The last model is also treated in Sect. 11.3.2. By inverting the potential (5.85) one also gets a metastable model (Sect. 5.8).

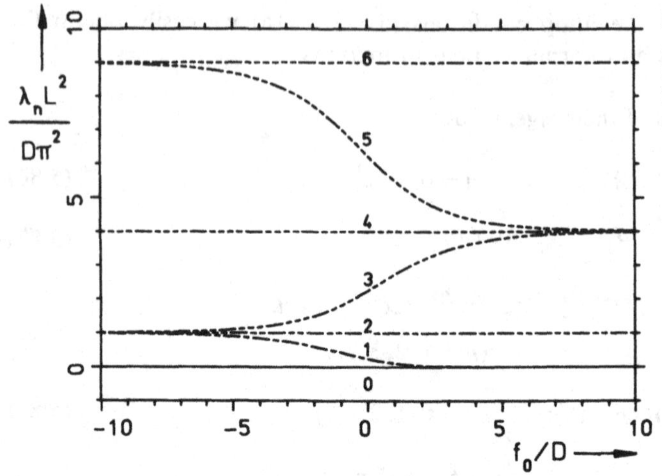

Fig. 5.4. The eigenvalues λ_n, $n = 0, 1, \ldots, 6$ of the bistable rectangular potential well as a function of f_0/D

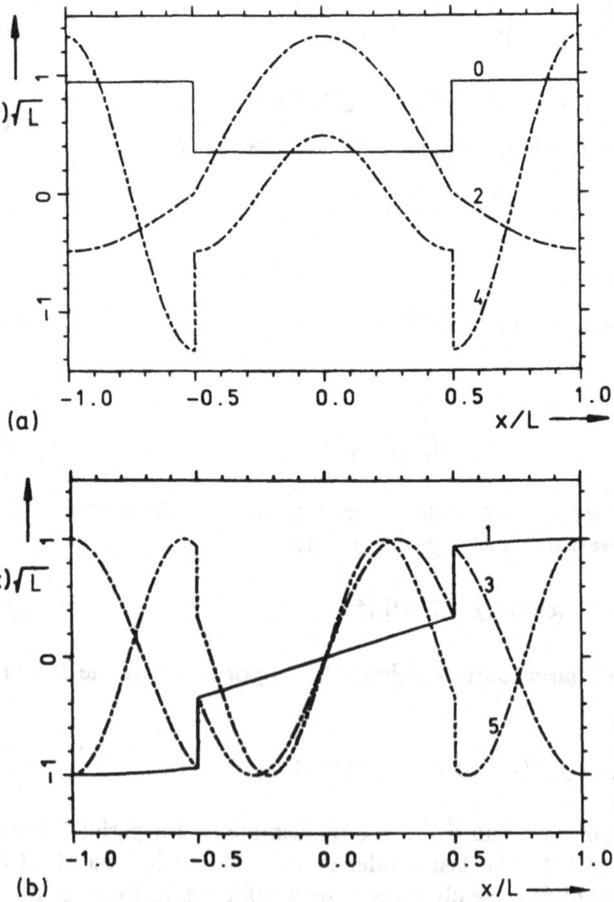

Fig. 5.5. The first three even **(a)** and odd **(b)** eigenfunctions of the bistable rectangular potential well for $f_0 = 2D$

5.8 Eigenfunctions and Eigenvalues of Inverted Potentials

In addition to the normalized Fokker-Planck equation (5.10) we consider this equation for the inverted (upside-down) potential

$$\bar{f}(x) = -f(x)\,, \tag{5.97}$$

but with the same x-independent diffusion constant D. As easily seen from (5.57), the operators $\bar{a}$ and $\hat{\bar{a}}$ for the inverted potential are connected with the operators a and $\hat{a}$ by the simple relations

$$\bar{a} = -\hat{a}\,; \quad \hat{\bar{a}} = -a\,. \tag{5.98}$$

Therefore the operator (5.48) for the inverted potential may be written as

$$\bar{L} = -\hat{\bar{a}}\bar{a} = -a\hat{a}\,. \tag{5.99}$$

We now apply the operator a to the eigenvalue equation $L\,\psi_n = -\lambda_n\psi_n$, i.e.,

$$aL\,\psi_n = -a\hat{a}\,a\,\psi_n = \bar{L}\,a\,\psi_n = -\lambda_n a\,\psi_n\,. \tag{5.100}$$

Thus if $a\,\psi_n$ is not identical to zero it is an eigenfunction of the operator belonging to the inverted problem.

The connection between the nth eigenfunction ψ_n of the original problem and the mth eigenfunction $\bar{\psi}_m$ of the inverted problem depends on the boundary conditions. Therefore we first discuss the transformation of the boundary conditions. Using $\varphi_n(x) = \exp[-(1/2)f(x)/D]\,\psi_n(x)$, $\bar{\varphi}_m(x) = \exp[-(1/2)\bar{f}(x)/D] \cdot \bar{\psi}_m(x)$, (5.57) and $\bar{\psi}_m \sim a\,\psi_n$ the probability current may be written in the form

$$\begin{aligned} S(x) &= -D\exp\left(-\frac{f(x)}{D}\right)\frac{\partial}{\partial x}\left[\exp\left(\frac{f(x)}{D}\right)\varphi_n(x)\right] \\ &= -D\exp\left(-\frac{f(x)}{2D}\right)\exp\left(-\frac{f(x)}{2D}\right)\frac{\partial}{\partial x}\left[\exp\left(\frac{f(x)}{2D}\right)\psi_n(x)\right] \\ &= -\sqrt{D}\exp\left(-\frac{f(x)}{2D}\right)a\,\psi_n(x) \\ &\sim \exp\left(-\frac{f(x)}{2D}\right)\bar{\psi}_m(x) = \exp\left(\frac{f(x)}{D}\right)\bar{\varphi}_m(x)\,. \end{aligned}$$

Thus the boundary condition

$$S(x_0) = -D\exp\left(-\frac{f(x_0)}{D}\right)\frac{\partial}{\partial x}\left[\exp\left(\frac{f(x)}{D}\right)\varphi_n(x)\right]\Bigg|_{x=x_0} = 0$$

for $\varphi_n(x)$ is transformed to the boundary condition

$$\exp\left(\frac{\bar{f}(x_0)}{D}\right)\bar{\varphi}_m(x_0) = 0$$

for the eigenfunction $\bar{\varphi}_m(x)$. Similarly the boundary condition

$$\exp\left(\frac{f(x_0)}{D}\right)\varphi_n(x_0) = 0$$

for $\varphi_n(x)$ is transformed to the boundary condition

$$\bar{S}(x_0) = -D\exp\left(-\frac{\bar{f}(x_0)}{D}\right)\frac{\partial}{\partial x}\left[\exp\left(\frac{\bar{f}(x)}{D}\right)\bar{\varphi}_m(x)\right]\Bigg|_{x=x_0} = 0$$

for the eigenfunctions $\bar{\varphi}_m$ of the inverted potential, in agreement with Table 5.1. (By inverting the potential a reflecting wall is transformed to an absorbing wall and vice versa). As may be checked the jump conditions for ψ_n and the jump conditions for $\bar{\psi}_m$ are connected according to

(5.81 a)		⇔ (5.82 a)	
(5.82 a)	for ψ_n	⇔ (5.81 a)	for $\bar{\psi}_m$.
(5.83 a)		⇔ (5.84 a)	
(5.84 a)		⇔ (5.83 a)	

(To see these connections $\psi_n = \bar{a}\bar{\psi}_m/\sqrt{\lambda_m} = -\hat{a}\bar{\psi}_m/\sqrt{\lambda_m}$ may be used; notice that normalized real eigenfunctions are defined only up to a factor ± 1.)

We now express the eigenvalues and normalized eigenfunctions of the operator $\bar{L}$ in terms of those of the operator L. To find the normalized functions

$$\int(a\psi_n)(a\psi_n)\,dx = \int\psi_n(\hat{a}a\psi_n)\,dx = \lambda_n$$

may be used. According to the various boundary conditions in Table 5.1 we have the following possibilities:

1) Boundary condition B1 (this includes natural boundary conditions) for the original problem, i.e. B4 for the inverted problem ($\lambda_0 = 0$; $a\psi_n = 0$)

$$\bar{\lambda}_n = \lambda_{n+1} > 0\,; \quad \bar{\psi}_n = a\psi_{n+1}/\sqrt{\lambda_{n+1}}\,, \quad n = 0, 1, 2, \ldots\,. \tag{5.101 a}$$

2) Boundary conditions B2 (B3) for the original problem, i.e. B3 (B2) for the inverted problem

$$\bar{\lambda}_n = \lambda_n > 0\,; \quad \bar{\psi}_n = a\psi_n/\sqrt{\lambda_n}\,, \quad n = 0, 1, 2, \ldots \tag{5.101 b}$$

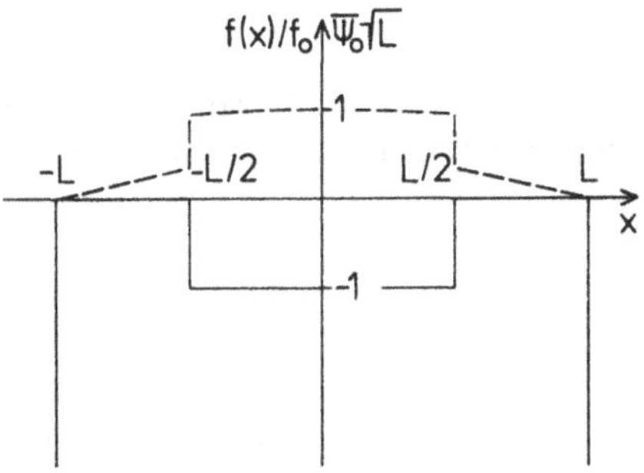

Fig. 5.6. Metastable potential model (*solid line*) and its lowest eigenfunction (*broken line*) for $f_0 = 2D$

3) Boundary condition B4 for the original problem, i.e. B1 for the inverted problem

$$\bar{\lambda}_0 = 0\,, \quad \bar{\lambda}_{n+1} = \lambda_n > 0\,, \quad \bar{\psi}_{n+1} = a\,\psi_n/\sqrt{\lambda_n}\,, \quad n = 0, 1, 2, \dots\,. \tag{5.101 c}$$

4) Periodic boundary conditions ($\lambda_0 = \bar{\lambda}_0 = 0$)

$$\bar{\lambda}_n = \lambda_n > 0\,, \qquad \bar{\psi}_n = a\,\psi_n/\sqrt{\lambda_n}\,, \qquad n = 1, 2, \dots\,. \tag{5.101 d}$$

The eigenfunction $\bar{\psi}_0$ in case 3 and 4 must be obtained from $\bar{a}\,\bar{\psi}_0 = 0$. The eigenvalues and eigenfunctions of the operator belonging to the parabolic potential and the inverted parabolic potential (Sects. 5.5.1, 2) are examples of (5.101 a, c).

Inverting the bistable potential model in Sect. 5.7 gives a metastable potential model. The lowest eigenvalue of this metastable model is then given by $\bar{\lambda}_0 = \lambda_1$ (5.96) and the corresponding even eigenfunction by (Fig. 5.6)

$$\begin{aligned} \bar{\psi}_0 &= L^{-1/2} \cos \nu \pi x/L\,, && 0 \leqq x < L/2 \\ \bar{\psi}_0 &= L^{-1/2} \sin \nu (L - x)\,\pi/L\,, && L/2 < x \leqq L\,. \end{aligned} \tag{5.92 a}$$

Here ν is defined by (5.95).

5.9 Approximate and Numerical Methods for Determining Eigenvalues and Eigenfunctions

Because the Fokker-Planck equation can be transformed to a Schrödinger equation, approximate and numerical methods used for solving the Schrödinger equation can also be used for solving the Fokker-Planck equation. We now want to discuss some of these methods which turn out to be quite effective. The Fokker-Planck equation is equivalent to a certain Langevin equation. The computer-simulation method for Langevin equations was already discussed in Sect. 3.6, therefore it will not be repeated here.

5.9.1 Variational Method

Assuming natural boundary conditions (for other types, see [5.2]) the Sturm-Liouville eigenvalue problem (5.41, 51, 52) is equivalent to the following variational problem. The function ψ which minimizes

$$\lambda = \frac{\int \left[\left(\frac{\partial \psi}{\partial x} \right)^2 D^{(2)}(x) + \psi^2 V_S(x) \right] dx}{\int \psi^2 dx} \tag{5.102}$$

leads to the eigenfunction ψ_0. The minimum of this expression is then the lowest eigenvalue λ_0. The next eigenfunction and eigenvalue are found by minimizing (5.102) subject to the auxiliary condition

$$\int \psi_0 \psi dx = 0 . \tag{5.103}$$

Higher eigenfunctions and eigenvalues are found similarly by adding the auxiliary condition that the function is orthogonal to all previous ones [5.2].

Approximate eigenvalues and eigenfunctions are obtained by the following procedure. One guesses some of the lowest eigenfunctions which in addition depend on certain parameters. One then minimizes (5.102) successively subject to $0, 1, 2, 3, \ldots$ auxiliary conditions. In this way, the parameters are determined, leading to approximate eigenfunctions and eigenvalues. From the practical point of view it is preferable to use such functions for the ansatz of the eigenfunctions so that the integral can be evaluated analytically. As for all variational methods, the results for the eigenvalues are much more accurate than those for the eigenfunctions and they are most effective for determining the lowest eigenvalues.

Lower and Upper Bounds

By the modified Ritz method of *Weinstein* [5.13] (see also [5.14]), one obtains lower and upper bounds for the eigenvalues. *Brand* et al. [5.15] applied this method to some Fokker-Planck equations.

5.9.2 Numerical Integration

Let us discuss the numerical integration method for the operator (5.54) of the Schrödinger equation. The method can also be applied to the Fokker-Planck operator (5.2) or to (5.51).

First assume that the potential (5.55) is symmetric, i.e. $V(x) = V(-x)$. The eigenfunctions ψ_s and ψ_a must then be either symmetric or antisymmetric. For the symmetric (antisymmetric) eigenfunction we start integrating at $x = 0$ with the initial condition $\psi(0) = A$; $\psi'(0) = 0$ ($\psi(0) = 0$, $\psi'(0) = A$) up to the boundary $x = x_B$ for some fixed value of λ. (If natural boundary conditions are considered x_B has to be chosen large enough consistent with the desired accuracy of the eigenvalues.) We then calculate the difference to the given boundary values and, by varying λ, determine the eigenvalues λ_n as the zeros of this difference. The eigenfunctions for λ_n are calculated in the above steps and can be normalized by choosing A suitably.

If the potential $V(x)$ is not symmetric but if x_{min}, x_{max} or both are finite and if no singularities in the differential equation occur, we start at $x = x_{min}$ (or at x_{max}) and integrate (5.41) up to x_{max} (x_{min}) using the boundary condition at $x = x_{min}$ (x_{max}). The eigenvalue λ is determined so that at $x = x_{max}$ (x_{min}) the boundary condition is also fulfilled. If x_{max} (x_{min}) is infinite we have to use an appropriately large $x_{max}^{(ap)}$ ($x_{min}^{(ap)}$). If there are singularities in the differential equation at some point, one should try an analytical power expansion around this point and use numerical integration in the other region.

If the potential $V(x)$ is not symmetric and if $x_{min} = -\infty$ and $x_{max} = \infty$, one may start at $x = 0$ with the initial condition $\psi(0) = A$, $\psi'(0) = B$ and integrate (5.41) in both directions to $-\infty$ and $+\infty$. By a proper choice of B and λ both boundary conditions can be fulfilled. In order to find these values of B and λ a regula falsi method for the two variables may be used. The constant A finally follows from the normalization (5.42). Numerical integration methods are usually very accurate even for higher eigenfunctions. In limiting cases, e.g., very high potentials or very small noise strength D, the numerical integration does not work. In these cases, however, analytical methods may be suitable, Sect. 5.10.

5.9.3 Expansion into a Complete Set

To solve the Fokker-Planck equation (5.1, 2) one may expand the probability density into a complete set $\varphi^q(x)$ satisfying the boundary conditions, i.e.,

$$W(x,t) = F(x) \sum_q c^q(t)\, \varphi^q(x) \, . \tag{5.104}$$

The choice of the arbitrary function $F(x)$ will be discussed below. For natural boundary conditions $x_{min} = -\infty$ and $x_{max} = \infty$ one may use for $\varphi^q(x)$ for instance Hermite functions $\sim \mathrm{H}_q(\alpha x) \exp(-\alpha^2 x^2/2)$, where α is a suitable scaling factor. Another possible choice for $\varphi^q(x)$ is the following. We may construct a system of polynomials orthogonal to a certain weight function [5.16]. As weight function we may use the stationary solution of the Fokker-Planck equation. With the latter choice one has the advantage that φ^q are adapted to the problem under consideration. If we use Hermite functions, only the scaling factor α can be adapted to the problem. The insertion of (5.104) into the Fokker-Planck equation leads to an infinite system of coupled differential equations for the expansion coefficients c^q. The truncated infinite system may then be solved. Sometimes the structure of the system of coupled differential equations may be such that only a finite number M of nearest-neighbor coefficients is coupled, i.e., of the form (9.17). Then one can cast the system into the form of the tridiagonal vector recurrence relation (9.10 or 121) which may be solved by matrix continued-fraction methods as discussed in Chap. 9. The matrix continued-fraction method has the advantage that a large number of expansion terms in (5.104) can be taken into account.

The M-nearest-neighbor coupling of the system seems at first glance to be valid only for very special Fokker-Planck operators. This is, however, not the case. If the drift and diffusion coefficients are rational functions of x, i.e.,

$$D^{(1)}(x) = \frac{P_1(x)}{P_2(x)} = \frac{a_0^{(1)} + a_1^{(1)}x + \ldots + a_{n_1}^{(1)}x^{n_1}}{a_0^{(2)} + a_1^{(2)}x + \ldots + a_{n_2}^{(2)}x^{n_2}},$$

$$D^{(2)}(x) = \frac{P_3(x)}{P_4(x)} = \frac{a_0^{(3)} + a_1^{(3)}x + \ldots + a_{n_3}^{(3)}x^{n_3}}{a_0^{(4)} + a_1^{(4)}x + \ldots + a_{n_4}^{(4)}x^{n_4}}, \tag{5.105}$$

and if there are natural boundary conditions, one can always find such a system with M-nearest-neighbor coupling by using Hermite functions and setting $F(x)$ equal to the product $P_2(x)P_4(x)$ of the denominators in (5.105). (In (5.105) it is tacitly assumed that the denominators are different from zero.) Generally, the function $F(x)$ should be chosen so that M is as small as possible. In [5.17] this method has been applied to the Fokker-Planck equation of a driven Josephson junction, where the drift coefficient is proportional to $a + x + (b + cx^2)^{-1}$ and the diffusion coefficient is a constant. An application to the laser Fokker-Planck equation, where an expansion into Laguerre functions has been made, will be discussed in Sect. 12.4.

5.10 Diffusion Over a Barrier

We first apply the Fokker-Planck equation (5.10) to calculate escape rates over a potential barrier, closely following the work of *Kramers* [1.17]. Then we want to calculate the lowest nonzero eigenvalue for a bistable potential and the lowest eigenvalue for a metastable potential. These types of problems have been extensively treated in the literature [1.6, 7, 5.18–30]. In this section we are mainly interested in the case where the diffusion coefficient D is small, or more precisely where the barrier height Δf is much larger than the diffusion coefficient D. As it turns out, one can get analytic expressions for the escape rate as well as for the lowest nonzero eigenvalue in a bistable potential in this limiting case. For smaller $\Delta f/D$ ratios, where no analytic expressions are generally available, one has to apply numerical methods, which, as discussed in the last section, work for not too large $\Delta f/D$ ratios.

For very low diffusion constants, the coefficient in front of the second derivative in (5.54) becomes very small. Therefore one may use singular perturbation methods [5.31] which have been applied to a bistable potential by *Larson* and *Kostin* [5.23] and *Dekker* [5.27]. In quantum mechanics, where the same problem occurs when one goes over to the classical limit, one uses the WKB method. This method was applied to a bistable potential by *Caroli* et al. [5.24]. More elaborate methods like the path integral method [5.28] and the Liouville projection operator method [5.29] have also been applied. In this chapter we are interested only in the quasi-stationary process. If one starts with a state where the particles are at the top of the barrier, being unstable, it will decay. The transients of such an unstable state will be discussed in Sect. 12.5 in connection with the transients of a laser model.

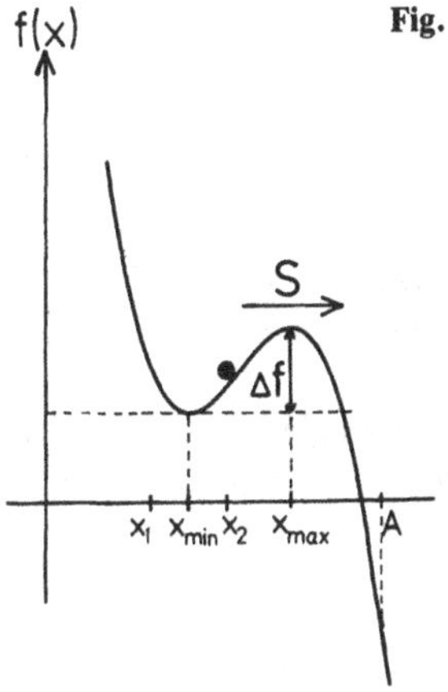

Fig. 5.7. Potential well for calculating the escape rate

5.10.1 Kramers' Escape Rate

We now want to calculate the escape rate for particles sitting in a deep well near $x = x_{min}$, Fig. 5.7. We assume that $\Delta f/D$ is very large. Furthermore, we restrict ourselves to a constant diffusion D, which, according to Sect. 5.1, can always be achieved by proper transformation. Then the probability current S over the top of the potential barrier near x_{max} is very small and the time change of the probability density $W(x,t)$ is also very small. For this quasi-stationary state the small probability current S must then be approximately independent of x (4.46). Integrating (5.15) with (5.56), i.e.,

$$-D\,e^{-f(x)/D}\frac{\partial}{\partial x}[e^{f(x)/D}W(x,t)] = S$$

between x_{min} and A we obtain

$$D[e^{f(x_{min})/D}W(x_{min},t) - e^{f(A)/D}W(A,t)] = S\int_{x_{min}}^{A} e^{f(x)/D}dx\,;$$

or if we assume that at $x = A$ the probability density is nearly zero (particles may for instance be taken away) we can express the probability current by the probability density at $x = x_{min}$, i.e.,

$$S = D\,e^{f(x_{min})/D}W(x_{min},t)\Big/\int_{x_{min}}^{A} e^{f(x)/D}dx\,. \tag{5.106}$$

If the barrier is high the distribution function near x_{min} will be given approximately by the stationary distribution

$$W(x,t) = W(x_{min},t)\,e^{-[f(x)-f(x_{min})]/D}\,. \tag{5.107}$$

The probability p to find the particle near x_{min} reads

$$p = \int_{x_1}^{x_2} W(x,t)\,dx = W(x_{min},t)\, e^{f(x_{min})/D} \int_{x_1}^{x_2} e^{-f(x)/D} dx\,. \tag{5.108}$$

Because for small D the probability density (5.107) becomes very small for x values appreciably different from x_{min}, the x_1, x_2 values need not be specified in detail.

The probability p times the escape rate r is the probability current S. Thus by using (5.106, 108) we get the following expression for the inverse of the escape rate:

$$\frac{1}{r} \equiv \frac{p}{S} = \frac{1}{D} \int_{x_1}^{x_2} e^{-f(x)/D} dx \int_{x_{min}}^{A} e^{f(x)/D} dx\,. \tag{5.109}$$

Whereas the main contribution to the first integral stems from the region around x_{min}, the main contribution to the second integral stems from the region around x_{max}. We therefore expand $f(x)$ for the first and second integrals according to

$$\begin{aligned} f(x) &\approx f(x_{min}) + \tfrac{1}{2} f''(x_{min})(x - x_{min})^2 \\ f(x) &\approx f(x_{max}) - \tfrac{1}{2} |f''(x_{max})| (x - x_{max})^2\,. \end{aligned} \tag{5.110}$$

Then we may extend the integration boundaries in both integrals to $\pm\infty$ and thus obtain the well-known Kramers' escape rate

$$r_K = (2\pi)^{-1} \sqrt{f''(x_{min})\,|f''(x_{max})|}\; e^{-[f(x_{max}) - f(x_{min})]/D}\,. \tag{5.111}$$

As shown by *Edholm* and *Leimar* [5.25], one can improve (5.111) by calculating the integrals in (5.109) more accurately. By using an expansion in (5.110) up to the fourth term and by evaluating the integrals according to

$$\begin{aligned} \int_{-\infty}^{\infty} e^{-ax^2 + bx^3 + cx^4} dx &\approx \int_{-\infty}^{\infty} \left(1 + bx^3 + cx^4 + \frac{1}{2} b^2 x^6\right) e^{-ax^2} dx \\ &= \sqrt{\frac{\pi}{a}} \left(1 + \frac{3}{4}\frac{c}{a^2} + \frac{15}{16}\frac{b^2}{a^3}\right), \end{aligned}$$

we get the improved escape rate

$$\begin{aligned} r = r_K \Bigg[1 - D \Bigg(&\frac{1}{8} \frac{f^{(IV)}(x_{max})}{[f''(x_{max})]^2} - \frac{1}{8} \frac{f^{(IV)}(x_{min})}{[f''(x_{min})]^2} \\ &+ \frac{5}{24} \frac{[f'''(x_{max})]^2}{|f''(x_{max})|^3} + \frac{5}{24} \frac{[f'''(x_{min})]^2}{[f''(x_{min})]^3} \Bigg) + O(D^2) \Bigg]\,. \end{aligned} \tag{5.112}$$

For the inverted potential $\bar{f}(x) = -f(x)$ we obtain exactly the same escape rates from the well of $\bar{f}$ at $\bar{x}_{min} = x_{max}$ over the barrier of $\bar{f}$ at $\bar{x}_{max} = x_{min}$.

5.10.2 Bistable and Metastable Potential

Let us now calculate the lowest nonvanishing eigenvalue for the symmetric bistable potential shown in Fig. 5.8a for small diffusion coefficients D. By inverting the potential we get the metastable potential in Fig. 5.8b. The lowest eigenvalue $\bar{\lambda}_0$ of the metastable potential agrees with the lowest nonvanishing eigenvalue $\bar{\lambda}_1$ of the bistable potential, Sect. 5.8. If the Fokker-Planck equation is interpreted as a Smoluchowski equation, the lowest eigenvalue of the metastable potential is the decay rate of particles in the well. In the bistable potential the lowest nonvanishing eigenvalue describes the transition rate between the left and right well.

We first look for the symmetric eigenfunction $\bar{\psi}_0$ and its lowest eigenvalue $\bar{\lambda}_0$ of the metastable potential. For reasons discussed below, we assume that at $x = \pm A$ the potential jumps to a negative infinite value (absorbing wall, Sect. 5.4), so that we have the jump condition (5.84). For further considerations it is useful to transform the eigenvalue equation (5.41) [see (5.35) with $D^{(2)} = D$ and (5.56)]

$$D\frac{\partial}{\partial x}\mathrm{e}^{-\bar{f}(x)/D}\frac{\partial}{\partial x}\mathrm{e}^{\bar{f}(x)/D}\bar{\varphi}_0 = -\bar{\lambda}_0\bar{\varphi}_0 \tag{5.113}$$

into an integral equation. Because of the symmetry of the potential and because the eigenfunction is symmetric $\bar{f}'$ and $\bar{\varphi}_0'$ must be zero at $x = 0$ (i.e., the probability current is zero at $x = 0$), and by integrating (5.113) we obtain

$$D\frac{\partial}{\partial x}\mathrm{e}^{\bar{f}(x)/D}\bar{\varphi}_0 = -\bar{\lambda}_0\mathrm{e}^{\bar{f}(x)/D}\int_0^x\bar{\varphi}_0(z)\,dz\,.$$

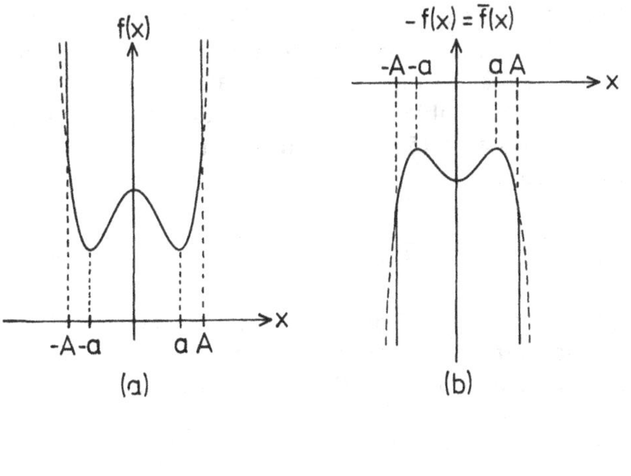

Fig. 5.8. Bistable (**a**) and metastable (**b**) potential

Integrating this equation once more we arrive at the integral equation

$$\bar{\varphi}_0(x) = \mathrm{e}^{-\bar{f}(x)/D}\left[\mathrm{e}^{\bar{f}(0)/D}\bar{\varphi}_0(0) - \frac{\bar{\lambda}_0}{D}\int_0^x \mathrm{d}y\, \mathrm{e}^{\bar{f}(y)/D}\int_0^y \mathrm{d}z\, \bar{\varphi}_0(z)\right]. \tag{5.114}$$

This equation together with the boundary condition

$$\bar{\varphi}_0(A) = 0 \tag{5.115}$$

determines the eigenvalue $\bar{\lambda}_0$ and the eigenfunction $\bar{\varphi}_0$.

For large barrier heights the eigenvalue $\bar{\lambda}_0/D$ will be very small. We may thus apply the following iteration procedure:
As zeroth approximation we use

$$\bar{\varphi}_0^{(0)}(x) = \mathrm{e}^{-\bar{f}(x)/D}\mathrm{e}^{\bar{f}(0)/D}\bar{\varphi}_0(0)\,; \qquad \bar{\lambda}_0^{(0)} = 0\,.$$

If we insert this zeroth approximation into the integral of (5.114) we obtain the first approximation for the eigenfunction

$$\bar{\varphi}_0^{(1)}(x) = \mathrm{e}^{-\bar{f}(x)/D}\mathrm{e}^{\bar{f}(0)/D}\bar{\varphi}_0(0)\left(1 - \frac{\bar{\lambda}_0^{(1)}}{D}\int_0^x \mathrm{d}y\, \mathrm{e}^{\bar{f}(y)/D}\int_0^y \mathrm{d}z\, \mathrm{e}^{-\bar{f}(z)/D}\right). \tag{5.116}$$

Because of (5.115) the eigenvalue $\bar{\lambda}_0$ in first approximation is given by

$$\bar{\lambda}_0^{(1)} = D\Big/\int_0^A \mathrm{d}y\, \mathrm{e}^{\bar{f}(y)/D}\int_0^y \mathrm{d}z\, \mathrm{e}^{-\bar{f}(z)/D}. \tag{5.117}$$

To obtain the eigenfunction and eigenvalue in second order we insert (5.116) in the integral of (5.114) and again use (5.115). Higher approximations are obtained similarly.

For small diffusion coefficients the double integral in (5.117) can be evaluated analytically. For $y = a$ and $z = 0$ there is a very sharp maximum of the integrand $\exp\{[\bar{f}(y) - \bar{f}(z)]/D\}$ for small D. The leading contribution to the double integral stems from the region near this maximum. We therefore expand $\bar{f}(y)$ and $\bar{f}(z)$ around this point ($y = a$, $z = 0$) up to second order, as in (5.110). The integration over y can then be taken from $-\infty$ to $+\infty$ and the integration over z from 0 to $+\infty$. Notice that the double integral factorizes in this approximation, leading to the same integrals as in (5.109) (up to a factor $\frac{1}{2}$ for the first integral). The eigenvalue finally reads

$$\bar{\lambda}_0^{(1)} = \pi^{-1}\sqrt{\bar{f}''(0)\,|\bar{f}''(a)|}\,\mathrm{e}^{-[\bar{f}(a)-\bar{f}(0)]/D} = 2r_{\mathrm{K}}\,. \tag{5.118}$$

Because we have two barriers in the potential of Fig. 5.8b, it is not surprising that the decay rate $\bar{\lambda}_0$ is twice the Kramers' escape rate over one barrier. We have chosen a finite A in Fig. 5.8b because otherwise the double integral in (5.117)

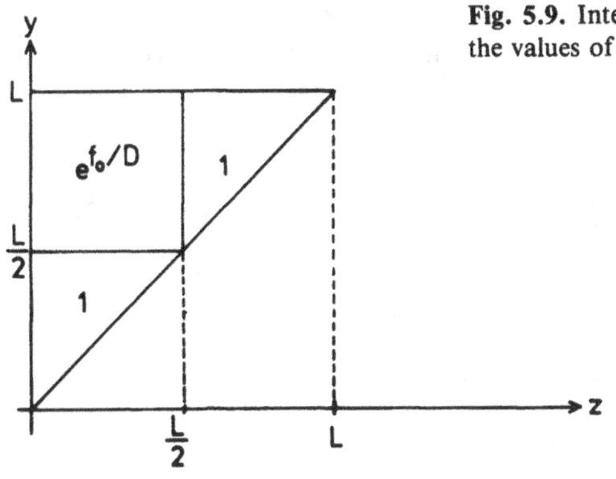

Fig. 5.9. Integration boundary of the double integral and the values of its integrand for the potential in Fig. 5.6

would diverge. As done for the escape rate, (5.118) can be improved by taking into account higher expansion terms of the potential near the maximum and minimum, leading to results in complete agreement with (5.112). (However, $\bar{f}'''(0)$ is now zero because $\bar{f}(x)$ was assumed to be symmetric.)

For the inverted potential of (5.85) (Fig. 5.6) the value of the integrand of the double integral is indicated in Fig. 5.9. The value of the double integral can immediately be read off Fig. 5.9, leading to

$$\bar{\lambda}_0^{(1)} = \frac{4D}{L^2(e^{f_0/D}+1)} \approx \frac{4D}{L^2} e^{-f_0/D}, \tag{5.119}$$

which agrees with (5.96) up to the order $\exp(-2f_0/D)$. Neglecting terms of the order $\exp(-3f_0/D)$ we get in second approximation

$$\bar{\lambda}_0^{(2)} = \frac{4D}{L^2}\left(e^{-f_0/D} - \frac{2}{3}e^{-2f_0/D}\right), \tag{5.120}$$

which again agrees with (5.96) but now up to the order $\exp(-3f_0/D)$.

Bistable Potential

The same method used for the metastable potential in Fig. 5.8b can be used for the bistable potential in Fig. 5.8a. Because at $x = A$ the probability current must now be zero (reflecting wall, Sect. 5.4), we obtain for the eigenfunction φ_1 the integral equation

$$\varphi_1(x) = e^{-f(x)/D}\left[e^{f(A)/D}\varphi_1(A) - \frac{\lambda_1}{D}\int_x^A dy\, e^{f(y)/D}\int_y^A dz\, \varphi_1(z)\right]. \tag{5.121}$$

The eigenfunction φ_1 belonging to the lowest nonvanishing eigenvalue must be an odd function for the bistable potential, i.e.,

$$\varphi_1(0) = 0\,. \tag{5.122}$$

The integral equation (5.121) together with (5.122) determine the eigenfunction φ_1 and the eigenvalue λ_1. We may now apply the same iteration procedure as before. In first order we have

$$\lambda_1^{(1)} = D\Big/\left(\int_0^A dy\, e^{f(y)/D} \int_y^A dz\, e^{-f(z)/D}\right). \tag{5.123}$$

It can be shown by partial integration that this expression agrees with (5.117) [notice $f(x) = -\bar{f}(x)$].

Asymmetric Metastable Potential

To treat the asymmetric metastable potential in Fig. 5.10 we need only minor modifications. Because the derivative $\bar{\varphi}_0'(0)$ is no longer zero [$\bar{f}'(0)$ is still zero] instead of the integral equation (5.114) we obtain

$$\bar{\varphi}_0(x) = e^{-\bar{f}(x)/D}\left[e^{\bar{f}(0)/D}\bar{\varphi}_0(0) + \int_0^x e^{\bar{f}(x)/D} dy\, \bar{\varphi}_0'(0) - \frac{\bar{\lambda}_0}{D}\int_0^x dy\, e^{\bar{f}(y)/D}\int_0^y dz\, \bar{\varphi}_0(z)\right]. \tag{5.124}$$

The eigenfunction $\bar{\varphi}_0$ must vanish at $x = A$ and $x = B$

$$\bar{\varphi}_0(A) = \bar{\varphi}_0(B) = 0\,. \tag{5.125}$$

To solve (5.124) we may apply the same iteration procedure as before. In zeroth approximation then

$$\bar{\varphi}_0^{(0)}(x) = e^{-[\bar{f}(x)-\bar{f}(0)]/D}\,\bar{\varphi}_0(0)$$

$$\bar{\varphi}_0^{(0)\prime}(0) = 0\,; \qquad \bar{\lambda}_0^{(0)} = 0\,.$$

Inserting this zeroth solution in the double integral gives

$$\bar{\varphi}_0^{(1)}(x) = e^{-\bar{f}(x)/D} e^{\bar{f}(0)/D}\bar{\varphi}_0(0)\left(1 + \int_0^x e^{\bar{f}(y)/D} dy\, \alpha - \frac{\bar{\lambda}_0^{(1)}}{D}\int_0^x dy\, e^{\bar{f}(y)/D}\int_0^y dz\, e^{-\bar{f}(z)/D}\right), \tag{5.126}$$

where α is given by

$$\alpha = \bar{\varphi}_0^{(1)\prime}(0)\, e^{-\bar{f}(0)/D}/\bar{\varphi}_0(0)\,. \tag{5.127}$$

The two conditions (5.125) then determine α and $\bar{\lambda}_0^{(1)}$. By expanding $f(x)$ near the maxima of the integrands up to second order [as in (5.110)] we obtain

$$\begin{aligned}\bar{\lambda}_0^{(1)} &= (2\pi)^{-1}\{\sqrt{\bar{f}''(0)\,|\bar{f}''(a)|}\,e^{-[\bar{f}(a)-\bar{f}(0)]/D} + \sqrt{\bar{f}''(0)\,|\bar{f}''(b)|}\,e^{-[\bar{f}(b)-\bar{f}(0)]/D}\}\\ &= r_{KR} + r_{KL}\,,\end{aligned} \tag{5.128}$$

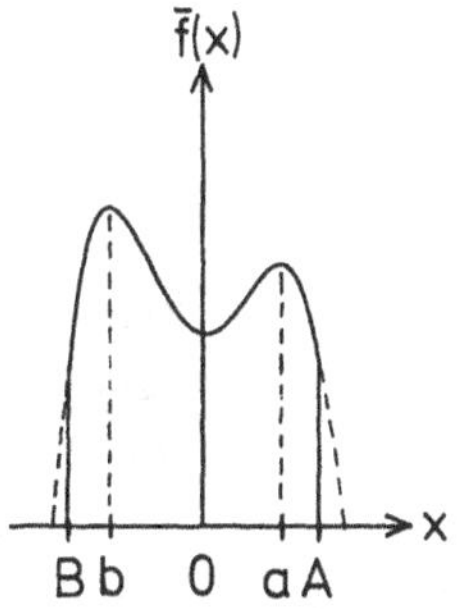

Fig. 5.10. Asymmetric metastable potential

i.e., the sum of the Kramers' escape rates (5.111) over the right and left barriers.

Transformation to a Homogeneous Fredholm Integral Equation

The integral equation (5.114) may be transformed into homogeneous Fredholm integral equation [5.32]. We show this for the metastable potential in Fig. 5.8b. For the bistable potential the expressions are more complicated because the stationary solution must be eliminated first by a projection formalism. Partial integration of (5.114) leads to

$$\bar{\varphi}_0(x) = \mathrm{e}^{-\bar{f}(x)/D}\left\{\mathrm{e}^{\bar{f}(0)/D}\bar{\varphi}_0(0) + \frac{\bar{\lambda}_0}{D}\left[u(x)\int_0^x \bar{\varphi}_0(y)\,\mathrm{d}y - \int_0^x u(y)\,\bar{\varphi}_0(y)\,\mathrm{d}y\right]\right\}, \tag{5.114a}$$

where we have defined $u(x)$ by

$$u(x) = \int_x^A \exp[\bar{f}(y)/D]\,\mathrm{d}y\,. \tag{5.129}$$

Because of the boundary condition (5.115), i.e.,

$$\bar{\varphi}_0(A) = \mathrm{e}^{-\bar{f}(A)/D}\left[\mathrm{e}^{\bar{f}(0)/D}\bar{\varphi}_0(0) - \frac{\bar{\lambda}_0}{D}\int_0^A u(y)\,\bar{\varphi}_0(y)\,\mathrm{d}y\right] = 0$$

we may write instead of (5.114a)

$$\bar{\varphi}_0(x) = \frac{\bar{\lambda}_0}{D}\,\mathrm{e}^{-\bar{f}(x)/D}\left[\int_0^x u(x)\,\bar{\varphi}_0(y)\,\mathrm{d}y + \int_x^A u(y)\,\bar{\varphi}_0(y)\,\mathrm{d}y\right].$$

Using instead of the eigenfunction $\bar{\varphi}_0(x)$ the function

$$\bar{\psi}_0(x) = \exp[\bar{f}(x)/(2D)]\,\bar{\varphi}_0(x) \tag{5.130}$$

we obtain the integral equation

$$\bar{\psi}_0(x) = \bar{\lambda}_0 \int_0^A K(x,y)\,\bar{\psi}_0(y)\,dy \tag{5.131}$$

with the symmetric kernel

$$K(x,y) = K(y,x) = D^{-1}\exp\{-[\bar{f}(x)+\bar{f}(y)]/(2D)\}\cdot\begin{cases} u(x) & \text{for} \quad y<x \\ u(y) & \quad\quad\;\; x<y \end{cases}. \tag{5.132}$$

Because we can express the second iterated kernel K_2 in terms of eigenvalues and eigenfunctions of (5.131) [5.2, Chap. III, (58)]

$$K_2(x,z) = \int_0^A K(x,y)\,K(y,z)\,dy = \sum_n \bar{\psi}_n(x)\,\bar{\psi}_n(z)/\bar{\lambda}_n^2$$

we obtain

$$\int_0^A K_2(x,x)\,dx = \int_0^A \int_0^A K(x,y)^2 dx\,dy = \sum_n 1/\bar{\lambda}_n^2 . \tag{5.133}$$

If we assume that $\bar{\lambda}_0$ is much smaller than the other eigenvalues

$$0 < \bar{\lambda}_0 \ll \bar{\lambda}_1 < \bar{\lambda}_2 < \ldots \tag{5.134}$$

we get

$$\bar{\lambda}_0 \approx \left[\int_0^A \int_0^A K(x,y)^2 dx\,dy\right]^{-1/2} . \tag{5.135}$$

For small D, $u(x)$ is approximately constant for $x<a$. Then the kernel $K(x,y)$ approximately factorizes and we finally obtain

$$\bar{\lambda}_0 \approx 1\Big/\int_0^A K(x,x)\,dx = D\Big/\int_0^A e^{-f(x)/D}\left(\int_x^A e^{f(y)/D} dy\right) dx . \tag{5.136}$$

As may be seen by using partial integration this expression agrees with (5.117).

Mean First-Passage Time for the Metastable Potential

The mean first-passage time $T_1(x')$ for a particle starting at $x = x'$ to leave the domain $|x| < A$ can either be obtained by [see (8.5, 9, 10)]

$$T_1(x') = \int_{-A}^{A} p_1(x,x')\,dx \, , \tag{5.137}$$

$$L_{FP}(x)\,p_1(x,x') = -\delta(x-x') \, , \tag{5.138}$$

$$p_1(\pm A, x') = 0 \tag{5.139}$$

or by, see (8.15a),

$$L_{FP}^{+}(x')\, T_1(x') = -1\,, \tag{5.140}$$

$$T_1(\pm A) = 0\,. \tag{5.141}$$

For the metastable potential in Fig. 5.8b we now calculate T_1 for $x' = 0$. Because the potential is symmetric and D is independent of x, $p_1(x, 0)$ and also $T_1(x')$ must be symmetric in x and x', respectively. Therefore, the first derivative of $p_1(x, 0)$ at $x = 0$ and of $T_1(x')$ at $x' = 0$ must vanish. Using

$$\begin{aligned} L_{FP}(x) &= D\frac{\partial}{\partial x}\, e^{-\bar{f}(x)/D} \frac{\partial}{\partial x}\, e^{\bar{f}(x)/D} \\ L_{FP}^{+}(x') &= D\, e^{\bar{f}(x')/D} \frac{\partial}{\partial x'}\, e^{-\bar{f}(x')/D} \frac{\partial}{\partial x'} \end{aligned} \tag{5.142}$$

it is easy to solve (5.137 – 139):

$$p_1(x, 0) = D^{-1} e^{-\bar{f}(x)/D} \int_x^A e^{\bar{f}(y)/D} \left[\int_0^y \delta(z)\, dz \right] dy\,. \tag{5.143}$$

The δ function in (5.138, 143) may be replaced by a sharp symmetric function of finite width. Then the integral over the δ function for $y > 0$ is 1/2, giving

$$\begin{aligned} T_1(0) &= \int_{-A}^{A} p_1(x, 0)\, dx = 2\int_0^A p_1(x, 0)\, dx \\ &= D^{-1} \int_0^A e^{-\bar{f}(x)/D} \left[\int_x^A e^{\bar{f}(y)/D} dy \right] dx\,. \end{aligned} \tag{5.144}$$

The solution of (5.140) with the boundary condition (5.141) and with $dT_1/dx'|_{x'=0} = 0$ reads ($|x'| \leqq A$)

$$T_1(x') = D^{-1} \int_{x'}^{A} e^{\bar{f}(y)/D} \left(\int_0^y e^{-\bar{f}(x)/D} dx \right) dy\,, \tag{5.145}$$

i.e., for $x' = 0$

$$\begin{aligned} T_1(0) &= D^{-1} \int_0^A e^{\bar{f}(y)/D} \left(\int_0^y e^{-\bar{f}(x)/D} dx \right) dy \\ &= 1/\bar{\lambda}_0^{(1)}\,. \end{aligned} \tag{5.146}$$

Thus, this expression is equal to the inverse of the first approximation of the eigenvalue, compare (5.117). It is, however, also equal to (5.136, 144) as may be seen by using partial integration.

Thus, for the mean first-passage time an exact expression valid for every potential height and arbitrary diffusion coefficients can be derived. For large potential heights the double integral can be evaluated analytically as done before. The inverse of the mean first-passage time is then given by the sum of the Kramer's escape rates (5.111) over the left and right barriers.

6. Fokker-Planck Equation for Several Variables; Methods of Solution

In this chapter we discuss methods of solution for the Fokker-Planck equation (4.94a, 95) for time-independent drift and diffusion coefficients, i.e., for

$$\partial W/\partial t = L_{\mathrm{FP}} W = -\partial S_i/\partial x_i \,, \tag{6.1}$$

$$L_{\mathrm{FP}} = -\frac{\partial}{\partial x_i} D_i(\{x\}) + \frac{\partial^2}{\partial x_i \partial x_j} D_{ij}(\{x\}) \,. \tag{6.2}$$

(With the exception of Sect. 6.6.5 the summation convention for Latin indices is used in this chapter.)

In (6.1) S_i is the probability current

$$S_i = D_i W - (\partial/\partial x_j) D_{ij} W \,. \tag{6.3}$$

In the stationary state and for one variable this probability current must be constant (and for natural boundary conditions it is zero). Therefore we could obtain the stationary solution in terms of an integral in Sect. 5.2. For N variables the probability current (6.3) is generally not constant in the stationary state and for natural boundary conditions it is generally not zero. Therefore, we can no longer proceed in the same way as for the one-variable Fokker-Planck equation. Only for certain conditions for the coefficients (potential conditions [6.1 – 4]) may the probability current be zero in the stationary state, and the procedure will then resemble the one-variable case, discussed in detail in Sect. 6.4 for a general diffusion matrix. In these introductory remarks we assume that the diffusion matrix is independent of $\{x\}$ and proportional to the unit matrix ($D_{ij} = D\delta_{ij}$). Then the probability current takes the simple form

$$S_i = W(D_i - D\,\partial \ln W/\partial x_i) \,. \tag{6.3a}$$

Obviously, the probability current can vanish only if D_i is the gradient of a potential Φ, i.e.,

$$D_i = -D\,\partial \Phi/\partial x_i \,. \tag{6.4}$$

Necessary and sufficient conditions for the existence of Φ are the potential conditions

$$\partial D_i/\partial x_j = \partial D_j/\partial x_i . \tag{6.5}$$

The stationary solution of (6.1, 2) then reads

$$W_{st} = N e^{-\Phi} , \tag{6.6}$$

where Φ is given by the line integral

$$\Phi(\{x\}) = -D^{-1} \int^{\{x\}} D_i(\{x'\}) \mathrm{d}x_i' . \tag{6.7}$$

The condition where the probability current vanishes is called the condition of detailed balance. (More precisely, if the irreversible part of it vanishes, Sect. 6.4.) If (6.5) is not fulfilled, the probability current cannot vanish everywhere and calculation of the stationary distribution is generally much more complicated.

If the potential conditions hold and if $D_{ij} = D\delta_{ij}$ (D independent of $\{x\}$), we may write the Fokker-Planck operator (6.2) in a form similar to the one-variable case (5.35), i.e.,

$$L_{FP} = D \frac{\partial}{\partial x_i} e^{-\Phi} \frac{\partial}{\partial x_i} e^{\Phi} . \tag{6.8}$$

From (6.8) we obtain the operator relation

$$L_{FP}^{+} e^{\Phi} = D e^{\Phi} \frac{\partial}{\partial x_i} e^{-\Phi} \frac{\partial}{\partial x_i} e^{\Phi} = e^{\Phi} L_{FP} . \tag{6.9}$$

Thus the operator

$$L = e^{\Phi/2} L_{FP} e^{-\Phi/2} = e^{-\Phi/2} L_{FP}^{+} e^{\Phi/2} = L^{+} \tag{6.10}$$

is an Hermitian operator. The problem of calculating the eigenvalues of L_{FP}, which are useful for obtaining nonstationary solutions, can therefore be reduced to an Hermitian eigenvalue problem if the potential condition (6.5) holds. The eigenvalues must then be real. Generally the Fokker-Planck operator cannot be brought to an Hermitian form and its eigenvalues and eigenfunctions may be complex.

6.1 Approach of the Solutions to a Limit Solution

To investigate the behavior of the solutions for large times we first compare two solutions of the Fokker-Planck equation. For certain conditions we show that every two solutions of the Fokker-Planck equation must agree for large times. (If the drift and diffusion coefficients do not depend on time this limit solution is the stationary solution of the Fokker-Planck equation.) This agreement will be proved by constructing a functional $H(t)$ of two functions which can decrease in time only if the two functions are different. Because this functional cannot reach negative values, one concludes that in the limit of large times these two functions must coincide.

Following *Lebowitz* and *Bergmann* [6.5], and *Graham* [4.8], we define the functional $H(t)$ by

$$\begin{aligned} H(t) &= \int W_1 \ln(W_1/W_2)\,\mathrm{d}^N x \\ &= \int (W_1 \ln W_1 - W_1 \ln W_2)\,\mathrm{d}^N x\,. \end{aligned} \tag{6.11}$$

The first term in the last line is Boltzmann's functional, for which he showed that it decreases in time if the solution W_1 of the Boltzmann equation did not agree with the stationary solution (H theorem) [1.21]. In a general analysis of thermodynamic nonequilibrium states, *Schlögl* [6.6] used a functional of the type (6.11) and he has shown that it serves as a Ljapunov function [6.7]. The information gain in probability theory [1.14, 6.8] is also of this form.

In our case, W_1 and W_2 are two positive solutions of the Fokker-Planck equation. We assume that there are natural boundary conditions and that both solutions are normalized to one, i.e.,

$$\int W_1 \mathrm{d}^N x = \int W_2 \mathrm{d}^N x = 1\,. \tag{6.12}$$

(We also assume that the integral (6.11) does exist; for instance, we exclude therefore δ functions for W_1, W_2.) Furthermore, we assume that the drift coefficients have no singularities and that they do not allow the solutions to run away to infinity. We first show that $H(t)$ cannot have negative values. By introducing the ratio

$$R = W_1/W_2\,, \tag{6.13}$$

using ($R \geqq 0$)

$$R \ln R - R + 1 = \int_1^R \ln x\,\mathrm{d}x \geqq 0\,, \tag{6.14}$$

and using the normalization for W_1 and W_2, we immediately obtain

$$\begin{aligned} H(t) &= \int W_1 \ln R\,\mathrm{d}^N x = \int (W_1 \ln R - W_1 + W_2)\,\mathrm{d}^N x \\ &= \int W_2 (R \ln R - R + 1)\,\mathrm{d}^N x \geqq 0\,. \end{aligned} \tag{6.15}$$

Next we derive an expression for $\dot{H}(t)$:

$$\begin{aligned} \dot{H}(t) &= \int [\dot{W}_1 \ln(W_1/W_2) + (W_1/W_1)\,\dot{W}_1 - (W_1/W_2)\,\dot{W}_2]\,\mathrm{d}^N x \\ &= \int (\dot{W}_1 \ln R - R\,\dot{W}_2)\,\mathrm{d}^N x = \end{aligned}$$

$$= \int [(L_{\mathrm{FP}} W_1) \ln R - R \dot{W}_2] \, \mathrm{d}^N x$$
$$= \int (W_1 L_{\mathrm{FP}}^{+} \ln R - R \dot{W}_2) \, \mathrm{d}^N x \tag{6.16}$$

[the integral $\int \dot{W}_1 \mathrm{d}^N x = (\mathrm{d}/\mathrm{d}t) \int W_1 \mathrm{d}^N x$ vanishes]. Because

$$L_{\mathrm{FP}}^{+} \ln R = \left(D_i + D_{ij} \frac{\partial}{\partial x_j} \right) \frac{\partial}{\partial x_i} \ln R$$
$$= \left(D_i + D_{ij} \frac{\partial}{\partial x_j} \right) \frac{1}{R} \frac{\partial R}{\partial x_i}$$
$$= \frac{1}{R} L_{\mathrm{FP}}^{+} R - D_{ij} \frac{1}{R^2} \frac{\partial R}{\partial x_j} \frac{\partial R}{\partial x_i} \tag{6.17}$$

we may write

$$\dot{H}(t) = \int \left(\frac{W_1}{R} L_{\mathrm{FP}}^{+} R - R \dot{W}_2 \right) \mathrm{d}^N x - \int W_1 D_{ij} \frac{1}{R^2} \frac{\partial R}{\partial x_i} \frac{\partial R}{\partial x_j} \mathrm{d}^N x$$
$$= \int (R L_{\mathrm{FP}} W_2 - R \dot{W}_2) \mathrm{d}^N x - \int W_1 D_{ij} \frac{1}{R} \frac{\partial R}{\partial x_i} \frac{1}{R} \frac{\partial R}{\partial x_j} \mathrm{d}^N x$$
$$= - \int W_1 D_{ij} \frac{\partial \ln R}{\partial x_i} \frac{\partial \ln R}{\partial x_j} \mathrm{d}^N x \leqq 0 \, . \tag{6.18}$$

If D_{ij} is positive definite, $\dot{H}(t)$ must always decrease for $\partial \ln R / \partial x_i \neq 0$. Because $H(t)$ is bounded from below, $H(t)$ cannot decrease indefinitely. Thus we conclude that finally both $\ln R$ and R must be independent of $\{x\}$. Because of the normalization, R must then be equal to 1 and $H(t)$ reaches its minimal value $H = 0$. Thus the two solutions W_1 and W_2 must coincide for large times. The same must, of course, be true if we take a third solution W_3, and so on. Therefore all solutions of the Fokker-Planck equation finally agree if we wait long enough. This result is valid regardless whether the drift and diffusion coefficients depend on time or not. If the drift and diffusion coefficients do not depend on time, a stationary solution W_{st}, i.e.,

$$L_{\mathrm{FP}} W_{\mathrm{st}} = 0 \, , \tag{6.19}$$

may exist. It then follows from $\dot{H} = 0$ that this solution is unique and that all other functions after some time T agree with it. It should be noted that this result is valid if D_{ij} is positive definite everywhere and if the drift coefficient has no singularities, though the time T may be very large, especially if high potential barriers occur (Sect. 5.10). It should further be noted that $W_1(\{x\}, t)$ is positive for every finite $\{x\}$ after a small time has elapsed. [Initially $W_1(\{x\}, t)$ may be zero in some regions, but because of (4.78, 109) it is positive everywhere after a small time τ.]

If, however, the drift coefficient has singularities (i.e., infinite high potential barriers), the problem may decompose into two or more separate problems, and the results are no longer valid. Furthermore, the results are no longer valid if the drift coefficients lead to run-away solutions of the deterministic equation (1.17). An example is the inverted harmonic potential in Sect. 5.5.2. If one solution starts with a sharp value at the left side of the potential and the other with a sharp value at the right side of the potential, both solutions will be driven apart. However, if we do not have any drift coefficient at all and if the diffusion coefficient is constant (Wiener process) the result is still valid. [Compare two solutions (5.20) with different x' and t' for large time t.]

Application to Kramers Equation

The results cannot be applied to the Kramers equation (4.112) because the diffusion matrix (3.131) is not positive definite. Here we can conclude from (6.18) only that the derivative with respect to v vanishes for large times

$$\partial \ln R/\partial v = 0\,. \tag{6.20}$$

If a stationary solution exists, it is given by the Boltzmann distribution $\exp\{-m[f(x)+v^2/2]/(kT)\}$, as may be seen by insertion. We conclude from (6.20) that for large times any distribution must have the form

$$W_1(x,v,t) = h(x,t)\exp[-mv^2/(2kT)]\,. \tag{6.21}$$

By inserting (6.21) into the Kramers equation (4.112) we obtain

$$\dot{h} = \left(-\frac{\partial}{\partial x} - \frac{mf'(x)}{kT}\right) v h\,.$$

Because h does not depend on the velocity v, then

$$\dot{h} = 0\,; \qquad h(x) = h_0 \exp[-mf(x)/(kT)]\,, \tag{6.22}$$

i.e., the stationary solution is also unique for the Kramers equation.

6.2 Expansion into a Biorthogonal Set

To solve the time-dependent Fokker-Planck equation (6.1, 2), one may use the separation ansatz

$$W(\{x\},t) = \varphi_\mu(\{x\})\,\mathrm{e}^{-\lambda_\mu t}\,, \tag{6.23}$$

which leads to the eigenvalue equation

$$L_{\mathrm{FP}}\varphi_\mu = -\lambda_\mu \varphi_\mu . \tag{6.24}$$

Because we use the summation convention for Latin indices we now indicate different eigenfunctions by Greek letters (no summation convention for Greek indices).

Because L_{FP} cannot be generally brought to an Hermitian form, we also need eigenfunctions φ_μ^+ of the adjoint operator [5.4]

$$L_{\mathrm{FP}}^+ \varphi_\mu^+ = -\lambda_\mu \varphi_\mu^+ . \tag{6.25}$$

If the eigenvalues are complex, the complex conjugate of an eigenvalue is also an eigenvalue because L_{FP} is a real operator. For the definition of φ_μ^+ we may use either the complex conjugate eigenvalue in (6.25) or the eigenvalue itself. The last choice requires that the scalar product must then be defined by

$$(\varphi_\mu^+, \varphi_\nu) = \int \varphi_\mu^+(\{x\})\, \varphi_\nu(\{x\})\, d^N x . \tag{6.26}$$

It is easy to show that the eigenvalues in (6.24, 25) are the same. Denoting for the moment the eigenvalue in (6.25) by λ_μ^+ we have

$$-\lambda_\mu(\varphi_\mu^+, \varphi_\mu) = (\varphi_\mu^+, L_{\mathrm{FP}}\varphi_\mu) = (L_{\mathrm{FP}}^+ \varphi_\mu^+, \varphi_\mu) = -\lambda_\mu^+(\varphi_\mu^+, \varphi_\mu) ,$$

i.e., $\lambda_\mu = \lambda_\mu^+$. Furthermore, it is easily seen that eigenfunctions $\varphi_\mu^+, \varphi_\nu$ for different eigenvalues are orthogonal

$$-\lambda_\nu(\varphi_\mu^+, \varphi_\nu) = (\varphi_\mu^+, L_{\mathrm{FP}}\varphi_\nu) = (L_{\mathrm{FP}}^+ \varphi_\mu^+, \varphi_\nu) = -\lambda_\mu(\varphi_\mu^+, \varphi_\nu) .$$

We may thus normalize the functions according to

$$(\varphi_\mu^+, \varphi_\nu) = \delta_{\mu\nu} . \tag{6.27}$$

Whereas for Hermitian operators a complete set always exists [5.4], it may not exist for non-Hermitian operators. If we were to use a suitable set we could transform the problem of diagonalizing the operator to the problem of diagonalizing non-Hermitian matrices. If the eigenvalues are all different, a non-Hermitian matrix can always be reduced to a diagonal form by a similarity transformation, otherwise it can be reduced to a Jordan canonical form only [6.9] [see also the remark following (6.120)].

In the following we assume that such a biorthogonal set does exist and that the completeness relation

$$\delta(\{x\} - \{x'\}) = \sum_\mu \varphi_\mu(\{x\})\, \varphi_\mu^+(\{x'\}) \tag{6.28}$$

is fulfilled.

If a stationary solution of the Fokker-Planck equation exists we obviously have

$$\lambda_0 = 0 ; \quad \varphi_0(\{x\}) = W_{\mathrm{st}}(\{x\}) ; \quad \varphi_0^+(\{x\}) = 1 . \tag{6.29}$$

The transition probability density is obtained by inserting in the formal solution of (6.1, 2) expression (6.28), i.e.,

$$P(\{x\},t|\{x'\},t') = e^{L_{FP}(\{x\})(t-t')}\delta(\{x\}-\{x'\}) = \sum_{\mu} e^{L_{FP}(\{x\})(t-t')}\varphi_\mu(\{x\})\varphi_\mu^+(\{x'\})$$
$$= \sum_{\mu} \varphi_\mu(\{x\})\varphi_\mu^+(\{x'\})\,e^{-\lambda_\mu(t-t')}. \tag{6.30}$$

6.3 Transformation of the Fokker-Planck Operator, Eigenfunction Expansions

We now assume that we have natural boundary conditions for all N variables. Other boundary conditions may, of course, also occur and some of the procedures and results of this chapter may also be applied to these boundary conditions. Generally, the boundary conditions must be given on an $(N-1)$-dimensional surface of the N-dimensional variable space. The variety of boundary conditions is much larger than in the one-variable case (see Table 5.1 for a summary of common boundary conditions). For instance, on some parts of the surface the probability density may vanish, whereas on other parts of the probability current may vanish. For Brownian motion in periodic potentials (Chap. 11) there are natural boundary conditions for velocity and a periodic boundary condition for position.

In addition to having natural boundary conditions we assume that a stationary solution $W_{st}(\{x\})$ of (6.1, 2) exists. Because the probability density must be positive, we may write

$$W_{st}(\{x\}) = N_{st}\,e^{-\Phi(\{x\})}; \qquad N_{st}^{-1} = \int e^{-\Phi} d^N x\,. \tag{6.31}$$

The function $\Phi(\{x\})$ is called a generalized potential. (If we use a proper normalization of this generalized potential the normalization constant N_{st} is unity.)

In this section we also assume that we already know the stationary solution, i.e., the generalized potential $\Phi(\{x\}) = -\ln W_{st} + \ln N_{st}$. To find $W_{st}(\{x\})$ for the general case we must solve the Fokker-Planck equation (6.1, 2) for the stationary state. As shown in Sect. 6.4, one can obtain $\Phi(\{x\})$ by quadratures if certain conditions, called potential conditions, are met by the drift and diffusion coefficients.

We now want to put the Fokker-Planck operator in a form similar to (5.35) for the one-variable case. Because now the probability current does not need to vanish in the stationary state even for natural boundary conditions we try the more general ansatz (this form was already used in [6.2])

$$L_{FP} = \frac{\partial}{\partial x_i} D_{ij} e^{-\Phi} \frac{\partial}{\partial x_j} e^{\Phi} - \frac{\partial}{\partial x_i} D_i^{(a)}$$
$$= \frac{\partial^2}{\partial x_i \partial x_j} D_{ij} - \frac{\partial}{\partial x_i}\left[e^{\Phi}\left(\frac{\partial}{\partial x_j} D_{ij} e^{-\Phi}\right) + D_i^{(a)}\right]. \tag{6.32}$$

(The operator in the bracket does not act on a function outside this bracket.) Because the operator (6.32) should agree with (6.2), the auxiliary drift coefficient $D_i^{(a)}$ in (6.32) is given by

$$D_i^{(a)} = D_i - \mathrm{e}^{\Phi}\left(\frac{\partial}{\partial x_j} D_{ij}\mathrm{e}^{-\Phi}\right) = D_i - \frac{\partial D_{ij}}{\partial x_j} + D_{ij}\frac{\partial \Phi}{\partial x_j}. \tag{6.33}$$

Since (6.31) is the stationary solution of (6.1, 2) then

$$\begin{aligned}\frac{\partial}{\partial x_i}(D_i^{(a)}\mathrm{e}^{-\Phi}) &= \frac{\partial}{\partial x_i}\left(D_i\mathrm{e}^{-\Phi} - \frac{\partial}{\partial x_j} D_{ij}\mathrm{e}^{-\Phi}\right)\\ &= -L_{\mathrm{FP}}\mathrm{e}^{-\Phi} = 0.\end{aligned} \tag{6.34}$$

The probability current in the stationary state, i.e.,

$$(S_{\mathrm{st}})_i = \left(D_i - \frac{\partial}{\partial x_j} D_{ij}\right) W_{\mathrm{st}} = D_i^{(a)} W_{\mathrm{St}}$$

vanishes only if $D_i^{(a)}$ is zero. (Because of (6.34) the divergence of the stationary current is always zero.)

As was done in the one-variable case (5.39), the first operator on the right-hand side in the first line of (6.32) can be brought to an Hermitian form for suitable boundary conditions by multiplying it from the left by $\exp(\Phi/2)$ and from the right by $\exp(-\Phi/2)$. By applying this transformation to (6.32) we obtain

$$L = \mathrm{e}^{\Phi/2} L_{\mathrm{FP}} \mathrm{e}^{-\Phi/2} = L_{\mathrm{H}} + L_{\mathrm{A}} \tag{6.35}$$

with

$$L_{\mathrm{H}} = \mathrm{e}^{\Phi/2}\frac{\partial}{\partial x_i} D_{ij}\mathrm{e}^{-\Phi}\frac{\partial}{\partial x_j}\mathrm{e}^{\Phi/2} = L_{\mathrm{H}}^{+} \tag{6.36}$$

and

$$L_{\mathrm{A}} = -\mathrm{e}^{\Phi/2}\frac{\partial}{\partial x_i} D_i^{(a)}\mathrm{e}^{-\Phi/2} = -L_{\mathrm{A}}^{+}. \tag{6.37}$$

Because of (6.34) it is easy to see that the last operator is an anti-Hermitian operator

$$\begin{aligned}L_{\mathrm{A}}^{+} &\equiv -\left[\mathrm{e}^{\Phi/2}\frac{\partial}{\partial x_i} D_i^{(a)}\mathrm{e}^{-\Phi}\mathrm{e}^{\Phi/2}\right]^{+} = -\left[\mathrm{e}^{\Phi/2} D_i^{(a)}\mathrm{e}^{-\Phi}\frac{\partial}{\partial x_i}\mathrm{e}^{\Phi/2}\right]^{+}\\ &= \mathrm{e}^{\Phi/2}\frac{\partial}{\partial x_i} D_i^{(a)}\mathrm{e}^{-\Phi/2} \equiv -L_{\mathrm{A}}.\end{aligned}$$

Any operator L can, of course, be decomposed into an Hermitian part $L_H = (L+L^+)/2$ and an anti-Hermitian part $L_A = (L-L^+)/2$. The decomposition (6.35), however, has the particularity that the operator L_A applied to the square root of the stationary distribution always leads to zero because of (6.34)

$$L_A \sqrt{W_{st}} = \sqrt{N_{st}} L_A e^{-\Phi/2} = -\sqrt{N_{st}} e^{\Phi/2} \left(\frac{\partial}{\partial x_i} D_i^{(a)} e^{-\Phi} \right) = 0 . \tag{6.38}$$

[Obviously, the same holds for the operator L_H as immediately seen from (6.36).]

The Fokker-Planck equation with the special drift coefficient

$$D_i^{(s)} = \frac{\partial D_{ij}}{\partial x_j} - D_{ij} \frac{\partial \Phi}{\partial x_j} \tag{6.39}$$

also has the stationary solution (6.31). In this special case, the anti-Hermitian part L_A vanishes. If an inverse of D_{ij} exists, then

$$\frac{\partial \Phi}{\partial x_i} = A_i \equiv (D^{-1})_{ij} \left(\frac{\partial D_{jk}}{\partial x_k} - D_j^{(s)} \right) . \tag{6.40}$$

This equation requires that $D_i^{(s)}$ and D_{ij} must satisfy the potential conditions

$$\partial A_i / \partial x_j = \partial A_j / \partial x_i . \tag{6.41}$$

Up to now, we have assumed that we know the stationary solution. Usually, the inverse problem arises, i.e., one is looking for the stationary solution for given drift and diffusion coefficients. If the drift and diffusion coefficients happen to obey the potential conditions (6.41) with $D_j = D_j^{(s)}$, the generalized potential Φ is according to (6.40) obtained from the line integral

$$\Phi(\{x\}) = \int_{\{x_0\}}^{\{x\}} A_i(\{x'\}) \, dx_i' \tag{6.42}$$

and the distribution function then follows from (6.31). If the drift and diffusion coefficients do not obey (6.41), one may try to split the drift coefficient according to (6.33) into a drift coefficient $D_i^{(s)}$ obeying (6.41) and a drift coefficient $D_i^{(a)}$, i.e.,

$$D_i = D_i^{(s)} + D_i^{(a)} . \tag{6.43}$$

Here, $D_i^{(a)}$ has to obey (6.34) with Φ given by (6.42). If a stationary solution exists, such a decomposition must always be possible. Without knowing the stationary distribution, however, this decomposition of the drift coefficient is generally hard to find. (In special cases, one may find the decomposition by some guesswork.) As shown in the next section, we always know this decomposition, if detailed balance is valid. Consequently, we can then obtain the stationary solution by evaluating only the line integral, which is, of course, much easier than solving a partial differential equation of second order.

The Hermitian part $L_{\rm H}$ may be cast in forms corresponding to the one-variable expressions (5.48 – 52). To do this we need the square root $(D^{1/2})_{ij} = (D^{1/2})_{ji}$ of the diffusion coefficient D_{ik}, which we assume to be positive definite. This square root $(D^{1/2})_{ij}$, defined by $(D^{1/2})_{ik}(D^{1/2})_{kj} = D_{ij}$, can also be assumed to be positive definite. (We can diagonalize the symmetric matrix D_{ij} by an orthogonal transformation, then take the positive square root of the positive eigenvalues and finally apply the inverse of the above orthogonal transformation.)

Thus we may write

$$L_{\rm H} = -a_i^+ a_i\,, \tag{6.44}$$

where a_i and a_i^+ are defined by

$$\begin{aligned} a_i &= (D^{1/2})_{ij} e^{-\Phi/2} \frac{\partial}{\partial x_j} e^{\Phi/2}\,, \\ a_i^+ &= -e^{\Phi/2} \frac{\partial}{\partial x_j} (D^{1/2})_{ij} e^{-\Phi/2}\,. \end{aligned} \tag{6.45}$$

The operator $L_{\rm H}$ has formally the same form as the Hamilton operator for a particle with a mass tensor $(D^{-1})_{ij}\hbar^2/2$ in quantum mechanics, i.e.,

$$L_{\rm H} = \frac{\partial}{\partial x_i} D_{ij} \frac{\partial}{\partial x_j} - V_{\rm S}\,, \tag{6.46}$$

where the potential $V_{\rm S}$ is given by ($\partial/\partial x_i$ does not act outside the brackets)

$$\begin{aligned} V_{\rm S}(\{x\}) &= e^{\Phi/2}\left[\frac{\partial}{\partial x_i} D_{ij} \frac{\partial}{\partial x_j} e^{-\Phi/2}\right] \\ &= \frac{1}{4} D_{ij} \frac{\partial \Phi}{\partial x_i}\frac{\partial \Phi}{\partial x_j} - \frac{1}{2}\left[\frac{\partial}{\partial x_i} D_{ij} \frac{\partial \Phi}{\partial x_j}\right] \\ &= \frac{1}{4}(D^{-1})_{ij}\left(\frac{\partial D_{ik}}{\partial x_k} - D_i^{(\rm s)}\right)\left(\frac{\partial D_{jl}}{\partial x_l} - D_j^{(\rm s)}\right) + \frac{1}{2}\frac{\partial D_i^{(\rm s)}}{\partial x_i} - \frac{1}{2}\frac{\partial^2 D_{ij}}{\partial x_i \partial x_j}\,. \end{aligned} \tag{6.47}$$

The separation 'ansatz' (6.23) for the time-dependent solutions now leads to the problem of finding eigenfunctions of L and L^+

$$\begin{aligned} L\,\psi_\mu &= (L_{\rm H} + L_{\rm A})\,\psi_\mu = -\lambda_\mu \psi_\mu \\ L^+ \psi_\mu^+ &= (L_{\rm H} - L_{\rm A})\,\psi_\mu^+ = -\lambda_\mu \psi_\mu^+\,. \end{aligned} \tag{6.48}$$

Because of the transformation (6.35) these eigenfunctions are connected to φ_μ and φ_μ^+ (Sect. 6.2) by

$$\varphi_\mu = \mathrm{e}^{-\Phi/2}\psi_\mu; \qquad \varphi_\mu^+ = \mathrm{e}^{\Phi/2}\psi_\mu^+ . \tag{6.49}$$

The completeness relation, the transition probability and the joint probability distribution $W_2 = P\,W_{\mathrm{st}}$ in the stationary state, expressed in terms of ψ_n and ψ_n^+ read for $t \geqq t'$ $(\psi_0 = \psi_0^+)$

$$\sum_\mu \psi_\mu(\{x\})\,\psi_\mu^+(\{x'\}) = \delta(\{x\} - \{x'\}), \tag{6.50}$$

$$P(\{x\}, t\,|\,\{x'\}, t') = [\psi_0(\{x\})/\psi_0(\{x'\})] \sum_\mu \psi_\mu(\{x\})\,\psi_\mu^+(\{x'\})\,\mathrm{e}^{-\lambda_\mu(t-t')}, \tag{6.51}$$

$$W_2(\{x\}, t; \{x'\}, t') = \psi_0(\{x\})\,\psi_0(\{x'\}) \sum_\mu \psi_\mu(\{x\})\,\psi_\mu^+(\{x'\})\,\mathrm{e}^{-\lambda_\mu(t-t')}. \tag{6.52}$$

The result for $t \leqq t'$ follows immediately by using the property of a joint distribution

$$W_2(\{x\}, t; \{x'\}, t') = W_2(\{x'\}, t'; \{x\}, t), \quad \text{i.e.,} \tag{6.53}$$

$$W_2(\{x\}, t; \{x'\}, t') = \psi_0(\{x\})\,\psi_0(\{x'\}) \sum_\mu \psi_\mu^+(\{x\})\,\psi_\mu(\{x'\})\,\mathrm{e}^{-\lambda_\mu|t-t'|}. \tag{6.54}$$

Positivity of the Real Part of the Eigenvalues

We first consider the real eigenvalues λ_μ^{H} and the real eigenfunctions ψ_μ^{H} of the Hermitian operator $\boldsymbol{L}_{\mathrm{H}}$, i.e.,

$$\boldsymbol{L}_{\mathrm{H}}\psi_\mu^{\mathrm{H}} = -\lambda_\mu^{\mathrm{H}}\psi_\mu^{\mathrm{H}}. \tag{6.55}$$

The eigenfunctions are assumed to be orthonormalized, i.e.,

$$\int \psi_\mu^{\mathrm{H}}\psi_\nu^{\mathrm{H}}\,\mathrm{d}^N x = \delta_{\mu\nu}. \tag{6.56}$$

(The eigenfunctions are orthogonal for different eigenvalues. Degenerate eigenfunctions may be chosen in such a way that they are mutually orthogonal [5.4].)

Because of (6.44) we may write

$$\lambda_\mu^{\mathrm{H}} = \int \psi_\mu^{\mathrm{H}} a_i^+ a_i \psi_\mu^{\mathrm{H}}\,\mathrm{d}^N x = \int (a_i\psi_\mu^{\mathrm{H}})(a_i\psi_\mu^{\mathrm{H}})\,\mathrm{d}^N x \geqq 0. \tag{6.57}$$

The equals sign requires that $a_i\psi_\mu^{\mathrm{H}} = 0$. Since $(D^{1/2})_{ij}$ was assumed to be positive definite, ψ_μ^{H} must be the square root of the stationary solution

$$\psi_0^{\mathrm{H}} = \sqrt{N_{\mathrm{st}}}\,\mathrm{e}^{-\Phi/2} = \sqrt{W_{\mathrm{st}}}. \tag{6.58}$$

All eigenvalues not belonging to the stationary solution $a_i \psi_0^{\mathrm{H}} = 0$ must be positive. They can be arranged in increasing order

$$0 = \lambda_0 < \lambda_1^{\mathrm{H}} \leqq \lambda_2^{\mathrm{H}} \leqq \lambda_3^{\mathrm{H}} \leqq \dots . \tag{6.59}$$

We now consider the generally complex eigenfunctions ψ_μ and the generally complex eigenvalues λ_μ of the non-Hermitian operator $\boldsymbol{L}$ (6.48). As a consequence of (6.38) the eigenfunction ψ_0 of $\boldsymbol{L}$ with stationary eigenvalue $\lambda_0 = 0$ is given by

$$\psi_0 = \psi_0^{\mathrm{H}} = \sqrt{N_{\mathrm{st}}}\, \mathrm{e}^{-\Phi/2} . \tag{6.60}$$

To prove that the lower bound of the real parts of the other eigenvalues is given by λ_1^{H} we expand the generally complex eigenfunctions ψ_μ into eigenfunctions of $\boldsymbol{L}_{\mathrm{H}}$

$$\psi_\mu = \sum_{\nu=1}^{\infty} c_{\mu\nu} \psi_\nu^{\mathrm{H}} \quad (\mu \geqq 1) . \tag{6.61}$$

The stationary solution ψ_0^{H} must be omitted in expansion (6.61) because the function ψ_μ should be orthogonal to $\psi_0^{+} = \psi_0 = \psi_0^{\mathrm{H}}$. (If the eigenvalue of ψ_μ is different from zero the eigenfunction ψ_μ must be orthogonal to ψ_0.) If we insert expansion (6.61) in the expression [this is not the scalar product defined in (6.26)]

$$\lambda_\mu = \frac{\int \psi_\mu^* (\boldsymbol{L}_{\mathrm{H}} + \boldsymbol{L}_{\mathrm{A}}) \psi_\mu \mathrm{d}^N x}{\int \psi_\mu^* \psi_\mu \mathrm{d}^N x} , \tag{6.62}$$

we obtain $(\mu \geqq 1)$

$$\mathrm{Re}\{\lambda_\mu\} = \frac{\sum_{\nu=1}^{\infty} |c_{\mu\nu}|^2 \lambda_\nu^{\mathrm{H}}}{\sum_{\nu=1}^{\infty} |c_{\mu\nu}|^2} \geqq \lambda_1^{\mathrm{H}} > 0 . \tag{6.63}$$

In deriving (6.63) we used (6.56, 59) and

$$\begin{aligned} I &= \sum_{\nu,\nu'} c_{\mu\nu}^* c_{\mu\nu'} \int \psi_\nu^{\mathrm{H}} \boldsymbol{L}_{\mathrm{A}} \psi_{\nu'}^{\mathrm{H}} \mathrm{d}^N x \\ &= -\sum_{\nu,\nu'} c_{\mu\nu}^* c_{\mu\nu'} \int (\boldsymbol{L}_{\mathrm{A}} \psi_\nu^{\mathrm{H}}) \psi_{\nu'}^{\mathrm{H}} \mathrm{d}^N x = -I^* , \end{aligned} \tag{6.64}$$

i.e., $\mathrm{Re}\{I\} = 0$.

Hence all solutions finally decay to the stationary solution, in agreement with the results of Sect. 6.1.

The results of this section were derived in [6.10] for the case where the coefficients satisfy the detailed balance condition (6.95 – 97). It was shown in [6.11] that these results are also valid if these conditions are not satisfied a priori.

Reduction to an Hermitian Problem by Analytic Continuation

The operator

$$\hat{L} = \hat{L}(\eta) = L_{\mathrm{H}} - \mathrm{i}\eta L_{\mathrm{A}} \tag{6.65}$$

with real η is a self-adjoint operator with respect to the complex scalar product:

$$\int \varphi^* \hat{L}\, \psi \,\mathrm{d}^N x = \int (\hat{L}^+ \varphi)^* \psi \,\mathrm{d}^N x . \tag{6.66}$$

We may thus look for real eigenvalues $\hat{\lambda}_\mu(\eta)$ and generally complex eigenfunctions $\hat{\psi}_\mu(\eta)$ of $\hat{L}$, i.e.,

$$\hat{L}\, \hat{\psi}_\mu = -\hat{\lambda}_\mu \hat{\psi}_\mu . \tag{6.67}$$

Having found analytic solutions of (6.67) we may then replace the parameter η by $\pm\mathrm{i}$ and in this way obtain eigenfunctions of L and L^+ with the eigenvalue $\lambda_\mu = \hat{\lambda}_\mu(\mathrm{i}) = [\hat{\lambda}_\mu(-\mathrm{i})]^*$, i.e.,

$$\psi_\mu = \hat{\psi}_\mu(\mathrm{i}) ; \quad \psi_\mu^+ = [\hat{\psi}_\mu(-\mathrm{i})]^* . \tag{6.68}$$

[In the last expression we must take the complex conjugate because our scalar product was defined by (6.26).] The complex conjugates of (6.68) are also eigensolutions of L and L^+, their eigenvalue is then λ_μ^*. For an example of this method see [6.11]. It should, however, be mentioned that this method seems to work only if analytic expressions for $\hat{\lambda}_\mu$ and $\hat{\psi}_\mu$ exist. In that case one may, of course, equally well directly solve the eigenvalue equation (6.48).

6.4 Detailed Balance

Detailed Balance for a Master Equation

Because it is easier to explain the principle of detailed balance [1.13, 14, 6.1–4, 6.10–13] for the master equation, we begin with the latter. A master equation is an equation of motion for the probability W_n of a state n. If $w(n \to m)$ are the transition rates from state n to state m, the master equation has the form [see also (1.34)]

$$\dot{W}_n = \sum_m [w(m \to n) W_m - w(n \to m) W_n] . \tag{6.69}$$

For a stationary or steady state solution the total number of transitions per time into state n must balance the total number of transitions per time out of state n, i.e., we have the

balance for steady state:

$$\sum_m w(m \to n) W_m = \sum_m w(n \to m) W_n . \tag{6.70}$$

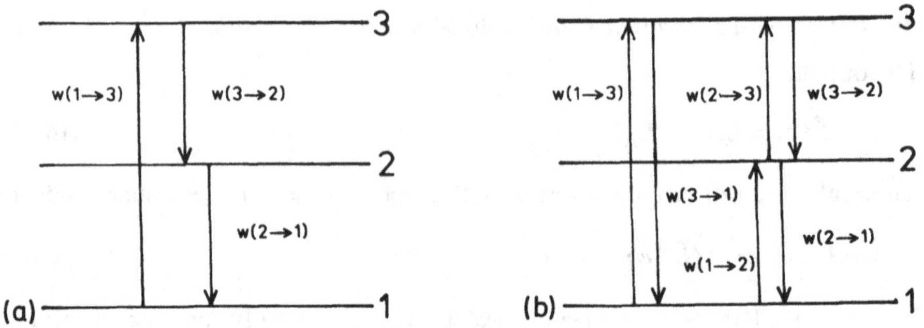

Fig. 6.1. Detailed balance is violated in (**a**). The probability current is given by $S = w(1\to 3)\, W_1 = w(3\to 2)\, W_3 = w(2\to 1)\, W_2$. If $w(i\to j) = w(j\to i)\, W_j/W_i$ detailed balance is fulfilled in (**b**)

One has detailed balance if each individual transition is balanced, i.e., if the number of transitions per time from state m into state n balances the number of transitions per time from state n to state m, i.e.,

detailed balance:

$$w(m\to n)\, W_m = w(n\to m)\, W_n\,. \tag{6.71}$$

In this case, stationary distribution is often called an equilibrium distribution.

If we have only two possible states, the steady-state condition (6.70) and the detailed balance condition (6.71) are the same. For three states the conditions may already be different (Fig. 6.1).

Detailed Balance of the Fokker-Planck Equation for Even Variables

First we consider only variables which do not change their sign if time reversal is considered. To find a connection with the detailed balance condition of the master equation we write the Fokker-Planck equation in the form of the continuous master equation

$$\frac{\partial W(\{x\},t)}{\partial t} = \int [w(\{x'\}\to\{x\})\, W(\{x'\},t) - w(\{x\}\to\{x'\})\, W(\{x\},t)]\, d^N x'\,. \tag{6.72}$$

Here the transition rate is given by [cf. (1.36) for the one-variable case]

$$\begin{aligned} w(\{x'\}\to\{x\}) &= \frac{d}{d\tau} P(\{x\},\tau\,|\,\{x'\},0)\bigg|_{\tau=0} \\ &= L_{FP}(\{x\})\, \delta(\{x\}-\{x'\})\,. \end{aligned} \tag{6.73}$$

We then have the following balance condition for the steady state [continuous analog to (6.70)]

$$\int w(\{x'\}\to\{x\})\, W_{st}(\{x'\})\, d^N x' = \int w(\{x\}\to\{x'\})\, W_{st}(\{x\})\, d^N x'$$

or by inserting (6.73)

$$L_{\mathrm{FP}}(\{x\}) \int W_{\mathrm{st}}(\{x'\})\, \delta(\{x\}-\{x'\})\, \mathrm{d}^N x'$$
$$= \int L_{\mathrm{FP}}(\{x'\})\, \delta(\{x\}-\{x'\})\, \mathrm{d}^N x'\, W_{\mathrm{st}}(\{x\}) \,.$$

Because the integration on the right-hand side leads to zero we obtain

balance for steady state:

$$L_{\mathrm{FP}}(\{x\})\, W_{\mathrm{st}}(\{x\}) = 0 \,. \tag{6.74}$$

Thus the probability density must satisfy the stationary Fokker-Planck equation.
The detailed balance condition (6.71) for the continuous case reads

$$w(\{x'\}\to\{x\})\, W_{\mathrm{st}}(\{x'\}) = w(\{x\}\to\{x'\})\, W_{\mathrm{st}}(\{x\}) \,, \tag{6.75}$$

or by inserting (6.73)

$$L_{\mathrm{FP}}(\{x\})\, \delta(\{x\}-\{x'\})\, W_{\mathrm{st}}(\{x'\}) = L_{\mathrm{FP}}(\{x'\})\, \delta(\{x\}-\{x'\})\, W_{\mathrm{st}}(\{x\}) \,. \tag{6.76}$$

Because of the δ function we can replace the argument $\{x'\}$ in the stationary distribution function on the left-hand side by $\{x\}$. If we write the distribution function on the right-hand side of (6.76) in front of the operator and if we then use (4.93) we obtain

$$L_{\mathrm{FP}}(\{x\})\, W_{\mathrm{st}}(\{x\})\, \delta(\{x\}-\{x'\}) = W_{\mathrm{st}}(\{x\}) L^{+}_{\mathrm{FP}}(\{x\})\, \delta(\{x\}-\{x'\}) \,.$$

This equation can be valid only if the operators in front of the δ function agree, i.e., we obtain the condition for

detailed balance:

$$L_{\mathrm{FP}}(\{x\})\, W_{\mathrm{st}}(\{x\}) = W_{\mathrm{st}}(\{x\}) L^{+}_{\mathrm{FP}}(\{x\}) \,. \tag{6.77}$$

This condition is an operator equation, i.e., it must be valid if it is applied to an arbitrary function. The steady-state condition (6.74) is not an operator equation, i.e., it cannot be applied to an arbitrary function. If we apply (6.77) to the function $f(\{x\}) = 1$ we recover (6.74) because $L^{+}_{\mathrm{FP}}(\{x\}) \cdot 1 = 0$. (Obviously, the detailed balance condition must contain the balance for the steady state.)

Detailed Balance for the Fokker-Planck Equation for Even and Odd Variables

The variables are classified according to their transformation with respect to time reversal $\{x\}\to\{\tilde{x}\}$. The even variables like position do not change their sign whereas the odd variables like velocity do change their sign

$$\left.\begin{array}{l} \tilde{x}_i = x_i \\ \tilde{x}_i = -x_i \end{array}\right\} \quad \tilde{x}_i = \varepsilon_i x_i \quad \left\{\begin{array}{ll} \varepsilon_i = 1 & \text{even variables} \\ \varepsilon_i = -1 & \text{odd variables} \end{array}\right. \tag{6.78}$$

or $\{\tilde{x}\} = \{\varepsilon x\}$. With respect to the summation convention we always disregard ε_i. If we have even and odd variables condition (6.75) for detailed balance must now be replaced by [1.13, 6.13]

$$w(\{x'\}\to\{x\})\, W_{\text{st}}(\{x'\}) = w(\{\varepsilon x\}\to\{\varepsilon x'\})\, W_{\text{st}}(\{\varepsilon x\}) \tag{6.79}$$

and

$$W_{\text{st}}(\{x\}) = W_{\text{st}}(\{\varepsilon x\})\,. \tag{6.80}$$

The left-hand side of (6.79) describes the number of transitions out of the state $\{x'\}$ into a state $\{x\}$. The right-hand side describes the number of transitions for the reverse process with reverse motion, i.e., odd variables (e.g., velocity) have to be replaced by their negative values. Equation (6.80) describes the fact that the stationary distribution must be invariant if time reversal is applied. If the transition probabilities depend on some external parameters (like the magnetic field) which also change their sign under time reversal, these transformed parameters have to be used on the right-hand side of (6.79, 80). If we insert in (6.79) the Fokker-Planck transition rate (6.73), by the same steps used to obtain (6.77) we then obtain the operator equation [6.10]

$$L_{\text{FP}}(\{x\})\, W_{\text{st}}(\{x\}) = W_{\text{st}}(\{\varepsilon x\})\, L_{\text{FP}}^{+}(\{\varepsilon x\})\,. \tag{6.81}$$

Detailed Balance Condition for Joint Distribution

Conditions (6.80, 81) lead to the usual conditions for detailed balance defined for the joint distributions [6.12]. If we apply (6.80, 81) once more we get

$$\begin{aligned}[L_{\text{FP}}(\{x\})]^2\, W_{\text{st}}(\{x\}) &= L_{\text{FP}}(\{x\})\, W_{\text{st}}(\{\varepsilon x\})\, L_{\text{FP}}^{+}(\{\varepsilon x\})\\ &= L_{\text{FP}}(\{x\})\, W_{\text{st}}(\{x\})\, L_{\text{FP}}^{+}(\{\varepsilon x\})\\ &= W_{\text{st}}(\{\varepsilon x\})\, [L_{\text{FP}}^{+}(\{\varepsilon x\})]^2\end{aligned}$$

or generally

$$[L_{\text{FP}}(\{x\})]^n\, W_{\text{st}}(\{x\}) = W_{\text{st}}(\{\varepsilon x\})\, [L_{\text{FP}}^{+}(\{\varepsilon x\})]^n\,. \tag{6.82}$$

By expanding the exponential function into a power series we thus obtain the operator relation

$$\mathrm{e}^{L_{\text{FP}}(\{x\})(t-t')}\, W_{\text{st}}(\{x\}) = W_{\text{st}}(\{\varepsilon x\})\, \mathrm{e}^{L_{\text{FP}}^{+}(\{\varepsilon x\})(t-t')}\,. \tag{6.83}$$

We now multiply (6.83) by the δ function $\delta(\{x\}-\{x'\}) = \delta(\{\varepsilon x\}-\{\varepsilon x'\})$ from the right-hand side. On the left-hand side we then replace the argument of the distribution function by $\{x'\}$ whereas on the right-hand side we use (4.93) and thus obtain

$$\mathrm{e}^{L_{\text{FP}}(\{x\})(t-t')}\, \delta(\{x\}-\{x'\})\, W_{\text{st}}(\{x'\}) = \mathrm{e}^{L_{\text{FP}}(\{\varepsilon x'\})(t-t')}\, \delta(\{\varepsilon x\}-\{\varepsilon x'\})\, W_{\text{st}}(\{\varepsilon x\})\,. \tag{6.84}$$

Because a formal solution of the transition probability has the form

$$P(\{x\},t|\{x'\},t') = e^{L_{FP}(\{x\})(t-t')}\delta(\{x\}-\{x'\})$$

and because the stationary joint distribution is given by

$$W_2(\{x\},t;\{x'\},t') = P(\{x\},t|\{x'\},t')\, W_{st}(\{x'\}),$$

we may write (6.84) in the form [6.12]

$$W_2(\{x\},t;\{x'\},t') = W_2(\{\varepsilon x'\},t;\{\varepsilon x\},t'). \tag{6.85}$$

If the joint probability distribution depends on external parameters which also change their sign under time reversal, then transformed parameters must be used on the right-hand side of (6.85).

Equation (6.85) may be interpreted in the following way. If we plot in some way the time dependence of W_2 on a movie we get the same time dependence on the reverse-running movie (velocities change their sign). If the movie is reversed at time $t=t_0$ we have

$$\begin{aligned} W_2(\{\varepsilon x\},t_0-t,\{\varepsilon x'\},t_0-t') &= W_2(\{\varepsilon x\},t';\{\varepsilon x'\},t) \\ &= W_2(\{\varepsilon x'\},t;\{\varepsilon x\},t') \\ &= W_2(\{x\},t;\{x'\},t'). \end{aligned} \tag{6.86}$$

In the first line we used the stationarity of the process (add $t+t'-t_0$ to the times on the left side) and in the second line the symmetry of a joint distribution. Conditions (6.79, 80) are recovered from (6.85) or its equivalent form (6.84). Differentiating (6.84) with respect to time we get for $t=t'$ the operator equation (6.81), which is equivalent to (6.79). Equation (6.80) follows from (6.84) by putting $t=t'$.

Consequence of the Operator Equation

To discuss the consequences of the operator equation (6.81) we first introduce the irreversible and reversible drift coefficients defined by

$$D_i^{ir}(\{x\}) = \tfrac{1}{2}[D_i(\{x\})+\varepsilon_i D_i(\{\varepsilon x\})], \tag{6.87a}$$

$$D_i^{rev}(\{x\}) = \tfrac{1}{2}[D_i(\{x\})-\varepsilon_i D_i(\{\varepsilon x\})], \tag{6.87b}$$

$$D_i(\{x\}) = D_i^{rev}(\{x\})+D_i^{ir}(\{x\}). \tag{6.87c}$$

Obviously the components of the irreversible and reversible parts are transformed according to

$$D_i^{ir}(\{x\}) = \varepsilon_i D_i^{ir}(\{\varepsilon x\}), \tag{6.88a}$$

$$D_i^{\rm rev}(\{x\}) = -\varepsilon_i D_i^{\rm rev}(\{\varepsilon x\}), \tag{6.88b}$$

i.e., the components of the reversible (irreversible) part are transformed in the same (opposite) way as the time derivative of x_i. The deterministic equation $\dot{x}_i = D_i^{\rm rev}(\{x\})$, for instance, is not changed by time reversal. This may be seen by multiplying the transformed equation $d(\varepsilon_i x_i)/d(-t) = D_i^{\rm rev}(\{\varepsilon x\})$ by ε_i and by using (6.88b) and $\varepsilon_i^2 = 1$. We may also split the Fokker-Planck operator into a reversible and an irreversible part

$$L_{\rm FP}(\{x\}) = L_{\rm rev}(\{x\}) + L_{\rm ir}(\{x\}) \quad \text{with} \tag{6.89}$$

$$L_{\rm rev}(\{x\}) = -\frac{\partial}{\partial x_i} D_i^{\rm rev}(\{x\}) = -L_{\rm rev}(\{\varepsilon x\}), \tag{6.90a}$$

$$\begin{aligned} L_{\rm ir}(\{x\}) &= -\frac{\partial}{\partial x_i} D_i^{\rm ir}(\{x\}) + \frac{\partial^2}{\partial x_i \partial x_j} D_{ij}^{\rm ir}(\{x\}) \\ &= L_{\rm ir}(\{\varepsilon x\}), \end{aligned} \tag{6.90b}$$

where the last result follows from (6.88, 95). The Fokker-Planck equation with only the reversible part is not changed by time reversal, i.e., it describes the reversible motion, whereas the Fokker-Planck equation with only the irreversible part changes its sign, i.e., it describes the irreversible motion. ($\partial W/\partial t$ changes its sign on time reversal.) The reversible operator $L_{\rm rev}$ is sometimes called a streaming operator and the irreversible operator $L_{\rm ir}$, the collison operator.

If the stationary distribution function is written in the form (6.31), we can define an irreversible probability current by [all arguments are $\{x\}$, compare (6.3)]

$$S_i^{\rm ir} = W_{\rm st}\left(D_i^{\rm ir} - \frac{\partial D_{ij}}{\partial x_j} + D_{ij}\frac{\partial \Phi}{\partial x_j}\right). \tag{6.91}$$

The total probability current may then be written as

$$S_i = S_i^{\rm ir} + S_i^{\rm rev}, \tag{6.92}$$

where the reversible probability current is defined by

$$S_i^{\rm rev} = W_{\rm st} D_i^{\rm rev}. \tag{6.93}$$

The operator equation (6.81) with $W_{\rm st}(\{\varepsilon x\})$ replaced by $W_{\rm st}(\{x\})$ reads explicitly

$$\begin{aligned} &\left[-\frac{\partial}{\partial x_i} D_i(\{x\}) + \frac{\partial^2}{\partial x_i \partial x_j} D_{ij}(\{x\})\right] W_{\rm st}(\{x\}) \\ &= W_{\rm st}(\{x\})\left\{D_i(\{\varepsilon x\})\frac{\partial}{\partial(\varepsilon_i x_i)} + D_{ij}(\{\varepsilon x\})\frac{\partial^2}{\partial(\varepsilon_i x_i)\,\partial(\varepsilon_j x_j)}\right\}. \end{aligned}$$

Considering the left-hand side we shift the operators $\partial/\partial x_i$ through from left to right. Then we write the terms of the right-hand side on the left-hand side. By using (6.3, 87b, 92, 93) and $D_{ij}=D_{ji}$ we thus obtain

$$-\frac{\partial(S_i^{\mathrm{rev}}(\{x\})+S_i^{\mathrm{ir}}(\{x\}))}{\partial x_i}-2S_i^{\mathrm{ir}}(\{x\})\frac{\partial}{\partial x_i}$$

$$+W_{\mathrm{st}}(\{x\})[D_{ij}(\{x\})-\varepsilon_i\varepsilon_j D_{ij}(\{\varepsilon x\})]\frac{\partial^2}{\partial x_i\partial x_j}=0\,. \tag{6.94}$$

This operator equation can be valid only if the coefficients in front of the zeroth, first and second derivatives vanish [6.2, 3, 10] (for Kramers-Moyal expansions see [6.13])

$$D_{ij}(\{x\})=\varepsilon_i\varepsilon_j D_{ij}(\{\varepsilon x\})\,, \tag{6.95}$$

$$S_i^{\mathrm{ir}}=0\,, \tag{6.96}$$

$$\partial S_i^{\mathrm{rev}}/\partial x_i=0\,. \tag{6.97}$$

Conditions (6.96, 97) read explicitly, see (6.91, 93),

$$D_i^{\mathrm{ir}}=\frac{\partial D_{ij}}{\partial x_j}-D_{ij}\frac{\partial\Phi}{\partial x_j}\,, \tag{6.96a}$$

$$\frac{\partial D_i^{\mathrm{rev}}}{\partial x_i}-D_i^{\mathrm{rev}}\frac{\partial\Phi}{\partial x_i}=0\,. \tag{6.97a}$$

Conditions (6.95 – 97) are the sufficient and necessary conditions for detailed balance. These conditions are sufficient as may be seen as follows. From (6.88a, 95, 96a) we conclude $\Phi(\{x\})=\Phi(\{\varepsilon x\})$, i.e., we get (6.80). Because (6.94 and 81) are equivalent, it follows that (6.79 and 85) must also be valid.

A necessary condition for the stationary solution to have a maximum or a minimum is obviously given by $\partial\Phi/\partial x_i=0$, i.e., by

$$D_i^{\mathrm{ir}}=\partial D_{ij}/\partial x_j\,. \tag{6.98}$$

If the diffusion coefficient does not depend on $\{x\}$ the maximum or minimum can occur only at those $\{x\}$ where the drift coefficient D_i^{ir} vanishes.

By comparing (6.87c, 96a, 97a) with (6.43, 39, 34) respectively, we see that the irreversible drift coefficient D_i^{ir} is now equal to the $D_i^{(\mathrm{s})}$ coefficient and that the reversible drift coefficient D_i^{rev} is equal to the $D_i^{(\mathrm{a})}$ coefficient. Hence, the decomposition of the drift vector in (6.43) is now a priori known, and the results of Sect. 6.3 can be used. For instance, the generalized potential and therefore the stationary distribution (6.31) follow from the line integral (6.42), if the inverse of D_{ij} exists. In this case D_i^{ir} must also obey (6.41), where A_i is given by (6.40) with $D_i^{(\mathrm{s})}$ replaced by D_i^{ir}.

If we apply the transformation (6.35), the key relation (6.81) for detailed balance then takes the form [6.10]

$$L(\{x\}) = L^+(\{\varepsilon x\}) . \tag{6.99}$$

The transformed reversible operator L_{rev} is then the anti-Hermitian operator (6.37) with $D_i^{(a)} = D_i^{\text{rev}}$ and the transformed irreversible operator L_{ir} is then the Hermitian operator (6.36) with $D_i^{(s)} = D_i^{\text{ir}}$. The eigenfunctions $\psi_\mu(\{x\})$ of $L(\{x\})$ and the eigenfunctions $\psi_\mu^+(\{x\})$ of the adjoint operator $L^+(\{x\})$ are therefore connected by

$$\psi_\mu(\{x\}) = \psi_\mu^+(\{\varepsilon x\}) ; \qquad \psi_\mu^+(\{x\}) = \psi_\mu(\{\varepsilon x\}) . \tag{6.100}$$

These connections can be used to simplify (6.51, 52) for P and W_2. The joint distribution (6.52), for instance, now reads ($t \geqq t'$)

$$W_2(\{x\}, t; \{x'\}, t') = \psi_0(\{x\})\, \psi_0(\{x'\}) \sum_\mu \psi_\mu(\{x\})\, \psi_\mu(\{\varepsilon x'\})\, \mathrm{e}^{-\lambda_\mu(t-t')} . \tag{6.101}$$

Because $\psi_0(\{x\}) = \psi_0(\{\varepsilon x\})$, it is explicitly seen that (6.85) is fulfilled.

Singular Diffusion Matrix

If the diffusion matrix is not positive definite it must then be semidefinite, i.e. $\operatorname{Det}\{D_{ij}\}$ is zero and the inverse of D_{ij} does not exist. If, for instance, D_{ij} has the special form

$$\boldsymbol{D} = \begin{pmatrix} D_{11}^{(\text{p})} \dots D_{1M}^{(\text{p})} & 0 \dots 0 \\ \vdots \qquad\quad \vdots & \vdots \quad\; \vdots \\ D_{M1}^{(\text{p})} \dots D_{MM}^{(\text{p})} & 0 \dots 0 \\ & \\ 0 \quad \dots 0 & 0 \dots 0 \\ \vdots \qquad\quad \vdots & \vdots \quad\; \vdots \\ 0 \quad \dots 0 & 0 \dots 0 \end{pmatrix} = \begin{pmatrix} \boldsymbol{D}^{(\text{p})} & \boldsymbol{0} \\ \boldsymbol{0} & \boldsymbol{0} \end{pmatrix} , \tag{6.102}$$

where the submatrix $D_{ij}^{(\text{p})}(i, j \leqq M)$ is positive definite, $\boldsymbol{D}^{-1}$ in (6.40–42) must be replaced by $(\boldsymbol{D}^{(\text{p})})^{-1}$, where the indices are restricted to $i, j \leqq M$. The dependence of the generalized potential on the first M variables $(x_1, \dots, x_M)$ can then be determined from (6.42). To determine the dependence of Φ on the other variables (6.96, 97) must be used.

Example: Kramers Equation

For Kramers equation (4.112) the drift and diffusion coefficients are given by (3.131). Obviously, the position is an even variable and the velocity an odd variable. The irreversible and the reversible drift coefficients read:

$$\begin{aligned} D_x^{\text{ir}} &= 0 , & D_v^{\text{ir}} &= -\gamma v , \\ D_x^{\text{rev}} &= v , & D_v^{\text{rev}} &= -f'(x) . \end{aligned} \tag{6.103}$$

The diffusion matrix is singular

$$D_{xx} = D_{xv} = D_{vx} = 0\,; \qquad D_{vv} \equiv D^{(p)} = \gamma kT/m\,. \tag{6.104}$$

The v dependence of Φ follows from [see (6.96a)]

$$\Phi(x,v) = \int \frac{1}{D_{vv}}(-D_v^{\text{ir}})\,\mathrm{d}v = \frac{m}{kT}\int v\,\mathrm{d}v = \frac{mv^2}{2kT} + h\,. \tag{6.105}$$

The integration constant h may depend on x. If detailed balance holds, this x dependence must be obtained from (6.97), i.e., from

$$\begin{aligned}\frac{\partial S_x^{\text{rev}}}{\partial x} + \frac{\partial S_v^{\text{rev}}}{\partial v} &= N_{\text{st}}\left(\frac{\partial(v\,\mathrm{e}^{-\Phi})}{\partial x} + \frac{\partial(-f'(x)\,\mathrm{e}^{-\Phi})}{\partial v}\right)\\ &= N_{\text{st}}\,\mathrm{e}^{-\Phi}\left[f'(x)\frac{\partial\Phi}{\partial v} - v\frac{\partial\Phi}{\partial x}\right]\\ &= N_{\text{st}}\,\mathrm{e}^{-\Phi}v\left(\frac{f'(x)m}{kT} - h'\right) = 0\end{aligned}$$

which immediately leads to

$$h(x) = mf(x)/(kT) + \text{const.} \tag{6.106}$$

Thus all the conditions (6.95 – 97) are fulfilled and therefore we have shown that detailed balance is valid.

6.5 Ornstein-Uhlenbeck Process

For the Ornstein-Uhlenbeck process [1.5, 1.7] the drift coefficient is linear and the diffusion coefficient constant, i.e.,

$$D_i = -\gamma_{ij}x_j\,; \qquad \gamma_{ij},\, D_{ij} = D_{ji} \text{ const. matrices}\,. \tag{6.107}$$

As discussed in Sect. 3.2, the corresponding Langevin equations can be solved and analytic expressions for the correlation functions can be obtained. Therefore it is not surprising that the Fokker-Planck equation can also be solved exactly for an Ornstein-Uhlenbeck process. The stochastic variables and the Langevin forces are coupled by the linear transformation (3.43), therefore the distribution functions of the stochastic variables must be Gaussian distributions, because the Langevin forces are Gaussian distributed (see the remark at the end of Sect. 2.3.3).

We now want to solve the Fokker-Planck equation (6.1, 2), where the drift and diffusion coefficients are given by (6.107). For the transition probability $P(\{x\}, t \mid \{x'\}, t')$ this equation reads

$$\frac{\partial P}{\partial t} = \gamma_{ij} \frac{\partial}{\partial x_i}(x_j P) + D_{ij} \frac{\partial^2 P}{\partial x_i \partial x_j}, \tag{6.108}$$

where P must satisfy the initial condition

$$P(\{x\}, t' \mid \{x'\}, t') = \delta(\{x\} - \{x'\}). \tag{6.109}$$

If we express P by its Fourier transform with respect to the variables $\{x\}$, i.e., by

$$P(\{x\}, t \mid \{x'\}, t') = (2\pi)^{-N} \int e^{i(k_1 x_1 + \ldots + k_N x_N)} \tilde{P}(\{k\}, t \mid \{x'\}, t') d^N k, \tag{6.110}$$

we obtain for the Fourier transform the first-order differential equation (one has to replace $\partial/\partial x_j$ by ik_j and x_j by $i\partial/\partial k_j$)

$$\frac{\partial \tilde{P}}{\partial t} = -\gamma_{ij} k_i \frac{\partial \tilde{P}}{\partial k_j} - D_{ij} k_i k_j \tilde{P}. \tag{6.111}$$

The initial condition (6.109) is transformed to

$$\tilde{P}(\{k\}, t' \mid \{x'\}, t') = \exp(-ik_j x_j'). \tag{6.112}$$

The first-order equation (6.111) may be solved by the method of characteristics [6.14]. Here we do not apply this method but proceed as follows. Because we already know that P and therefore $\tilde{P}$ must be Gaussian functions, we make the 'ansatz' ($\sigma_{ij} = \sigma_{ji}$)

$$\tilde{P}(\{k\}, t \mid \{x'\}, t') = \exp[-ik_i M_i(t-t') - \tfrac{1}{2} k_i k_j \sigma_{ij}(t-t')]. \tag{6.113}$$

Inserting this 'ansatz' into (6.111) leads to

$$\begin{aligned} &\dot{\tilde{P}} + \gamma_{ij} k_i \frac{\partial \tilde{P}}{\partial k_j} + D_{ij} k_i k_j \tilde{P} \\ &\quad = (-ik_i \dot{M}_i - \tfrac{1}{2} k_i k_j \dot{\sigma}_{ij} - \gamma_{ij} k_i i M_j - \gamma_{ij} k_i \sigma_{jl} k_l + D_{ij} k_i k_j) \tilde{P} \\ &\quad = 0. \end{aligned}$$

This equation requires that M_i and σ_{ij} must obey the differential equations

$$\dot{M}_i = -\gamma_{ij} M_j, \tag{6.114}$$

$$\dot{\sigma}_{ij} = -\gamma_{il} \sigma_{lj} - \gamma_{jl} \sigma_{li} + 2D_{ij}. \tag{6.115}$$

[To obtain the last equation one has to change indices and observe that in $A_{ij}k_ik_j$ the antisymmetric part of the matrix drops out. Therefore only the symmetric part of $\boldsymbol{\gamma\sigma}$ appears in (6.115).] The initial condition (6.112) requires the following initial conditions for M_i and σ_{ij}:

$$M_i(0) = x_i' \; ; \qquad \sigma_{ij}(0) = 0 \, . \tag{6.116}$$

The solution of (6.114) with (6.116) can be written as

$$M_i(t-t') = G_{ij}(t-t')x_j' \, , \tag{6.117}$$

where $G_{ij}(t)$ is the Green's function of the homogeneous Langevin equation (3.31), i.e., it obeys (3.36) with the initial condition (3.35). A formal solution is given by (3.37). In terms of the Green's function G_{ij} the solution of (6.115) with the initial condition (6.116) reads (3.45, 46)

$$\sigma_{ij}(\tau) = \int_0^\tau G_{ik}(\tau')G_{js}(\tau')\,\mathrm{d}\tau' \, 2D_{ks} \, . \tag{6.118}$$

Expansion into a Biorthogonal Set

We now assume that a complete biorthogonal set of the matrix $\boldsymbol{\gamma}$ exists, i.e.,

$$\gamma_{ij}u_j^{(\alpha)} = \lambda_\alpha u_i^{(\alpha)} ; \qquad v_i^{(\alpha)}\gamma_{ij} = \lambda_\alpha v_j^{(\alpha)} \tag{6.119}$$

with the orthonormality and completeness relation

$$\sum_\alpha v_i^{(\alpha)}u_j^{(\alpha)} = \delta_{ij} ; \qquad u_i^{(\alpha)}v_i^{(\beta)} = \delta_{\alpha\beta} \tag{6.120}$$

(summation convention for Latin indices but not for Greek indices). Such a complete biorthogonal set exists if the N eigenvalues are all different. Otherwise the matrix $\boldsymbol{\gamma}$ can be reduced to a Jordan canonical form only [6.9]. To avoid these forms we may change the matrix γ_{ij} by adding terms $\varepsilon\hat{\gamma}_{ij}$ so that all eigenvalues are different. In the final results we may then take the limit $\varepsilon \to 0$. The spectral decomposition of the matrix $\boldsymbol{\gamma}$ reads

$$\gamma_{ij} = \sum_\alpha \lambda_\alpha u_i^{(\alpha)}v_j^{(\alpha)} \tag{6.121}$$

and we have

$$G_{ij}(t) = [\exp(-\boldsymbol{\gamma}t)]_{ij} = \sum_\alpha \mathrm{e}^{-\lambda_\alpha t}u_i^{(\alpha)}v_j^{(\alpha)} \, . \tag{6.122}$$

By inserting this expression in (6.118) we can perform the integration, obtaining

$$\sigma_{ij}(t) = 2\sum_{\alpha,\beta} \frac{1-\mathrm{e}^{-(\lambda_\alpha+\lambda_\beta)t}}{\lambda_\alpha+\lambda_\beta} D^{(\alpha,\beta)}u_i^{(\alpha)}u_j^{(\beta)} \, ,$$

$$D^{(\alpha,\beta)} = v_k^{(\alpha)} D_{kl} v_l^{(\beta)} . \tag{6.123}$$

For symmetric matrices γ_{ij} all the eigenvalues are real and the eigenvectors $u_i^{(\alpha)}$ and $v_i^{(\alpha)}$ coincide.

Transition Probability Density

If we now insert (6.113) into (6.110) and perform the integration (the integration is explained in Sect. 2.3.3), we arrive at

$$\begin{aligned} P(\{x\},t\,|\,\{x'\},t') = {} & (2\pi)^{-N/2}[\mathrm{Det}\,\sigma(t-t')]^{-1/2} \\ & \times \exp\{-\tfrac{1}{2}[\sigma^{-1}(t-t')]_{ij}[x_i - G_{ik}(t-t')x'_k] \\ & \times [x_j - G_{jl}(t-t')x'_l]\} . \end{aligned} \tag{6.124}$$

Stationary Distribution

If all real parts of the eigenvalues of the damping matrix γ_{ij} are larger than zero, a stationary solution exists. For large $t-t'$ the Green's function G_{ik} will be zero and (6.124) reduces to

$$W_{\mathrm{st}}(\{x\}) = (2\pi)^{-N/2}[\mathrm{Det}\,\sigma(\infty)]^{-1/2}\exp\{-\tfrac{1}{2}[\sigma^{-1}(\infty)]_{ij}x_i x_j\} . \tag{6.125}$$

The matrix $\sigma_{ij}(\infty)$ is determined from [see (6.115)]

$$\gamma_{il}\sigma_{lj}(\infty) + \gamma_{jl}\sigma_{li}(\infty) = 2D_{ij} \tag{6.126}$$

or from (6.118). If the biorthogonal set (6.119) is used, we have from (6.123)

$$\sigma_{ij}(\infty) = 2\sum_{\alpha,\beta} D^{(\alpha,\beta)} u_i^{(\alpha)} u_j^{(\beta)}/(\lambda_\alpha + \lambda_\beta) . \tag{6.127}$$

With the help of this expression and because of (6.120, 122), (6.123) becomes

$$\sigma_{ij}(t) = [\delta_{is}\delta_{jr} - G_{is}(t)G_{jr}(t)]\,\sigma_{sr}(\infty) . \tag{6.128}$$

This relation can also be proved by inserting (6.126) into (6.118) and by using $-\gamma_{ik}G_{kl} = \dot{G}_{il}$ (3.36).

Joint Probability Density

In the stationary state the joint probability density is given by $P(\{x\},t\,|\,\{x'\},t') \times W_{\mathrm{st}}(\{x'\})$. With the help of this joint probability density any two-time expectation value can be calculated by integration.

Potential Condition

In accordance with (6.87–88) we split the matrix γ into an irreversible and reversible matrix

$$\gamma_{ij} = \gamma_{ij}^{\text{ir}} + \gamma_{ij}^{\text{rev}}, \qquad \gamma_{ij}^{\substack{\text{ir}\\ \text{rev}}} = \tfrac{1}{2}[\gamma_{ij} \pm \varepsilon_i \gamma_{ij} \varepsilon_j] . \tag{6.129}$$

Then $D_i^{\text{ir}} = -\gamma_{ij}^{\text{ir}} x_j$ is the irreversible and $D_i^{\text{rev}} = -\gamma_{ij}^{\text{rev}} x_j$ the reversible drift coefficient. If the inverse of D_{ij} exists we can apply the potential condition (6.96a) which requires that

$$(D^{-1})_{ij} \gamma_{jk}^{\text{ir}} = (D^{-1})_{kj} \gamma_{ji}^{\text{ir}} . \tag{6.130}$$

From (6.96a) we obtain

$$\frac{\partial^2 \Phi}{\partial x_i \partial x_k} = [\sigma^{-1}(\infty)]_{ik} = (D^{-1})_{ij} \gamma_{jk}^{\text{ir}} . \tag{6.131}$$

[If (6.130) is not fulfilled, (6.131) is not a solution because the right-hand side is then no longer symmetric as it should be.] If the inverse of D_{ij} does not exist, we obtain from (6.96a) the condition

$$\gamma_{ik}^{\text{ir}} = D_{ij} [\sigma^{-1}(\infty)]_{jk} . \tag{6.132}$$

It follows from (6.97, 126, 131) that γ_{ik}^{rev} must obey the relations

$$\gamma_{ii}^{\text{rev}} = 0 ; \qquad \gamma_{il}^{\text{rev}} \sigma_{lj}(\infty) + \gamma_{jl}^{\text{rev}} \sigma_{li}(\infty) = 0 . \tag{6.133}$$

Expectation Values

Expectation values are obtained from (6.124) or from the joint probability density by integration. For the calculation of simple forms of expectation values it is – because of the linearity of the corresponding Langevin equation – easier to treat the Ornstein-Uhlenbeck process by the Langevin equation method. If we need, for instance, the expectation value $\langle x_i(t) \rangle$, we may directly integrate (3.31) and take the expectation value as in Sect. 3.2. One may also derive an equation of motion for expectation values directly from the Fokker-Planck equation (6.108). Multiplying, for instance, (6.108) by x_i and integrating the resulting equation, i.e.,

$$\frac{\partial}{\partial t} \int x_i P d^N x = \int x_i \gamma_{jk} \frac{\partial}{\partial x_j} (x_k P) d^N x + \int x_i D_{jk} \frac{\partial^2 P}{\partial x_j \partial x_k} d^N x ,$$

by using partial integration we obtain

$$\frac{d}{dt} \langle x_i(t) \rangle = -\gamma_{ik} \langle x_k(t) \rangle . \tag{6.134}$$

This equation must then be solved. This method is usually simpler than working with the general solution (6.124) of the Fokker-Planck equation.

Applications of the results of this subsection to the Kramers equation (4.112) for a harmonic potential are given in Sect. 10.2.1.

Time-Dependent Matrices

If the matrix γ_{ij} and the diffusion matrix D_{ik} depend on time we still obtain for P a Gaussian distribution function. The moments $M_i(t,t')$ follow from the solution of (6.114) which can be found with the help of the Green's function $G_{ij}(t,t')$ of (6.114). The matrix σ follows from (6.115) [1.24].

6.6 Further Methods for Solving the Fokker-Planck Equation

If no analytic expressions can be found the difficulty of solving a Fokker-Planck equation usually increases with increasing number of variables. At first, one may therefore try to eliminate some variables. If some variables decay very rapidly (fast variables) they may be eliminated adiabatically. If the drift and diffusion coefficients do not depend on some variables, the latter may also be eliminated. These methods are treated in Sects. 8.2, 3 and will therefore not be discussed here.

6.6.1 Transformation of Variables

One may transform the variables in such a way that the transformed Fokker-Planck equation can be solved analytically. The transformed coefficients are given by (4.131, 132). The problem of finding such a transformation is, however, as hard as solving the Fokker-Planck equation. The method is usually applied in the opposite way. One starts with a Fokker-Planck equation whose solution is known, e.g., the Fokker-Planck equation for an Ornstein-Uhlenbeck process. If one makes a nonlinear transformation of variables, one obtains a complicated Fokker-Planck equation which, of course, can then be solved.

6.6.2 Variational Method

If the anti-Hermitian operator L_{A} in (6.35) is absent, the eigenvalues λ and the eigenfunctions ψ of L_{H} given by (6.46) can be obtained from a variational problem. The function ψ which minimizes

$$\lambda = \frac{\int \left(D_{ij} \dfrac{\partial \psi}{\partial x_i} \dfrac{\partial \psi}{\partial x_j} + V_{\mathrm{S}} \psi^2 \right) d^N x}{\int \psi^2 d^N x} \tag{6.135}$$

leads to the lowest eigenfunction ψ_0. The minimum of this expression is then the lowest eigenvalue λ_0. The next eigenfunction ψ and eigenvalue are found by minimizing (6.135) subject to the auxiliary condition

$$\int \psi_0 \psi \, d^N x = 0 , \tag{6.136}$$

and so on, see the discussion in Sect. 5.9.1 for the one-variable case. Similar to the one-variable case, lower bounds for the eigenvalues can also be found.

6.6.3 Reduction to an Hermitian Problem

The problem of finding the stationary solution of the Fokker-Planck equation can always be reduced to a real Hermitian eigenvalue problem. (More and better techniques exist for solving Hermitian eigenvalue problems than for solving non-Hermitian problems.) Obviously, the operator $L_{\mathrm{FP}}^{+}L_{\mathrm{FP}}$ is an Hermitian operator. If we have found a solution of $L_{\mathrm{FP}}^{+}L_{\mathrm{FP}}W=0$ we have also found a solution of $L_{\mathrm{FP}}W=0$. This may be seen as follows. Using the notation

$$\int W_1 L_{\mathrm{FP}}^{+}L_{\mathrm{FP}}W_2 d^N x = (W_1, L_{\mathrm{FP}}^{+}L_{\mathrm{FP}}W_2),$$

we have the following steps of conclusions

$$\begin{array}{ccc} L_{\mathrm{FP}}^{+}L_{\mathrm{FP}}W=0 & \Rightarrow & (W, L_{\mathrm{FP}}^{+}L_{\mathrm{FP}}W)=0 \\ \Uparrow & & \Downarrow \\ L_{\mathrm{FP}}W=0 & \Leftarrow & (L_{\mathrm{FP}}W, L_{\mathrm{FP}}W)=0\,. \end{array}$$

Seybold [6.15] has applied this method to calculate the stationary distribution of a Fokker-Planck equation where the detailed balance condition is not valid.

6.6.4 Numerical Integration

One way of performing a numerical integration of the Fokker-Planck equation (6.1, 2) is to use instead of the continuous variables $\{x\}$ the discrete variables $\{n\}$, defined by $x_i = \Delta x_i n_i$ (no summation convention), with discrete times $t_m = \Delta t m$. If the differentials are then approximated by differences in a consistent way, solving the Fokker-Planck equation is reduced to iterating a difference equation. The difference equations must be stable in the sense that the probability error does not increase faster than the probability itself, otherwise one does not obtain an approximate solution to the continuous Fokker-Planck equation. As discussed in the literature [6.16–18], it is crucial for this stability that the differentials are approximated by appropriate differences.

Another difficulty arises for natural boundary conditions where the variables extend to infinity. Such problems for infinite regions must be approximated by a problem for a finite region.

6.6.5 Expansion into Complete Sets

The method will be explained for a Fokker-Planck equation with two variables x and y. Though the method may in principle be applied to a Fokker-Planck equation with an arbitrary number of variables, it becomes less practical for a greater number of variables. We first assume that we have two complete orthonormalized sets $\varphi^q(x)$ and $\psi_n(y)$ which satisfy the boundary conditions for x and y. If x and y extend from minus infinity to plus infinity (natural boundary conditions) we may use, for instance, Hermite functions $\varphi^q(x) = N^q \mathrm{H}_q(\alpha x) \exp(-\alpha^2 x^2/2)$ and $\psi_n(y) = \mathrm{N}_n \mathrm{H}_n(\beta y) \exp(-\beta^2 y^2/2)$ with suitable scaling factors α and β. One may, of course, use complete sets which are more adapted to the problem, as proposed in the one-variable case (Sect. 5.9.3). Because the sets are complete the probability density may be expanded similarly to the one-variable case (5.104)

$$W(x,y,t) = F(x,y) \sum_{q,n} c_n^q(t)\, \varphi^q(x)\, \psi_n(y)\,. \tag{6.137}$$

The choice of the function $F(x,y)$ will be discussed later on. (It should be mentioned that the expansion functions need not necessarily be of a product form.) If we insert (6.137) into the Fokker-Planck equation, we obtain an infinite system of coupled differential equations for the expansion coefficients c_n^q. One choice for this system is given by (assuming $F(x,y) \neq 0$)

$$\dot{c}_n^q = \sum_{p,m} [F^{-1} L_{\mathrm{FP}} F]_{nm}^{qp} c_m^p\,. \tag{6.138}$$

Here we denoted the matrix element of an operator A by

$$A_{nm}^{qp} = \iint [\varphi^q(x)\, \psi_n(y)]^* A\, \varphi^p(x)\, \psi_m(y)\, dx\, dy\,. \tag{6.139}$$

If we do not divide by $F(x,y)$ we obtain the more complicated system

$$\sum_{p,m} F_{nm}^{qp} \dot{c}_m^p = \sum_{p,m} [L_{\mathrm{FP}} F]_{nm}^{qp} c_m^p\,, \tag{6.140}$$

which may, however, have certain advantages (see below).

If the infinite system is truncated at $q = Q$ and $n = N$ we may solve it numerically, leading to approximate solutions of the Fokker-Planck equation. Because the number of equations in the truncated system is of the order QN, we can use only low Q and N values. As an example, in Sect. 12.5.2 we obtain the transient of laser oscillation by solving the truncated coupled equations (6.138).

6.6.6 Matrix Continued-Fraction Method

Sometimes the structure of the system (6.138) is such that only nearest neighbors with respect to the lower index are coupled. (If such a coupling occurs for the upper indices, we may of course change notation.) If we introduce the column vector

$$\boldsymbol{c}_n = \begin{pmatrix} c_n^1 \\ \vdots \\ c_n^Q \end{pmatrix} \tag{6.141}$$

the system (6.138) then has the form of the vector recurrence relation (9.10), which can be solved by matrix continued fractions, as explained in Chap. 9. The advantage of the matrix continued-fraction method is that we now only have to invert $Q \times Q$ matrices N times. This is much easier to perform than to invert the $Q \cdot N \times Q \cdot N$ matrices in (6.138) once. Therefore much higher Q and N values can be treated by this method.

If M nearest neighbors with respect to the lower index are coupled, we have to use a $Q \cdot M$ dimensional column vector, so may again cast (6.138) into a tridiagonal vector recurrence relation, as explained in Chap. 9.

As already discussed for the one-variable case, M nearest-neighbor coupling is valid for a large class of Fokker-Planck operators. If, for instance, we have natural boundary conditions, if we use Hermite functions and if D_i and D_{ij} are given by polynomials in y of finite order, we obtain a finite number of nearest-neighbor couplings in the lower index for $F = 1$. (The coefficients of the polynomial may depend on x.) By using a suitable notation the equation of motion can then be cast into the tridiagonal vector recurrence relation (9.10). If the drift and diffusion coefficients are rational functions in y (i.e., quotient of two polynomials in y) and if we choose F to be the common denominator, we can then cast the equation for the coefficients into the form (9.121), which may also be solved by matrix continued fractions. Whether the method will actually work depends on the dimension of matrices to be inverted. This dimension in turn depends on the number of expansion terms which must be used to approximate the distribution function (6.137) fairly well.

General applications of the matrix continued-fraction method to the Kramers equation are given in Chap. 10. Explicit numerical results for Brownian motion in a periodic potential are presented in Chap. 11. For a simple laser model, a Fokker-Planck equation for two variables (intensity, inversion) without detailed balance was solved by *Mörsch* [6.19] using this method. In optical bistability a Fokker-Planck equation for two variables (intensity, phase) has to be solved. *Haug* et al. [6.20] have investigated the optical bistability due to a two-photon absorption resonance also using this method. We conclude from these examples that the matrix continued-fraction method seems to be very effective for treating some Fokker-Planck equations for two variables without detailed balance, which can hardly be solved by other methods. The method is also useful for obtaining eigenvalues and eigenfunctions in those cases where the problem cannot be reduced to an Hermitian one. As shown in App. A1, certain expectation values of distribution functions, which obey Fokker-Planck equations corresponding to linear differential equations with multiplicative colored noise, can also be obtained by the matrix continued-fraction method.

One restriction of the matrix continued-fraction method, however, is the following: for very small diffusion coefficients the distribution function gets more peaked and therefore more coefficients in the expansion have to be taken into account. Thus, the dimension of the matrices to be inverted also increases. For numerical calculations the method is then no longer tractable.

6.6.7 WKB Method

For very small diffusion coefficients one may use a WKB method. This method has been applied to diffusion in one-dimensional and multi-dimensional bistable potentials by *Caroli* et al. [5.24, 6.21]. It essentially consists in the following. After indicating the smallness of the diffusion coefficient by a parameter $\varepsilon > 0$ we insert in the Fokker-Planck equation

$$\dot{W} = \left(-\frac{\partial}{\partial x_i} D_i + \varepsilon \frac{\partial^2}{\partial x_i \partial x_j} D_{ij} \right) W \tag{6.142}$$

the 'ansatz'

$$W = A \exp\left(-\frac{1}{\varepsilon} w \right) \tag{6.143}$$

and obtain in lowest order

$$\dot{w} = -D_i \frac{\partial w}{\partial x_i} - D_{ij} \frac{\partial w}{\partial x_i} \frac{\partial w}{\partial x_j} + O(\varepsilon) \, . \tag{6.144}$$

This first-order nonlinear partial differential equation is usually easier to solve than the Fokker-Planck equation. It can be treated by the method of characteristics [6.14]. The WKB method is well known in quantum mechanics, where it is useful for describing the transition to classical mechanics, and in optics, where it is useful for describing the transition from wave optics to ray optics. The method is also called the ray method [6.22], where equations for the higher-order terms can also be found. (In higher order A also depends on variables.) Some care has to be taken for the application of the WKB method. We know from quantum mechanics [6.23] that at a classical turning point the WKB solution is not valid. Near these points the full Schrödinger equation must be used and this solution must then be matched with the WKB solutions. Because in certain cases (e.g. detailed balance) the Fokker-Planck equation can be transformed to a Schrödinger equation, the same should also be true for the Fokker-Planck equation, see [5.24, 6.21] for an application to bistable systems.

7. Linear Response and Correlation Functions

We consider a system in a stable steady state or in equilibrium. If we disturb the system by applying some external fields or by changing some parameter the system will be driven away from its former steady state. The external fields or the changes of the parameters are usually small. Then we only need to take into account those deviations from the steady state which are linear in the external fields (linear response). The deviations of expectation values from their steady-state values also depend linearly on the fields. This dependence can be described by a response function. If the external fields are switched off, the deviations from the steady state decay or dissipate (in the physical literature the word 'dissipate' is usually used for the decay of energy).

The response function describes the decay of an expectation value to its steady-state value, thus describing the dissipation of the deviation of an expectation value from its steady-state value.

In a steady state the variables are not at rest but fluctuating. The fluctuations of two variables $A(t)$ and $B(t)$ are described by their mutual correlation function $\langle A(t_1) B(t_2) \rangle$. Similar to an external field the fluctuations also drive the system away from its steady state and then the system decays to its steady state again. Therefore one expects that these correlation functions (describing fluctuations) are in some way connected to the response functions (describing dissipation).

These connections are called dissipation-fluctuation or fluctuation-dissipation theorems. They have been derived for a large class of classical and quantum mechanical systems [7.1 – 4]. The relations in which the transport coefficients are expressed in terms of time integrals over correlation functions are the well-known *Green-Kubo* expressions [7.1, 3]. Dissipation-fluctuation theorems for systems in nonthermal equilibrium, which are described by Fokker-Planck equations, have been derived by *Agarwal* [7.5].

For linear systems, i.e., linear Langevin equations, one can calculate the response function of the variable x by a deterministic equation, because the Langevin force drops out. One can then use the dissipation-fluctuation theorem to obtain the correlation function from this deterministic equation.

In this chapter we first derive a general expression of the linear response function for systems described by Fokker-Planck equations. We shall then see that this expression can always be written as a correlation function, i.e., dissipation-fluctuation theorems exist. Next we discuss for some cases the connection of different correlation functions and thus obtain different forms of the dissipation-fluctuation theorem. Finally, it is shown how the Fourier transform of the response function, i.e., the susceptibility, is connected to the Fourier transform of the correlation function, i.e., to the spectral density.

7.1 Linear Response Function

If our system is described by a Fokker-Planck equation and if some external fields or some parameters depend on time, the Fokker-Planck operator generally depends on time. It is written in the form

$$L_{FP}(\{x\},t) = L_{FP}(\{x\}) + L_{ext}(\{x\},t) . \tag{7.1}$$

We assume that the time-independent Fokker-Planck operator L_{FP} has the stationary solution W_{st}, i.e.,

$$L_{FP}(\{x\})\, W_{st}(\{x\}) = 0 . \tag{7.2}$$

Any solution of the Fokker-Planck equation can be split into the stationary solution and a time-dependent solution

$$W(\{x\},t) = W_{st}(\{x\}) + w(\{x\},t) . \tag{7.3}$$

If the external fields or the changes of external parameters are small L_{ext} and w will also be small. In the Fokker-Planck equation

$$\begin{aligned} \dot{W}(\{x\},t) &= \dot{w}(\{x\},t) = L_{FP}(\{x\},t)\, W(\{x\},t) \\ &= [L_{FP}(\{x\}) + L_{ext}(\{x\},t)]\,[W_{st}(\{x\}) + w(\{x\},t)] \end{aligned}$$

we may therefore neglect the term $L_{ext}(\{x\},t) \cdot w(\{x\},t)$ and retain only the linear terms, i.e.,

$$\dot{w}(\{x\},t) = L_{FP}(\{x\})\, w(\{x\},t) + L_{ext}(\{x\},t)\, W_{st}(\{x\}) . \tag{7.4}$$

A formal solution of this equation is given by

$$w(\{x\},t) = \int_{-\infty}^{t} e^{L_{FP}(\{x\})\cdot(t-t')} L_{ext}(\{x\},t')\, W_{st}(\{x\})\, dt' , \tag{7.5}$$

as is easily seen by insertion.

The Fokker-Planck operator $L_{ext}(\{x\},t)$ which describes the time variations of the external fields or of some parameters generally reads (summation convention)

$$L_{ext}(\{x\},t) = -\frac{\partial}{\partial x_i} D_i^{(ext)}(\{x\},t) + \frac{\partial^2}{\partial x_i \partial x_j} D_{ij}^{(ext)}(\{x\},t) . \tag{7.6}$$

We now assume that the drift and diffusion coefficients depend on time in a multiplicative way, i.e., (no summation convention)

$$\begin{aligned} D_i^{(\text{ext})}(\{x\},t) &= D_i^{(\text{ext})}(\{x\}) \cdot F_i(t)\,, \\ D_{ij}^{(\text{ext})}(\{x\},t) &= D_{ij}^{(\text{ext})}(\{x\}) \cdot F_{ij}(t)\,. \end{aligned} \tag{7.7}$$

The operator in (7.6) is then a sum of operators of the form

$$L_{\text{ext}}(\{x\},t) = L_{\text{ext}}(\{x\}) \cdot F(t)\,, \tag{7.8}$$

where $L_{\text{ext}}(\{x\})$ stands for (no summation convention)

$$-\frac{\partial}{\partial x_i} D_i^{(\text{ext})}(\{x\}) \quad \text{or} \quad \frac{\partial^2}{\partial x_i \partial x_j} D_{ij}^{(\text{ext})}(\{x\})$$

and $F(t)$ for $F_i(t)$ or $F_{ij}(t)$, respectively. Because of the linearity it is therefore sufficient to use (7.8) in (7.5) to calculate $w(\{x\},t)$. The deviation ΔA of any expectation value of $A(\{x\})$ from its stationary value then takes the form

$$\begin{aligned} \Delta A(t) &\equiv \int A(\{x\})\, w(\{x\},t)\, d^N x \\ &= \int_{-\infty}^{\infty} R_{A,L}(t-t')F(t')dt'\,, \end{aligned} \tag{7.9}$$

where the response function $R_{A,L}(t)$ describing the response of A to L_{ext} is defined by

$$\begin{aligned} R_{A,L}(t) &= \int A(\{x\})\, e^{L_{\text{FP}}(\{x\})\cdot t} L_{\text{ext}}(\{x\})\, W_{\text{st}}(\{x\})\, d^N x \quad \text{for} \quad t \geqq 0\,, \\ R_{A,L}(t) &= 0 \quad \text{for} \quad t<0\,. \end{aligned} \tag{7.10}$$

The following three linear response functions are used.

a) Pulse-Response Function

The response of A to a δ-function force $F(t) = \delta(t)$ is called the pulse-response function

$$\Delta A^{(\text{p})}(t) = \int_{-\infty}^{\infty} R_{A,L}(t-t')\,\delta(t')dt' = R_{A,L}(t) \tag{7.11a}$$

and is just given by (7.10) [It should be mentioned that the force must be small for the linear response theory to be valid. It would be better to write $F(t) = \varepsilon\delta(t)$, where ε is small, so $\Delta A^{(\text{p})}(t)$ is then ε times $R_{A,L}(t)$.]

b) Step-Response Function or Excitation Function

The step-response function $\Delta A^{(\text{s})}(t)$ [7.6, 7] or the excitation function [7.3] is the linear response of the system to an external field, which is proportional to the step function $\Theta(t)$, i.e., the constant external field is switched on at $t=0$. Inserting the step function in (7.9) we obtain

$$\Delta A^{(s)}(t) = \int_{-\infty}^{\infty} R_{A,L}(t-t')\,\Theta(t')\,\mathrm{d}t' = \int_{0}^{t} R_{A,L}(t-t')\,\mathrm{d}t'$$

$$= \int_{0}^{t} R_{A,L}(t')\,\mathrm{d}t' \,, \tag{7.11 b}$$

i.e., the time derivative of the step-response function $\Delta A^{(s)}(t)$ is equal to the pulse-response function $\Delta A^{(p)}(t) = R_{A,L}(t)$. At $t=0$ $\Delta A^{(s)}$ is zero.

c) After-Effect-Response Function or Relaxation Function

The after-effect-response function $\Delta A^{(a)}(t)$ [7.6, 7] or the relaxation function [7.3] is the linear response of the system to the step function $\Theta(-t)$, i.e., the constant external field is switched off at $t=0$. We have

$$\Delta A^{(a)}(t) = \int_{-\infty}^{\infty} R_{A,L}(t-t')\,\Theta(-t')\,\mathrm{d}t' = \int_{-\infty}^{0} R_{A,L}(t-t')\,\mathrm{d}t'$$

$$= \int_{t}^{\infty} R_{A,L}(t')\,\mathrm{d}t' \,, \tag{7.11 c}$$

i.e., the negative derivative of the after-effect-response function $\Delta A^{(a)}(t)$ is equal to the pulse-response function $\Delta A^{(p)}(t) = R_{A,L}(t)$. At $t=0$, $\Delta A^{(a)}$ describes the static linear response to a unit force. Obviously, for all t the sum of the step-response function and the after-effect-response function is equal to the static linear response to a unit force.

7.2 Correlation Functions

We now consider correlation functions of $A(\{\xi(t_1)\})$ and $B(\{\xi(t_2)\})$ for the different times t_1 and t_2 in the stationary state. Because of the stationarity, the correlation function depends only on the time difference $t = t_1 - t_2$. The correlation function may then be defined by

$$K_{A,B}(t) = \langle A(\{\xi(t)\})\,B(\{\xi(0)\})\rangle$$

$$= \iint A(\{x\})\,B(\{x'\})\,W_2(\{x\},t;\{x'\},0)\,\mathrm{d}^N x\,\mathrm{d}^N x' \,. \tag{7.12}$$

If we express W_2 by the product of the transition probability times the stationary distribution and if we insert the formal solution for the transition probability, we have $(t \geqq 0)$

$$K_{A,B}(t) = \iint A(\{x\})\, P(\{x\},t\,|\,\{x'\},0)\, B(\{x'\})\, W_{\rm st}(\{x'\})\, d^N x\, d^N x'$$
$$= \iint A(\{x\})\, e^{L_{\rm FP}(\{x\})t} \delta(\{x\}-\{x'\})\, B(\{x'\})\, W_{\rm st}(\{x'\})\, d^N x\, d^N x' \,.$$

By performing the integration over $\{x'\}$ we obtain the following result for the stationary correlation function for

$t \geqq 0$:

$$K_{A,B}(t) = \int A(\{x\})\, e^{L_{\rm FP}(\{x\})t} B(\{x\})\, W_{\rm st}(\{x\})\, d^N x \,. \tag{7.13}$$

For $t \leqq 0$ we interchange the factors in (7.12), using the stationarity of the process (subtract t in both time arguments),

$$\langle A(\{\xi(t)\})\, B(\{\xi(0)\})\rangle = \langle B(\{\xi(-t)\})\, A(\{\xi(0)\})\rangle \tag{7.12a}$$

and thus obtain for

$t \leqq 0$:

$$K_{A,B}(t) = \int B(\{x\})\, e^{L_{\rm FP}(\{x\})(-t)} A(\{x\})\, W_{\rm st}(\{x\})\, d^N x \,. \tag{7.13a}$$

Dissipation-Fluctuation Theorem

By comparing (7.10) with (7.13) we see that the response function (7.10) for $t \geqq 0$ is given by the correlation (7.13) if the function $B(\{x\})$ is defined by

$$B(\{x\}) = [W_{\rm st}(\{x\})]^{-1} L_{\rm ext}(\{x\})\, W_{\rm st}(\{x\})$$
$$= e^{\Phi(\{x\})} L_{\rm ext}(\{x\})\, e^{-\Phi(\{x\})} \,. \tag{7.14}$$

In the last line we have introduced the generalized potential, see (6.31). The dissipation-fluctuation theorem may thus be written as

$$\begin{aligned} R_{A,L}(t) &= K_{A,B}(t) \quad &\text{for} \quad t \geqq 0\,,\\ R_{A,L}(t) &= 0 \quad &\text{for} \quad t < 0\,, \end{aligned} \tag{7.15}$$

where $B(\{x\})$ has to be calculated according to (7.14). Sometimes we add the index B to the response function to indicate the external field.

Connection Between Correlation Functions

The correlation function (7.13) may be expressed by time derivatives of other correlation functions. Thus one obtains different forms of the dissipation-fluctuation theorems. To derive these different forms one has to specify $L_{\rm FP}$ and $L_{\rm ext}$.

Example I

In the first example we make the following assumptions: the diffusion matrix is independent of the variables $\{x\}$ and it is positive definite. Furthermore, the detailed balance condition (6.41) should be valid and the reversible drift coefficient (6.88 b) is zero. We further assume that the external force acts on the ith variable in an additive way, i.e., $L_{\text{ext}}(\{x\}, t)$ is given by

$$L_{\text{ext}}^{(i)}(\{x\}, t) = L_{\text{ext}}^{(i)}(\{x\}) \cdot F(t) , \quad L_{\text{ext}}^{(i)}(\{x\}) = -\partial/\partial x_i . \tag{7.16}$$

For the Langevin equation this means that we add the force $F(t)$ on the right-hand side of (3.110) at the ith component. Using (6.96 a) we then get for the function $B(\{x\})$ in (7.14)

$$B_i(\{x\}) = -\mathrm{e}^{\Phi}\frac{\partial}{\partial x_i}\mathrm{e}^{-\Phi} = \frac{\partial \Phi}{\partial x_i} = -(D^{-1})_{ik}D_k . \tag{7.17}$$

Furthermore, we have for the above assumptions

$$\begin{aligned}(L_{\text{FP}}x_i W_{\text{st}}) &= x_i(L_{\text{FP}}W_{\text{st}}) - \left(D_k\frac{\partial x_i}{\partial x_k} + 2D_{kj}\frac{\partial x_i}{\partial x_k}\,\frac{\partial \Phi}{\partial x_j}\right)W_{\text{st}} \\ &= -[D_i - 2D_{ij}(D^{-1})_{jl}D_l]\,W_{\text{st}} = D_i W_{\text{st}} .\end{aligned} \tag{7.18}$$

Therefore we obtain the following connection between the correlation functions K_{A,x_i} and K_{A,D_i}:

$$\begin{aligned}\frac{\mathrm{d}}{\mathrm{d}t}K_{A,x_i}(t) &= \int A\,\mathrm{e}^{L_{\text{FP}}t}(L_{\text{FP}}x_i W_{\text{st}})\,\mathrm{d}^N x \\ &= \int A\,\mathrm{e}^{L_{\text{FP}}t}D_i W_{\text{st}}\,\mathrm{d}^N x = K_{A,D_i}(t) .\end{aligned} \tag{7.19}$$

The response function $R_{A,L}(t)$ then takes the form [7.5]

$$\begin{aligned}R_{A,L}(t) &= K_{A,B_i}(t) = -(D^{-1})_{ik}K_{A,D_k}(t) \\ &= -(D^{-1})_{ik}\frac{\mathrm{d}}{\mathrm{d}t}K_{A,x_k}(t) .\end{aligned} \tag{7.20}$$

Example II

We assume that the process is described by the Kramers equation (4.112). For an external force F in direction x we have

$$L_{\text{ext}}(x, v) = -\frac{1}{m}\,\frac{\partial}{\partial v} . \tag{7.21}$$

In the Langevin equations (3.130) this would mean that we add the force $F(t)/m$ on the right-hand side of the first equation. Because the generalized potential is given by (6.105) we get for B:

$$B = \mathrm{e}^{\Phi}\left(-\frac{1}{m}\frac{\partial}{\partial v}\right)\mathrm{e}^{-\Phi} = \frac{v}{kT}. \tag{7.22}$$

The response of the velocity thus reads ($t \geqq 0$)

$$R_{v,L}(t) = K_{v,B}(t) = \frac{1}{kT}K_{v,v}(t) = \frac{1}{kT}\langle v(t)\,v(0)\rangle . \tag{7.23}$$

For the Kramers equation this velocity correlation function can be expressed by the correlation function of $f'(x)$. This is seen as follows: we obviously have

$$\begin{aligned}\left(\frac{\mathrm{d}}{\mathrm{d}t}+\gamma\right)^2 K_{v,v}(t) &= \iint v(L_{\mathrm{FP}}+\gamma)^2 \mathrm{e}^{L_{\mathrm{FP}}t} v\, W_{\mathrm{st}} \mathrm{d}x\,\mathrm{d}v \\ &= \iint [(L_{\mathrm{FP}}^{+}+\gamma)\,v]\,\mathrm{e}^{L_{\mathrm{FP}}t}(L_{\mathrm{FP}}+\gamma)\,v\,W_{\mathrm{st}}\mathrm{d}x\,\mathrm{d}v .\end{aligned}$$

Because of

$$\begin{aligned}(L_{\mathrm{FP}}^{+}+\gamma)\,v &= \left\{v\frac{\partial}{\partial x} - [\gamma v + f'(x)]\frac{\partial}{\partial v} + \frac{\gamma kT}{m}\frac{\partial^2}{\partial v^2} + \gamma\right\} v \\ &= -f'(x)\end{aligned}$$

and

$$\begin{aligned}(L_{\mathrm{FP}}+\gamma)\,v\,W_{\mathrm{st}} &= \left[2\gamma v + f'(x)\frac{\partial v}{\partial v}\right] W_{\mathrm{st}} + 2\gamma\frac{kT}{m}\frac{\partial v}{\partial v}\frac{\partial W_{\mathrm{st}}}{\partial v} \\ &= f'(x)\,W_{\mathrm{st}} ,\end{aligned}$$

we obtain

$$\left(\frac{\mathrm{d}}{\mathrm{d}t}+\gamma\right)^2 K_{v,v}(t) = -K_{f'(x),f'(x)}(t) . \tag{7.24}$$

This relation can also be derived by using the corresponding Langevin equations.

Example III

Here we consider again the Kramers equation (4.112), but we now assume that the temperature T changes in time, i.e.,

$$T = T[1+\Delta T(t)/T] = T[1+F(t)] , \qquad F(t) = \Delta T(t)/T , \tag{7.25}$$

$$L_{\mathrm{ext}}(x,v) = \frac{\gamma kT}{m}\frac{\partial^2}{\partial v^2} . \tag{7.26}$$

The B function in (7.14) then reads

$$B = e^{\Phi} \frac{\gamma kT}{m} \frac{\partial^2}{\partial v^2} e^{-\Phi} = \gamma \left(\frac{m v^2}{kT} - 1 \right). \tag{7.27}$$

Using the energy $E = \frac{1}{2} m v^2 + m f(x)$ and

$$\begin{aligned} L_{\mathrm{FP}} E W_{\mathrm{st}} &= L_{\mathrm{FP}} \left(\frac{m v^2}{2} + m f \right) W_{\mathrm{st}} \\ &= -\gamma (m v^2 - kT) W_{\mathrm{st}} \end{aligned} \tag{7.28}$$

we have

$$K_{A,B}(t) = -\frac{1}{kT} \frac{\mathrm{d}}{\mathrm{d}t} K_{A,E}(t) . \tag{7.29}$$

Thus the response of the energy ($A = E$) is in this case given by the negative time derivative of the energy correlation function divided by kT, i.e., ($t \geqq 0$)

$$R_{E,L}(t) = -\frac{1}{kT} \frac{\mathrm{d}}{\mathrm{d}t} K_{E,E}(t) = -\frac{1}{kT} \frac{\mathrm{d}}{\mathrm{d}t} \langle E(t) E(0) \rangle . \tag{7.30}$$

By using $\langle E(\infty) E(0) \rangle = \langle E(0) \rangle^2$ we get for the static response

$$\Delta E_{\mathrm{s}} = \int_0^\infty R_{E,L}(t) \mathrm{d}t = \frac{\langle (E(0)^2 \rangle - \langle E(0) \rangle^2}{kT} . \tag{7.31}$$

This result also follows from

$$\Delta E_{\mathrm{s}} = \lim_{F \to 0} \frac{1}{F} \left[\frac{\int E \exp\left(-\frac{E}{kT(1+F)} \right) \mathrm{d}v \, \mathrm{d}x}{\int \exp\left(-\frac{E}{kT(1+F)} \right) \mathrm{d}v \, \mathrm{d}x} - \frac{\int E \exp\left(-\frac{E}{kT} \right) \mathrm{d}v \, \mathrm{d}x}{\int \exp\left(-\frac{E}{kT} \right) \mathrm{d}v \, \mathrm{d}x} \right] .$$

The application of l'Hôpital's rule leads to

$$\begin{aligned} \Delta E_{\mathrm{s}} &= \frac{1}{kT} \left[\frac{\int E^2 \exp\left(-\frac{E}{kT} \right) \mathrm{d}v \, \mathrm{d}x}{\int \exp\left(-\frac{E}{kT} \right) \mathrm{d}v \, \mathrm{d}x} - \left(\frac{\int E \exp\left(-\frac{E}{kT} \right) \mathrm{d}v \, \mathrm{d}x}{\int \exp\left(-\frac{E}{kT} \right) \mathrm{d}v \, \mathrm{d}x} \right)^2 \right] \\ &= \frac{1}{kT} (\langle E^2 \rangle - \langle E \rangle^2) . \end{aligned}$$

Example IV

Here we consider an Ornstein-Uhlenbeck process with an external field acting on the ith coordinate,

$$L_{\text{ext}}(\{x\}) = -\frac{\partial}{\partial x_i}.$$

In terms of the Langevin equation (3.31) we have to add the force $F(t)$ on the right-hand side of the ith component.

Because of (6.125) we obtain for the B function (7.14)

$$B_i = \partial \Phi/\partial x_i = [\sigma^{-1}(\infty)]_{ij} x_j. \tag{7.32}$$

The response function of the coordinate x_k can therefore be expressed by the correlation function $\langle x_k(t) x_j(0) \rangle$,

$$R_{x_k,L}(t) = [\sigma^{-1}(\infty)]_{ij} \langle x_k(t) x_j(0) \rangle. \tag{7.33}$$

Because the Ornstein-Uhlenbeck process is linear, the Langevin force drops out by averaging (3.31). The response function $R_{x_k,L}(t)$ is therefore identical to the Green's function $G_{ki}(t)$ of the Langevin equation (3.31). This also follows explicitly by inserting the correlation function for the stationary state (3.56a) into (7.33).

Sum Rules

A relation of the form (7.19) where one correlation function is connected to the time derivative of another correlation function leads to a relation for eigenvalues and eigenfunctions. This may be seen as follows: for $t = 0$ (7.19) may be written as

$$\begin{aligned}
&\int A(\{x\}) L_{\text{FP}}(\{x\}) x_i W_{\text{st}}(\{x\}) d^N x \\
&\quad = \int\int A(\{x\}) L_{\text{FP}}(\{x\}) \delta(\{x\} - \{x'\}) x_i' W_{\text{st}}(\{x'\}) d^N x \, d^N x' \\
&\quad = \int A(\{x\}) D_i(\{x\}) W_{\text{st}}(\{x\}) d^N x .
\end{aligned}$$

If we insert here (6.50) for the δ function we thus obtain

$$\sum_\mu \lambda_\mu \int A(\{x\}) \psi_\mu(\{x\}) d^N x \int \psi_\mu^+(\{x\}) x_i W_{\text{st}}(\{x\}) d^N x = \int A(\{x\}) D_i(\{x\}) W_{\text{st}}(\{x\}) d^N x . \tag{7.34}$$

(A very well known sum rule of this type is the sum rule for oscillator strength in quantum mechanics [Ref. 5.11, p. 1318].) Sum rules are good checks for the accuracy of numerically computed eigenvalues and eigenfunctions. Sum rules for susceptibility are discussed in Sect. 7.3.

7.3 Susceptibility

The convolution in (7.9) reduces to a multiplication for the corresponding Fourier transforms. Introducing the Fourier transforms of $\Delta A(t)$, $F(t)$ and $R_{A,L}(t)$,

$$\Delta\tilde{A}(\omega) = \int_{-\infty}^{\infty} \Delta A(t)\, e^{-i\omega t} dt\,, \tag{7.35}$$

$$\tilde{F}(\omega) = \int_{-\infty}^{\infty} F(t)\, e^{-i\omega t} dt\,, \tag{7.36}$$

$$\begin{aligned} \chi_A(\omega) &= \int_{-\infty}^{\infty} R_{A,L}(t)\, e^{-i\omega t} dt \\ &= \int_{0}^{\infty} R_{A,L}(t)\, e^{-i\omega t} dt\,, \end{aligned} \tag{7.37}$$

(7.9) transforms to [5.1, p. 113]

$$\Delta\tilde{A}(\omega) = \chi_A(\omega)\tilde{F}(\omega)\,. \tag{7.38}$$

(In (7.37) we used (7.10).) The generally complex quantity $\chi_A(\omega)$ is called the susceptibility. The real part is usually denoted by $\chi_A'(\omega)$ and the imaginary part by $-\chi_A''(\omega)$,

$$\chi_A(\omega) = \chi_A'(\omega) - i\chi_A''(\omega)\,. \tag{7.39}$$

The minus sign is used if the Fourier transforms (7.35 – 37) are defined with a minus sign. The real part is an even function in ω,

$$\chi_A'(\omega) = \int_{0}^{\infty} R_{A,L}(t) \cos \omega t\, dt = \chi_A'(-\omega) \tag{7.40}$$

and the imaginary part an odd function in ω,

$$\chi_A''(\omega) = \int_{0}^{\infty} R_{A,L}(t) \sin \omega t\, dt = -\chi_A''(-\omega)\,. \tag{7.41}$$

(We assume that $R_{A,L}(t)$ is a real function.)

Because $R_{A,L}(t)$ decays in time all the poles of $\chi_A(\omega)$ must lie in the upper half of the complex ω plane. It then follows that $\chi_A'(\omega)$ and $\chi_A''(\omega)$ are connected by the Kramers-Kronig relations [7.8]

$$\chi_A''(\omega) = -\frac{1}{\pi} P \int_{-\infty}^{\infty} \frac{\chi'(\nu)}{\nu - \omega} d\nu\,, \tag{7.42a}$$

$$\chi_A'(\omega) = \frac{1}{\pi} P \int_{-\infty}^{\infty} \frac{\chi''(\nu)}{\nu - \omega} \mathrm{d}\nu \,, \tag{7.42b}$$

where the principal value is denoted by P.

For Brownian motion the energy dissipation is proportional to

$$\Delta = \gamma \int_{-\infty}^{\infty} \langle v(t) \rangle F(t) \mathrm{d}t \,. \tag{7.43}$$

If we insert here the inverse Fourier transforms of (7.35, 36) with $A = v$

$$\langle v(t) \rangle = (2\pi)^{-1} \int_{-\infty}^{\infty} \langle \tilde{v}(\omega) \rangle \mathrm{e}^{\mathrm{i}\omega t} \mathrm{d}\omega \,, \tag{7.35a}$$

$$F(t) = (2\pi)^{-1} \int_{-\infty}^{\infty} \tilde{F}(\omega') \mathrm{e}^{\mathrm{i}\omega' t} \mathrm{d}\omega' \,, \tag{7.36a}$$

by using (2.84, 7.38) we obtain

$$\Delta = \gamma (2\pi)^{-1} \int_{-\infty}^{\infty} \chi_v(\omega) \tilde{F}(\omega) \tilde{F}(-\omega) \mathrm{d}\omega \,.$$

For real fields we have $\tilde{F}(\omega) = \tilde{F}^*(-\omega)$. Because $\chi_v'(\omega)$ is even and $\chi_v''(\omega)$ is odd, we finally get

$$\Delta = \gamma (2\pi)^{-1} \int_{-\infty}^{\infty} \chi_v'(\omega) |\tilde{F}(\omega)|^2 \mathrm{d}\omega \,. \tag{7.43a}$$

Hence, absorption at frequency ω is proportional to $\chi_v'(\omega)$. Had we used x instead of v, the absorption at the frequency ω would be proportional to $\omega \cdot \chi_x''(\omega)$.

Connection to Spectral Density

The spectral density $S_{A,B}(\omega)$ is the Fourier transform of the correlation function multiplied by 2, see (2.86),

$$S_{A,B}(\omega) = 2 \int_{-\infty}^{\infty} K_{A,B}(t) \mathrm{e}^{-\mathrm{i}\omega t} \mathrm{d}t \,. \tag{7.44}$$

Because of (7.15) the susceptibility is the half-sided Fourier transform of this correlation function:

$$\chi_A(\omega) = \int_{0}^{\infty} K_{A,B}(t) \mathrm{e}^{-\mathrm{i}\omega t} \mathrm{d}t \,. \tag{7.45}$$

If the correlation function is symmetric (antisymmetric) we can express $\chi'_A(\chi''_A)$ by $S_{A,B}$. The other part $\chi''_A(\chi'_A)$ then follows from the Kramers-Kronig relation (7.42a) (Eq. (7.42b)).

Symmetric Correlation Function

$$K_{A,B}(t) = K_{A,B}(-t) \Rightarrow S_{A,B}(\omega) = S_{A,B}(-\omega)$$

$$\chi'_A(\omega) = \int_0^\infty K_{A,B}(t)\cos\omega t\,\mathrm{d}t = \frac{1}{2}\int_{-\infty}^{\infty} K_{A,B}(t)\,\mathrm{e}^{-\mathrm{i}\omega t}\mathrm{d}t$$

$$= \frac{1}{4}S_{A,B}(\omega)\,. \tag{7.46}$$

Antisymmetric Correlation Function

$$K_{A,B}(t) = -K_{A,B}(-t) \Rightarrow S_{A,B}(\omega) = -S_{A,B}(-\omega)$$

$$\chi''_A(\omega) = \int_0^\infty K_{A,B}(t)\sin\omega t\,\mathrm{d}t = -\frac{1}{2\mathrm{i}}\int_{-\infty}^{\infty} K_{A,B}(t)\,\mathrm{e}^{-\mathrm{i}\omega t}\mathrm{d}t$$

$$= -\frac{1}{4\mathrm{i}}S_{A,B}(\omega)\,. \tag{7.47}$$

Because of the stationarity (7.12a) we have

$$K_{A,B}(t) = K_{B,A}(-t)$$

and therefore a correlation function for $A = B$ is always symmetric in the stationary state

$$K_{A,A}(t) = K_{A,A}(-t)\,.$$

In Examples II and III we expressed the response function by the $\langle v(t)v(0)\rangle$ (7.23) and the $\langle E(t)E(0)\rangle$ (7.30) correlation functions. Therefore we can immediately express the real part of the susceptibility in Example II and the imaginary part of the susceptibility in Example III by the corresponding spectral densities:

$$\chi'_v(\omega) = \frac{1}{kT}\,\frac{1}{4}S_{v,v}(\omega) \quad \text{for Example II}\,, \tag{7.48}$$

$$\chi''_E(\omega) = \frac{1}{kT}\,\frac{\omega}{4}S_{E,E}(\omega) \quad \text{for Example III}\,. \tag{7.49}$$

Connection to the Einstein Relation

Equation (7.48) is the dissipation-fluctuation theorem for Brownian motion in the frequency domain. If applied to free Brownian motion, for $\omega = 0$ (7.48) is identical to the Einstein relation (3.19). This is seen as follows. If the response function for free Brownian motion $R_{v,L}(t) = \exp(-\gamma t)/m$ is integrated over time, we get

$$\chi_v'(0) = \int_0^\infty R_{v,L}(t)\,\mathrm{d}t = 1/(m\gamma)\,. \tag{7.50}$$

On the other hand, the diffusion constant, defined by (3.18), can be expressed by

$$\begin{aligned} D &= \lim_{t\to\infty} \frac{1}{2}\frac{\mathrm{d}}{\mathrm{d}t}\langle [x(t)-x(0)]^2\rangle = \lim_{t\to\infty} \langle v(t)[x(t)-x(0)]\rangle \\ &= \lim_{t\to\infty}\int_0^t \langle v(t)\,v(t')\rangle \mathrm{d}t' = \lim_{t\to\infty}\int_0^t \langle v(t)\,v(t-\tau)\rangle \mathrm{d}\tau\,, \end{aligned}$$

or, after using the stationarity (add $\tau - t$ to both arguments), by

$$D = \int_0^\infty \langle v(\tau)\,v(0)\rangle \mathrm{d}\tau = \frac{1}{4} S_{v,v}(0)\,. \tag{7.51}$$

By inserting (7.50, 51) into (7.48) we thus obtain the Einstein relation (3.19).

Application to Example I

If we have even and odd variables and if detailed balance is valid the correlation function has the following property [cf. (6.85 and 7.12a)]

$$\begin{aligned} \langle x_j(t)x_i(0)\rangle &= \varepsilon_j\varepsilon_i\langle x_j(0)x_i(t)\rangle \\ &= \varepsilon_j\varepsilon_i\langle x_j(-t)x_i(0)\rangle\,. \end{aligned} \tag{7.52}$$

Thus for $\varepsilon_i\varepsilon_j = 1$ the correlation function is even and for $\varepsilon_i\varepsilon_j = -1$ the correlation function is odd. Because of (7.20) the response function of the variable x_j to L_{ext} in Example I is

$$R_{x_j,L}(t) = -(D^{-1})_{ik}\frac{\mathrm{d}}{\mathrm{d}t}\langle x_j(t)x_k(0)\rangle\,. \tag{7.53}$$

The Fourier transform of this expression leads to the susceptibility

$$\chi_{ji}(\omega) = -(D^{-1})_{ik}\int_0^\infty \mathrm{e}^{-\mathrm{i}\omega t}\frac{\mathrm{d}}{\mathrm{d}t}\langle x_j(t)x_k(0)\rangle \mathrm{d}t\,.$$

After using partial integration, we thus obtain for the real and imaginary parts of the susceptibility

$$\chi'_{ji}(\omega) = (D^{-1})_{ik}\left[\langle x_j(0)\,x_k(0)\rangle - \omega\int_0^\infty \langle x_j(t)\,x_k(0)\rangle \sin\omega t\,\mathrm{d}t\right]$$

$$\chi''_{ji}(\omega) = (D^{-1})_{ik}\,\omega\int_0^\infty \langle x_j(t)\,x_k(0)\rangle \cos\omega t\,\mathrm{d}t\,.$$

The diffusion matrix in Example I was independent of $\{x\}$ and had to satisfy (6.95). If follows from this relation and from the symmetry of D_{ij} that we have

$$(D^{-1})_{ij} = 0 \quad \text{for} \quad \varepsilon_i\varepsilon_j = -1 \tag{7.54}$$

and we can express x'_{ji} and χ''_{ji} by the spectral density [7.5]

$$\chi''_{ji} = (D^{-1})_{ik}\frac{\omega}{4}S_{kj}(\omega) \quad \text{for} \quad \varepsilon_i\varepsilon_j = 1\,, \tag{7.55}$$

$$\chi'_{ji} = (D^{-1})_{ik}\frac{\mathrm{i}\omega}{4}S_{kj}(\omega) \quad \text{for} \quad \varepsilon_i\varepsilon_j = -1 \tag{7.56}$$

($\langle x_i(0)\,x_j(0)\rangle$ must vanish for $\varepsilon_i\varepsilon_j = -1$).

Sum Rules

An identity in which an integration over the susceptibility appears is called a sum rule for the susceptibility. Because χ'_A is the cos-transform of the correlation function $K_{A,B}(t)$, we have, for example,

$$\begin{aligned}\frac{2}{\pi}\int_0^\infty \chi'_A(\omega)\,\mathrm{d}\omega &= \frac{2}{\pi}\int_0^\infty\int_0^\infty K_{A,B}(t)\cos\omega t\,\mathrm{d}t\,\mathrm{d}\omega\\ &= 2\int_0^\infty K_{A,B}(t)\lim_{\Omega\to\infty}\frac{\sin\Omega t}{\pi t}\\ &= 2\int_0^\infty K_{A,B}(t)\,\delta(t)\,\mathrm{d}t = K_{A,B}(0)\,.\end{aligned} \tag{7.57}$$

Using (7.23, 57), the susceptibilities χ'_v and χ''_E in Examples II and III become

$$\frac{2}{\pi}\int_0^\infty \chi'_v(\omega)\,\mathrm{d}\omega = \frac{1}{kT}\langle v^2(0)\rangle = 1/m\,, \tag{7.58}$$

and similarly, using (7.41, 30),

$$\frac{2}{\pi}\int_0^\infty \chi''_E(\omega)\frac{\mathrm{d}\omega}{\pi} = \frac{2}{\pi}\int_0^\infty\int_0^\infty\left[-\frac{\mathrm{d}}{\mathrm{d}t}\langle E(t)\,E(0)\rangle\right]\frac{1}{kT}\,\frac{\sin\omega t}{\omega}\,\mathrm{d}t\,\mathrm{d}\omega$$

$$= \frac{2}{\pi}\int_0^\infty\int_0^\infty \frac{1}{kT}\langle E(t)E(0)\rangle \cos\omega t\,\mathrm{d}\omega\,\mathrm{d}t$$

$$- \lim_{t\to\infty}\frac{1}{kT}\langle E(t)E(0)\rangle \frac{2}{\pi}\int_0^\infty \frac{\sin\omega t}{\omega}\mathrm{d}\omega$$

$$= \frac{1}{kT}[\langle (E(0))^2\rangle - \langle E(0)\rangle^2]\,. \tag{7.59}$$

In deriving (7.59) we have used an integration by parts, $\lim_{t\to\infty}\langle E(t)E(0)\rangle = \langle E(0)\rangle^2$, $\int_0^\infty (\sin x/x)\,\mathrm{d}x = \frac{\pi}{2}$ and evaluated the integral according to (7.57). The last expression in (7.59) is equal to the static response (7.31).

By expressing $K_{A,B}(t)$ as time derivatives of other correlation functions, further sum rules may be derived. If the susceptibilities are obtained numerically, the sum rules are useful for checking the accuracy of the calculations.

Application to Brownian Motion in a Harmonic Potential

Brownian motion in a harmonic potential with an external force $F(t)$ is described by the Langevin equation

$$\ddot{x} + \gamma\dot{x} + \omega_0^2 x = F(t)/m + \Gamma(t)\,. \tag{7.60}$$

Because the system is linear, the response function $R_{x,L}(t)$ can be obtained by averaging (7.60). Hence, it is the Green's function of (7.60) without the Langevin force, i.e., a solution of the harmonic oscillator equation with $F(t) = \delta(t)$. This Green's function reads ($t \geqq 0$)

$$R_{x,L}(t) = \frac{1}{\omega_1 m}\mathrm{e}^{-\gamma t/2}\sin\omega_1 t\,; \qquad \omega_1 = \sqrt{\omega_0^2 - \gamma^2/4}\,. \tag{7.61}$$

The response of the velocity follows from (7.61) ($t \geqq 0$)

$$R_{v,L}(t) = \dot{R}_{x,L}(t) = \frac{1}{m}\mathrm{e}^{-\gamma t/2}\left(\cos\omega_1 t - \frac{\gamma}{2\omega_1}\sin\omega_1 t\right)\,. \tag{7.62}$$

The dissipation-fluctuation theorem (7.23) then immediately leads to the velocity autocorrelation function

$$\langle v(t)\,v(0)\rangle = \frac{kT}{m}\mathrm{e}^{-\gamma t/2}\left(\cos\omega_1 t - \frac{\gamma}{2\omega_1}\sin\omega_1 t\right)\,. \tag{7.63}$$

The susceptibilities $\chi_x(\omega)$ and $\chi_v(\omega)$ take the form

$$\chi_x(\omega) = \int_0^\infty R_{x,L}(t)\,\mathrm{e}^{-\mathrm{i}\omega t}\mathrm{d}t$$

$$= \frac{1}{m}\,\frac{1}{2\omega_1 \mathrm{i}}\left(\frac{1}{\frac{1}{2}\gamma+\mathrm{i}(\omega-\omega_1)} - \frac{1}{\frac{1}{2}\gamma+\mathrm{i}(\omega+\omega_1)}\right)$$

$$= \frac{1}{m}\,\frac{\omega_0^2-\omega^2-\mathrm{i}\gamma\omega}{(\omega_0^2-\omega^2)^2+\gamma^2\omega^2}\,, \tag{7.64}$$

$$\chi_v(\omega) = \int_0^\infty \dot{R}_{x,L}(t)\,\mathrm{e}^{-\mathrm{i}\omega t}\mathrm{d}t = R_{x,L}(0) + \mathrm{i}\omega\int_0^\infty R_{x,L}(t)\,\mathrm{e}^{-\mathrm{i}\omega t}\mathrm{d}t = \mathrm{i}\omega\chi_x(\omega)\,. \tag{7.65}$$

Thus we finally have

$$\omega\chi_x' = -\chi_v'' = \frac{1}{m}\,\frac{(\omega_0^2-\omega^2)\,\omega}{(\omega_0^2-\omega^2)^2+\gamma^2\omega^2} \approx \frac{1}{2m}\,\frac{\omega_0-\omega}{(\omega_0-\omega)^2+(\gamma/2)^2}\,, \tag{7.66}$$

$$\omega\chi_x'' = \chi_v' = \frac{1}{m}\,\frac{\gamma\omega^2}{(\omega_0^2-\omega^2)^2+\gamma^2\omega^2} \approx \frac{1}{4m}\,\frac{\gamma}{(\omega_0-\omega)^2+(\gamma/2)^2}\,. \tag{7.67}$$

The last expressions in (7.66, 67) are valid only for positive ω and small damping constant $\gamma \ll \omega_0$. The dissipation of energy is described by $\chi_v' = \omega\chi_x''$ (7.43 a).

8. Reduction of the Number of Variables

Usually, the difficulty of solving the Fokker-Planck equation like any other partial differential equation increases with increasing number of independent variables. It is therefore advisable to eliminate as many variables as possible, so we discuss below three cases where the number of independent variables can be reduced.

In Sect. 8.1 we treat first-passage time problems. As shown, the mean first-passage time can be obtained by solving an equation where the time variable no longer appears. Though in some experiments one may really measure the first-passage times, one usually uses the mean first-passage time to obtain approximate lifetimes for problems, where the boundary condition is slightly different from that for the first-passage time problem.

In Sect. 8.2 we look for solutions of those Fokker-Planck equations where the drift and diffusion coefficients do not depend on some variables. The Fourier transform of the probability density for these variables can then be obtained by an equation where these variables no longer appear. This equation is applied to calculate distribution functions for variables which are time integrals of other stochastic Markovian variables, in Sect. 8.2.1.

Finally, in Sect. 8.3 we assume that the decay constants for some variables are much larger than those for other ones. These "fast" variables can then be eliminated (adiabatic elimination procedure).

8.1 First-Passage Time Problems

First-passage time is the time at which the stochastic variable $\xi(t)$ first leaves a given domain [8.1, 2, 1.12]. In the following we restrict ourselves to the one-variable case.

If we start with the realizations at $t=0$ with $\xi(0)=x'$ (Fig. 8.1), the first-passage time T is the time when $\xi(t)$ reaches a boundary for the first time. If both boundaries are absorbing, then either $\xi(T)=x_2$ or $\xi(T)=x_1$. If one of the boundaries is reflecting we only have to take into account the other boundary. Obviously the time T varies from realization to realization, i.e., the first-passage time is a random variable.

We now want to calculate the distribution function for these first-passage times. We are especially interested in the moments of first-passage times, because

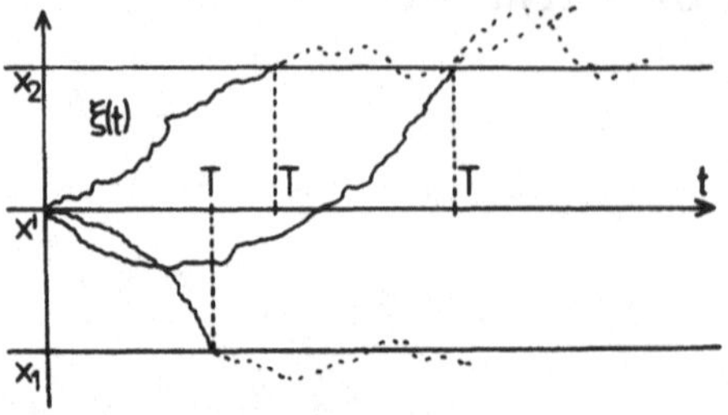

Fig. 8.1. The first-passage times T of the realizations $\xi(t)$ to leave the domain $x_1 < x < x_2$

these moments can be obtained by solving an inhomogeneous differential equation where only the variable x will enter. We first ask for the probability density $P(x,t|x',0)$ for the stochastic variable $\xi(t)$ starting at $t=0$ with $\xi(0)=x'$ to reach x at time t. If the stochastic variable reaches either x_2 or x_1 the first time we no longer count these realizations. Therefore P must be zero for $x \geqq x_2$ and $x \leqq x_1$. In other words, the prescription that the particles are no longer counted if they have passed a boundary is substituted by an absorbing wall. For $x_1 < x < x_2$ P must satisfy the Fokker-Planck equation (5.1, 2),

$$\frac{\partial P}{\partial t} = L_{\mathrm{FP}}(x)P; \quad P(x,0|x',0) = \delta(x-x') \quad \text{for} \quad x_1 < x < x_2,$$
$$P(x,t|x',0) = 0 \quad \text{for} \quad x = x_2 \quad \text{or} \quad x = x_1. \tag{8.1}$$

The probability $W(x',t)$ of realizations which have started at x' and which have not yet reached either one of the boundaries up to the time t is given by

$$W(x',t) = \int_{x_1}^{x_2} P(x,t|x',0)\,\mathrm{d}x. \tag{8.2}$$

The probability $-\mathrm{d}W$ of those realizations which reach one of the boundaries in the time interval $(t, t+\mathrm{d}t)$ thus reads

$$-\mathrm{d}W(x',t) = -\int_{x_1}^{x_2} \dot{P}(x,t|x',0)\,\mathrm{d}x\,\mathrm{d}t. \tag{8.3}$$

The distribution function $w(T)$ for the first-passage time T is therefore given by

$$w(x',T) = -\frac{\mathrm{d}W(x',T)}{\mathrm{d}T} = -\int_{x_1}^{x_2} \dot{P}(x,T|x',0)\,\mathrm{d}x. \tag{8.4}$$

The moments of the first-passage time distribution are

$$T_n(x') = \int_0^\infty T^n w(x',T)\,\mathrm{d}T = \int_{x_1}^{x_2} p_n(x,x')\,\mathrm{d}x, \tag{8.5}$$

where $p_n(x,x')$ is defined by

$$p_n(x,x') = -\int_0^\infty T^n \dot{P}(x,T|x',0)\,\mathrm{d}T\,. \tag{8.6}$$

We obviously have

$$p_0(x,x') = -\int_0^\infty \dot{P}(x,T|x',0)\,\mathrm{d}T = P(x,0|x',0) = \delta(x-x')\,. \tag{8.7}$$

Performing a partial integration gives

$$p_n(x,x') = n\int_0^\infty T^{n-1}P(x,T|x',0)\,\mathrm{d}T\,, \qquad n \geqq 1\,. \tag{8.8}$$

By applying the operator $L_{\mathrm{FP}}(x)$ to (8.8) and using (8.1, 6), we obtain the following system of coupled differential equations

$$L_{\mathrm{FP}}(x)\,p_n(x,x') = -np_{n-1}(x,x') \qquad n \geqq 1\,,$$

i.e.,

$$\begin{aligned} L_{\mathrm{FP}}(x)\,p_1(x,x') &= -\delta(x-x') \\ L_{\mathrm{FP}}(x)\,p_2(x,x') &= -2p_1(x,x') \\ L_{\mathrm{FP}}(x)\,p_3(x,x') &= -3p_2(x,x') \\ \dots \qquad & \qquad \dots\,. \end{aligned} \tag{8.9}$$

From this system we can obtain $p_n(x,x')$ by solving the equations successively starting with the first one. The boundary conditions for p_n must be the same as for P, i.e.,

$$p_n(x,x') = 0 \qquad \text{for} \qquad x = x_1 \qquad \text{or} \qquad x = x_2\,. \tag{8.10}$$

The first equation of the system in (8.9) describes the stationary probability density, if at x' a unit rate of probability is injected into the system. [Integrating the first equation from $x'-\varepsilon$ to $x'+\varepsilon$ leads to $S(x'+\varepsilon) - S(x'-\varepsilon) = 1$, where S is the probability current (4.47).]

Formal Solution

A formal solution of (8.9) is given by

$$p_n(x,x') = n!\,[-L_{\mathrm{FP}}(x)]^{-n}\delta(x-x')\,. \tag{8.11}$$

This formal solution also follows by integration of the formal solution of P, i.e.,

$$p_n(x,x') = -\int_0^\infty t^n L_{FP}(x)\, e^{L_{FP}(x)t}\delta(x-x')\,dt$$
$$= n!\,[-L_{FP}(x)]^{-n}\delta(x-x'). \quad (8.12)$$

Solution in Terms of the Adjoint Operator

Because $P(x,t\,|\,x',0) = P(x,0\,|\,x',-t)$ also satisfies the backward Kolmogorov equation (4.96),

$$\frac{\partial P}{\partial t} = L_{FP}^{+}(x')P, \quad (8.13)$$

in (8.9, 11) we can replace the operator $L_{FP}(x)$ by its adjoint acting on x', i.e., by $L_{FP}^{+}(x')$. If x' lies outside the interval (x_1, x_2) the probability P must vanish, i.e., we may equally well solve (8.13) with the boundary condition

$$P(x,t|x',0) = 0 \quad \text{for} \quad x' = x_1 \quad \text{and} \quad x' = x_2. \quad (8.14)$$

Hence, to obtain the moments (8.5) we may also use the adjoint equation (8.13). We then have [8.1, 2]

$$L_{FP}^{+}(x')\,T_n(x') = -n\,T_{n-1}(x'); \quad n \geqq 1,$$
$$T_0(x') = 1. \quad (8.15)$$

For two absorbing boundaries $T_n(x')$ must satisfy (8.14). The equation $(n=1)$ for the mean first-passage time $T_1(x')$

$$L_{FP}^{+}(x')\,T_1(x') = -1 \quad (8.15\text{a})$$

was derived in [8.1], the system (8.15) in [8.2]. At first glance system (8.15) seems somewhat easier to solve than system (8.9), because here only one variable occurs. However, to obtain for instance $T_1(x')$ for a certain value of $x' = a$, one has to solve (8.15a) for all x' in the interval $x_1 \leqq x' \leqq x_2$, whereas the first equation of (8.9) needs to be solved only for $x' = a$. As shown in Sect. 5.10.2 for a metastable potential, both (8.15a, 9) can be used to calculate T_1 leading to the same result if the potential difference is large compared to the diffusion constant D. It was also shown there that the inverse of the first-passage time agrees approximately with the first eigenvalue and that the precise value of the boundaries $x_1 = -A$ and $x_2 = A$ are not important for $A > x_{max}$ (Fig. 5.8a).

It follows from the form (5.35) of the Fokker-Planck operator that the solutions of (8.9, 15) can be obtained by quadratures for the one-variable case see for instance (5.143). For several variables one may also define a mean first-passage time when a particle leaves a certain domain. However, the equations for the moments of the first-passage times can then no longer be given in terms of quadratures in the general case.

Boundary Conditions for the Kramers Equation

As we have seen the boundary condition for the first-passage time problem is the boundary of an absorbing wall. For Brownian motion of particles, whose probability density $P(x,v,t|x',v',0)$ in phase space is a solution of the Kramers equation (4.112), this boundary condition is more complicated than (8.14). If, for instance, we have an absorbing wall at the left side of the domain at $x_1 = x_{\min}$ we require that probability current in x-direction must vanish for those particles leaving the wall into the domain i.e. for the particles with positive velocities. Because of (3.131, 4.104) we have $S_x = D_x P = vP$. Therefore we must require that the probability density for positive velocities is zero at $x_1 = x_{\min}$

$$P(x_1,v,t|x',v',0) = 0 \quad \text{for} \quad v > 0\,. \tag{8.16a}$$

If we have an absorbing wall at $x_2 = x_{\max}$ we have similarly

$$P(x_2,v,t|x',v',0) = 0 \quad \text{for} \quad v < 0\,. \tag{8.16b}$$

The initial condition of P is given by $P(x,v,0|x',v',0) = \delta(x-x')\,\delta(v-v')$. Here we assume that particles with the velocity v' are injected into the system at x'. If the particles are injected at x' with a velocity distribution $g(v)$ we have to replace the δ function by $g(v)$.

Burschka and *Titulaer* [8.3, 4] calculated probability densities for the Kramers equation with the boundary condition (8.16a) and similar conditions. In these works further discussions and various references may also be found.

8.2 Drift and Diffusion Coefficients Independent of Some Variables

If the drift and diffusion coefficients do not depend on some variables, let say on $x_1,\dots,x_n$, we can reduce the problem of solving the Fokker-Planck equation to a problem where only the other variables $x_{n+1},\dots,x_N$ occur. If we make a Fourier transform with respect to the first n variables

$$\begin{aligned} W(x_1,\dots,x_N,t) = (2\pi)^{-n} \int \hat{W}(k_1,\dots,k_n,x_{n+1},\dots,x_N,t) \\ \times \mathrm{e}^{\mathrm{i}(k_1x_1+\dots+k_nx_n)}\mathrm{d}k_1\dots\mathrm{d}k_n \end{aligned} \tag{8.17}$$

it is easily seen by inserting (8.17) into the Fokker-Planck equation (4.94a, 95) and by performing partial integrations that the following equation for $\hat{W}$ with the variables $x_{n+1},\dots,x_N$ must then be solved:

$$\partial \hat{W}/\partial t = \hat{L}_{\mathrm{FP}} \hat{W},$$

$$\begin{aligned} \hat{L}_{\mathrm{FP}} &= \hat{L}_{\mathrm{FP}}(x_{n+1},\dots,x_N) \\ &= -\mathrm{i}\sum_{i=1}^{n} k_i D_i - \sum_{i=n+1}^{N} \frac{\partial}{\partial x_i} D_i - \sum_{i,j=1}^{n} k_i k_j D_{ij} \\ &\quad + 2\mathrm{i}\sum_{i=1}^{n}\sum_{j=n+1}^{N} k_i \frac{\partial}{\partial x_j} D_{ij} + \sum_{i,j=n+1}^{N} \frac{\partial^2}{\partial x_i \partial x_j} D_{ij} \, . \end{aligned} \tag{8.18}$$

Generally, (8.18) must be solved for every $k_1, \dots, k_n$. Sometimes, however, one is interested only in the distribution function for certain k_i. If, for instance, $x_1 = \varphi$ is the angle variable and one is looking only for periodic solutions in the angle variable, k_1 must be an integer number and the integral must be replaced by a sum over these integer numbers. Furthermore, if one is interested only in some expectation values of the form $\langle \exp \mathrm{i} m\varphi(t)\rangle$, only the solution of (8.18) with $k_1 = -m$ needs to be calculated. An application will be given in Sect. 12.3.

If the laser Fokker-Planck equation for the complex amplitude is transformed to intensity and angle variable, the drift and diffusion coefficients do not depend on the angle variable and therefore the reduction above may be used. For a class of Fokker-Planck equations with two variables where the drift and diffusion coefficients do not depend on one variable and where the solutions are given in terms of hypergeometric functions, see [8.5, 6] and App. A6.

8.2.1 Time Integrals of Markovian Variables

For a stochastic variable $\xi(t)$ one may not be interested in the properties of the stochastic variable itself but rather in the properties of the time integral $\int_{t_0}^{t} \xi(t')\,\mathrm{d}t'$ of the stochastic variable $\xi(t)$. Usually, the properties of the time-integrated variable can be easier measured than the properties of the variable $\xi(t)$. For instance, if $\xi(t) = v(t)$ is the velocity of a Brownian particle, one usually measures the position at different times $x(t) - x(t_0) = \int_{t_0}^{t} v(t')\,\mathrm{d}t'$ and not the velocity itself. Actually every measuring process needs a certain time. Instead of measuring the statistical properties of the velocity, one measures the properties of the time-integrated quantity

$$\bar{v}(t_0, \tau) = \frac{x(t_0+\tau) - x(t_0)}{\tau} = \frac{1}{\tau}\int_{t_0}^{t_0+\tau} v(t')\,\mathrm{d}t' \, . \tag{8.19}$$

Only if τ can be made very small (8.19) has the properties of the velocity measured, otherwise the properties of $\bar{v}$ are different from those of v.

For N stochastic variables $\{\xi(t)\}$ one may be interested in the stochastic properties of

$$I(t) = \int_{t_0}^{t} f(\{\xi(t')\}, t')\,\mathrm{d}t' \, . \tag{8.20}$$

To find the stochastic properties of $I(t)$ we first assume that $\{\xi(t)\}$ are Markovian variables which obey the Langevin equations (3.110, 111). Because of (8.20), the time derivative and initial condition of $I(t)$ are given by

$$\dot{I}(t) = f(\{\xi(t)\}, t)\,, \tag{8.21}$$

$$I(t_0) = 0\,. \tag{8.22}$$

We may add (8.21) to the Langevin equations (3.110). These combined equations are then Langevin equations for the $N+1$ Markovian variables $\xi_1(t), \ldots, \xi_N(t), I(t)$.

The corresponding Fokker-Planck equation for the distribution function $W(I, \{x\}, t)$ then takes the form

$$\partial W/\partial t = \boldsymbol{L}\, W\,, \tag{8.23}$$

where the operator $\boldsymbol{L}$ is given by

$$\boldsymbol{L} = \boldsymbol{L}_{\mathrm{FP}}(\{x\}) - f(\{x\}, t)\, \partial/\partial I\,. \tag{8.24}$$

Here $\boldsymbol{L}_{\mathrm{FP}}$ is the Fokker-Planck operator of the Fokker-Planck equation which corresponds to the Langevin equations (3.110). Because of (8.22), W has the initial condition

$$W(I, \{x\}, t_0) = \delta(I)\, W_{\mathrm{i}}(\{x\})\,, \tag{8.25}$$

where $W_{\mathrm{i}}(\{x\})$ is the initial distribution of the variables $\{x\}$. In the stationary state W_{i} is the stationary solution $W_{\mathrm{st}}(\{x\})$ of the Fokker-Planck equation, i.e., $\boldsymbol{L}_{\mathrm{FP}}\, W_{\mathrm{st}} = 0$.

The variable I does not appear in any drift or diffusion coefficient. Therefore we can apply the method discussed in Sect. 8.2. (Because no second derivative with respect to I occurs, the method simplifies somewhat.) Introducing the Fourier transform with respect to I

$$W(I, \{x\}, t) = (2\pi)^{-1} \int \mathrm{e}^{\mathrm{i}kI}\, \hat{W}(k, \{x\}, t)\, \mathrm{d}k\,, \tag{8.26}$$

this Fourier transform $\hat{W}$ now obeys the equation

$$\partial \hat{W}/\partial t = \hat{\boldsymbol{L}}\, \hat{W} \tag{8.27}$$

with

$$\hat{\boldsymbol{L}} = \boldsymbol{L}_{\mathrm{FP}}(\{x\}) - \mathrm{i}k f(\{x\}, t)\,. \tag{8.28}$$

The initial condition (8.25) for W transforms to

$$\hat{W}(k, \{x\}, 0) = W_{\mathrm{i}}(\{x\})\,. \tag{8.29}$$

If we are interested in expectation values of the form

$$\begin{aligned}\langle e^{i\alpha I(t)}\rangle &= \int\int e^{i\alpha I} W(I,\{x\},t)\,d^N x\,dI \\ &= (2\pi)^{-1}\int\int\int e^{i(\alpha+k)I}\hat{W}(k,\{x\},t)\,d^N x\,dI\,dk \\ &= \int \hat{W}(-\alpha,\{x\},t)\,d^N x\,,\end{aligned} \tag{8.30}$$

we need to calculate $\hat{W}$ only for $k = -\alpha$. However, if we are interested in the distribution function of the variable I

$$\begin{aligned}W(I,t) &= \int W(I,\{x\},t)\,d^N x \\ &= (2\pi)^{-1}\int\int e^{ikI}\hat{W}(k,\{x\},t)\,d^N x\,dk\end{aligned} \tag{8.31}$$

we must find the solution of (8.27–29) for every k. The result (8.30) was derived by *Lax* [1.11c]. He wrote it in the form

$$\langle e^{i\alpha I(t)}\rangle = \int M_0(\{x_0\},t,t_0)\,W_i(\{x_0\})\,d^N x_0\,, \tag{8.32}$$

where M_0 is given by

$$M_0(\{x_0\},t,t_0) = \int \hat{P}(-\alpha,\{x\},t\,|\,\{x_0\},t_0)\,d^N x\,. \tag{8.33}$$

Here, $\hat{P}$ is the Green's function of (8.27) with $k = -\alpha$ in (8.28), i.e., it is the solution of (8.27) with the initial condition

$$\hat{P}(\alpha,\{x\},t_0\,|\,\{x_0\},t_0) = \delta(\{x\}-\{x_0\})\,. \tag{8.34}$$

Because the solution $\hat{W}$ can be expressed by

$$\hat{W}(\alpha,\{x\},t) = \int \hat{P}(-\alpha,\{x\},t\,|\,\{x_0\},t_0)\,W_i(\{x_0\})\,d^N x_0\,, \tag{8.35}$$

the equivalence of (8.32–33) to (8.30) is easily seen.

The quantity M_0 may be considered as a kind of normalization. Because of the additional term on the right-hand side of (8.28) the usual normalization is no longer conserved. We have, for instance,

$$\frac{dM_0}{dt} = \int \hat{L}\hat{P}\,d^N x = i\alpha\int f(\{x\},t)\hat{P}\,d^N x\,, \tag{8.36}$$

which reduces for $t = t_0$ to

$$\left.\frac{dM_0}{dt}\right|_{t=t_0} = i\alpha f(\{x_0\},t_0)\,. \tag{8.37}$$

To solve (8.27, 28) we may apply some of the methods of Chaps. 5 and 6, respectively, for solving the Fokker-Planck equation. Obviously, methods requiring that L_{FP} be brought to an Hermitian form do not work for the non-Hermitian operator $\hat{L}$, but other methods like eigenfunction expansions, numerical integration and matrix continued-fraction methods may be applied to solve (8.27, 28).

Example

As a rather trivial example we ask for the stationary distribution function $W(\bar{v})$ for the time-averaged velocity (8.19) for a Brownian particle obeying the Langevin equation (3.1). [It is trivial because the integrated velocity is the position coordinate $x(t)$ and thus the problem may be reduced to solving the Kramers equation for free Brownian motion, i.e., (4.112) with $f'(x) = 0$.] In this case, $f = v/\tau$, and in order not to mix up the wave vector k with the Boltzmann constant, the latter is denoted by k_B. Then $\hat{L}$ reads

$$\hat{L} = \gamma \frac{\partial}{\partial v}\left(v + \frac{k_B T}{m}\frac{\partial}{\partial v}\right) - \frac{ik}{\tau} v . \tag{8.38}$$

The solution of (8.27) with $\hat{L}$ given by (8.38) and with the initial distribution

$$W_{st}(v) = W_{\text{Maxwell}} = \sqrt{\frac{m}{2\pi k_B T}} \exp\left(-\frac{mv^2}{2k_B T}\right) \tag{8.39}$$

is easily obtained by making a further Fourier transform with respect to v. As may be checked by insertion, $\hat{W}(k, v, t)$ is given by

$$\begin{aligned}\hat{W}(k,v,t) = \frac{1}{2\pi}\int \exp\Bigg\{ &-\frac{k_B T}{2m} q^2 - \left[\frac{k}{\gamma\tau}\frac{k_B T}{m}(1 - e^{-\gamma(t-t_0)}) - iv\right] q \\ &- \frac{k^2}{(\gamma\tau)^2}\frac{k_B T}{m}[\gamma(t-t_0) - 1 + e^{-\gamma(t-t_0)}]\Bigg\} dq .\end{aligned} \tag{8.40}$$

Here, t is equal to $t_0 + \tau$ (8.19).

The integration of $\hat{W}$ over v leads to

$$\int \hat{W}(k, v, t_0 + \tau)\, dv = \exp[-k^2 g(\gamma\tau) k_B T/(2m)] \tag{8.41}$$

with

$$g(\gamma\tau) = 2(\gamma\tau - 1 + e^{-\gamma\tau})/(\gamma\tau)^2 . \tag{8.42}$$

Thus we finally obtain the following distribution function for $\bar{v}$ in the stationary state

$$\begin{aligned}W(\bar{v}) &= \frac{1}{2\pi}\int \exp[ik\bar{v} - k^2 g(\gamma\tau) k_B T/(2m)]\, dk \\ &= \sqrt{\frac{m}{2\pi k_B T g(\gamma\tau)}} \exp\left(-\frac{m\bar{v}^2}{2k_B T g(\gamma\tau)}\right).\end{aligned} \tag{8.43}$$

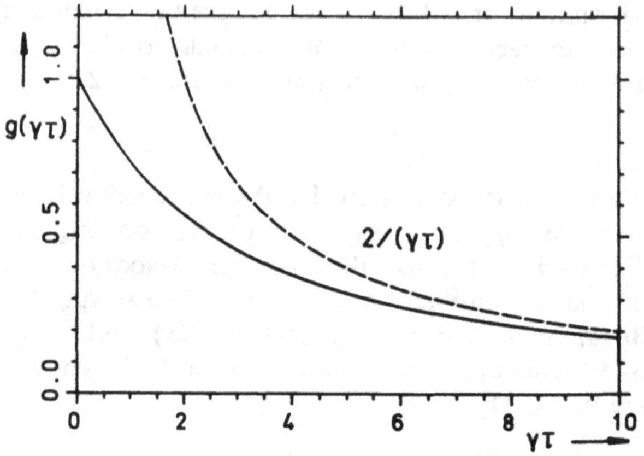

Fig. 8.2. Expression (8.42) as a function of $\gamma\tau$ (*full line*) and the approximation $2/(\gamma\tau)$ for $\gamma\tau \gg 1$ (*broken line*)

The function $g(\gamma\tau)$ is shown in Fig. 8.2. For $\gamma\tau \ll 1$ we thus recover the Maxwell distribution (8.39), whereas for $\gamma\tau \gg 1$ we obtain a very sharp distribution around $\bar{v} = 0$ with width $\sim(\gamma\tau)^{-1/2}$.

8.3 Adiabatic Elimination of Fast Variables

For simplicity we restrict ourselves to two variables x and y. The generalization to more variables is easy in principle though the explicit calculation may become quite complicated. If the y variable decays much faster than the x variable, the y variable is called a fast variable and the x variable a slow variable. We write the Langevin equations for the two variables in the form

$$\dot{x} = h_x(x,y) + g_x(x,y)\Gamma_x, \tag{8.44}$$

$$\dot{y} = \gamma h_y(x,y) + \sqrt{\gamma}\, g_y(x,y)\Gamma_y, \tag{8.45}$$

where Γ_x and Γ_y are Langevin forces with the correlation functions

$$\begin{aligned} &\langle \Gamma_x(t)\,\Gamma_x(t')\rangle = \langle \Gamma_y(t)\,\Gamma_y(t')\rangle = 2\delta(t-t'),\\ &\langle \Gamma_x(t)\,\Gamma_y(t')\rangle = 0\,. \end{aligned} \tag{8.46}$$

In (8.45) we introduced a parameter γ to indicate the different time scales. We are looking for solutions of (8.44, 45) in the limit $\gamma \to \infty$, consistent with the assumption that y will decay very rapidly to an equilibrium value y_{eq}. We assume that (8.45) without noise has a stable equilibrium value y_{eq}, which is determined by $h_y(x, y_{\text{eq}}) = 0$. The equilibrium value y_{eq} generally depends on the slow variable x. If the noise term in (8.45) is taken into account, we then get an equilibrium distribution for the y variable, which is independent of γ because of the factor $\sqrt{\gamma}$.

If one is interested only in a time scale large compared to the decay time ($\sim \gamma^{-1}$) of the fast variable y, the process described by (8.44, 45) is then mainly described by the motion of the slow variable x. Hence we may say that the slow variable slaves the fast variable [1.14]. The slow variable is the important or relevant variable, since the fast variable becomes irrelevant for the above time scale because it can be expressed by the slow variable.

We now want to derive an equation of motion for the distribution function of the relevant variable x. Such an elimination method was derived by *Haken* [Ref. 1.14, p. 202 ff.]. In our procedure we follow closely the work of *Kaneko* [8.7]. Our starting point is the Fokker-Planck equation corresponding to (8.44, 45). This Fokker-Planck equation for the distribution function $W(x,y,t)$ is written in the form

$$\partial W/\partial t = [L_x(x,y) + \gamma L_y(y,x)]\, W\,, \tag{8.47}$$

where L_x and L_y are given by

$$L_x(x,y) = -\frac{\partial}{\partial x} D_x(x,y) + \frac{\partial^2}{\partial x^2} D_{xx}(x,y)\,, \tag{8.48}$$

$$L_y(y,x) = -\frac{\partial}{\partial y} D_y(x,y) + \frac{\partial^2}{\partial y^2} D_{yy}(x,y)\,, \tag{8.49}$$

and where the drift and diffusion coefficients read (3.95)

$$\begin{aligned} D_x &= h_x(x,y) + g_x(x,y)\,\partial g_x(x,y)/\partial x\,,\\ D_y &= h_y(x,y) + g_y(x,y)\,\partial g_y(x,y)/\partial y\,. \end{aligned} \tag{8.50}$$

$$D_{xx} = g_x^2\,, \qquad D_{yy} = g_y^2\,. \tag{8.51}$$

We first look for eigenfunctions of the operator $L_y(y,x)$. Here the variable x appears only as a parameter. We assume that for every parameter x a stationary solution and discrete eigenvalues and eigenfunctions exist ($n \geqq 0$),

$$L_y(y,x)\,\varphi_n(y,x) = -\lambda_n(x)\,\varphi_n(y,x)\,. \tag{8.52}$$

The eigenvalues as well as the eigenfunctions generally depend on the parameter x. For $n = 0$ we have the stationary solution

$$\varphi_0(y,x) = W_{st}(y,x)\,; \qquad \lambda_0 = 0\,. \tag{8.53}$$

The eigenfunctions φ_n and the eigenfunctions φ_n^+ of the adjoint operator L_y^+, i.e.,

$$L_y^+(y,x)\,\varphi_n^+(y,x) = -\lambda_n(x)\,\varphi_n^+(y,x)\,, \tag{8.54}$$

may be expressed in terms of the eigenfunctions ψ_n of the corresponding Sturm-Liouville problem according to (Sect. 5.4)

$$\begin{aligned} \varphi_n(y,x) &= \sqrt{\varphi_0(y,x)}\, \psi_n(y,x)\,, \\ \varphi_n^+(y,x) &= \psi_n(y,x)/\sqrt{\varphi_0(y,x)} = \varphi_n(y,x)/\varphi_0(y,x)\,. \end{aligned} \tag{8.55}$$

Obviously then

$$\varphi_0^+(y,x) = 1\,. \tag{8.56}$$

The orthonormality and completeness relations read

$$\int \varphi_n^+(y,x)\, \varphi_m(y,x)\, dy = \int \psi_n(y,x)\, \psi_m(y,x)\, dy = \delta_{nm}\,, \tag{8.57}$$

$$\sum_{n=0}^{\infty} \varphi_n^+(y,x)\, \varphi_n(y',x) = \sum_{n=0}^{\infty} \psi_n(y,x)\, \psi_n(y',x) = \delta(y-y')\,. \tag{8.58}$$

The idea of solving (8.52) with x as a parameter is taken from the Born-Oppenheimer approximation in quantum mechanics. If applied to the H_2 molecule, the Born-Oppenheimer approximation consists in first solving the Schrödinger equation for the two electrons (fast variables), thereby keeping the variables for the two H nuclei (slow variables) fixed. After making a certain ansatz for the wave function, one then solves the Schrödinger equation for the two H nuclei [8.8].

We now expand the distribution function $W(x,y,t)$ into the complete set φ_n of the operator $\boldsymbol{L}_y$:

$$W(x,y,t) = \sum_{m=0}^{\infty} c_m(x,t)\, \varphi_m(y,x)\,. \tag{8.59}$$

We insert this expansion into the Fokker-Planck equation (8.47), multiply the resulting equation by φ_n^+ and then integrate over the y variable, thus obtaining

$$\left[\frac{\partial}{\partial t} + \gamma\lambda_n(x)\right] c_n = \sum_{m=0}^{\infty} \boldsymbol{L}_{n,m} c_m \tag{8.60}$$

with

$$\boldsymbol{L}_{n,m} = \int \varphi_n^+(y,x)\, \boldsymbol{L}_x(x,y)\, \varphi_m(y,x)\, dy\,. \tag{8.61}$$

The $\boldsymbol{L}_{n,m}$ are operators with respect to x. In contrast to the ansatz one usually makes in the Born-Oppenheimer approximation, the infinite set of equations (8.60) is still exact. For large γ, however, we may use the following approximation procedure. Because we are interested only in the time scale large compared to the decay constant $(\gamma\lambda_1)^{-1}$ of the fast variable, in (8.60) we neglect the time derivative in the equation with $n \geqq 1$. We thus have

$$\dot{c}_0 = \sum_{m=0}^{\infty} L_{0,m} c_m \,, \tag{8.62}$$

$$\begin{aligned} c_n &= [\gamma \lambda_n(x)]^{-1} \sum_{m=0}^{\infty} L_{n,m} c_m \\ &= [\gamma \lambda_n(x)]^{-1} L_{n,0} c_0 + O(\gamma^{-2}) \,, \quad n \geqq 1 \,. \end{aligned} \tag{8.63}$$

Because the c_n for $n \geqq 1$ are of the order γ^{-1} and c_0 is of the order γ^{-0}, we have taken into account only the term with $m = 0$ for the last line of (8.63). By inserting (8.63) into (8.62) we get

$$\dot{c}_0 = L_0 c_0 \,, \tag{8.64}$$

$$L_0 = L_{0,0} + \gamma^{-1} \sum_{n=1}^{\infty} L_{0,n} [\lambda_n(x)]^{-1} L_{n,0} + O(\gamma^{-2}) \,. \tag{8.65}$$

I should like to mention that (8.65) has the same form as the expression for energy corrections in second-order perturbation theory in quantum mechanics.

It follows from (8.56, 57, 59) that c_0 is the integral of $W(x,y,t)$ over the y variable

$$c_0(x,t) = \int W(x,y,t)\,dy = W(x,t) \,. \tag{8.66}$$

Thus (8.64) is the desired equation of motion for the distribution function of the relevant variable x. The operator $L_{0,0}$ has the form

$$L_{0,0} = -\frac{\partial}{\partial x} \bar{D}_x(x) + \frac{\partial^2}{\partial x^2} \bar{D}_{xx}(x) \,, \tag{8.67}$$

with drift and diffusion coefficients defined by

$$\begin{aligned} \bar{D}_x(x) &= \int D_x(x,y)\, \varphi_0(y,x)\,dy \,, \\ \bar{D}_{xx}(x) &= \int D_{xx}(x,y)\, \varphi_0(y,x)\,dy \,, \end{aligned} \tag{8.68}$$

i.e., one only has to take the average with respect to the stationary distribution of the y variable. The term proportional to γ^{-1} is more complicated. It is easy to see that it starts with a $\partial/\partial x$ term [because of (8.56, 61)], but in general it contains derivatives with respect to x of the order higher than two. If, however, $D_{xx}(x,y)$ is independent of y, one may show [8.7] that then only derivatives with respect to x of the order two occur, i.e., L_0 can be brought to the form of a Fokker-Planck operator, up to terms of the order γ^{-2}. To solve (8.64) explicitly, the operator L_0 should be given analytically. This is the case only if the eigenvalues and eigenfunctions of $L_y(y,x)$ are known analytically and if the matrix elements occurring in (8.65) can be calculated analytically. An application of this procedure will be given below.

8.3.1 Linear Process with Respect to the Fast Variable

We now apply the adiabatic elimination method to the case where the functions $h_{\substack{x\\y}}$ depend linearly on y and where $g_x = \sqrt{D_{xx}}$, $g_y = \sqrt{D_{yy}}$ are constants. Then the Langevin equations (8.44, 45) take the form

$$\dot{x} = h_0(x) + h_1(x)y + \sqrt{D_{xx}}\,\Gamma_x, \tag{8.69}$$

$$\dot{y} = -\gamma a(x)[y - \alpha(x)] + \sqrt{\gamma}\sqrt{D_{yy}}\,\Gamma_y, \tag{8.70}$$

where the correlation function of the Langevin forces is still given by (8.46).

A usual adiabatic elimination procedure consists in neglecting the time derivative in (8.70). Inserting

$$y = \alpha(x) + \frac{\sqrt{D_{yy}}}{\sqrt{\gamma a(x)}}\Gamma_y \tag{8.71}$$

into (8.69) yields

$$\dot{x} = h_0(x) + h_1(x)\,\alpha(x) + h_1(x)\frac{\sqrt{D_{yy}}}{\sqrt{\gamma a(x)}}\Gamma_y + \sqrt{D_{xx}}\,\Gamma_x. \tag{8.72}$$

According to (3.95) this leads to the following Fokker-Planck equation for $W = W(x,t)$

$$\dot{W} = L\,W, \tag{8.73}$$

$$L = -\frac{\partial}{\partial x}\left[h_0 + h_1\alpha + \frac{D_{yy}}{\gamma}\left(\frac{h_1}{a}\right)'\frac{h_1}{a}\right] + \frac{\partial^2}{\partial x^2}\left(D_{xx} + \frac{D_{yy}}{\gamma}\frac{h_1^2}{a^2}\right). \tag{8.74}$$

To apply the elimination procedure discussed previously, we notice that the operator (8.49) now reduces to

$$L_y(y,x) = \frac{\partial}{\partial y}a(x)[y - \alpha(x)] + D_{yy}\frac{\partial^2}{\partial y^2}. \tag{8.75}$$

Introducing the shifted variable

$$\xi = \sqrt{a(x)/(2D_{yy})}\,[y - \alpha(x)] \tag{8.76}$$

and the boson operators (5.62), operator (8.75) may be cast into the form

$$L_y(y,x) = -\mathrm{e}^{-\xi^2/2}a(x)\,b^+b\,\mathrm{e}^{\xi^2/2}. \tag{8.75a}$$

The eigenvalues and eigenfunctions φ_n and φ_n^+ then read (compare Sect. 5.5.1)

$$\lambda_n(x) = n a(x) ,$$

$$\varphi_n(y,x) = \sqrt{\frac{a(x)}{2\pi D_{yy}}}\ \frac{1}{2^n n!} \mathrm{H}_n(\xi)\,\mathrm{e}^{-\xi^2} , \tag{8.77}$$

$$\varphi_n^+(y,x) = \mathrm{H}_n(\xi) .$$

The operator $L_x(x,y)$ now specializes to

$$L_x(x,y) = -\frac{\partial}{\partial x}[h_0(x) + h_1(x)y] + D_{xx}\frac{\partial^2}{\partial x^2} . \tag{8.78}$$

We first calculate $L_{0,n}$. We easily find using (8.56, 57, 61) and $\xi = \frac{1}{2}\mathrm{H}_1(\xi) = \frac{1}{2}\varphi_1^+$

$$L_{0,n} = -\frac{\partial}{\partial x}\left[h_0 \int \varphi_n \mathrm{d}y + h_1 \int\left(\sqrt{\frac{2D_{yy}}{a}}\,\xi + \alpha\right)\varphi_n \mathrm{d}y\right] + D_{xx}\frac{\partial^2}{\partial x^2}\int \varphi_n \mathrm{d}y$$

$$= \left[-\frac{\partial}{\partial x}(h_0 + h_1\alpha) + D_{xx}\frac{\partial^2}{\partial x^2}\right]\delta_{0,n} - \frac{\partial}{\partial x}\sqrt{\frac{D_{yy}}{2a}}\,h_1\,\delta_{1,n} . \tag{8.79}$$

Because of (8.79), we need to calculate only

$$L_{1,0} = \int 2\xi\left\{-\frac{\partial}{\partial x}\left[h_0 + h_1\left(\sqrt{\frac{2D_{yy}}{a}}\,\xi + \alpha\right)\right] + D_{xx}\frac{\partial^2}{\partial x^2}\right\}\varphi_0 \mathrm{d}y$$

in order to write down (8.65). Using $2\xi^2 = \frac{1}{2}\mathrm{H}_2 + \mathrm{H}_0$ and $2\xi' = (a'/a)\xi - \sqrt{2a/D_{yy}}\,\alpha'$, $2\xi'' = (a''/a)\xi - (a'^2/(2a^2))\xi - 2a'\alpha'/\sqrt{D_{yy}2a} - \sqrt{2a/D_{yy}}\,\alpha''$ (a prime denotes a derivative with respect to x) we obtain

$$L_{1,0} = \frac{1}{2}\,\frac{a'}{a}h_1\sqrt{\frac{2D_{yy}}{a}} - \frac{\partial}{\partial x}\sqrt{\frac{2D_{yy}}{a}}\,h_1 - \sqrt{\frac{2a}{D_{yy}}}\,\alpha' h_0$$

$$-\sqrt{\frac{2a}{D_{yy}}}\,\alpha' h_1 \alpha + 2D_{xx}\sqrt{\frac{2a}{D_{yy}}}\,\alpha'\frac{\partial}{\partial x} + D_{xx}\sqrt{\frac{2a}{D_{yy}}}\,\alpha'' \tag{8.80}$$

and thus finally arrive at [8.7]

$$L_0 = -\frac{\partial}{\partial x}(h_0 + h_1\alpha) + \frac{\partial^2}{\partial x^2}D_{xx}$$

$$+\frac{1}{\gamma}\,\frac{\partial}{\partial x}\left(D_{yy}\frac{h_1}{a}\,\frac{\partial}{\partial x}\,\frac{h_1}{a} + \frac{h_0 h_1 \alpha'}{a} + \frac{h_1^2 \alpha\alpha'}{a}\right.$$

$$\left. - \frac{2D_{xx}h_1\alpha'}{a}\,\frac{\partial}{\partial x} - D_{xx}\frac{h_1}{a}\alpha''\right) . \tag{8.81}$$

If α is independent of x, (8.81) agrees with (8.74). Otherwise the elimination procedure with $\dot{y}=0$ in the Langevin equation does not lead to the same result in the terms of the order γ^{-1}. If, however, $\gamma\alpha$ is of the order $\gamma^0[\alpha \sim O(\gamma^{-1})]$ the adiabatic elimination procedure in the Langevin equation is correct in the order γ^{-1}.

For the special case

$$\begin{aligned} &y=v\,; \quad h_0=0\,; \quad h_1=1\,; \quad a(x)=1\,; \\ &\gamma\alpha=-\tilde{f}'/m=-f'\,; \quad D_{xx}=0\,; \quad D_{yy}=kT/m \end{aligned} \tag{8.82}$$

system (8.69) is identical to (3.130) describing Brownian motion of particles in the potential $\tilde{f}(x)$. Up to terms of the order γ^{-1}, the operator L_0 is then identical to the operator (5.2) of the Smoluchowski equation with $D^{(1)}$, $D^{(2)}$ given by (5.3, 4). In Chap. 10 we treat the Fokker-Planck or Kramers equation for Brownian motion in a potential. Especially in Sect. 10.4 we derive the corrections of L_0 up to terms of the order γ^{-5}. A solution of this two-variable Fokker-Planck equation in terms of matrix continued fractions is derived in Sect. 10.3.

If the operators $L_{n,m}$ (8.61) are such that

$$L_{n,m}=0 \quad \text{for} \quad |m-n| \geqq L\,, \tag{8.83}$$

where L is a finite number, system (8.60) may then – after a further expansion of the expansion coefficients $c_n(x,t)$ into a complete set $\varphi^p(x)$ is made – be cast into a tridiagonal vector recurrence relation which may also be solved by the matrix continued-fraction method, as discussed in Chap. 9. For problems where L_y is of the form (8.75) and where $h_x(x,y)$ is a polynomial in y, (8.83) is valid.

8.3.2 Connection to the Nakajima-Zwanzig Projector Formalism

Quite often the elimination of one or more variables is done with the *Nakajima-Zwanzig* projector formalism [1.26, 27, 8.9]. This formalism can of course also be applied to the present elimination problem, whereby a projection operator $\boldsymbol{P}$ is defined by

$$\boldsymbol{P}\ldots=(\textstyle\int\ldots dy)\,\varphi_0=(\textstyle\int\varphi_0^+\ldots dy)\,\varphi_0\,. \tag{8.84}$$

Using the normalization (8.57), it is easy to see that the relation $\boldsymbol{P}^2=\boldsymbol{P}$ for a projection operator holds. Because the system φ_n, φ_n^+ is complete, the operator $\boldsymbol{1}-\boldsymbol{P}$ may be cast in the form

$$(\boldsymbol{1}-\boldsymbol{P})\ldots=\sum_{n=1}^{\infty}(\textstyle\int\varphi_n^+\ldots dy)\,\varphi_n\,. \tag{8.85}$$

In the projector formalism the equation of motion $\dot{W}=\boldsymbol{L}W$ is split up into two coupled equations for $\boldsymbol{P}W$ and $(\boldsymbol{1}-\boldsymbol{P})W$, i.e., into

$$P\dot{W} = PLPW + PL(1-P)W\,, \tag{8.86}$$

$$(1-P)\dot{W} = (1-P)LPW + (1-P)L(1-P)W\,. \tag{8.87}$$

If we use the expansion (8.59), then (8.86) is identical to (8.60) with $n = 0$, whereas (8.87) is identical to (8.60) with $n \geqq 1$. The usual Markov approximation to the formal solution of (8.87) consists in neglecting the time derivative, as in (8.63).

9. Solutions of Tridiagonal Recurrence Relations, Application to Ordinary and Partial Differential Equations

As shown in the next chapter, the Fokker-Planck equation describing the Brownian motion in arbitrary potentials, i.e., the Kramers equation, can be cast into a tridiagonal vector recurrence relation by suitable expansion of the distribution function. In this chapter we shall investigate the solutions of tridiagonal vector recurrence relations. As it turns out, the Laplace transform of these solutions as well as the eigenvalues and eigenfunctions can be obtained in terms of matrix continued fractions. Therefore, the corresponding solutions of the Kramers equation can also be given in terms of matrix continued fractions. This method has the advantage that no detailed balance condition is needed for its application. This matrix continued-fraction method is especially suitable for numerical calculations and for some problems it seems to be the most accurate and fastest method, as will be discussed in other chapters. Besides its advantage for numerical purposes, the matrix continued-fraction solutions are also very useful for analytical evaluations. By a proper Taylor series expansion of the matrix continued fractions we obtain, for instance, in Sect. 10.4 the high-friction limit solutions of the Kramers equation.

Though in this book tridiagonal recurrence relations are applied mainly to the Fokker-Planck equation, it should be mentioned that other linear ordinary as well as partial differential equations and difference equations occurring in physics and other fields may also be solved by this method. As will be shown, the one-dimensional Schrödinger equation with an anharmonic potential also leads to a tridiagonal vector recurrence relation. Also master equations lead to tridiagonal scalar or vector recurrence relations. In Sect. 9.1 the forms and the applications of tridiagonal scalar, and vector recurrence relations are given and discussed. It is now appropriate to give a short list of some references.

Ordinary Continued Fractions

The standard text books on ordinary continued fractions are those by *Perron* [9.1], *Wall* [9.2], and *Jones* and *Thron* [9.3]; for a connection of ordinary continued fractions with *Padé* expansions, consult [9.4]. Many functions appearing in mathematical physics can be expressed by continued fractions [9.1 – 3]. Some applications of ordinary continued fractions to problems in physics may be found in [9.5], further applications are discussed in connection with: (i) determination of eigenvalues of the Schrödinger equation for an anharmonic potential [9.6]; (ii) pocket-calculator determination of eigenvalues of the Schrödinger equation [9.7]; (iii) master equations [9.8]; (iv) differential equations with harmonic coefficients, especially the Mathieu equation [Ref. 5.1, Chaps. 7–5] and the Bloch equations [9.9–11]. In some applications [9.5] continued fractions are used to improve the convergence of poorly or only asymptotic convergent Taylor series, similarly to how one uses Padé approximants for improving poorly convergent Taylor series.

Matrix Continued Fractions

An application of matrix continued fractions to the Schrödinger equation of an anharmonic oscillator was given in [9.12]. (There, however, the eigenvalues are not determined by calculating numerically matrix continued fractions.) Also Padé approximants were generalized to matrices (see the short chapter on matrix Padé approximants in [9.4]). *Allegrini* et al. [9.13] used 2×2 matrix continued fractions for solving the Bloch equations in connection with saturation effects in spin – 1/2 radio frequency spectroscopy. *Vollmer* and I have used matrix continued fractions for solving the stationary Brownian motion problem in periodic potentials with an external force [9.14, 15], as well as for the time-dependent Brownian motion problem in periodic potentials [9.16, 17], and we applied them to the laser Fokker-Planck equation [9.18], the general solution of the Kramers equation [9.19], and the Boltzmann equation with a BGK collision operator [9.16]. (In [9.18, 19] the general method presented in Sect. 9.3 for solving vector recurrence relations was given.) It should be mentioned that an iteration procedure similar to that for determining the matrix continued fractions has been applied by *Dieterich* et al. [9.20] as a trick to solve coupled differential equations for the Brownian motion problem in periodic potentials. For other applications of matrix continued fractions to the solution of Fokker-Planck equations, see [6.19, 20, 9.21]. The stochastic optical Bloch equations, where the Gaussian noise has a finite coherence time, was also treated by matrix continued-fraction methods, [9.22, 23] and App. 1. Whereas a lot of convergence theorems exist for ordinary continued fractions [9.1–3], we only know the method in [9.24] where a mathematical proof of the convergence of matrix continued fractions used in [9.14–19, 22, 23] was given.

9.1 Applications and Forms of Tridiagonal Recurrence Relations

9.1.1 Scalar Recurrence Relation

A scalar tridiagonal recurrence relation has the form

$$Q_n^- c_{n-1} + Q_n c_n + Q_n^+ c_{n+1} = 0\,, \tag{9.1}$$

where $Q_n^\pm$, Q_n are some given constants generally depending on n. We are looking for solutions of the unknown coefficients c_n, which usually depend on time in the applications. Instead of (9.1) we then consider the time-dependent recurrence relation

$$\dot{c}_n = Q_n^- c_{n-1} + Q_n c_n + Q_n^+ c_{n+1}, \tag{9.2}$$

which is a system of tridiagonal coupled differential equations. If the index n is not restricted ($-\infty < n < \infty$), we speak of a two-sided recurrence relation; if the coefficients start with $n = 0$ ($0 \leqq n < \infty, Q_0^- = 0$), we speak of a one-sided recurrence relation. Sometimes the index n may be restricted from below and above ($0 \leqq n \leqq N$) and we call (9.2) a finite tridiagonal recurrence relation.

Master Equation with Nearest-Neighbor Coupling

A master equation (1.34) with nearest-neighbor coupling is given by

$$\begin{aligned} \dot{p}_n = {} & w(n-1 \to n) p_{n-1} + w(n+1 \to n) p_{n+1} \\ & - [w(n \to n+1) + w(n \to n-1)] p_n . \end{aligned} \tag{9.3}$$

In (9.3) p_n are the probabilities for the nth state (e.g., number of photons in a cavity mode) and $w(n \to n+1)$ is the transition probability from state n to $n+1$. If $p_n = 0$ for $n<0$, (9.3) has the form of the one-sided tridiagonal recurrence relation (9.2) with

$$\begin{aligned} Q_n^\pm &= w(n \pm 1 \to n) \\ Q_n &= -[w(n \to n+1) + w(n \to n-1)] \\ &= -(Q_{n+1}^- + Q_{n-1}^+) . \end{aligned} \tag{9.4}$$

As seen from the last expression, the coefficients $Q_n^\pm, Q_n$ are not independent for a master equation. Generally, however, the constants Q_n, $Q_n^\pm$ are assumed to be independent.

Schrödinger Equation for a Discrete Variable

If instead of the continuous variable x the discrete variable $n = x/\Delta$ is used, the Schrödinger equation for the wave function $\psi(x)$

$$\mathrm{i}\hbar\dot{\psi} = -(\hbar^2/2m)\psi'' + V(x)\psi \tag{9.5}$$

is transformed into the tridiagonal recurrence relation (9.2) by setting

$$\begin{aligned} &\psi(n\Delta) = c_n, \quad \psi''(x)\big|_{x=n\Delta} \approx (c_{n+1} + c_{n-1} - 2c_n)/\Delta^2, \\ &Q_n^\pm = \mathrm{i}\hbar/(2m\Delta^2), \quad Q_n = -\mathrm{i}\hbar/(m\Delta^2) - \mathrm{i}V(n\Delta)/\hbar . \end{aligned} \tag{9.6}$$

The same procedure can also be applied to other differential equations with a derivative up to second order in x, e.g., a Fokker-Planck equation with one variable.

Moment Equation for Laser Intensity

As shown in Sect. 12.1.3, the moments of intensity for a laser near threshold are given by, see (12.43),

$$(\mathrm{d}/\mathrm{d}t)\langle I^n\rangle = -2n\langle I^{n+1}\rangle + 2na\langle I^n\rangle + 4n^2\langle I^{n-1}\rangle\,. \tag{9.7}$$

Thus, by setting

$$\begin{aligned} &c_n = \langle I^n\rangle\,, \quad (n \geqq 0)\,,\\ &Q_n^- = 4n^2\,, \quad Q_n = 2na\,, \quad Q_n^+ = -2n\,, \end{aligned} \tag{9.8}$$

we again recover the tridiagonal recurrence relation (9.2).

Higher-Order Time Derivatives

In the following we treat equations of the form (9.2) with a first-order time derivative. With slight modifications the methods of this chapter will, however, also hold for higher-order time derivatives. For instance, the equation of motion for a linear chain

$$m_n\ddot{q}_n = \lambda_{n+1}(q_{n+1}-q_n) - \lambda_n(q_n - q_{n-1}) \tag{9.9}$$

can be treated by this method. The spring constants λ_n and the masses m_n of the particles may depend on the lattice site n.

9.1.2 Vector Recurrence Relation

A tridiagonal vector recurrence relation with a first-order time derivative is of the form

$$\dot{\boldsymbol{c}}_n = \boldsymbol{Q}_n^- \boldsymbol{c}_{n-1} + \boldsymbol{Q}_n \boldsymbol{c}_n + \boldsymbol{Q}_n^+ \boldsymbol{c}_{n+1}\,. \tag{9.10}$$

In (9.10) $\boldsymbol{c}_n$ is a time-dependent column vector with M components

$$\boldsymbol{c}_n = (c_n^p) = \begin{pmatrix} c_n^1 \\ c_n^2 \\ \vdots \\ c_n^M \end{pmatrix} \tag{9.11}$$

and $\boldsymbol{Q}_n^\pm, \boldsymbol{Q}_n$ are time-independent $M \times M$ matrices with the form for $\boldsymbol{Q}_n$

$$\boldsymbol{Q}_n = ((Q_n)^{pq}) = \begin{pmatrix} Q_n^{11} & Q_n^{12} & \dots & Q_n^{1M} \\ Q_n^{21} & Q_n^{22} & \dots & Q_n^{2M} \\ \vdots & \vdots & & \vdots \\ Q_n^{M1} & Q_n^{M2} & \dots & Q_n^{MM} \end{pmatrix} \tag{9.12}$$

and similar forms for $\boldsymbol{Q}_n^\pm$.

We want to show that recurrence relations with L nearest-neighbor coupling can always be reduced to tridiagonal vector recurrence relations, first for $L=2$ and then give the result for arbitrary L.

We consider the one-sided pentadiagonal recurrence relation ($c_n=0$ for $n<0$)

$$\dot{c}_n = A_n^{-2}c_{n-2}+A_n^{-1}c_{n-1}+A_n^0 c_n+A_n^1 c_{n+1}+A_n^2 c_{n+2} \tag{9.13}$$

and write it down for even and odd indices n

$$\begin{aligned}\dot{c}_{2n} &= A_{2n}^{-2}c_{2n-2}+A_{2n}^{-1}c_{2n-1}+A_{2n}^0 c_{2n}+A_{2n}^1 c_{2n+1}+A_{2n}^2 c_{2n+2},\\ \dot{c}_{2n+1} &= A_{2n+1}^{-2}c_{2n-1}+A_{2n+1}^{-1}c_{2n}+A_{2n+1}^0 c_{2n+1}+A_{2n+1}^1 c_{2n+2}+A_{2n+1}^2 c_{2n+3}.\end{aligned} \tag{9.14}$$

If we now introduce the column vectors

$$\mathbf{c}_n=\begin{pmatrix} c_{2n} \\ c_{2n+1}\end{pmatrix}, \quad \mathbf{c}_{n-1}=\begin{pmatrix} c_{2n-2} \\ c_{2n-1}\end{pmatrix}, \quad \mathbf{c}_{n+1}=\begin{pmatrix} c_{2n+2} \\ c_{2n+3}\end{pmatrix}, \tag{9.15}$$

we can cast (9.14) into the form (9.10), where the matrices $\mathbf{Q}_n^{\pm}, \mathbf{Q}_n$ are given by

$$\begin{aligned}\mathbf{Q}_n^- &= \begin{pmatrix} A_{2n}^{-2} & A_{2n}^{-1} \\ 0 & A_{2n+1}^{-2}\end{pmatrix}, \quad \mathbf{Q}_n = \begin{pmatrix} A_{2n}^{0} & A_{2n}^{1} \\ A_{2n+1}^{-1} & A_{2n+1}^{0}\end{pmatrix},\\ \mathbf{Q}_n^+ &= \begin{pmatrix} A_{2n}^{2} & 0 \\ A_{2n+1}^{1} & A_{2n+1}^{2}\end{pmatrix}.\end{aligned} \tag{9.16}$$

Generally the recurrence relation with L nearest-neighbor coupling

$$\dot{c}_n = \sum_{l=-L}^{L} A_n^l c_{n+l} \tag{9.17}$$

is cast into the form of (9.10) by using the column vectors with L components

$$\mathbf{c}_n = \begin{pmatrix} c_{Ln} \\ c_{Ln+1} \\ \vdots \\ c_{Ln+L-1}\end{pmatrix} \tag{9.18}$$

and the matrices $\mathbf{Q}_n^{\pm}, \mathbf{Q}_n$ with the matrix elements

$$\begin{aligned}(\mathbf{Q}_n^{\pm})^{qr} &= A_{Ln+q-1}^{r-q\pm L},\\ (\mathbf{Q}_n)^{qr} &= A_{Ln+q-1}^{r-q}.\end{aligned} \tag{9.19}$$

In (9.19) one has to set $A_n^l=0$ for $|l|>L$.

Master Equation with 2 Nearest-Neighbor Coupling

As a first example for (9.13) we consider the master equation with 2 nearest-neighbor coupling

$$\begin{aligned}\dot{p}_n = & w(n-2\to n)p_{n-2} + w(n-1\to n)p_{n-1} \\ & + w(n+1\to n)p_{n+1} + w(n+2\to n)p_{n+2} \\ & - [w(n\to n+2) + w(n\to n+1) + w(n\to n-1) \\ & + w(n\to n-2)]p_n ,\end{aligned} \tag{9.20}$$

which for instance describes two photon absorption and emission processes. Obviously, (9.20) is of the form (9.17), where the constants A_n^l are immediately obtained by comparison with (9.17).

Schrödinger Equation for an Anharmonic Potential

The Schrödinger equation for the anharmonic potential

$$\begin{aligned} & \mathrm{i}\hbar\dot{\psi} = H\psi \\ & H = -\frac{\hbar^2}{2m}\frac{\mathrm{d}^2}{\mathrm{d}x^2} + \frac{1}{2}m\omega_0^2x^2 + a[(2m\omega_0)^2/\hbar]x^4 \\ & \quad = \hbar\omega_0(b^+b+\tfrac{1}{2}) + \hbar a(b^+ + b)^4 \end{aligned} \tag{9.21}$$

may be reduced to the form (9.17) by expanding the wave function $\psi(x,t)$ into eigenfunctions $|n\rangle$ of the number operator b^+b ($b^+b|n\rangle = n|n\rangle$, $\langle n|m\rangle = \delta_{nm}$, $b^+|n\rangle = \sqrt{n+1}\,|n+1\rangle$, $b|n\rangle = \sqrt{n}\,|n-1\rangle$ [5.3]), i.e.,

$$\psi = \sum_{n=0}^{\infty} c_n(t)|n\rangle \tag{9.22}$$

and by inserting (9.22) into (9.21). This yields again (9.17) with ($n \geqq 0$) $M=4$ and

$$\begin{aligned} A_n^{-4} &= -\mathrm{i}a\sqrt{(n-3)(n-2)(n-1)n}\,, \\ A_n^{-2} &= -\mathrm{i}a2(2n-1)\sqrt{(n-1)n}\,, \\ A_n^{0} &= -\mathrm{i}\omega_0(n+\tfrac{1}{2}) - \mathrm{i}a3(1+2n+2n^2) \\ A_n^{2} &= -\mathrm{i}a2(2n+3)\sqrt{(n+1)(n+2)}\,, \\ A_n^{4} &= -\mathrm{i}a\sqrt{(n+1)(n+2)(n+3)(n+4)}\,, \end{aligned} \tag{9.23}$$

and where all other A_n^l are zero. The same method can be applied to an anharmonicity of the form $\sim x^{2n}$ [9.12] or more generally to a potential of the form

$$V = \sum_{\nu=0}^{2n} V_\nu x^\nu, \quad V_{2n} > 0,$$

leading to a recurrence relation of length $2n+1$.

A Schrödinger equation may also be brought to the form (9.17) not only by using a discrete variable as in (9.6), but also by using higher-order difference approximations of the second derivative like ($c_n = \psi(n\Delta)$)

$$\psi''(x)|_{x=n\Delta} = (-c_{n-2} + 16c_{n-1} - 30c_n + 16c_{n+1} - c_{n+2})/(12\Delta^2). \tag{9.24}$$

As shown in Sect. 12.4, the laser Fokker-Planck equation can also be written in the form (9.17) with $M = 2$, if the distribution function is expanded into a suitable set of Laguerre functions. A linear chain with more than nearest-neighbor coupling leads to an equation of the type (9.17), where the first derivative is replaced by the second time derivative. Obviously, such an equation can also be cast into a tridiagonal recurrence relation of the form (9.10), where the first derivative is replaced by the second time derivative.

Application to Partial Differential Equations with More Variables

As we have seen, some partial differential equations like the Schrödinger equation and the laser Fokker-Planck equation can be brought to the tridiagonal vector recurrence relation (9.10). These partial differential equations essentially depend only on two variables, i.e., x and t. More complicated linear partial differential equations like a Fokker-Planck equation with an x, y and t variable may, however, also be cast into the tridiagonal vector recurrence relation. By expanding the distribution function into a complete set with respect to the y variable being consistent with the boundary condition (Sect. 6.6.5)

$$W(x,y,t) = \sum_{n=0}^{\infty} c_n(x,t)\,\psi_n(y), \tag{9.25}$$

one may obtain an equation of the form (9.17), where A_n^l are operators with respect to x. By using a further expansion into a complete set with respect to the x variable truncated at $q = M$

$$c_n(x,t) = \sum_{q=1}^{M} c_n^q \varphi^q(x), \tag{9.26}$$

(9.17) then becomes a recurrence relation for the column vectors $\boldsymbol{c}_n = (c_n^q)$, where A_n^l are now $M \times M$ matrices with matrix elements given by

$$(A_n^l)^{pq} = \int [\varphi^p(x)]^* A_n^l \varphi^q(x)\,dx. \tag{9.27}$$

The complete set $\psi_n(y)$ and $\varphi^q(x)$ should fulfill the boundary condition for the x and y variables. The set should be chosen in such a way that expansion (9.25) leads to (9.17) with a low L. The Kramers equation, where $\psi_n(v)$ are the Hermite functions with the velocity v as y variable, is an example where one obtains the form (9.17) with $L=1$ (Chap. 10).

For $L \geqq 2$, one may reduce the system to the tridiagonal vector recurrence relation (9.10) by the same procedure as applied in the beginning of this section. For $L=2$, for instance, the matrices $\boldsymbol{Q}_n^{\pm}, \boldsymbol{Q}_n$ are then given by (9.16), where the A_n^l are $M \times M$ submatrices. For numerical purposes it is very convenient if the two complete sets $\psi_n(y)$ and $\varphi^q(x)$ are chosen so that the matrix elements (9.27) can be calculated analytically.

In Sect. 9.4 it is shown briefly how tridiagonal recurrence relations can be used to solve linear differential equations with parameters which harmonically depend on time.

9.2 Solutions of Scalar Recurrence Relations

9.2.1 Stationary Solution

We first discuss the stationary solution of (9.2), i.e., the solution of (9.1). If we divide (9.1) by c_n and introduce the ratio

$$S_n = \frac{c_{n+1}}{c_n}, \tag{9.28}$$

the recurrence relation (9.1) takes the form

$$\frac{Q_n^-}{S_{n-1}} + Q_n + Q_n^+ S_n = 0 . \tag{9.1 a}$$

Thus, instead of the tridiagonal recurrence relation (9.1) for the coefficients c_n, we now get a recurrence relation for the ratios S_n, where only two adjacent indices are coupled. We can therefore express S_n by S_{n+1} [changing $n \rightarrow n+1$ in (9.1 a)]

$$S_n = -\frac{Q_{n+1}^-}{Q_{n+1} + Q_{n+1}^+ S_{n+1}} \tag{9.29}$$

or, if we use (9.29) again and again we obtain the following continued fraction for the ratios S_n

$$S_n = \frac{c_{n+1}}{c_n} = -\cfrac{Q_{n+1}^-}{Q_{n+1} - \cfrac{Q_{n+1}^+ Q_{n+2}^-}{Q_{n+2} - \cfrac{Q_{n+2}^+ Q_{n+3}^-}{Q_{n+3} - \dots}}} . \tag{9.30}$$

By iteration of (9.28) we can then express c_n by c_0

$$c_n = S_{n-1} S_{n-2} \dots S_0 c_0 \,. \tag{9.28a}$$

Sometimes instead of writing the continued fraction with the long fraction strokes the notation given in [9.1] is used

$$K = \cfrac{a_1}{b_1 + \cfrac{a_2}{b_2 + \cfrac{a_3}{b_3 + \dots}}}$$

$$\equiv \frac{a_1|}{|b_1} + \frac{a_2|}{|b_2} + \frac{a_3|}{|b_3} + \dots \,. \tag{9.31}$$

A continued fraction may have either a finite or an infinite number of terms. An infinite continued fraction may be approximated by a finite continued fraction where we stop after the Nth term:

$$K_N = \frac{a_1|}{|b_1} + \frac{a_2|}{|b_2} + \dots + \frac{a_N|}{|b_N} \,. \tag{9.32}$$

We call K_N the Nth approximant of the infinite continued fraction K. The infinite continued fraction (9.31) is said to be convergent if

$$K = \lim_{N \to \infty} K_N \tag{9.33}$$

exists. The Nth approximant of S_0 in (9.30) may be obtained if we truncate the system (9.1) after the Nth term, i.e., if we omit the equations for $n \geqq N$ and if we put

$$c_{N+1} = c_{N+2} = \dots = 0 \,. \tag{9.34}$$

We then have the recurrence relation (9.29) for $n \leqq N-1$ with

$$S_N = 0 \,. \tag{9.35}$$

For a theory of continued fractions and especially for various convergence theorems the reader should consult [9.1 – 3].

Ambiguity of the Solution of the Tridiagonal Recurrence Relation and Uniqueness of the Continued-Fraction Solution

Let us consider the recurrence relation (9.1) with $Q_0 = Q_0^{\pm} = 0$, see (9.8) for an example. Then the recurrence relation (9.1) has two solutions whereas the infinite continued fraction – assuming it converges – leads to only one solution. This is seen as follows. Starting with the second equation of (9.1), both coefficients c_0

and c_1 can be chosen arbitrarily and all other coefficients c_n with $n \geqq 2$ follow by upiteration according to (9.1). The continued fraction (9.30), however, gives a unique S_0 and thus c_0 and $c_1 = S_0 c_0$ are no longer independent. To gain some insight into this phenomenon we first treat the case where $Q_n^{\pm}$ and Q_n are independent of n for $n \geqq 1$. Then the S_n will also be independent of n and we obtain from (9.1 a) for S the quadratic equation

$$Q^+ S^2 + QS + Q^- = 0$$

with the two solutions

$$S = x_{\substack{1\\2}} = (2Q^+)^{-1}(-Q \pm \sqrt{Q^2 - 4Q^- Q^+}) \,. \tag{9.36}$$

The infinite continued fraction (9.30) becomes the periodic continued fraction

$$S_c = -\frac{Q^-|}{|Q} - \frac{Q^+ Q^-|}{|Q} - \frac{Q^+ Q^-|}{|Q} - \ldots \,. \tag{9.37}$$

If we assume that $Q^{\pm}, Q$ are real and that two different real roots of (9.36) exist, it can be proved [9.1] that the continued fraction (9.37) converges to that solution where $|x_{\substack{1\\2}}|$ is a minimum,

$$S_c = x_1 \quad \text{if} \quad |x_1| < |x_2|$$

or (9.38)

$$S_c = x_2 \quad \text{if} \quad |x_1| > |x_2| \,.$$

(If there are two different complex roots the continued fraction (9.37) cannot converge.) For instance, if we have $Q^- = 2$; $Q^+ = 1$, $Q = 3$ we have the two solutions $x_1 = -1$; $x_2 = -2$, whereas the continued fraction (9.37) leads to only one solution with the value

$$-\frac{2|}{|3} - \frac{2|}{|3} - \frac{2|}{|3} - \ldots = -1$$

as may be easily checked on a programmable pocket calculator.

The coefficients c_n may be expressed by c_0 according to

$$c_n = S_c^n c_0 \,. \tag{9.39}$$

We see that the continued fraction singles out that solution where, for increasing n, the absolute amount of the coefficients either decreases in the fastest way for $|S_c| < 1$ or increases in the slowest way for $|S_c| > 1$.

We may also look at this selection process from a different point of view. Because of the linearity of (9.1) the general solution of (9.1) may be written as

$$c_n = [A x_1^n + (1 - A) x_2^n] c_0 \,, \tag{9.40}$$

where A is an arbitrary constant. The solution with the special value $c_{N+1}=0$ indicated by the index N has the form

$$c_{n,N}=\frac{x_2^{N+1}x_1^n-x_1^{N+1}x_2^n}{x_2^{N+1}-x_1^{N+1}}c_0\,. \tag{9.41}$$

If we assume that two different real values x_1 and x_2 exist and if we take the limit $N\to\infty$ we get exactly the solution (9.39).

Another very instructive example is the recurrence relation (9.7) for the moments of laser intensity. The stationary solution obviously follows from the one-sided recurrence relation

$$\begin{aligned}&c_0=1\\&2nc_{n-1}+ac_n-c_{n+1}=0\quad\text{for}\quad n\geqq 1\,.\end{aligned} \tag{9.42}$$

With the help of (9.30) it is an easy matter to obtain an expression for the first moment

$$\begin{aligned}\langle I\rangle=c_1=S_0c_0=S_0&=-\frac{2|}{|a}+\frac{4|}{|a}+\frac{6|}{|a}+\dots\\&=-a\frac{2/a^2|}{|1}+\frac{4/a^2|}{|1}+\frac{6/a^2|}{|1}+\dots\end{aligned} \tag{9.43}$$

and for all higher moments

$$\begin{aligned}&\langle I^{n+1}\rangle=S_n\langle I^n\rangle=S_nS_{n-1}\dots S_0\\&S_n=-a\frac{2(n+1)/a^2|}{|1}+\frac{2(n+2)/a^2|}{|1}+\dots\,.\end{aligned} \tag{9.44}$$

On the other hand, c_n can be calculated exactly by Laplace's method [9.25]. We do not, however, need this method here because we know the stationary distribution of the laser intensity, Sect. 12.2. A solution of (9.42) is given by, see (12.46, 48),

$$c_n^{(1)}(a)=F_n(a)/F_0(a) \tag{9.45}$$

with

$$F_n(a)=\int_0^\infty I^n\exp\left(-\frac{1}{4}I^2+\frac{aI}{2}\right)\mathrm{d}I\,. \tag{9.46}$$

It is easily checked by partial integration that (9.45) fulfills (9.42). As clearly seen, another independent solution $c_n^{(2)}$ is obtained by

$$c_n^{(2)}(a)=(-1)^nc_n^{(1)}(-a)\,. \tag{9.47}$$

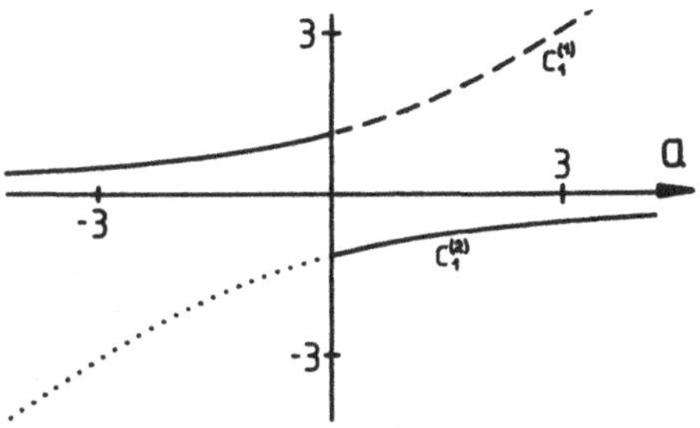

Fig. 9.1. Solution $c_1^{(1)}$ (*upper curve*), solution $c_1^{(2)}$ (*lower curve*) and the continued-fraction solution (9.43) (*solid line*) as a function of the parameter a

Though this solution must be rejected on physical grounds (because the intensity moments cannot be negative) it is nevertheless a possible solution of (9.42).

Therefore a more general solution (probably the complete solution) of (9.42) has the form

$$c_n(a) = A\,c_n^{(1)}(a) + (1-A)\,c_n^{(2)}(a)\,, \tag{9.48}$$

where again A is an arbitrary constant. The solutions $c_1^{(1)}(a)$ and $c_1^{(2)}(a)$ are given in Fig. 9.1. For positive a solution $c_n^{(1)}$ increases with increasing n more rapidly than solution $c_n^{(2)}$. This may be shown by first expressing $F_n(a)$ by the parabolic cylinder functions [9.26]

$$F_n(a) = n!\,2^{(n+1)/2}\mathrm{e}^{a^2/8}\mathrm{D}_{-(n+1)}(-a/\sqrt{2}) \tag{9.49}$$

and then using the asymptotic expressions for large n and bounded a, yielding from [Ref. 9.26, last equation of Sect. 8.1.6]

$$F_n(a) = \sqrt{2\pi}\,\exp\{a^2/8 + \tfrac{1}{2}n(\ln n + \ln 2 - 1) + a\sqrt{n/2}\}(1 + O(1/\sqrt{n}))\,. \tag{9.50}$$

Thus the ratio of $c_n^{(1)}(a)$ and $c_n^{(2)}(a)$ is given for large n by

$$\left|\frac{c_n^{(1)}(a)}{c_n^{(2)}(a)}\right| = \frac{F_0(-a)}{F_0(a)}\exp(\sqrt{2n}\,a)\,. \tag{9.51}$$

Thus, $c_n^{(1)}$ increases for $a>0$ with increasing n faster than $c_n^{(2)}$, whereas for $a<0$ the reverse is true. The continued fraction (9.43) agrees with $c_1^{(1)}(a)$ for $a<0$ and with $c_1^{(2)}(a)$ for $a>0$ (Fig. 9.1). [At $a=0$ the continued fraction (9.43) does not exist.] This may be shown numerically by evaluating (9.43) on a programmable pocket calculator. It may be proved analytically as follows: by using (16) of [Ref. 9.1, Vol. II, Sect. 20] we obtain (after making the substitution $I-a=2t$ and putting $a=-2\xi$) for $a<0$

$$\langle I\rangle = \frac{2}{F_0(a)} + a = \frac{F_1(a)}{F_0(a)} = c_1^{(1)}(a)\,.$$

Had we chosen the coefficients

$$\hat{c}_n = c_n/n! \tag{9.52}$$

instead of the coefficients c_n, the recurrence relation (9.42) would read

$$2\hat{c}_{n-1} + a\hat{c}_n - (n+1)\hat{c}_{n+1} = 0 \tag{9.42a}$$

and both solutions $\hat{c}_n^{(1)}$ and $\hat{c}_n^{(2)}$ would tend to zero for $n \to \infty$. Solution $\hat{c}_n^{(1)}$ then decreases less rapidly than $\hat{c}_n^{(2)}$ for $a > 0$ and vice versa for $a < 0$.

Thus, we see that also in this more complicated example the continued-fraction method singles out that solution for which $|c_n|$ increases with increasing n in the slowest way or decreases with increasing n in the fastest way. This seems to be a general feature of the continued-fraction method. The explanation runs as follows. The Nth approximant $S_0(N) = c_1/c_0$ could also be obtained by truncating the system (9.2) after the Nth term, as discussed in the beginning of this section. If we then start with a value c_N and calculate c_n for low n iteratively, only that solution will survive in the limit of large N, where the $|c_n|$ grow with decreasing n in the fastest way (or decrease with decreasing n in the slowest way). If we perform the reverse iteration, i.e., if we start with the two coefficients c_0 and $c_1 = S_0(\infty)c_0$ and calculate coefficients with larger n, $|c_n|$ must then decrease with increasing n in the fastest way (or increase with increasing n in the slowest way). This solution is called a minimal solution of the recurrence relation (9.1). If the minimal solution is denoted by $c_n^{\min}$ and another solution by c_n, we then have [9.3] $\lim\limits_{n\to\infty} c_n^{\min}/c_n = 0$.

As explained in [9.3] the upiteration of the recurrence relation (9.1) is numerically unstable for a minimal solution whereas the upiteration according to (9.28a) is stable.

If the coefficients c_n are the expansion coefficients of a function W into a complete set $\psi_n(x)$, i.e.,

$$W(x) = \sum_n c_n \psi_n(x), \tag{9.53}$$

this would mean that the continued-fraction method singles out that solution where expansion (9.53) converges in the best possible way. If for instance $W(x)$ is determined by a differential equation (which may lead to the recurrence relations (9.1) for a proper chosen set $\psi_n(x)$) one is usually looking for a solution where the expansion converges in the best possible way. Thus, the continued-fraction method automatically leads to this desired solution. If, however, the coefficients are some other quantities like the moments in the example just discussed, one has to check very carefully whether the solution obtained by the continued-fraction method is the desired one.

Approximate Determination of S_N

If the coefficients Q_n, $Q_n^{\pm}$ are slowly dependent on n for large n so that $Q_n \approx Q_{n+1}$, $Q_n^{\pm} \approx Q_{n+1}^{\pm}$ for large n the ratio S_{N-1} may be approximated by S_N for large N. We thus obtain from (9.1a) the quadratic equation

$$Q_N^+ S_N^2 + Q_N S_N + Q_N^- = 0 .$$

For S_N we then get the roots (9.36) where $Q^\pm$, Q and S has to be replaced by $Q_N^\pm$, Q_N and S_N respectively. As explained before we must take that root where $|x_{N\frac{1}{2}}|$ is smallest, i.e. we use

$$\begin{aligned} S_N &= x_{N1} \quad \text{if} \quad |x_{N1}| < |x_{N2}| \\ S_N &= x_{N2} \quad \text{if} \quad |x_{N1}| > |x_{N2}| \\ x_{N\frac{1}{2}} &= (2Q_N^+)^{-1}\left(-Q_N \pm \sqrt{Q_N^2 - 4Q_N^- Q_N^+}\right) \end{aligned} \tag{9.35a}$$

instead of (9.35) as the starting value for obtaining S_n with $n \leqq N-1$ according to (9.29). For the example (9.42), for instance, with $a<0$ we have

$$S_N = \sqrt{2N + a^2/4} + a/2 .$$

For $a = -1$ only $N \approx 150$ is needed to obtain $\langle I \rangle = S_0$ accurate to 9 digits with the above starting value whereas with the starting value $S_N = 0$ $N \approx 300$ must be used to obtain the same accuracy.

By using

$$S_{N-1} = S_N - \mathrm{d}S_N/\mathrm{d}N + \dots$$

more accurate solutions of (9.1a) may be obtained for large N. (In the above example one can derive an expansion of S_N in powers of $(2N)^{-1/2}$ which further reduces the number N of iterations, which are necessary to achieve the same accuracy.)

9.2.2 Initial Value Problem

To find the solution of the one-sided recurrence relation (9.2) with the initial value $c_m(0)$, we first observe that the solution $c_n(t)$ can be expressed by

$$c_n(t) = \sum_{m=0}^{\infty} G_{n,m}(t)\, c_m(0) , \tag{9.54}$$

where the Green's function $G_{n,m}(t)$ is a solution of

$$\dot{G}_{n,m} = Q_n^- G_{n-1,m} + Q_n G_{n,m} + Q_n^+ G_{n+1,m} \tag{9.55}$$

with the initial condition

$$G_{n,m}(0) = \delta_{nm} . \tag{9.56}$$

Next we make a Laplace transform

$$\tilde{G}_{n,m}(s) = \int_0^\infty e^{-st} G_{n,m}(t)\,dt \tag{9.57}$$

and obtain from (9.55) the following recurrence relation for the Laplace-transformed Green's function $\tilde{G}_{n,m}(s)$

$$Q_n^- \tilde{G}_{n-1,m} + \hat{Q}_n \tilde{G}_{n,m} + Q_n^+ \tilde{G}_{n+1,m} = -\delta_{nm}, \tag{9.58}$$

where $\hat{Q}_n$ is given by

$$\hat{Q}_n = \hat{Q}_n(s) = Q_n - s. \tag{9.59}$$

If we disregard the inhomogeneity in (9.58), it is the same as (9.1), whose solutions we just discussed. We therefore proceed now in a similar manner. If we truncate the infinite system after the Nth term, i.e., if we neglect the equations (9.58) for $n \geqq N+1$ and if we put

$$\tilde{G}_{N+1,m} = \tilde{G}_{N+2,m} = \dots = 0 \quad \text{for all} \quad m \geqq 0, \tag{9.60}$$

system (9.58) reads explicitly ($Q_0^- = 0$)

$$\begin{array}{llll}
 & \hat{Q}_0 \tilde{G}_{0,m} & + Q_0^+ \tilde{G}_{1,m} & = 0 \\
Q_1^- \tilde{G}_{0,m} & + \hat{Q}_1 \tilde{G}_{1,m} & + Q_1^+ \tilde{G}_{2,m} & = 0 \\
\dots & \dots & \dots & \dots \\
Q_{m-1}^- \tilde{G}_{m-2,m} & + \hat{Q}_{m-1} \tilde{G}_{m-1,m} & + Q_{m-1}^+ \tilde{G}_{m,m} & = 0 \\
Q_m^- \tilde{G}_{m-1,m} & + \hat{Q}_m \tilde{G}_{m,m} & + Q_m^+ \tilde{G}_{m+1,m} & = -1 \\
Q_{m+1}^- \tilde{G}_{m,m} & + \hat{Q}_{m+1} \tilde{G}_{m+1,m} & + Q_{m+1}^+ \tilde{G}_{m+2,m} & = 0 \\
\dots & \dots & \dots & \dots \\
Q_{N-1}^- \tilde{G}_{N-2,m} & + \hat{Q}_{N-1} \tilde{G}_{N-1,m} & + Q_{N-1}^+ \tilde{G}_{N,m} & = 0 \\
Q_N^- \tilde{G}_{N-1,m} & + \hat{Q}_N \tilde{G}_{N,m} & & = 0\,.
\end{array} \tag{9.61}$$

We now introduce the two ratios

$$\tilde{S}_n^+(s) = \frac{\tilde{G}_{n+1,m}}{\tilde{G}_{n,m}}, \quad \tilde{S}_n^-(s) = \frac{\tilde{G}_{n-1,m}}{\tilde{G}_{n,m}}. \tag{9.62}$$

According to the last equation in (9.61) we may express $\tilde{G}_{N,m}$ in terms of $\tilde{G}_{N-1,m}$

$$\tilde{G}_{N,m} = \tilde{S}_{N-1}^{+}(s)\,\tilde{G}_{N-1,m}\,, \qquad \tilde{S}_{N-1}^{+}(s) = -\frac{Q_N^-}{\hat{Q}_N}\,. \tag{9.63}$$

Inserting this $\tilde{G}_{N,m}$ into the second-last equation we may express $\tilde{G}_{N-1,m}$ in terms of $\tilde{G}_{N-2,m}$

$$\tilde{G}_{N-1,m} = \tilde{S}_{N-2}^{+}\tilde{G}_{N-2,m}\,, \qquad \tilde{S}_{N-2}^{+}(s) = -\frac{Q_{N-1}^-}{\hat{Q}_{N-1} - \dfrac{Q_{N-1}^+ Q_N^-}{\hat{Q}_N}}\,. \tag{9.64}$$

We proceed in this way till we reach the equation with the inhomogeneity -1:

$$\tilde{G}_{m+1,m} = \tilde{S}_m^+ \tilde{G}_{m,m}$$

$$\begin{aligned} \tilde{S}_m^+(s) \;&= -\frac{Q_{m+1}^-|}{|\hat{Q}_{m+1}} - \frac{Q_{m+1}^+ Q_{m+2}^-|}{|\hat{Q}_{m+2}} - \dots \\ &\quad - \frac{Q_{N-2}^+ Q_{N-1}^-|}{|\hat{Q}_{N-1}} - \frac{Q_{N-1}^+ Q_N^-|}{|\hat{Q}_N}\,. \end{aligned} \tag{9.65}$$

If we let N tend to infinity we obtain the infinite continued fraction

$$\tilde{S}_m^+(s) = -\frac{Q_{m+1}^-|}{|Q_{m+1}-s} - \frac{Q_{m+1}^+ Q_{m+2}^-|}{|Q_{m+2}-s} - \dots\,. \tag{9.66}$$

According to the first equation in (9.61) we may express $\tilde{G}_{0,m}$ in terms of $\tilde{G}_{1,m}$

$$\tilde{G}_{0,m} = \tilde{S}_1^-(s)\,\tilde{G}_{1,m}\,, \qquad \tilde{S}_1^-(s) = -\frac{Q_0^+}{\hat{Q}_0}\,. \tag{9.67}$$

Inserting this $\tilde{G}_{0,m}$ in the second equation we may express $\tilde{G}_{1,m}$ in terms of $\tilde{G}_{2,m}$

$$\tilde{G}_{1,m} = \tilde{S}_2^-(s)\,\tilde{G}_{2,m}\,, \qquad \tilde{S}_2^-(s) = -\frac{Q_1^+}{\hat{Q}_1 - \dfrac{Q_1^- Q_0^+}{\hat{Q}_0}} \tag{9.67a}$$

and so on, till we reach the equation with the inhomogeneity -1

$$\tilde{G}_{m-1,m} = \tilde{S}_m^- \tilde{G}_{m,m}$$

$$\begin{aligned} \tilde{S}_m^-(s) \;&= -\frac{Q_{m-1}^+|}{|Q_{m-1}-s} - \frac{Q_{m-1}^- Q_{m-2}^+|}{|Q_{m-2}-s} - \dots \\ &\quad - \frac{Q_2^- Q_1^+|}{|Q_1-s} - \frac{Q_1^- Q_0^+|}{|Q_0-s}\,. \end{aligned} \tag{9.68}$$

For the one-sided recurrence relation (9.55) $S_m^-(s)$ is a finite continued fraction. For a two-sided recurrence relation, (9.68) does not stop at the last term but continues to negative indices, i.e., S_m^- will then also be an infinite continued fraction. Finally, from the equation with the inhomogeneity -1 we have

$$(Q_m^- \tilde{S}_m^- + \hat{Q}_m + Q_m^+ \tilde{S}_m^+)\,\tilde{G}_{m,m} = -1$$

or if we introduce the abbreviation

$$\tilde{K}_m(s) = Q_m^- \tilde{S}_m^-(s) + Q_m^+ \tilde{S}_m^+(s) \tag{9.69}$$

and if we observe (9.59) we have

$$[Q_m - s + \tilde{K}_m(s)]\,\tilde{G}_{m,m} = -1\,. \tag{9.70}$$

By this equation $\tilde{G}_{m,m}$ can be calculated

$$\tilde{G}_{m,m}(s) = [s - Q_m - \tilde{K}_m(s)]^{-1}\,. \tag{9.71}$$

Thus the essential function is

$$\tilde{K}_m(s) = \tilde{K}_m^-(s) + \tilde{K}_m^+(s)\,, \tag{9.72}$$

where $\tilde{K}_m^\pm(s)$ are given by the infinite and finite continued fractions

$$\tilde{K}_m^+(s) = -\frac{Q_m^+ Q_{m+1}^- |}{|Q_{m+1} - s} - \frac{Q_{m+1}^+ Q_{m+2}^- |}{|Q_{m+2} - s} - \dots, \quad \text{for} \quad m \geqq 0 \tag{9.73}$$

$$\tilde{K}_m^-(s) = -\frac{Q_m^- Q_{m-1}^+ |}{|Q_{m-1} - s} - \frac{Q_{m-1}^- Q_{m-2}^+ |}{|Q_{m-2} - s} - \dots$$

$$-\frac{Q_2^- Q_1^+ |}{|Q_1 - s} - \frac{Q_1^- Q_0^+ |}{|Q_0 - s}, \quad \text{for} \quad m \geqq 1 \tag{9.74}$$

$$\tilde{K}_0^-(s) = 0\,.$$

The other Laplace-transformed Green's function then follows by iteration according to (9.65, 68) from $\tilde{G}_{m,m}(s)$

$$\tilde{G}_{n,m}(s) = \tilde{U}_{n,m}(s)\,\tilde{G}_{m,m}(s)\,, \tag{9.75}$$

$$\tilde{U}_{n,m}(s) = \tilde{S}_{n-1}^+(s)\,\tilde{S}_{n-2}^+(s)\dots\tilde{S}_m^+(s) \quad \text{for} \quad n \geqq m+1\,, \tag{9.76}$$

$$\tilde{U}_{m,m}(s) = 1\,, \tag{9.77}$$

$$\tilde{U}_{n,m}(s) = \tilde{S}_{n+1}^-(s)\,\tilde{S}_{n+2}^-(s)\dots\tilde{S}_m^-(s) \quad \text{for} \quad 0 \leqq n \leqq m-1\,. \tag{9.78}$$

Thus the Laplace transform of Green's function and therefore also the Laplace transform of the general initial value problem is expressed in terms of continued fractions. [We have tacitly assumed that only the minimal solutions of (9.61) are of interest.]

Equation (9.70) is the Laplace transform of the integrodifferential equation

$$\dot{G}_{m,m}(t) = Q_m G_{m,m}(t) + \int_0^t K_m(t-\tau)\, G_{m,m}(\tau)\,d\tau \tag{9.79}$$

with the initial condition (9.56). The memory kernel $K_m(t)$ is the inverse Laplace transform of $\tilde{K}_m(s)$. Thus the exact elimination of the functions $G_{n,m}(t)$ with $n \neq m$ in the infinite system (9.55) of coupled differential equations is completely taken into account by the memory function $K_m(t)$. Because of (9.75) the other Green's functions are expressed by the convolution ($n \neq m$)

$$G_{n,m}(t) = \int_0^t U_{n,m}(t-\tau)\, G_{m,m}(\tau)\,d\tau\,, \tag{9.80}$$

where $U_{n,m}(t)$ is the inverse Laplace transform of $\tilde{U}_{n,m}(s)$. A solution of the general initial value problem is then given by (9.54).

A somewhat different method to solve the initial value problem of (9.2) would be the following. Making the Laplace transform

$$\tilde{c}_n(s) = \int_0^\infty c_n(t)\,e^{-st}\,dt\,,$$

we obtain the inhomogeneous recurrence relation ($\hat{Q}_n = Q_n - s$)

$$Q_n^-\,\tilde{c}_{n-1} + \hat{Q}_n \tilde{c}_n + Q_n^+\,\tilde{c}_{n+1} = -c_n(0)\,. \tag{9.81}$$

If we truncate the system after the Nth term ($\tilde{c}_{N+1} = \tilde{c}_{N+2} = \ldots = 0$ and omit the equations with $n \geqq N+1$) we obtain by inserting the 'ansatz'

$$\tilde{c}_{n+1} = \tilde{S}_n^+\,\tilde{c}_n + \tilde{a}_{n+1} \quad \text{for} \quad n \geqq 0\,, \quad \tilde{c}_0 = \tilde{a}_0 \tag{9.82}$$

the following recursions

$$\begin{aligned}
\tilde{S}_{N-1}^+(s) &= -\frac{Q_N^-}{\hat{Q}_N}\\
\tilde{a}_N(s) &= -\frac{c_N(0)}{\hat{Q}_N}\\
\tilde{S}_n^+(s) &= -\frac{Q_{n+1}^-}{\hat{Q}_{n+1} + Q_{n+1}^+ \tilde{S}_{n+1}^+(s)} \quad \text{for} \quad 0 \leqq n \leqq N-2\\
\tilde{a}_n(s) &= -\frac{c_n(0) + Q_n^+\,\tilde{a}_{n+1}}{\hat{Q}_n + Q_n^+\,\tilde{S}_n^+(s)} \quad \text{for} \quad 0 \leqq n \leqq N-1\,.
\end{aligned} \tag{9.83}$$

Thus all $\tilde{c}_n$ with $n \geqq 1$ can then be obtained by iteration. Finally $\tilde{c}_0$ follows from (9.81) with $n=0$, i.e. from

$$\tilde{c}_0(s) = \tilde{a}_0(s)\,. \tag{9.84}$$

This iteration procedure may be better suited than (9.75 – 78) for those cases for which the continued fraction (9.74) becomes numerically unstable for large m.

Taylor Expansion Method

The foregoing continued-fraction method should not be mixed up with the following expansion method in which a continued fraction may be used to improve convergence. If we insert the 'ansatz'

$$\begin{aligned} c_n(t) &= \sum_{\nu=0}^{\infty} a_{n,\nu} t^{\nu}/\nu! \\ a_{n,0} &= c_n(0) \end{aligned} \tag{9.85}$$

into (9.81) or if we insert its Laplace transform (or its one-sided Fourier transform $s = \mathrm{i}\omega$)

$$\tilde{c}_n(s) = \sum_{\nu=0}^{\infty} a_{n,\nu} s^{-(\nu+1)} \tag{9.85a}$$

into (9.2), we can determine the expansion coefficients $a_{n,\nu}$ of the Taylor series by iteration according to ($\nu \geqq 0$)

$$a_{n,\nu+1} = Q_n^- a_{n-1,\nu} + Q_n a_{n,\nu} + Q_n^+ a_{n+1,\nu}\,. \tag{9.86}$$

(This method is, of course, not restricted to a tridiagonal recurrence relation.)

Usually the convergence of the above Taylor series is not very good especially for large t or small s; the series may even converge only asymptotically for $t \to 0$ or $s \to \infty$, i.e., diverge for every $t > 0$ or finite s. The convergence may be improved or the divergent series may become convergent when Padé approximants or continued fractions, which are a special form of Padé approximants, are used [9.4]. Depending on the form of these continued fractions they may or may not agree with (9.65, 68). In any case the advantage of (9.65, 68) is that (9.65, 68) are directly given in terms of $Q_n^{\pm}$, Q_n, whereas the coefficients of the continued fractions (or more generally the coefficients of Padé approximants) for the Taylor series (9.85, 85a) are expressed in a more complicated way by determinants or recurrence relations [9.1 – 4].

9.2.3 Eigenvalue Problem

Because $Q_n^{\pm}$ and Q_n in (9.2) do not depend on time, we can make the separation 'ansatz'

$$c_n(t) = \hat{c}_n e^{-\lambda t} \tag{9.87}$$

and obtain the homogeneous recurrence relation

$$Q_n^- \hat{c}_{n-1} + (Q_n + \lambda)\hat{c}_n + Q_n^+ \hat{c}_{n+1} = 0 \,. \tag{9.88}$$

Because (9.88) is a homogeneous linear equation, the following infinite determinant must be zero

$$\begin{vmatrix} Q_0+\lambda & Q_0^+ & & & \\ Q_1^- & Q_1+\lambda & Q_1^+ & & \\ & Q_2^- & Q_2+\lambda & Q_2^+ & \\ & & \dots & \dots & \dots \end{vmatrix} = 0 \,. \tag{9.89}$$

Determinants of this form, where all elements are zero except the diagonal elements and the elements in the two adjacent lines, are called continuants [Ref. 2.8, Chap. 13]. Here we do not need the theory of continuants. To find an equation for the eigenvalues we eliminate all coefficients $\hat{c}_n$ with $n \neq m$ by the same method used for $\tilde{G}_{n,m}$ in (9.61). Instead of (9.70) we now obtain ($\lambda = -s$)

$$[Q_m + \lambda + \tilde{K}_m(-\lambda)]\,\hat{c}_m = 0 \,. \tag{9.90}$$

From this equation we immediately get the equation

$$Q_m + \lambda + \tilde{K}_m(-\lambda) = 0 \,, \tag{9.91}$$

from which the eigenvalues λ can be calculated. Obviously, the Green's function $G_{m,m}(s)$ (9.71) has poles at $s = -\lambda$. Usually the index $m \geqq 0$ is arbitrary. For those eigenvalues which are close to $-Q_n$ it is, however, advisable to use the nth equation. If one or more of the coefficients are zero ($\hat{c}_l = 0$), we cannot use (9.91) for $m = l$.

The other coefficients then follow from $\hat{c}_m$ by [compare (9.75)]

$$\hat{c}_n = \tilde{U}_{n,m}(-\lambda)\,\hat{c}_m \,. \tag{9.92}$$

If the stationary solution of (9.2) exists, one eigenvalue λ must be zero. For $Q_0 = 0$ only $\tilde{K}_0$ exists and we must have $\tilde{K}_0(0) = 0$. The other coefficients then follow from (9.92) with $m = 0$, which agrees with (9.28a).

Example

As an example we look for the eigenvalues of the intensity moments equation (9.7). Because in this case $\hat{c}_0$ must be zero for $\lambda \neq 0$, we cannot use (9.91) for $m = 0$. For $m = 1$, (9.91) then reads

$$2a + \lambda + \tilde{K}_1(-\lambda) = 0 \,, \tag{9.93}$$

where $\tilde{K}_1(-\lambda)$ is given by, see (9.8, 72, 73),

$$\begin{aligned}\tilde{K}_1(-\lambda) &= \tilde{K}_1^+(-\lambda)\\ &= \frac{2\cdot 4\cdot 2^2|}{|2\cdot 2\cdot a+\lambda} + \frac{4\cdot 4\cdot 3^2|}{|2\cdot 3\cdot a+\lambda} + \frac{6\cdot 4\cdot 4^2|}{|2\cdot 4\cdot a+\lambda} + \dots\\ &= 2\cdot \frac{4|}{|a+\lambda/4} + \frac{6|}{|a+\lambda/6} + \frac{8|}{|a+\lambda/8} + \dots \,.\end{aligned} \tag{9.94}$$

Continued fractions of the above form can easily be evaluated on a programmable pocket calculator. The roots of (9.93) can then be obtained by some root-finding technique, for example by a regula falsi method. Thus, for this example one can calculate with a programmable pocket calculator the eigenvalues below threshold $a<0$, otherwise obtained only by more elaborate methods. Above threshold ($a>0$), however, the method is not applicable because the continued-fraction method singles out an unphysical solution, as was already discussed for the stationary problem in Sect. 9.2.1. If c_n are expansion coefficients of some function into a complete set, we may truncate system (9.2) at large N, leading to the Nth approximant of the continued-fraction. If we then take the limit $N\to\infty$ we can always use (9.91) to determine the eigenvalues. (For an application of this method to the calculation of the eigenvalues of the Schrödinger equation with discrete variables (9.6), see [9.7].)

9.3 Solutions of Vector Recurrence Relations

We now discuss the solution of the vector recurrence relation (9.10). As will become evident, we can also apply the methods used for the scalar recurrence relation (9.2) for (9.10). The decisive difference will be, however, that for non-commutative matrices $Q_n^{\pm}$, Q_n the order of multiplications and inversions will be essential. The multiplications and inversions now become matrix multiplications and matrix inversions. Thus the ordinary continued fractions of (9.30) will now become matrix continued fractions.

If, for the moment, we neglect the first ($n=0$) equation in (9.10) and if we put $\dot{c}_n=0$, the tridiagonal vector recurrence relation (9.10) has $2M$ independent solutions. We may choose for instance the two vectors $\boldsymbol{c}_0$ and $\boldsymbol{c}_1$ arbitrarily ($2M$ arbitrary constants) and obtain all $\boldsymbol{c}_n$ with $n\geqq 2$ by upiteration of (9.10). The matrix continued-fraction solution of (9.10) leads – if it converges – to a matrix S_0^+ which connects $\boldsymbol{c}_1$ and $\boldsymbol{c}_0$, i.e. $\boldsymbol{c}_1 = S_0^+\boldsymbol{c}_0$ and thus both $\boldsymbol{c}_1$ and $\boldsymbol{c}_0$ can no longer be chosen arbitrarily. We may choose an arbitrary $\boldsymbol{c}_0$, still leading to M arbitrary constants. In the examples we discuss in the following chapters the first equation of (9.10) together with some normalization or proper initial conditions then determine all these M coefficients and thus give a unique solution. If matrix continued fractions are not used M arbitrary constants cannot be determined. Similar to the scalar case, the matrix continued-fraction method seems to single out that solution where the absolute amount of the coefficients c_n^p decreases with increasing n in the fastest way or increases with increasing n in the slowest way. If the c_n^p are the expansion coefficients of some function into two complete sets, the matrix continued-fraction solution is just that solution which converges in the best possible way, which usually is the desired solution.

Because the stationary solution is the special eigenvalue problem with eigenvalue $\lambda = 0$, we do not treat the stationary solution separately as in Sect. 9.2.1.

9.3.1 Initial Value Problem

We proceed here in the same way as in Sect. 9.2.2. The general solution of (9.10) is expressed by the Green's function matrix $\boldsymbol{G}_{n,m}(t)$, i.e.,

$$\boldsymbol{c}_n(t) = \sum_{m=0}^{\infty} \boldsymbol{G}_{n,m}(t)\,\boldsymbol{c}_m(0)\,, \tag{9.95}$$

with the initial value

$$\boldsymbol{G}_{n,m}(0) = \boldsymbol{I}\,\delta_{nm}\,. \tag{9.96}$$

Here $\boldsymbol{I}$ is the unit matrix. After making the Laplace transform

$$\tilde{\boldsymbol{G}}_{n,m}(s) = \int_0^{\infty} \mathrm{e}^{-st}\boldsymbol{G}_{n,m}(t)\,\mathrm{d}t\,, \tag{9.97}$$

we obtain as a sufficient condition for the solution of (9.10)

$$\boldsymbol{Q}_n^-\,\tilde{\boldsymbol{G}}_{n-1,m} + \hat{\boldsymbol{Q}}_n\tilde{\boldsymbol{G}}_{n,m} + \boldsymbol{Q}_n^+\,\tilde{\boldsymbol{G}}_{n+1,m} = -\boldsymbol{I}\,\delta_{nm} \tag{9.98}$$

with

$$\hat{\boldsymbol{Q}}_n(s) = \boldsymbol{Q}_n - s\boldsymbol{I}\,. \tag{9.99}$$

To solve (9.98) we introduce two matrices $\tilde{\boldsymbol{S}}_n^{\pm}$ which connect $\tilde{\boldsymbol{G}}_{n,m}$ with $\tilde{\boldsymbol{G}}_{n\pm1,m}$

$$\tilde{\boldsymbol{G}}_{n\pm1,m} = \tilde{\boldsymbol{S}}_n^{\pm}\,\tilde{\boldsymbol{G}}_{n,m}\,. \tag{9.100}$$

We may now use the same elimination procedure as in Sect. 9.2.2. To shorten the derivation, we proceed as follows. Neglecting for the moment the inhomogeneous term, we have either

$$\boldsymbol{Q}_n^-\,\tilde{\boldsymbol{G}}_{n-1,m} + (\hat{\boldsymbol{Q}}_n + \boldsymbol{Q}_n^+\,\tilde{\boldsymbol{S}}_n^+)\,\tilde{\boldsymbol{G}}_{n,m} = \boldsymbol{0} \tag{9.101}$$

or

$$(\boldsymbol{Q}_n^-\,\tilde{\boldsymbol{S}}_n^- + \hat{\boldsymbol{Q}}_n)\,\tilde{\boldsymbol{G}}_{n,m} + \boldsymbol{Q}_n^+\,\tilde{\boldsymbol{G}}_{n+1,m} = \boldsymbol{0}\,. \tag{9.102}$$

By multiplying both equations with the inverse of the matrix in the parenthesis and by comparing the results with (9.100), we immediately obtain

$$\tilde{S}^{\pm}_{n\mp1} = -(\hat{Q}_n + Q^{\mp}_n \tilde{S}^{\pm}_n)^{-1} Q^{\mp}_n$$

or, if we change the index n and insert (9.99)

$$\tilde{S}^{\pm}_n = (sI - Q_{n\pm1} - Q^{\pm}_{n\pm1} \tilde{S}^{\pm}_{n\pm1})^{-1} Q^{\mp}_{n\pm1} . \tag{9.103}$$

For the upper sign in (9.103) we get by iteration the infinite continued fraction

$$\begin{aligned}\tilde{S}^{+}_n &= (sI - Q_{n+1} - Q^{+}_{n+1}(sI - Q_{n+2} - Q^{+}_{n+2}(sI - Q_{n+3} - \dots)^{-1} \\ &\quad \times Q^{-}_{n+3})^{-1} Q^{-}_{n+2})^{-1} Q^{-}_{n+1} .\end{aligned} \tag{9.104}$$

For a one-sided vector recurrence relation the iteration of (9.103) for the lower sign leads to the finite continued fraction ($n \geqq 3$)

$$\begin{aligned}\tilde{S}^{-}_0 &= 0 \\ \tilde{S}^{-}_1 &= (sI - Q_0)^{-1} Q^{+}_0 \\ \tilde{S}^{-}_2 &= (sI - Q_1 - Q^{-}_1 (sI - Q_0)^{-1} Q^{+}_0)^{-1} Q^{+}_1 \\ \tilde{S}^{-}_n &= (sI - Q_{n-1} - Q^{-}_{n-1}(sI - \dots - Q^{-}_1 (sI - Q_0)^{-1} Q^{+}_0 \dots)^{-1} Q^{+}_{n-2})^{-1} Q^{+}_{n-1} .\end{aligned} \tag{9.105}$$

If the inhomogeneous term $-I$ in (9.98) is taken into account, the index n in (9.101) must be restricted to $n \geqq m+1$ and in (9.102) to $n \leqq m-1$. Because of the change of index in (9.103), the index n in (9.104) must be restricted to $n \geqq m$ and in (9.105) to $n \leqq m$. Considering the equation with $n = m$ of (9.98), we may express $\tilde{G}_{m-1,m}$ by $\tilde{S}^{-}_m \tilde{G}_{m,m}$ and $\tilde{G}_{m+1,m}$ by $\tilde{S}^{+}_m \tilde{G}_{m,m}$, thus obtaining

$$(Q^{-}_m \tilde{S}^{-}_m + \hat{Q}_m + Q^{+}_m \tilde{S}^{+}_m) \tilde{G}_{m,m} = -I .$$

Introducing

$$\tilde{K}_m(s) = \tilde{K}^{-}_m(s) + \tilde{K}^{+}_m(s) \tag{9.106}$$

with

$$\tilde{K}^{\pm}_m(s) = Q^{\pm}_m \tilde{S}^{\pm}_m \tag{9.107}$$

we have

$$[Q_m - sI + \tilde{K}_m(s)] \tilde{G}_{m,m} = -I . \tag{9.108}$$

Thus $\tilde{G}_{m,m}$ is given by

$$\tilde{G}_{m,m}(s) = [sI - Q_m - \tilde{K}_m(s)]^{-1} . \tag{9.109}$$

The other $\tilde{G}_{n,m}$ follow by iteration of (9.100)

$$\tilde{G}_{n,m}(s) = \tilde{U}_{n,m}(s)\,\tilde{G}_{m,m}(s)\,; \tag{9.110}$$

$$\begin{aligned} \tilde{U}_{n,m}(s) &= \tilde{S}^+_{n-1}(s)\tilde{S}^+_{n-2}(s)\dots\tilde{S}^+_m(s) \quad \text{for} \quad n \geqq m+1 \\ \tilde{U}_{m,m}(s) &= I \\ \tilde{U}_{n,m}(s) &= \tilde{S}^-_{n+1}(s)\tilde{S}^-_{n+2}(s)\dots\tilde{S}^-_m(s) \quad \text{for} \quad 0 \leqq n \leqq m-1\,. \end{aligned} \tag{9.111}$$

In particular, we have for $m = 0$

$$\begin{aligned} \tilde{K}_0(s) &= \tilde{K}_0^+(s) = Q_0^+\tilde{S}_0^+ \\ &= Q_0^+(sI - Q_1 - Q_1^+(sI - Q_2 - Q_2^+(sI - Q_3 - \dots)^{-1}Q_3^-)^{-1}Q_2^-)^{-1}Q_1^-\,. \end{aligned} \tag{9.112}$$

If we write the matrix inversions by fraction lines, (9.112) takes the form

$$\tilde{K}_0(s) = Q_0^+ \cfrac{I}{sI - Q_1 - Q_1^+ \cfrac{I}{sI - Q_2 - Q_2^+ \cfrac{I}{sI - Q_3 - \dots} Q_3^-} Q_2^-} Q_1^-\,. \tag{9.112a}$$

Equation (9.108) is the Laplace transform of

$$\dot{G}_{m,m}(t) = Q_m G_{m,m}(t) + \int_0^t K_m(t-\tau)\,G_{m,m}(\tau)\,d\tau \tag{9.113}$$

with the initial condition (9.96). The memory matrix-kernel $K_m(t)$ is the inverse Laplace transform of $\tilde{K}_m(s)$. The other solutions $G_{n,m}(t)$ then follow by convolutions with the inverse Laplace transform of $\tilde{U}_{n,m}(s)$ according to (9.110).

It should be noted that for a numerical evaluation of $\tilde{G}_{n,m}$ one does not need (9.110, 111) for all n. It is necessary to determine only two $\tilde{G}_{n,m}$ with adjacent indices n by (9.110, 111). The other $\tilde{G}_{n,m}$ may then be determined by iteration according to (9.98). By this iteration procedure, however, one may encounter numerical instabilities. If for instance we have obtained $G_{0,0}$ from (9.109) and $G_{1,0}$ from (9.110, 111) the other $G_{n,0}$ may then be obtained by iteration according to (9.98). As already mentioned in [9.3] for the scalar recurrence relation (9.1) this upiteration is numerically unstable if $G_{0,0}$ is a minimal solution of the recurrence relation. No such instability seems to occur if $G_{n,0}$ with $n \geqq 1$ are determined by (9.110, 111).

As in Sect. 9.2.2 for the scalar case, it is also possible to find the general solution of (9.10) by the following iteration. Making the 'ansatz' ($n \geqq 0$, $\tilde{c}_n = 0$ for $n \geqq N+1$)

$$\begin{aligned} \tilde{c}_{n+1} &= \tilde{S}_n^+\,\tilde{c}_n + \tilde{a}_{n+1}\,, \\ \tilde{c}_0 &= \tilde{a}_0\,, \end{aligned} \tag{9.114}$$

for the Laplace transform of the solution $c_n(t)$ of (9.10) with the initial value $c_n(0)$, we obtain in analogy to (9.83)

$$\begin{aligned}
\tilde{S}^+_{N-1} &= [sI - Q_N]^{-1} Q^-_N \\
\tilde{a}_N(s) &= [sI - Q_N]^{-1} c_N(0) \\
\tilde{S}^+_n(s) &= [sI - Q_{n+1} - Q^+_{n+1}\tilde{S}^+_{n+1}(s)]^{-1} Q^-_{n+1} \quad \text{for} \quad 0 \leqq n \leqq N-2 \\
\tilde{a}_n(s) &= [sI - Q_n - Q^+_n \tilde{S}^+_n(s)]^{-1} [c_n(0) + Q^+_n \tilde{a}_{n+1}] \quad \text{for} \quad 0 \leqq n \leqq N-1 \,.
\end{aligned} \tag{9.115}$$

This iteration procedure may be better suited than (9.110, 111) for those cases for which the continued fraction (9.105) becomes unstable for large n.

Taylor Expansion Method

In complete analogy to the scalar case it is also possible to solve (9.10) by a Taylor series

$$c^p_n(t) = \sum_{\nu=0}^{\infty} a^p_{n,\nu} t^\nu / \nu! \,.$$

The coefficients are found by iteration according to

$$a^p_{n,\nu+1} = \sum_q [(Q^-_n)^{pq} a^q_{n-1,\nu} + (Q_n)^{pq} a^q_{n,\nu} + (Q^+_n)^{pq} a^q_{n+1,\nu}] \,,$$

starting with

$$a^p_{n,0} = c^p_n(0) \,.$$

The convergence of the above Taylor series may again be improved by an ordinary continued fraction. If the matrices Q^+_n, Q_n do not commute, these ordinary continued fractions have nothing to do with the foregoing matrix continued fractions (9.104, 105).

9.3.2 Eigenvalue Problem

For the determination of the eigenvalues of (9.10), we may proceed in a similar way as in Sect. 9.2.3 for the scalar case. By inserting the separation 'ansatz'

$$c_n(t) = \hat{c}_n e^{-\lambda t} \tag{9.116}$$

into (9.10) we get the homogeneous recurrence relation

$$Q^-_n \hat{c}_{n-1} + (Q_n + \lambda I)\hat{c}_n + Q^+_n \hat{c}_{n+1} = 0 \,. \tag{9.117}$$

With the exception of the equation $n = m$, (9.117) agrees with (9.98) ($s = -\lambda$). We may therefore eliminate all $\boldsymbol{c}_n$ for $n \neq m$ by the same procedure used in Sect. 9.3.1. For the equation with $n = m$ we then have

$$[\boldsymbol{Q}_m + \lambda \boldsymbol{I} + \tilde{\boldsymbol{K}}_m(-\lambda)]\,\hat{\boldsymbol{c}}_m = \boldsymbol{0}\,. \tag{9.118}$$

With the exception of the inhomogeneous term in (9.108) equation (9.118) agrees with (9.108). Because (9.118) is a homogeneous equation the determinant

$$D_m(\lambda) = \operatorname{Det}\,[\boldsymbol{Q}_m + \lambda \boldsymbol{I} + \tilde{\boldsymbol{K}}_m(-\lambda)] = 0 \tag{9.119}$$

must be zero. This is the desired equation for determining the eigenvalues. Because of (9.109), the determinant of the Green's function matrix $\tilde{\boldsymbol{G}}_{m,m}(s)$ has poles at $s = -\lambda$. The eigenvectors $\hat{\boldsymbol{c}}_n$ then follow from the eigenvectors $\hat{\boldsymbol{c}}_m$ of (9.118) by the relation

$$\hat{\boldsymbol{c}}_n = \tilde{\boldsymbol{U}}_{n,m}(-\lambda)\,\hat{\boldsymbol{c}}_m\,, \tag{9.120}$$

see (9.110). If the continued fractions (9.104, 105) exist any of the equations (9.119) for $m = 0, 1, 2, \ldots$ may be used to calculate the eigenvalues. For those eigenvalues where $\tilde{\boldsymbol{K}}_m(-\lambda)$ is small compared to $\boldsymbol{Q}_m$ it is advisable to use $D_m(\lambda)$.

Concerning the calculation of eigenvalues the following remark is pertinent. To calculate the infinite continued fraction $\tilde{\boldsymbol{K}}_m^+$ we have to approximate it by some approximant of finite order, e.g., by the Nth approximant. As discussed in Sect. 9.2.1, this is equivalent to truncating the recurrence relation (9.117) at the index N. Equation (9.117) is then equivalent to a linear system for the $(N+1)M$ coefficients $\hat{c}_n^p$ and the eigenvalue problem would be equivalent to diagonalizing an $(N+1)M \times (N+1)M$ matrix. The matrix continued-fraction method requires the calculation of N inversions and multiplications of $M \times M$ matrices, which is, especially for large N, much easier to perform than diagonalizing the $(N+1)M \times (N+1)M$ matrix. To obtain the eigenvalues by the matrix continued-fraction method, the roots of (9.119) have to be found with some root-finding technique.

As already mentioned, the stationary solution does not require any special treatment. Here, the determinant (9.119) must be zero for $\lambda = 0$. The vector $\hat{\boldsymbol{c}}_m$ then follows from (9.118) and the other $\hat{\boldsymbol{c}}_n$ from (9.120). Vector $\hat{\boldsymbol{c}}_m$ and therefore also vectors $\hat{\boldsymbol{c}}_n$ contain a multiplicative arbitrary constant, which must be determined by the normalization condition. If $\operatorname{Det} \boldsymbol{Q}_0 = 0$ the matrix continued fractions (9.105) do not exist and we must use (9.119) for $m = 0$.

Generalizations

(i) The continued fractions were written down for a one-sided tridiagonal vector recurrence relation. For a two-sided tridiagonal vector recurrence relation similar expressions are valid. The only difference will be that the continued fractions $\tilde{\boldsymbol{S}}_n^-$ and $\tilde{\boldsymbol{K}}_n^-$ will not terminate at $n = 1$ but will extend to n equal to minus infinity and therefore will also become infinite continued fractions.

(ii) A more general tridiagonal recurrence relation has the form

$$A_n^- \dot{c}_{n-1} + A_n \dot{c}_n + A_n^+ \dot{c}_{n+1} = Q_n^- c_{n-1} + Q_n c_n + Q_n^+ c_{n+1} . \tag{9.121}$$

An equation of this form, for instance, may be obtained from a two-variable Fokker-Planck equation, if the drift and diffusion coefficients are rational functions, Sect. 6.6.6. To calculate the eigenvalues, (9.117) has to be replaced by

$$(Q_n^- + \lambda A_n^-) \hat{c}_{n-1} + (Q_n + \lambda A_n) \hat{c}_n + (Q_n^+ + \lambda A_n^+) \hat{c}_{n+1} = 0 , \tag{9.122}$$

which may be solved similarly to (9.117) by matrix continued fractions. The initial value problem is now more complicated because the inhomogeneity $A_n^\pm, A_n$ occurs for $n = m \mp 1$ and $n = m$ in (9.98). Of course, $Q_n^\mp$ and $\hat{Q}_n$ in (9.98) must be replaced by $Q_n^\pm - sA_n^\pm$ and $Q_n - sA_n$ in this case. The elimination of $\tilde{G}_{n\pm l,m}$ with $l \geqq 2$ can be done as in Sect. 9.3.1. Only the remaining three equations for $\tilde{G}_{n\pm 1,m}$ and $\tilde{G}_{n,m}$ must then be solved.

(iii) As mentioned in the introduction, higher derivatives may also be treated in the same way, i.e., one may solve the tridiagonal recurrence relation

$$\sum_{l=0}^{L} [A_n^{-(l)} (\mathrm{d}/\mathrm{d}t)^l c_{n-1} + A_n^{(l)} (\mathrm{d}/\mathrm{d}t)^l c_n + A_n^{+(l)} (\mathrm{d}/\mathrm{d}t)^l c_{n+1}] = 0 \tag{9.123}$$

by the procedures given in this chapter.

9.4 Ordinary and Partial Differential Equations with Multiplicative Harmonic Time-Dependent Parameters

We now discuss briefly the application of ordinary and matrix continued fractions to differential equations, where some multiplicative parameters depend harmonically on time. In Sect. 9.4.1 we show the application to ordinary differential equations, especially to the Mathieu and the Bloch equations, and in Sect. 9.4.2 to partial differential equations, especially to the one-variable Fokker-Planck equation.

9.4.1 Ordinary Differential Equations

Mathieu Equation

One of the simplest equations with a multiplicative harmonic time-dependent parameter is the Mathieu equation [Ref. 5.1, Chap. 7 – 5]. It reads in normalized variables

$$\mathrm{d}^2 x/\mathrm{d}\tau^2 + (\Omega - 2q \cos 2\tau) x = 0 . \tag{9.124}$$

Because of Floquet's theorem [5.1], a solution of (9.124) can be written in the form

$$x(\tau) = e^{\mu\tau} u_\mu(\tau) , \quad u_\mu(\tau+\pi) = u_\mu(\tau) , \tag{9.125}$$

where u_μ is a periodic function with the same period π as the time-dependent parameter $\cos 2\tau$ in (9.124). If the characteristic exponent μ is purely imaginary, solution (9.125) is stable; for $\mathrm{Re}\{\mu\}>0$ it will become unstable. Because of the periodicity of u_μ we can make the Fourier expansion

$$u_\mu(\tau) = \sum_{n=-\infty}^{\infty} c_n e^{i2n\tau} . \tag{9.126}$$

By inserting this expansion into (9.125, 124), we obtain the two-sided tridiagonal recurrence relation

$$[(\mu+2in)^2 + \Omega] c_n - q(c_{n+1}+c_{n-1}) = 0 . \tag{9.127}$$

The elimination of all c_n with $n \neq 0$ leads to

$$\left[\mu^2+\Omega - \frac{q^2|}{|(\mu+2i)^2+\Omega} - \frac{q^2|}{|(\mu+4i)^2+\Omega} - \dots \right.$$
$$\left. - \frac{q^2|}{|(\mu-2i)^2+\Omega} - \frac{q^2|}{|(\mu-4i)^2+\Omega} - \dots \right] c_0 = 0 . \tag{9.128}$$

If $c_0 \neq 0$, the characteristic exponents μ can be calculated from the condition that the bracket in (9.128) vanishes. For $\mu = 0$ the solution $x(\tau)$ is periodic in τ. For solutions with $c_0 \neq 0$, (9.128) then gives a relation between Ω and q for which only such a periodic solution can exist [5.1].

Generalizations

(i) If higher harmonics occur, i.e.,

$$d^2x/d\tau^2 + (a_0 + a_2\cos 2\tau + a_4 \cos 4\tau + \dots + a_{2M} \cos 2M\tau$$
$$+ b_2 \sin 2\tau + b_4 \sin 4\tau + \dots + b_{2M} \sin 2M\tau) x = 0 , \tag{9.129}$$

the 'ansatz' (9.126) then leads to a recurrence relation of the form

$$\sum_{l=-M}^{M} A_{n,l} c_{n+l} = 0 , \tag{9.130}$$

which can be cast into a tridiagonal vector recurrence relation (Sect. 9.1.2).

(ii) Coupled equations of the form

$$\mathrm{d}^2x^p/\mathrm{d}\tau^2 + \sum_q (a_0^{pq} + a_2^{pq}\cos 2\tau + \ldots + a_{2M}^{pq}\cos 2M\tau + b_2^{pq}\sin 2\tau + \ldots + b_{2M}^{pq}\sin 2M\tau)x^q = 0 \tag{9.129a}$$

lead to

$$\sum_q \sum_{l=-M}^{M} A_{n,l}^{pq} c_{n+l}^q = 0\,, \tag{9.130a}$$

which, by using a suitable vector notation, can also be cast into a tridiagonal vector recurrence relation.

(iii) If a damping term $\gamma\mathrm{d}x/\mathrm{d}\tau$ occurs in (9.124, 129), the same method should also work.

Bloch Equations

The Bloch equations are equations of motion for the matrix elements ρ_{ij} of the density operator for a two-level system. For an external cosine field $E = A\cos\nu t$ they take the form, see (12.11, 12), ($\rho_{12} = \rho_{21}^*$)

$$\begin{aligned} \dot\rho_{22} - \dot\rho_{11} &= \lambda - \gamma_1(\rho_{22} - \rho_{11}) + \mathrm{i}(2\mathrm{e}/\hbar)x_{12}(\rho_{12} - \rho_{21})A\cos\nu t\,, \\ \dot\rho_{12} &= (\mathrm{i}\,\omega_0 - \gamma_2)\rho_{12} + \mathrm{i}(\mathrm{e}/\hbar)x_{12}(\rho_{22} - \rho_{11})A\cos\nu t\,. \end{aligned} \tag{9.131}$$

The constant term λ describes the pumping. We look for periodic solutions with period $2\pi/\nu$. Because of the special form of the equations, $\rho_{22} - \rho_{11}$ can have Fourier terms with even n only and ρ_{12} have those with odd n only. The Fourier expansions

$$\begin{aligned} \rho_{22} - \rho_{11} &= \sum_{n=\mathrm{even}} d_n \mathrm{e}^{\mathrm{i}n\nu t}\,, \\ \rho_{12} - \rho_{21} &= \sum_{n=\mathrm{odd}} s_n \mathrm{e}^{\mathrm{i}n\nu t}\,, \\ \rho_{12} + \rho_{21} &= \sum_{n=\mathrm{odd}} c_n \mathrm{e}^{\mathrm{i}n\nu t} \end{aligned} \tag{9.132}$$

lead to [9.11]

$$\begin{aligned} n\nu d_n &= \omega_1(s_{n+1} + s_{n-1}) + \mathrm{i}\gamma_1 d_n - \mathrm{i}\lambda\delta_{n0}\,, && (n\ \text{even}) \\ n\nu s_n &= \omega_0 c_n + \omega_1(d_{n+1} + d_{n-1}) + \mathrm{i}\gamma_2 s_n\,, && (n\ \text{odd}) \\ n\nu c_n &= \omega_0 s_n + \mathrm{i}\gamma_2 c_n\,, && (n\ \text{odd})\,, \end{aligned} \tag{9.133}$$

where ω_1 is given by

$$\omega_1 = (\mathrm{e}/\hbar)x_{12}A\,. \tag{9.134}$$

By eliminating c_n from the last two equations in (9.133) we get (n odd)

$$s_n = \frac{\mathrm{i}}{2}\left(\frac{1}{\gamma_2+\mathrm{i}(n\nu-\omega_0)} + \frac{1}{\gamma_2+\mathrm{i}(n\nu+\omega_0)}\right)\omega_1(d_{n+1}+d_{n-1})\,; \tag{9.135}$$

and from the first equation of (9.133) we have (n even)

$$d_n = \frac{\mathrm{i}\,\omega_1}{\gamma_1+\mathrm{i}\,n\,\nu}(s_{n+1}+s_{n-1}) + \frac{\lambda}{\gamma_1}\delta_{n0}\,. \tag{9.136}$$

If we introduce

$$x_n = \begin{cases} d_n\gamma_1/\lambda & \text{for} \quad n \text{ even} \\ s_n\gamma_1/\lambda & \text{for} \quad n \text{ odd} \end{cases} \tag{9.137}$$

and

$$D_n = \begin{cases} (\gamma_1+\mathrm{i}\,n\,\nu)^{-1} & \text{for} \quad n \text{ even} \\ \frac{1}{2}\{[\gamma_2+\mathrm{i}(n\nu-\omega_0)]^{-1} + [\gamma_2+\mathrm{i}(n\nu+\omega_0)]^{-1}\} & \text{for} \quad n \text{ odd} \end{cases}$$

both recurrence relations (9.135, 136) can be cast into the following inhomogeneous two-sided tridiagonal scalar recurrence relation

$$x_n - \mathrm{i}\,\omega_1 D_n(x_{n+1}+x_{n-1}) = \delta_{n,0}\,. \tag{9.138}$$

This relation can be solved in terms of ordinary continued fractions similar to the one-sided tridiagonal recurrence relation in Sect. 9.2.2. The continued fractions and some results are given by *Stenholm* [9.11], see also [9.9, 10]. *Allegrini* et al. [9.13] have solved a similar problem for a spin-1/2 system, by 2×2 matrix continued fractions. Obviously, the method can also be generalized to multilevel systems leading to matrix continued fractions of higher dimensions.

9.4.2 Partial Differential Equations

The procedure of Sect. 9.4.1 can also be applied to partial differential equations. Here we discuss only the application to a one-variable Fokker-Planck equation. (A one-variable Schrödinger equation with a harmonic time-dependent potential may be solved similarly.) Assuming time dependence of a cosine form, this Fokker-Planck equation reads

$$\dot{W}(x,t) = L_{\mathrm{FP}}(x,t)\,W(x,t)\,, \tag{9.139}$$

$$L_{\mathrm{FP}}(x,t) = L_0(x) + L_1(x)\cos\nu t\,. \tag{9.140}$$

According to Floquet's theorem we have multiplicative solutions

$$W(x,t) = \mathrm{e}^{\mu t}u_\mu(x,t)\,, \tag{9.141}$$

where $u_\mu(x,t)$ is a periodic function in t with period $2\pi/\nu$. Therefore we may expand $u_\mu(x,t)$ into a Fourier series with respect to t. We furthermore expand the Fourier coefficients into a complete set $\varphi^p(x)$ satisfying the boundary conditions for x. Hence, the total expansion of the probability density reads

$$W(x,t) = F(x) \sum_p \sum_n c_n^p \mathrm{e}^{(\mathrm{i}n\nu+\mu)t} \varphi^p(x) . \qquad (9.142)$$

In (9.142) we have added a function $F(x) \neq 0$ which may be useful to simplify the final recurrence relation (Sects. 5.9.3, 6.6.5). Inserting (9.142) into (9.139, 140) leads to

$$\sum_q [(L_0^{pq} - \mathrm{i}n\nu - \mu) c_n^q + \tfrac{1}{2} L_1^{pq} (c_{n+1}^q + c_{n-1}^q)] = 0 , \qquad (9.143)$$

where $L_{\substack{0\\1}}^{pq}$ are defined by

$$L_{\substack{0\\1}}^{pq} = \int F(x)^{-1} [\varphi^p(x)]^* L_{\substack{0\\1}} F(x) [\varphi^q(x)] \,\mathrm{d}x . \qquad (9.144)$$

By truncating the upper indices, we thus obtain a two-sided tridiagonal vector recurrence relation for the vector $\boldsymbol{c}_n = (c_n^q)$. The characteristic exponent can then be determined in a similar way to the eigenvalue λ in Sect. 9.3.2. If the Fokker-Planck operator has the more general time dependence

$$\begin{aligned} L_{\mathrm{FP}}(x) &= L_0(x) + L_{c1}(x) \cos \nu t + L_{c2}(x) \cos 2\nu t + \ldots + L_{cM}(x) \cos M\nu t \\ &\quad + L_{s1}(x) \sin \nu t + L_{s2}(x) \sin 2\nu t + \ldots + L_{sM}(x) \sin M\nu t \end{aligned} \qquad (9.145)$$

the 'ansatz' (9.142) leads to a recurrence relation of the form

$$\sum_{l=-M}^{M} \sum_q A_{n,l}^{pq} c_{n+l}^q = 0 , \qquad (9.146)$$

which can – by a suitable notation – also be cast into a tridiagonal vector recurrence relation. It may also happen that for a suitable choice of $F(x)$ and $\varphi^p(x)$ the $A_{n,l}^{pq}$ are of a tridiagonal form in the upper index, which then can also be solved by a matrix continued fraction. By using a similar method *Breymayer* et al. [9.21] have solved the problem of harmonic mixing in cosine potential for large damping and arbitrary field strength.

9.5 Methods for Calculating Continued Fractions

9.5.1 Ordinary Continued Fractions

To calculate the Nth approximant of the continued fraction

$$K_N = b_0 + \frac{a_1|}{|b_1} + \frac{a_2|}{|b_2} + \ldots \frac{a_N|}{|b_N} \qquad (9.147)$$

we may start with a_N/b_N, then add b_{N-1}, take the inverse and multiply it with a_{N-1} and proceed in this way till we reach b_0. This downward or tail to head iteration is very simple. If the a_n and b_n have simple forms, the evaluation can easily be performed with a programmable pocket calculator. To check the convergence of (9.147) for $N\to\infty$ one has to start the whole procedure again with larger N. This disadvantage of the downward iteration is avoided by upward or head to tail iteration. Here we first calculate A_n and B_n by the recurrence relations [9.1 – 3]

$$\begin{aligned} A_n &= b_n A_{n-1} + a_n A_{n-2} \\ B_n &= b_n B_{n-1} + a_n B_{n-2}, \end{aligned} \tag{9.148}$$

starting with

$$\begin{aligned} A_{-1} &= 1 \quad & A_0 &= b_0 \\ B_{-1} &= 0 \quad & B_0 &= 1\,. \end{aligned} \tag{9.149}$$

The Nth approximant (9.147) is then given by

$$K_N = A_N/B_N\,. \tag{9.150}$$

If we now increase N we need not calculate the first N, A_n and B_n again. We now, however, have a more complicated algorithm than for the downward iteration.

9.5.2 Matrix Continued Fractions

Downward Iteration

The Nth approximant of the infinite continued fraction (9.112),

$$\begin{aligned} \tilde{\boldsymbol{K}}_{0N} = \boldsymbol{Q}_0^+ (s\boldsymbol{I} - \boldsymbol{Q}_1 - \boldsymbol{Q}_1^+ (s\boldsymbol{I} - \boldsymbol{Q}_2 - \ldots \boldsymbol{Q}_{N-1}^+ (s\boldsymbol{I} - \boldsymbol{Q}_N)^{-1} \boldsymbol{Q}_N^- \ldots)^{-1} \\ \times \boldsymbol{Q}_2^-)^{-1} \boldsymbol{Q}_1^-\,, \end{aligned} \tag{9.151}$$

may be calculated by starting with the inversion of $s\boldsymbol{I} - \boldsymbol{Q}_N$, multiplying it with $\boldsymbol{Q}_{N-1}^+$ from the left and with $\boldsymbol{Q}_N^-$ from the right and subtracting it from $s\boldsymbol{I} - \boldsymbol{Q}_{N-1}$. The inversion of the resulting matrix is then the beginning of the next step, and we thus finally obtain $\tilde{\boldsymbol{K}}_{0N}$. In this way approximately $2N$ matrix multiplications, $2N$ matrix additions and N matrix inversions are necessary. For large matrices of dimension $M\times M$ the number of arithmetic operations is then essentially given by $3NM^3$.

If a new approximant with a larger N is needed, we have to start the whole process again with this larger N. To check for convergence we may increase the number N proportional to 2^k ($k = 1, 2, 3, \ldots$). In this way about $6NM^3$ arithmetic operations are necessary.

Upward Iteration

To perform the upward iteration we first consider a matrix continued fraction where the $\boldsymbol{Q}_n^-$ matrices are normalized to the unit matrix. That this can be done is seen from the recurrence relation (9.117) which may be multiplied with the inverse of $\boldsymbol{Q}_n^{-1}$, if it exists. Thus we consider the matrix continued fraction

$$\boldsymbol{K} = \boldsymbol{b}_0 + \boldsymbol{a}_1(\boldsymbol{b}_1 + \boldsymbol{a}_2(\boldsymbol{b}_2 + \boldsymbol{a}_3(\boldsymbol{b}_3 + \dots)^{-1})^{-1})^{-1} . \tag{9.152}$$

If we introduce the matrices $\boldsymbol{A}_n, \boldsymbol{B}_n$ by the recursion relations

$$\begin{aligned} \boldsymbol{A}_n &= \boldsymbol{A}_{n-1}\boldsymbol{b}_n + \boldsymbol{A}_{n-2}\boldsymbol{a}_n , \\ \boldsymbol{B}_n &= \boldsymbol{B}_{n-1}\boldsymbol{b}_n + \boldsymbol{B}_{n-2}\boldsymbol{a}_n , \end{aligned} \tag{9.153}$$

with the initial values

$$\begin{aligned} \boldsymbol{A}_{-1} &= \boldsymbol{I} , \qquad \boldsymbol{A}_0 = \boldsymbol{b}_0 , \\ \boldsymbol{B}_{-1} &= \boldsymbol{0} , \qquad \boldsymbol{B}_0 = \boldsymbol{I} , \end{aligned} \tag{9.154}$$

the Nth approximant of (9.152) is given by

$$\boldsymbol{K}_N = \boldsymbol{A}_N \boldsymbol{B}_N^{-1} . \tag{9.155}$$

To check for convergence, the recursion (9.153) can be continued. For this downward iteration $2N$ matrix additions and $4N$ matrix multiplications are necessary, leading approximately to $4NM^3$ arithmetic operations. It is important to notice that depending on the magnitude of the matrix elements in $\boldsymbol{a}_n$ and $\boldsymbol{b}_n$, the magnitude of $\boldsymbol{A}_n$ and $\boldsymbol{B}_n$ may increase or decrease, leading to arithmetic overflow or underflow. This must be avoided by rescaling the matrices $\boldsymbol{A}_n, \boldsymbol{B}_n, \boldsymbol{A}_{n-1}, \boldsymbol{B}_{n-1}$ by a constant factor after an appropriate number of steps has been made.

A comparison of the upward and downward iteration methods shows that upward iteration is somewhat faster (roughly by a factor of 2), but downward iteration is simpler to put into program statements and has less storage requirements for matrices (roughly by a factor of 2). A similar upward iteration was used in [9.22].

10. Solutions of the Kramers Equation

The Kramers equation is a special Fokker-Planck equation describing the Brownian motion in a potential. For a one-dimensional problem it is an equation for the distribution function in position and velocity space. This Kramers equation was derived and used by *Kramers* [1.17] to describe reaction kinetics. Later on it turned out that it had more general applicability, e.g., to such different fields as superionic conductors, Josephson tunneling junction, relaxation of dipoles, second-order phase-locked loops. These applications will be discussed in Chap. 11. For large damping constants the Kramers equation reduces to the Smoluchowski equation which is a special Fokker-Planck equation for the distribution function for the position coordinate only. In this chapter some of the well-known solutions for linear forces are presented. Next we shall derive a general solution of the Kramers equation in terms of matrix continued fractions for arbitrary forces. Expansion of these matrix continued-fraction solutions for large damping constants into powers of the inverse friction constant gives the Smoluchowski equation and its different correction terms. Whereas the position will become a slow variable and the velocity a fast variable in the high-friction limit, the energy will become a slow variable and the position (or velocity) a fast variable in the low-friction limit (see Sect. 8.3 for a discussion of slow and fast variables). In the low-friction limit the procedure depends on the topology of the energy surface in phase space, which in turn depends on the specific form of the potential. With the exception of the linear force, special potentials are not treated in this chapter. Therefore, the low-friction limit is treated in Chap. 11 (Sects. 4, 6.3, 8.1, 9.1), where Brownian motion in a periodic potential is investigated.

10.1 Forms of the Kramers Equation

The Langevin equation describing the Brownian motion of particles with mass m in a potential $mf(x)$ reads [cf. (3.132)]

$$\begin{aligned} & m\ddot{x} + m\gamma\dot{x} + mf'(x) = m\Gamma(t)\,, \\ & \langle \Gamma(t)\rangle = 0\,, \qquad \langle \Gamma(t)\Gamma(t')\rangle = 2\gamma(kT/m)\,\delta(t-t')\,, \end{aligned} \tag{10.1}$$

Γ Gaussian distributed.

Here γ is the damping constant ($\tau = 1/\gamma$ is the relaxation time), $-f'(x) = -\mathrm{d}f/\mathrm{d}x$ is the force per mass m due to the potential $mf(x)$, k is Boltzmann's constant and T is the temperature of the surrounding heat bath. The stochastic Langevin force $m\Gamma(t)$ is assumed to be a Gaussian random process with δ correlation. To obtain the Fokker-Planck equation, (10.1) is first written as a system of two first-order equations

$$\begin{aligned} \dot{x} &= v \\ \dot{v} &= -\gamma v - f'(x) + \Gamma(t) . \end{aligned} \tag{10.2}$$

Using the results of Chap. 4, the special Fokker-Planck equation for this process then reads (4.112)

$$\partial W/\partial t = L_{\mathrm{K}} W \tag{10.3a}$$

$$L_{\mathrm{K}} = L_{\mathrm{K}}(x,v) = -\frac{\partial}{\partial x} v + \frac{\partial}{\partial v}[\gamma v + f'(x)] + \gamma v_{\mathrm{th}}^2 \frac{\partial^2}{\partial v^2} . \tag{10.3b}$$

In (10.3) $W(x,v,t)$ is the distribution function in position and velocity space, $v_{\mathrm{th}} = \sqrt{kT/m}$ is the thermal velocity (10.9). Equation (10.3) may be written as a continuity equation (6.1)

$$\frac{\partial W}{\partial t} + \frac{\partial S_x}{\partial x} + \frac{\partial S_v}{\partial v} = 0 , \tag{10.4a}$$

where the x and v components of the probability current are defined by (6.3)

$$S_x = vW , \quad S_v = -[\gamma v + f'(x)]W - \gamma v_{\mathrm{th}}^2 \partial W/\partial v . \tag{10.4b}$$

For three dimensions the Langevin equation (10.1) and the corresponding Fokker-Planck operator L_{K} of the Kramers equation are given by [1.5–7] [$\boldsymbol{x} = (x_1, x_2, x_3)$; $\boldsymbol{v} = (v_1, v_2, v_3)$]

$$\begin{aligned} &m\ddot{x}_i + m\gamma\dot{x}_i + m\partial f(\boldsymbol{x})/\partial x_i = m\Gamma_i(t) \\ &\langle \Gamma_i(t)\Gamma_j(t)\rangle = 2\gamma(kT/m)\,\delta_{ij}\,\delta(t-t') \end{aligned} \tag{10.5}$$

$$L_{\mathrm{K}}(\boldsymbol{x},\boldsymbol{v}) = \sum_{i=1}^{3}\left[-\frac{\partial}{\partial x_i} v_i + \frac{\partial}{\partial v_i}\left(\gamma v_i + \frac{\partial f}{\partial x_i}\right) + \gamma v_{\mathrm{th}}^2 \frac{\partial^2}{\partial v_i \partial v_i}\right] . \tag{10.6}$$

10.1.1 Normalization of Variables

For numerical calculations it is convenient to use the Kramers equation in normalized form. By introducing the following variables and parameters

$$\begin{aligned} t_n &= v_{th} t\,, \quad v_n = v/v_{th}\,, \quad x_n = x\,, \\ f_n &= f/v_{th}^2 = mf/(kT)\,, \quad \gamma_n = \gamma/v_{th}\,, \\ W_n &= W v_{th}\,, \end{aligned} \tag{10.7}$$

the Kramers equation (10.3) transforms to

$$\frac{\partial W_n}{\partial t_n} = \left[-\frac{\partial}{\partial x_n} v_n + \frac{\partial}{\partial v_n}(\gamma_n v_n + f_n') + \gamma_n \frac{\partial^2}{\partial v_n^2} \right] W_n\,. \tag{10.8}$$

The thermal velocity v_{th} of a particle with mass m is defined by the square root of the averaged velocity v^2 without potential in the stationary state (3.11, 12)

$$v_{th} = \sqrt{\langle v^2 \rangle_{f=0}} = \sqrt{kT/m}\,. \tag{10.9}$$

The velocity v_n is a dimensionless variable whereas the time t_n has the dimension of length. The inverse of the normalized friction constant is the mean free path

$$1/\gamma_n = l = \tau v_{th}\,, \tag{10.10}$$

i.e., the distance which a particle with velocity v_{th} would reach in the relaxation time τ. (The variable x may be of course an angle variable. Then v will become the angle velocity.) The normalization (10.7) corresponds to the normalization $kT/m = 1$ in the Langevin equation (10.1). The normalized from (10.8) is not very convenient if the zero temperature limit $T \to 0$ (i.e., $v_{th} \to 0$) is considered.

10.1.2 Reversible and Irreversible Operators

The operator L_K of the Kramers equation may be split into a reversible or streaming operator L_{rev} and an irreversible or collision operator L_{ir} (6.90, 103, 104)

$$L_K = L_{rev} + L_{ir} \tag{10.11}$$

$$L_{rev} = -v\,\partial/(\partial x) + f'\,\partial/(\partial v) \tag{10.12}$$

$$L_{ir} = \gamma\,\partial/(\partial v)\,[v + v_{th}^2\,\partial/(\partial v)]\,. \tag{10.13}$$

The operator L_{rev} describes the motion of an ensemble obeying the reversible equations

$$\begin{aligned} \dot{x} &= v \\ \dot{v} &= -f'(x)\,. \end{aligned} \tag{10.14}$$

The total derivative of the distribution for such an ensemble is given by

$$\frac{\mathrm{d}W}{\mathrm{d}t}=\frac{\partial W}{\partial t}+\frac{\partial W}{\partial x}\dot{x}+\frac{\partial W}{\partial v}\dot{v}=\frac{\partial W}{\partial t}-L_{\text{rev}}W=0\,. \tag{10.15}$$

The solution of this equation with the initial distribution $W(x_0, v_0, 0)$ reads

$$W(x,v,t)=\iint\delta(x-X(x_0,v_0,t))\,\delta(v-V(x_0,v_0,t))\,W(x_0,v_0,0)\,\mathrm{d}x_0\mathrm{d}v_0\,, \tag{10.16}$$

where

$$\begin{aligned} x &= X(x_0,v_0,t) \\ v &= V(x_0,v_0,t) \end{aligned} \tag{10.17}$$

is the solution of (10.14) with initial values $x = x_0$, $v = v_0$ at $t = 0$. Because of the special form of (10.14) the motion is reversible. Therefore, starting with the x and v coordinates, the initial coordinates are reached after time $-t$, or starting with x and $-v$ after time t

$$\begin{aligned} x_0 &= X(x,v,-t) = X(x,-v,t) \\ v_0 &= V(x,v,-t) = -V(x,-v,t)\,. \end{aligned} \tag{10.18}$$

The solution $W(x,v,t)$ of (10.15) with initial condition $W(x,v,0)$ may then by written as

$$W(x,v,t)=W(X(x,v,-t),\,V(x,v,-t),0)\,. \tag{10.19}$$

The derivation of (10.19) from (10.16) runs as follows. We first use X and V as new coordinates. Next we note that the Jacobian

$$J=\begin{vmatrix} \dfrac{\partial X}{\partial x_0} & \dfrac{\partial X}{\partial v_0} \\ \dfrac{\partial V}{\partial x_0} & \dfrac{\partial V}{\partial v_0} \end{vmatrix}=1 \tag{10.20}$$

is equal to one. This may be shown by differentiating J with respect to time. It then follows from (10.14) that $\dot{J}=0$, i.e., $J(t)=J(0)=1$. Because the inverse relation of (10.17) is given by (10.18) we obtain (10.19). We finally remark that L_{rev} is an anti-Hermitian operator.

The irreversible operator L_{ir} has a second derivative if the temperature is not zero. [For zero temperature but finite friction constant γ the solution of (10.1) is no longer reversible and (10.19) no longer holds.] The operator L_{ir} is neither an anti-Hermitian operator nor an Hermitian operator.

10.1.3 Transformation of the Operators

The stationary solution of $\boldsymbol{L}_{\mathrm{ir}}$ is proportional to $\exp[-v^2/(2v_{\mathrm{th}}^2)]$. By multiplying $\boldsymbol{L}_{\mathrm{ir}}$ from the right with the square root of the stationary solution and from the left with the inverse of the square root of the stationary solution, $\boldsymbol{L}_{\mathrm{ir}}$ can be brought to an Hermitian form (Sect. 5.4):

$$\begin{aligned}\bar{\boldsymbol{L}}_{\mathrm{ir}} &= \exp\left[\frac{1}{4}\left(\frac{v}{v_{\mathrm{th}}}\right)^2\right]\boldsymbol{L}_{\mathrm{ir}}\exp\left[-\frac{1}{4}\left(\frac{v}{v_{\mathrm{th}}}\right)^2\right] \\ &= \gamma\left[v_{\mathrm{th}}^2\frac{\partial^2}{\partial v^2}-\frac{1}{4}\left(\frac{v}{v_{\mathrm{th}}}\right)^2+\frac{1}{2}\right]=\bar{\boldsymbol{L}}_{\mathrm{ir}}^{+}\,. \end{aligned} \tag{10.21}$$

The operator $\bar{\boldsymbol{L}}_{\mathrm{ir}}$ has the same form as the Hamilton operator for the harmonic oscillator in quantum mechanics. We therefore employ this connection with quantum mechanics and introduce the annihilation and creation operators b and b^+

$$b = v_{\mathrm{th}}\frac{\partial}{\partial v}+\frac{1}{2}\,\frac{v}{v_{\mathrm{th}}}\,,\qquad b^+ = -v_{\mathrm{th}}\frac{\partial}{\partial v}+\frac{1}{2}\,\frac{v}{v_{\mathrm{th}}}\,, \tag{10.22}$$

where the commutator for these boson operators is equal to one

$$[b, b^+] = 1\,. \tag{10.23}$$

The operator $\bar{\boldsymbol{L}}_{\mathrm{ir}}$ is then

$$\bar{\boldsymbol{L}}_{\mathrm{ir}} = -\gamma b^+ b\,. \tag{10.24}$$

To take advantage of (10.24) we have to transform the Fokker-Planck operator $\boldsymbol{L}_{\mathrm{K}}$ according to (10.21). Instead of (10.21) we use the more general transformation

$$\begin{aligned}&\bar{\boldsymbol{L}}_{\mathrm{K}} = \bar{\boldsymbol{L}}_{\mathrm{rev}}+\bar{\boldsymbol{L}}_{\mathrm{ir}} \\ &\bar{\boldsymbol{L}}_{\substack{\mathrm{rev}\\ \mathrm{ir}}} = \exp\left[\frac{1}{4}\left(\frac{v}{v_{\mathrm{th}}}\right)^2+\varepsilon\frac{f(x)}{v_{\mathrm{th}}^2}\right]\boldsymbol{L}_{\substack{\mathrm{rev}\\ \mathrm{ir}}}\exp\left[-\frac{1}{4}\left(\frac{v}{v_{\mathrm{th}}}\right)^2-\varepsilon\frac{f(x)}{v_{\mathrm{th}}^2}\right],\end{aligned} \tag{10.25}$$

where ε is an arbitrary constant. Because $\boldsymbol{L}_{\mathrm{ir}}$ does not act on x, $\bar{\boldsymbol{L}}_{\mathrm{ir}}$ is still given by (10.21, 24). For the reversible operator $\bar{\boldsymbol{L}}_{\mathrm{rev}}$ we get

$$\begin{aligned}\bar{\boldsymbol{L}}_{\mathrm{rev}} &= -v\,\partial/\partial x+f'\,\partial/\partial v+(\varepsilon-\tfrac{1}{2})f'\,v/v_{\mathrm{th}}^2 \\ &= \boldsymbol{L}_{\mathrm{rev}}+(\varepsilon-\tfrac{1}{2})f'\,v/v_{\mathrm{th}}^2 = -bD-b^+\hat{D}\,. \end{aligned} \tag{10.26}$$

In the last expression we have introduced the differential operators D and $\hat{D}$ defined by

$$D = v_{\text{th}}\,\partial/\partial x - \varepsilon f'/v_{\text{th}}\,, \qquad \hat{D} = v_{\text{th}}\,\partial/\partial x + (1-\varepsilon) f'/v_{\text{th}}\,. \tag{10.27}$$

The commutator between D and $\hat{D}$ is

$$[D, \hat{D}] = f''\,. \tag{10.28}$$

For $\varepsilon = \frac{1}{2}$ the differential operators are the negative adjoints of each other, i.e.,

$$D^+ = -\hat{D} \quad \text{for} \quad \varepsilon = \tfrac{1}{2} \tag{10.29}$$

and the transformation (10.25) is then the transformation with the square root of the stationary solution (if it exists), according to (6.35). For $\varepsilon = \frac{1}{2}$ the operator $\bar{L}_{\text{rev}}$ is still an anti-Hermitian operator

$$\bar{L}_{\text{rev}}^+ = -b^+ D^+ - b\hat{D}^+ = b^+\hat{D} + bD = -\bar{L}_{\text{rev}}\,, \quad \text{for} \quad \varepsilon = \tfrac{1}{2}\,. \tag{10.30}$$

For other values of the parameter ε this is no longer true. As will be shown below, the choice $\varepsilon = 0$ is also very convenient.

In the three-dimensional case we may proceed in the same manner. By introducing the boson and differential operators

$$b_i = v_{\text{th}}\frac{\partial}{\partial v_i} + \frac{1}{2}\frac{v_i}{v_{\text{th}}}\,, \qquad b_i^+ = -v_{\text{th}}\frac{\partial}{\partial v_i} + \frac{1}{2}\frac{v_i}{v_{\text{th}}}\,, \tag{10.31}$$

$$D_i = v_{\text{th}}\frac{\partial}{\partial x_i} - \frac{\varepsilon}{v_{\text{th}}}\frac{\partial f}{\partial x_i}\,, \qquad \hat{D}_i = v_{\text{th}}\frac{\partial}{\partial x_i} + \frac{(1-\varepsilon)}{v_{\text{th}}}\frac{\partial f}{\partial x_i} \tag{10.32}$$

the Fokker-Planck operator may be written as

$$L_{\text{K}} = \exp\left(-\frac{1}{4}\frac{v^2}{v_{\text{th}}^2} - \varepsilon\frac{f(\boldsymbol{x})}{v_{\text{th}}^2}\right)\bar{L}_{\text{K}}\exp\left(\frac{1}{4}\frac{v^2}{v_{\text{th}}^2} + \varepsilon\frac{f(\boldsymbol{x})}{v_{\text{th}}^2}\right), \tag{10.33}$$

$$\bar{L}_{\text{K}} = \bar{L}_{\text{rev}} + \bar{L}_{\text{ir}}\,, \tag{10.34}$$

$$\bar{L}_{\text{rev}} = -\sum_{i=1}^{3}(b_i D_i + b_i^+\hat{D}_i)\,, \tag{10.35}$$

$$\bar{L}_{\text{ir}} = -\gamma\sum_{i=1}^{3} b_i^+ b_i\,. \tag{10.36}$$

10.1.4 Expansion into Hermite Functions

By expanding the velocity part of the distribution function $W(x, v, t)$ of the Fokker-Planck equation into Hermite functions we obtain a tridiagonal coupled system of partial differential equations for the position and time-dependent expansion coefficients. As shown in Sect. 10.3, this coupled system may be solved by matrix continued-fraction methods. We choose Hermite functions for the following reasons: (i) they are eigenfunctions of $\bar{L}_{\text{ir}}$, (ii) they form a complete system, (iii) they have the correct natural boundary conditions in velocity space $-\infty < v < \infty$, (iv) they lead to the tridiagonal structure of the coupling mentioned above.

The normalized eigenfunctions of the operator $\bar{L}_{\text{ir}} = -\gamma b^+ b$, i.e.,

$$\bar{L}_{\text{ir}} \psi_n(v) = -\gamma n \psi_n(v), \quad b^+ b \psi_n(v) = n \psi_n(v) \tag{10.37}$$

are given by

$$\begin{aligned} \psi_n(v) &= (b^+)^n \psi_0(v)/\sqrt{n!} \\ \psi_0(v) &= \exp[-\tfrac{1}{4}(v/v_{\text{th}})^2]/\sqrt{v_{\text{th}}\sqrt{2\pi}}\,. \end{aligned} \tag{10.38}$$

In terms of the Hermite polynomials $\text{H}_n(x)$ usually used in quantum mechanics or in terms of the Hermite polynomials $\text{He}_n(x) = 2^{-n/2}\text{H}_n(x/\sqrt{2})$ (see [Ref. 9.26, p. 250] for a definition) the functions $\psi_n(v)$ read

$$\psi_n(v) = \text{H}_n(v/(\sqrt{2}\, v_{\text{th}})) \exp[-v^2/(4v_{\text{th}}^2)]/\sqrt{n!\, 2^n v_{\text{th}}\sqrt{2\pi}}\,, \tag{10.39}$$

$$\psi_n(v) = \text{He}_n(v/v_{\text{th}}) \exp[-v^2/(4v_{\text{th}}^2)]/\sqrt{n!\, v_{\text{th}}\sqrt{2\pi}}\,. \tag{10.40}$$

They agree with those used in Sect. 5.5.1, if we replace v by x and v_{th} by $\sqrt{D/\gamma}$. The variable ξ used in that section is then connected to v by $\sqrt{2}\,\xi = v/v_{\text{th}}$.

The eigenfunctions $\psi_n(v)$ are orthonormalized

$$\int_{-\infty}^{\infty} \psi_n(v)\, \psi_m(v)\, dv = \delta_{nm}\,. \tag{10.41}$$

Because the Fokker-Planck operator L_{K} is of the form [compare (10.25, 38)]

$$\begin{aligned} L_{\text{K}} = &-\psi_0(v) \exp[-\varepsilon f(x)/v_{\text{th}}^2](\gamma b^+ b + bD + b^+ \hat{D}) \\ &\times \exp[\varepsilon f(x)/v_{\text{th}}^2][\psi_0(v)]^{-1} \end{aligned} \tag{10.42}$$

we expand $W(x, v, t)$ in the following way:

$$W(x, v, t) = \psi_0(v) \exp[-\varepsilon f(x)/v_{\text{th}}^2] \sum_{n=0}^{\infty} c_n(x, t)\, \psi_n(v)\,. \tag{10.43}$$

The distribution function in position only is then given by the first expansion coefficient

$$\int_{-\infty}^{\infty} W(x, v, t)\, dv = \exp[-\varepsilon f(x)/v_{\text{th}}^2]\, c_0(x, t) \tag{10.44}$$

and the x component of the probability current (10.4b) integrated over the velocities is given by the next expansion coefficient $[v\,\psi_0(v) = v_{\text{th}}\,\psi_1(v)]$

$$\begin{aligned}\int S_x(x,v,t)\,\mathrm{d}v &= \int v\,W(x,v,t)\,\mathrm{d}v \\ &= v_{\text{th}}\exp[-\varepsilon f(x)/v_{\text{th}}^2]\,c_1(x,t)\,.\end{aligned} \tag{10.44a}$$

Equation of Motion for the Expansion Coefficients

To obtain an equation of motion for the expansion coefficients, we insert (10.43) into the Fokker-Planck equation $\dot{W} = L_{\text{K}} W$. Because of (10.26, 37) and because

$$\begin{aligned} b^+\,\psi_n(v) &= \sqrt{n+1}\,\psi_{n+1}(v) \\ b\,\psi_n(v) &= \sqrt{n}\quad\psi_{n-1}(v)\,, \end{aligned} \tag{10.45}$$

we easily obtain the following hierarchy for the expansion coefficients $c_n(x,t)$ ($c_n = 0$ for $n<0$)

$$\partial c_n/\partial t = -\sqrt{n}\hat{D}c_{n-1} - n\gamma c_n - \sqrt{n+1}\,D c_{n+1}\,. \tag{10.46}$$

If instead of the coefficients c_n the coefficients

$$\bar{c}_n = c_n/\sqrt{n!} \tag{10.47}$$

are used, the square roots in the recurrence relations (10.46) disappear

$$\partial \bar{c}_n/\partial t = -\hat{D}\bar{c}_{n-1} - n\gamma\bar{c}_n - (n+1)D\bar{c}_{n+1}\,. \tag{10.48}$$

[For numerical purposes one should use the coefficients c_n, because otherwise the large factors $\sqrt{n!}$ occur in the expansion (10.43).]

The Laplace transform of (10.48) for $\varepsilon = 0$ was first derived by *Brinkman* [10.1], therefore (10.46, 48) are sometimes called Brinkman's hierarchy. This hierarchy of equations is equivalent to the Kramers equation. Equations (10.46) read explicitly

$$\begin{aligned} &(\partial/\partial t)c_0 + \sqrt{1}Dc_1 &&= 0 \\ &\sqrt{1}\hat{D}c_0 + (\partial/\partial t + 1\gamma)c_1 + \sqrt{2}Dc_2 &&= 0 \\ &\qquad\sqrt{2}\hat{D}c_1 + (\partial/\partial t + 2\gamma)c_2 + \sqrt{3}Dc_3 &&= 0 \\ &\qquad\qquad\sqrt{3}\hat{D}c_2 + (\partial/\partial t + 3\gamma)c_3 + \sqrt{4}Dc_4 &&= 0 \\ &\qquad\qquad\qquad\ldots \qquad \ldots \qquad \ldots &&= 0\,. \end{aligned} \tag{10.46a}$$

In the stationary state, a general solution of (10.46a) is easily obtained for the case where the probability current in x direction integrated over the velocities (10.44a) vanishes, i.e., $c_1 = 0$. This solution is given by

$$\begin{aligned} \hat{D}c_0 &= 0\,, \quad \text{i.e.,} \quad c_0(x) \sim \exp[-(1-\varepsilon)f(x)/v_{\text{th}}^2)] \\ c_n &= 0 \quad \text{for} \quad n \geqq 1 \end{aligned} \tag{10.49}$$

and the corresponding distribution (10.43) is the Boltzmann distribution

$$\begin{aligned} W_{\text{st}} &= N\psi_0^2(v)\exp[-f(x)/v_{\text{th}}^2] = [N/(\sqrt{2\pi}\,v_{\text{th}})]\exp[-E/(kT)]\,, \\ N &= \{\textstyle\int \exp[-f(x)/v_{\text{th}}^2]\,dx\}^{-1}\,. \end{aligned} \tag{10.49a}$$

Here $E = mv^2/2 + mf(x)$ is the energy. If the probability current in x direction integrated over the velocities does not vanish or if the instationary solution is considered, the coefficients c_n are generally not zero for $n \geqq 1$.

In the three-dimensional case we may use the expansion

$$\begin{aligned} W(\boldsymbol{x}, \boldsymbol{v}, t) &= \psi_0(v_1)\,\psi_0(v_2)\,\psi_0(v_3)\exp[-\varepsilon f(\boldsymbol{x})/v_{\text{th}}^2] \\ &\quad \times \sum_{n_1,n_2,n_3} c_{n_1,n_2,n_3}(\boldsymbol{x},t)\,\psi_{n_1}(v_1)\,\psi_{n_2}(v_2)\,\psi_{n_3}(v_3) \end{aligned} \tag{10.50}$$

and obtain the hierarchy

$$\begin{aligned} \dot{c}_{n_1,n_2,n_3} = &-\sqrt{n_1}\,\hat{D}_1 c_{n_1-1,n_2,n_3} - \sqrt{n_2}\,\hat{D}_2 c_{n_1,n_2-1,n_3} - \sqrt{n_3}\,\hat{D}_3 c_{n_1,n_2,n_3-1} \\ &- \gamma(n_1+n_2+n_3)\,c_{n_1,n_2,n_3} - \sqrt{n_1+1}\,D_1 c_{n_1+1,n_2,n_3} \\ &- \sqrt{n_2+1}\,D_2 c_{n_1,n_2+1,n_3} - \sqrt{n_3+1}\,D_3 c_{n_1,n_2,n_3+1}\,, \end{aligned} \tag{10.51}$$

where D_i and $\hat{D}_i$ are given by (10.32).

In this case, the operator $\bar{L}_{\text{ir}}$ is essentially the same as the Hamilton operator for a spherical harmonic oscillator.

10.2 Solutions for a Linear Force

For a linear force per mass $-f'(x) = ax + b$ the Fokker-Planck equation can be solved exactly. We first note that the constant b is taken into account by a proper shift in x ($\bar{x} = x + b/a$) so that without loss of generality b can be put to zero. First we treat the case of a harmonically bound particle [1.5–7]

$$-f'(x) = -\omega_0^2 x\,, \quad f(x) = \omega_0^2 x^2/2\,, \quad \omega_0^2 > 0 \tag{10.52}$$

and later the case of a repulsive linear force.

10.2.1 Transition Probability

The Langevin equations (10.2), which may be written in the form

$$\frac{\mathrm{d}}{\mathrm{d}t}\begin{pmatrix} x \\ v \end{pmatrix} = -\begin{pmatrix} 0 & -1 \\ \omega_0^2 & \gamma \end{pmatrix}\begin{pmatrix} x \\ v \end{pmatrix} + \begin{pmatrix} 0 \\ \Gamma(t) \end{pmatrix}, \tag{10.53}$$

describe a two-dimensional Ornstein-Uhlenbeck process. Therefore the general solutions in Sect. 6.5 can be applied. The drift matrix $\boldsymbol{\gamma}$ and the diffusion matrix $\boldsymbol{D}$ then read

$$\boldsymbol{\gamma} = \begin{pmatrix} 0 & -1 \\ \omega_0^2 & \gamma \end{pmatrix}; \quad \boldsymbol{D} = \begin{pmatrix} 0 & 0 \\ 0 & \gamma v_{\text{th}}^2 \end{pmatrix}. \tag{10.54}$$

According to (6.124) the transition probability $P(x,v,t|x',v',0)$ is given by the two-variable Gaussian distribution ($t \geqq 0$)

$$\begin{aligned} P(x,v,t|x',v',0) = (2\pi)^{-1}(\operatorname{Det}\boldsymbol{\sigma})^{-1/2}\exp\{-\tfrac{1}{2}[\boldsymbol{\sigma}^{-1}(t)]_{xx}[x-x(t)]^2 \\ -[\boldsymbol{\sigma}^{-1}(t)]_{xv}[x-x(t)][v-v(t)] - \tfrac{1}{2}[\boldsymbol{\sigma}^{-1}(t)]_{vv}[v-v(t)]^2\} \end{aligned} \tag{10.55}$$

with mean values (6.117, 122)

$$\begin{aligned} \langle x\rangle = x(t) = [\exp(-\boldsymbol{\gamma}t)]_{xx}x' + [\exp(-\boldsymbol{\gamma}t)]_{xv}v' \\ \langle v\rangle = v(t) = [\exp(-\boldsymbol{\gamma}t)]_{vx}x' + [\exp(-\boldsymbol{\gamma}t)]_{vv}v' . \end{aligned} \tag{10.56}$$

By using the spectral decomposition (6.121) ($*$ indicates the dyadic product)

$$\boldsymbol{\gamma} = \lambda_1 \boldsymbol{u}^{(1)} * \boldsymbol{v}^{(1)} + \lambda_2 \boldsymbol{u}^{(2)} * \boldsymbol{v}^{(2)}, \tag{10.57}$$

$$\boldsymbol{u}^{(1)} = \begin{pmatrix} -1 \\ \lambda_1 \end{pmatrix}, \quad \boldsymbol{u}^{(2)} = \begin{pmatrix} 1 \\ -\lambda_2 \end{pmatrix}, \tag{10.58}$$

$$\boldsymbol{v}^{(1)} = \frac{1}{\lambda_1-\lambda_2}(\lambda_2, 1), \quad \boldsymbol{v}^{(2)} = \frac{1}{\lambda_1-\lambda_2}(\lambda_1, 1), \tag{10.59}$$

$$\lambda_{1,2} = \tfrac{1}{2}(\gamma \pm \sqrt{\gamma^2 - 4\omega_0^2}), \quad \lambda_1+\lambda_2 = \gamma, \quad \lambda_1\lambda_2 = +\omega_0^2 \tag{10.60}$$

we get, see (6.122),

$$G_{xx}(t) = [\exp(-\boldsymbol{\gamma}t)]_{xx} = \frac{\lambda_1 \mathrm{e}^{-\lambda_2 t} - \lambda_2 \mathrm{e}^{-\lambda_1 t}}{\lambda_1 - \lambda_2}$$

$$G_{xv}(t) = [\exp(-\boldsymbol{\gamma}t)]_{xv} = \frac{\mathrm{e}^{-\lambda_2 t} - \mathrm{e}^{-\lambda_1 t}}{\lambda_1 - \lambda_2}$$

$$G_{vx}(t) = [\exp(-\gamma t)]_{vx} = \omega_0^2 \frac{e^{-\lambda_1 t} - e^{-\gamma_2 t}}{\lambda_1 - \lambda_2}$$

$$G_{vv}(t) = [\exp(-\gamma t)]_{vv} = \frac{\lambda_1 e^{-\lambda_1 t} - \lambda_2 e^{-\lambda_2 t}}{\lambda_1 - \lambda_2}. \tag{10.61}$$

The symmetry matrix σ^{-1} follows from the symmetric matrix $\sigma(\sigma_{xv} = \sigma_{vx})$

$$\begin{aligned} (\sigma^{-1})_{xx} &= \sigma_{vv}/\mathrm{Det}\,\sigma \\ (\sigma^{-1})_{xv} &= (\sigma^{-1})_{vx} = -\sigma_{xv}/\mathrm{Det}\,\sigma \\ (\sigma^{-1})_{vv} &= \sigma_{xx}/\mathrm{Det}\,\sigma \\ \mathrm{Det}\,\sigma &= \sigma_{xx}\sigma_{vv} - (\sigma_{xv})^2 . \end{aligned} \tag{10.62}$$

Finally, the expressions for the σ matrix, which may be obtained from (6.123) or (6.118), are

$$\begin{aligned} \sigma_{xx}(t) = \frac{\gamma v_{\mathrm{th}}^2}{(\lambda_1 - \lambda_2)^2} \Bigg[& \frac{\lambda_1 + \lambda_2}{\lambda_1 \lambda_2} + \frac{4}{\lambda_1 + \lambda_2} (e^{-(\lambda_1 + \lambda_2)t} - 1) \\ & - \frac{1}{\lambda_1} e^{-2\lambda_1 t} - \frac{1}{\lambda_2} e^{-2\lambda_2 t} \Bigg], \end{aligned} \tag{10.63}$$

$$\sigma_{xv}(t) = \frac{\gamma v_{\mathrm{th}}^2}{(\lambda_1 - \lambda_2)^2} (e^{-\lambda_1 t} - e^{-\lambda_2 t})^2 ,$$

$$\sigma_{vv}(t) = \frac{\gamma v_{\mathrm{th}}^2}{(\lambda_1 - \lambda_2)^2} \left[\lambda_1 + \lambda_2 + \frac{4\lambda_1 \lambda_2}{\lambda_1 + \lambda_2} (e^{-(\lambda_1 + \lambda_2)t} - 1) - \lambda_1 e^{-2\lambda_1 t} - \lambda_2 e^{-2\lambda_2 t} \right].$$

These expressions may be found for the three-dimensional case in [1.6, 7], together with expressions for (10.61, 63) in terms of hyperbolic and trigonometric functions (overdamped or underdamped case).

Stationary Solution

For $\omega_0^2 > \gamma^2/4$ the real parts of the eigenvalues $\lambda_{1,2}$ and for $\gamma^2/4 \geqq \omega_0^2 > 0$ the eigenvalues are larger than zero. Therefore (10.61) vanishes for $t \to \infty$ and for (10.63) we obtain

$$\begin{aligned} \sigma_{xx}(\infty) &= \frac{v_{\mathrm{th}}^2}{\lambda_1 \lambda_2} = \frac{v_{\mathrm{th}}^2}{\omega_0^2} \\ \sigma_{xv}(\infty) &= 0 \\ \sigma_{vv}(\infty) &= v_{\mathrm{th}}^2 . \end{aligned} \tag{10.64}$$

Thus we have

$$
\begin{aligned}
&(\sigma^{-1})_{xx} = \omega_0^2 v_{\mathrm{th}}^{-2}, \quad (\sigma^{-1})_{xv} = 0, \quad (\sigma^{-1})_{vv} = v_{\mathrm{th}}^{-2}, \\
&\quad \mathrm{Det}\,\sigma = v_{\mathrm{th}}^4/\omega_0^2
\end{aligned} \tag{10.65}
$$

and the stationary distribution is the Boltzmann distribution

$$
\begin{aligned}
W_{\mathrm{st}}(x,v) &= P(x,v,\infty\,|x',v',0) \\
&= \frac{\omega_0}{2\pi v_{\mathrm{th}}^2} \exp\left(-\frac{1}{2}\frac{v^2}{v_{\mathrm{th}}^2} - \frac{\omega_0^2 x^2}{2 v_{\mathrm{th}}^2}\right) \\
&= \frac{m\omega_0}{2\pi kT} \exp\left(-\frac{mv^2 + m\omega_0^2 x^2}{2kT}\right) \\
&= \frac{m\omega_0}{2\pi kT} \exp\left(-\frac{E}{kT}\right).
\end{aligned} \tag{10.66}
$$

Free Brownian Motion

For Brownian motion without an external force, we obtain the corresponding expression by taking the limit $\omega_0^2 \to 0$, i.e., $\lambda_1 \to \gamma$, $\lambda_2 \to 0$.

Expressions (10.56, 63) then simplify to

$$
\begin{aligned}
x(t) &= x' + \gamma^{-1}(1 - \mathrm{e}^{-\gamma t})v' \\
v(t) &= \mathrm{e}^{-\gamma t} v'
\end{aligned} \tag{10.67}
$$

$$
\begin{aligned}
\sigma_{xx}(t) &= v_{\mathrm{th}}^2 \gamma^{-2}(2\gamma t - 3 + 4\mathrm{e}^{-\gamma t} - \mathrm{e}^{-2\gamma t}) \\
\sigma_{xv}(t) &= v_{\mathrm{th}}^2 \gamma^{-1}(1 - \mathrm{e}^{-\gamma t})^2 \\
\sigma_{vv}(t) &= v_{\mathrm{th}}^2 (1 - \mathrm{e}^{-2\gamma t}).
\end{aligned} \tag{10.68}
$$

Inverted Parabolic Potential

For the inverted parabolic potential

$$
f(x) = -\tfrac{1}{2}\omega_1^2 x^2, \quad \omega_1^2 > 0 \tag{10.69}
$$

all the transition-probability expressions remain valid. The only exception is that the eigenvalue equation (10.60) has to be replaced by

$$
\lambda_{1,2} = \tfrac{1}{2}\left(\gamma \pm \sqrt{\gamma^2 + 4\omega_1^2}\right). \tag{10.70}
$$

As it is immediately seen, the second eigenvalue λ_2 then has a negative value. For large times $x(t)$ and $v(t)$ grow exponentially if $x(0) \neq 0$ or $v(0) \neq 0$. Also the width of the distribution grows exponentially. No stationary solution exists in this case.

10.2.2 Eigenvalues and Eigenfunctions

To find the eigenvalues and eigenfunctions of the operator (10.3 b)

$$L_{\mathrm{K}}\psi(x,v) = -\lambda\,\psi(x,v) \tag{10.71}$$

we first transform the Fokker-Planck operator according to (10.25) with $\varepsilon = \frac{1}{2}$. The eigenfunctions $\bar{\psi}$ of the transformed operator $\bar{L}_{\mathrm{K}}$

$$\bar{L}_{\mathrm{K}}\bar{\psi}(x,v) = -\lambda\,\bar{\psi}(x,v) \tag{10.72}$$

and those of L_{K} are connected by

$$\bar{\psi}(x,v) = \exp\left[\frac{1}{4}\left(\frac{v}{v_{\mathrm{th}}}\right)^2 + \frac{1}{2}\,\frac{f(x)}{v_{\mathrm{th}}^2}\right]\psi(x,v)\,. \tag{10.73}$$

For a harmonic potential (10.52), the commutator of D and $\hat{D}$ is equal to the constant ω_0^2 (10.28). We therefore introduce the boson operators

$$\begin{aligned} a &= \frac{\hat{D}}{\omega_0} = \frac{v_{\mathrm{th}}}{\omega_0}\,\frac{\partial}{\partial x} + \frac{\omega_0}{2v_{\mathrm{th}}}\,x \\ a^+ &= -\frac{D}{\omega_0} = -\frac{v_{\mathrm{th}}}{\omega_0}\,\frac{\partial}{\partial x} + \frac{\omega_0}{2v_{\mathrm{th}}}\,x\,, \end{aligned} \tag{10.74}$$

which fulfill the boson commutation relation $[a,a^+]=1$. The operator $\bar{L}_{\mathrm{K}}$ then takes the form

$$\bar{L}_{\mathrm{K}} = -\gamma b^+ b - \omega_0(ab^+ - a^+ b)\,. \tag{10.75}$$

The stationary ($\lambda = 0$) solution of (10.72) which we call $\bar{\psi}_{0,0}$ is obtained from

$$a\,\bar{\psi}_{0,0} = b\,\bar{\psi}_{0,0} = 0\ , \tag{10.76}$$

leading to

$$\bar{\psi}_{0,0}(x,v) = \sqrt{\frac{\omega_0}{2\pi v_{\mathrm{th}}^2}}\,\exp\left[-\frac{1}{4}\left(\frac{v}{v_{\mathrm{th}}}\right)^2 - \frac{1}{4}\,\frac{\omega_0^2 x^2}{v_{\mathrm{th}}^2}\right], \tag{10.77}$$

which is the square root of (10.66). (The normalization will be discussed later.)

To transform (10.75) to a simpler form we introduce new operators by the transformations

$$\begin{aligned} c_{1+} &= \delta^{-1/2}(\sqrt{\lambda_1}\,b^+ - \sqrt{\lambda_2}\,a^+) \\ c_{1-} &= \delta^{-1/2}(\sqrt{\lambda_1}\,b + \sqrt{\lambda_2}\,a) \\ c_{2+} &= \delta^{-1/2}(-\sqrt{\lambda_2}\,b^+ + \sqrt{\lambda_1}\,a^+) \\ c_{2-} &= \delta^{-1/2}(\sqrt{\lambda_2}\,b + \sqrt{\lambda_1}\,a)\,. \end{aligned} \tag{10.78}$$

In (10.78) λ_1 and λ_2 are the eigenvalues of the equation of motion without noise [see (10.60)] and the square root is denoted by δ

$$\lambda_{\substack{1\\2}} = (\gamma \pm \delta)/2\,, \quad \delta = \sqrt{\gamma^2 - 4\omega_0^2} = \lambda_1 - \lambda_2\,. \tag{10.79}$$

The inverse relations to (10.78) read

$$\begin{aligned} b^+ &= \delta^{-1/2}(\sqrt{\lambda_1}c_{1+} + \sqrt{\lambda_2}c_{2+}) \\ b &= \delta^{-1/2}(\sqrt{\lambda_1}c_{1-} - \sqrt{\lambda_2}c_{2-}) \\ a^+ &= \delta^{-1/2}(\sqrt{\lambda_2}c_{1+} + \sqrt{\lambda_1}c_{2+}) \\ a &= \delta^{-1/2}(-\sqrt{\lambda_2}c_{1-} + \sqrt{\lambda_1}c_{2-})\,. \end{aligned} \tag{10.80}$$

The operator $\bar{L}_\mathrm{K}$ is transformed to

$$\bar{L}_\mathrm{K} = -\lambda_1 c_{1+}c_{1-} - \lambda_2 c_{2+}c_{2-} \tag{10.75a}$$

and the commutation relations are

$$\begin{aligned} &[c_{1-}, c_{1+}] = [c_{2-}, c_{2+}] = 1 \\ &[c_{1-}, c_{2+}] = [c_{2-}, c_{1+}] = [c_{1-}, c_{2-}] = [c_{1+}, c_{2+}] = 0\,. \end{aligned} \tag{10.81a}$$

$$[(-\bar{L}_\mathrm{K}), c_{i\pm}] = \pm\lambda_i c_{i\pm}\,; \quad (i = 1, 2)\,. \tag{10.81b}$$

Though c_{i+} is not the adjoint operator of c_{i-}, eigenfunctions of $\bar{L}_\mathrm{K}$ can be constructed in the same way as for the harmonic oscillator in quantum mechanics

$$\bar{\psi}_{n_1,n_2}(x, v) = (n_1!\, n_2!)^{-1/2}(c_{1+})^{n_1}(c_{2+})^{n_2}\bar{\psi}_{0,0}(x, v)\,. \tag{10.82}$$

The eigenvalues of $\bar{L}_\mathrm{K}$ are therefore given by

$$\begin{aligned} \lambda_{n_1,n_2} &= \lambda_1 n_1 + \lambda_2 n_2 \\ &= \tfrac{1}{2}\gamma(n_1 + n_2) + \tfrac{1}{2}\delta(n_1 - n_2)\,. \end{aligned} \tag{10.83}$$

For the underdamped case $\gamma < 2\omega_0$ these eigenvalues are complex

$$\delta = \mathrm{i}\sqrt{4\omega_0^2 - \gamma^2} \equiv 2\mathrm{i}\omega\,, \tag{10.84}$$

whereas for the overdamped case all the eigenvalues are real. For large friction $\gamma \gg 2\omega_0$ the eigenvalues are grouped together according to

$$\lambda_{n_1,n_2} = \gamma n_1 + (\omega_0^2/\gamma)(n_2 - n_1) + O(\gamma^{-3})\,. \tag{10.85}$$

Some patterns of eigenvalues are shown in Fig. 10.1.

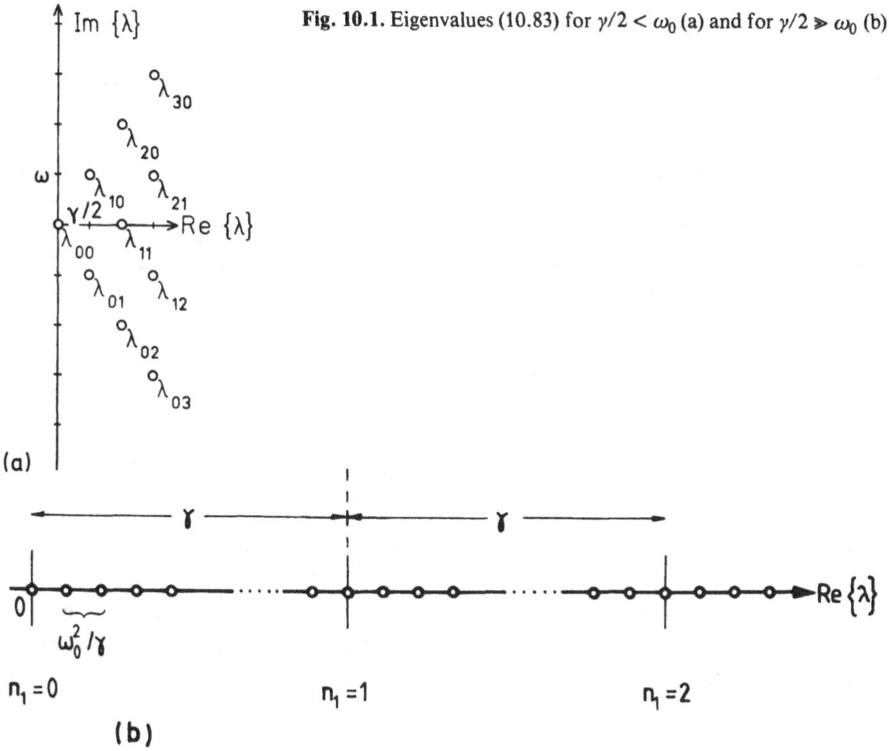

Fig. 10.1. Eigenvalues (10.83) for $\gamma/2 < \omega_0$ (a) and for $\gamma/2 \gg \omega_0$ (b)

The eigenfunctions $\bar{\psi}^+$ of the adjoint operator

$$\bar{L}_K^+ = -\lambda_1 c_{1-}^+ c_{1+}^+ - \lambda_2 c_{2-}^+ c_{2+}^+ \tag{10.86}$$

with the eigenvalues λ_{n_1,n_2} are

$$\bar{\psi}_{n_1,n_2}^+(x,v) = (n_1!\, n_2!)^{-1/2}(c_{1-}^+)^{n_1}(c_{2-}^+)^{n_2}\bar{\psi}_{0,0}(x,v)\,. \tag{10.87}$$

The adjoint operator is defined by

$$\int (A^+\varphi)\,\psi\,\mathrm{d}x\,\mathrm{d}v = \int \varphi A\,\psi\,\mathrm{d}x\,\mathrm{d}v\,. \tag{10.88}$$

Notice that we do not take the conjugate complex definition of the scalar product as used in quantum mechanics.

The eigenfunctions $\bar{\psi}^+$ and $\bar{\psi}$ form a biorthogonal set, i.e.,

$$\int \bar{\psi}_{n_1,n_2}^+(x,v)\,\bar{\psi}_{n_1',n_2'}(x,v)\,\mathrm{d}x\,\mathrm{d}v = \delta_{n_1,n_1'}\delta_{n_2,n_2'}\,. \tag{10.89}$$

As seen from (10.30, 22) or (10.26), the adjoint operator of $\bar{L}_K$ for $\varepsilon = \frac{1}{2}$ is obtained from $\bar{L}_K$ by replacing v by $-v$

$$\bar{L}_K^+(x,v) = \bar{L}_K(x,-v)\,. \tag{10.90}$$

Because $c_{1-}^+(x,v) = -c_{1+}(x,-v)$ and $c_{2-}^+(x,v) = c_{2+}(x,-v)$, the eigenfunctions (10.87, 82) are connected by

$$\bar{\psi}_{n_1,n_2}^+(x,v) = (-1)^{n_1}\,\bar{\psi}_{n_1,n_2}(x,-v)\,. \tag{10.91}$$

Because $\bar{\psi}_{n_1,n_2}^+$ and $\bar{\psi}_{n_1,n_2}$ form a complete set and because the prefactor of ψ^+ is the inverse of the corresponding one for ψ (10.73), the transition probability (10.55) in terms of eigenfunctions is given by

$$P(x,v,t\,|\,x',v',0) = \exp\left(-\frac{1}{4}\,\frac{v^2-v'^2}{v_{\text{th}}^2} - \frac{\omega_0^2}{4}\,\frac{x^2-x'^2}{v_{\text{th}}^2}\right)$$
$$\times \sum_{n_1,n_2=0}^{\infty} \bar{\psi}_{n_1,n_2}^+(x',v')\,\bar{\psi}_{n_1,n_2}(x,v)\exp(-\lambda_{n_1,n_2}t)\,. \tag{10.92}$$

(For $t=0$ one obtains the completeness relation.) A similar derivation to that here is in the appendix of [10.2].

Another method for determining the eigenvalues of (10.75) is the following. We expand the eigenfunction $\bar{\psi}$ into eigenfunctions $\psi_n(v)$ of b^+b and $\varphi_n(x)$ of a^+a in the form

$$\bar{\psi}(x,v) = \sum_{n,m=0}^{\infty} (c_{n,m}/\sqrt{n!\,m!})\,\psi_n(v)\,\varphi_m(x)\,. \tag{10.93}$$

Inserting (10.93) into $\bar{L}_K\bar{\psi} = -\lambda\bar{\psi}$ leads to ($c_{n,m}=0$ for $n<0$ or $m<0$)

$$(\gamma n-\lambda)c_{n,m} + \omega_0 n c_{n-1,m+1} - \omega_0 m c_{n+1,m-1} = 0\,. \tag{10.94}$$

By this relation only a finite number of coefficients with $n+m=N$ are coupled. Changing the notation of the coefficients to

$$y_n^{(N)} = c_{n,N-n} \tag{10.95}$$

we get the tridiagonal coupled relation ($0 \leqq n \leqq N$)

$$n\,\omega_0 y_{n-1}^{(N)} + (n\gamma-\lambda)\,y_n^{(N)} - (N-n)\,\omega_0 y_{n+1}^{(N)} = 0\,. \tag{10.96}$$

The eigenvalues for $N=3$, for instance, can be determined from the special determinant (= continuant)

$$\begin{vmatrix} -\lambda & -3\omega_0 & 0 & 0 \\ 1\omega_0 & \gamma-\lambda & -2\omega_0 & 0 \\ 0 & 2\omega_0 & 2\gamma-\lambda & -\omega_0 \\ 0 & 0 & 3\omega_0 & 3\gamma-\lambda \end{vmatrix} = 0\,. \tag{10.97}$$

It follows from (10.82, 83) that the four eigenvalues must have the form $(n_1+n_2=N=3)$

$$\lambda = 3\gamma/2 + \sqrt{\gamma^2/4-\omega_0^2}\times\begin{cases}3\\1\\-1\\-3\end{cases}, \tag{10.98}$$

which may be checked by insertion.

Velocity Correlation Function

For this linear process, the stationary velocity correlation function $\langle v(t)\,v(0)\rangle$ can best be calculated using the Langevin equation or the fluctuation-dissipation theorem, see (7.63). As an application of the eigenfunction expansion (10.92), however, we determine the correlation function using this expansion. The correlation function is defined by

$$\begin{aligned}\langle v(t)\,v(0)\rangle &= \int v\,v'\,P(x,v,t\,|\,x',v',0)\,W_{\rm st}(x',v')\,dx\,dv\,dx'\,dv'\\ &= \sum_{n_1 n_2}\int v\,\bar\psi_{0,0}\,\bar\psi_{n_1,n_2}dx\,dv\int v\,\bar\psi_{0,0}\,\bar\psi^+_{n_1,n_2}dx\,dv\,\exp(-\lambda_{n_1,n_2}t)\,.\end{aligned} \tag{10.99}$$

From $v=v_{\rm th}(b^+ + b)$ and (10.77, 80, 82, 87) we have

$$\begin{aligned}v\,\bar\psi_{0,0} &= v_{\rm th}b^+\,\bar\psi_{0,0} = v_{\rm th}\delta^{-1/2}(\sqrt{\lambda_1}c_{1+}+\sqrt{\lambda_2}c_{2+})\,\bar\psi_{0,0}\\ &= v_{\rm th}\delta^{-1/2}(\sqrt{\lambda_1}\,\bar\psi_{1,0}+\sqrt{\lambda_2}\,\bar\psi_{0,1})\\ &= v_{\rm th}\delta^{-1/2}(\sqrt{\lambda_1}c^+_{1-}-\sqrt{\lambda_2}c^+_{2-})\,\bar\psi_{0,0}\\ &= v_{\rm th}\delta^{-1/2}(\sqrt{\lambda_1}\,\bar\psi^+_{1,0}-\sqrt{\lambda_2}\,\bar\psi^+_{0,1})\,.\end{aligned} \tag{10.100}$$

The biorthogonality relation (10.89) leads to

$$\begin{aligned}&\langle v(t)\,v(0)\rangle = v_{\rm th}^2\frac{[\lambda_1\exp(-\lambda_{1,0}t)-\lambda_2\exp(-\lambda_{0,1}t)]}{\lambda_1-\lambda_2}\,,\\ &\lambda_{1,0}=\lambda_1\,,\qquad \lambda_{0,1}=\lambda_2\,.\end{aligned} \tag{10.101}$$

With a little algebra it is seen that for the underdamped case (10.101) reads

$$\begin{aligned}\langle v(t)\,v(0)\rangle = v_{\rm th}^2\exp(-\gamma t/2)\{&\cos\sqrt{\omega_0^2-\gamma^2/4}\,t\\ &-[\gamma/(2\sqrt{\omega_0^2-\gamma^2/4})]\sin\sqrt{\omega_0^2-\gamma^2/4}\,t\}\,.\end{aligned} \tag{10.102}$$

Inverted Parabolic Potential

For the inverted parabolic potential (10.69) one of the two eigensolutions of the noise-free equation increases exponentially and no stationary solution of the

Kramers equation exists. Nevertheless, we may look for decaying solutions of the Kramers equation, which can be normalized according to $\int \psi^+ \psi \,\mathrm{d}x\,\mathrm{d}v = 1$, where ψ^+ is the eigenfunction of the adjoint operator L^+.

A similar result was obtained in Sect. 5.5.2 for the solution of the Smoluchowski equation with an inverted potential. One may use the same expressions as for the parabolic potential (10.52), but replace ω_0 by $\mathrm{i}\,\omega_1$:

$$\begin{aligned} a^+ &= -\frac{D}{\mathrm{i}\omega_1} = -\frac{v_{\mathrm{th}}}{\mathrm{i}\omega_1}\frac{\partial}{\partial x} + \frac{\mathrm{i}\omega_1}{2v_{\mathrm{th}}}x \equiv \mathrm{i}\hat{a} \\ a &= \frac{\hat{D}}{\mathrm{i}\omega_1} = +\frac{v_{\mathrm{th}}}{\mathrm{i}\omega_1}\frac{\partial}{\partial x} + \frac{\mathrm{i}\omega_1}{2v_{\mathrm{th}}}x \equiv \mathrm{i}\hat{a}^+ . \end{aligned} \tag{10.103}$$

For the operators $\hat{a}$ and $\hat{a}^+$ one has formally the same relation as (10.74) with ω_0 replaced by ω_1. Therefore these operators should now be used. The transformed operator (10.75) then reads

$$\bar{L}_{\mathrm{K}} = -\gamma b^+ b - \omega_1(\hat{a}b - \hat{a}^+ b^+) \tag{10.104}$$

which cannot be brought to the form (10.81) by the transformation (10.80).

The easiest way to obtain the eigenvalues seems to be the following. We first introduce the new variables

$$\begin{aligned} z_1 &= v + \lambda_2 x \\ z_2 &= v + \lambda_1 x . \end{aligned} \tag{10.105}$$

The Langevin equations (10.2) are then transformed to

$$\dot{z}_i = -\lambda_i z_i + \Gamma(t) \quad \text{for} \quad i = 1,2 \tag{10.106}$$

and the corresponding Fokker-Planck equation reads

$$\begin{aligned} \frac{\partial W}{\partial t} &= L\,W, \\ L &= \lambda_1 \frac{\partial}{\partial z_1} z_1 + \lambda_2 \frac{\partial}{\partial z_2} z_2 + \gamma v_{\mathrm{th}}^2 \left(\frac{\partial}{\partial z_1} + \frac{\partial}{\partial z_2}\right)^2 . \end{aligned} \tag{10.107}$$

This equation was used in [1.6, 7] to determine the transition probability (10.55).

For the inverted potential (10.69), the first eigenvalue λ_1 is positive but the other is negative

$$\begin{aligned} \lambda_1 &= \tfrac{1}{2}\left(\sqrt{\gamma^2 + 4\omega_1^2} + \gamma\right) = \mu_1 > 0 \\ \lambda_2 &= -\tfrac{1}{2}\left(\sqrt{\gamma^2 + 4\omega_1^2} - \gamma\right) = -\mu_2 < 0 . \end{aligned} \tag{10.108}$$

Next we introduce the boson operators

$$b_i = \alpha_i \frac{\partial}{\partial z_i} + \frac{1}{2}\frac{z_i}{\alpha_i}, \qquad b_i^+ = -\alpha_i \frac{\partial}{\partial z_i} + \frac{1}{2}\frac{z_i}{\alpha_i},$$
$$[b_i, b_j^+] = \delta_{ij} \tag{10.109}$$

with

$$\alpha_i^2 = \gamma v_{\text{th}}^2/\mu_i . \tag{10.110}$$

In terms of these operators the operator L has the form

$$L = \exp(-\Phi)\bar{L}\exp(\Phi)$$
$$\bar{L} = -\mu_1 b_1^+ b_1 - \mu_2 b_2 b_2^+ - 2\omega_1 b_1^+ b_2 \tag{10.111}$$
$$\Phi = z_1^2/(4\alpha_1^2) - z_2^2/(4\alpha_2^2) .$$

Finally we expand the eigenfunction ψ of (10.107) into eigenfunctions $\psi_n^{(1)}$ and $\psi_n^{(2)}$ of the number operators $b_i^+ b_i$

$$b_i^+ b_i \psi_n^{(i)}(z_i) = n\,\psi_n^{(i)}(z_i) \tag{10.112}$$

in the following way:

$$\psi(x,v) = \exp(-\Phi)\,\bar{\psi}(x,v)$$
$$\bar{\psi}(x,v) = \sum_{n_1,n_2=0}^{\infty} c_{n_1,n_2}\psi_{n_1}^{(1)}(z_1)\,\psi_{n_2}^{(2)}(z_2) . \tag{10.113}$$

Then the eigenvalue equation $\bar{L}\,\bar{\psi} = -\lambda\,\bar{\psi}$ leads to the following coupled system for the expansion coefficients c_{n_1,n_2} ($c_{n_1,n_2} = 0$ for $n_1 < 0$ or $n_2 < 0$)

$$[\mu_1 n_1 + \mu_2(n_2+1) - \lambda]\,c_{n_1,n_2} + 2\omega_1\sqrt{n_1}\sqrt{n_2+1}\,c_{n_1-1,n_2+1} = 0 . \tag{10.114}$$

By this relation only a finite number of coefficients with $n_1 + n_2 = N$ are coupled. Changing the notation of the coefficients to

$$y_n^{(N)} = c_{N-n,n} , \tag{10.115}$$

we get the recursion relation ($n_1 = N-n$, $n_2 = n$)

$$[\mu_1(N-n) + \mu_2(n+1) - \lambda]\,y_n^{(N)} + 2\omega_1\sqrt{N-n}\sqrt{n+1}\,y_{n+1}^{(N)} = 0 . \tag{10.116}$$

We immediately obtain the expression for the eigenvalues by requiring that the recursion terminates at some finite $n \leqq N$ ($n = 0, 1, \ldots, N \geqq 0$; $n_1 \geqq 0$ $n_2 \geqq 0$)

$$\begin{aligned}\lambda_n^{(N)} &= \mu_1(N-n)+\mu_2(n+1)\\ \lambda_{n_1,n_2} &= \mu_1 n_1+\mu_2(n_2+1)\\ &= \tfrac{1}{2}\sqrt{\gamma^2+4\omega_1^2}(n_1+n_2+1)+\tfrac{1}{2}\gamma(n_1-n_2-1)\,.\end{aligned} \tag{10.117}$$

The eigenvalues are all real and positive. The coefficients $y_m^{(N)}$ with $m \leqq n$ are obtained from (10.116) by iteration.

For the eigenfunctions ψ^+ of the adjoint operator L^+ we may proceed in the same way. Writing in analogy to (10.111)

$$\begin{aligned}L^+ &= \exp(\Phi)\bar{L}^+\exp(-\Phi)\,,\\ \bar{L}^+ &= -\mu_1 b_1^+ b_1-\mu_2 b_2 b_2^+ -2\omega_1 b_1 b_2^+\,,\end{aligned} \tag{10.111 a}$$

we can make the same expansion (10.113) for the eigenfunctions $\bar{\psi}^+ = \exp(-\Phi)\psi^+$ of $\bar{L}^+$. These two expansions then guarantee that the integral $\int \psi^+ \psi \mathrm{d}x\mathrm{d}v = \int \bar{\psi}^+ \bar{\psi} \mathrm{d}x\mathrm{d}v$ exists. We now want to show this explicitly for the eigenfunctions ψ and ψ^+ with the lowest eigenvalue of (10.117).

Lowest Eigenvalue and Eigenfunction

The lowest eigenvalue is obtained by setting $n_1 = n_2 = 0$

$$\lambda_{0,0} = \tfrac{1}{2}(\sqrt{\gamma^2+4\omega_1^2}-\gamma)\,. \tag{10.118}$$

The eigenfunction for the eigenvalue (10.118) is given by

$$\begin{aligned}\psi &= N\exp(-\Phi)\,\psi_0^{(1)}(z_1)\,\psi_0^{(2)}(z_2)\\ &= N\exp[-z_1^2/(2\alpha_1^2)] = N\exp[-\mu_1(v-\mu_2 x)^2/(2\gamma v_{\mathrm{th}}^2)]\,.\end{aligned} \tag{10.119}$$

The eigenfunction ψ^+ of the adjoint operator L^+ for the lowest eigenvalue (10.117) is

$$\begin{aligned}\psi^+ &= N\exp(\Phi)\,\psi_0^{(1)}(z_1)\,\psi_0^{(2)}(z_2)\\ &= N\exp[-z_2^2/(2\alpha_2^2)] = N\exp[-\mu_2(v+\mu_1 x)^2/(2\gamma v_{\mathrm{th}}^2)]\,.\end{aligned} \tag{10.120}$$

The integral of the product of (10.119, 120) exists and the normalization constant N reads

$$N = \left(\int\!\!\int \psi^+ \psi \mathrm{d}x\mathrm{d}v\right)^{-1/2} = [(\mu_1+\mu_2)\,\omega_1/(2\pi\gamma v_{\mathrm{th}}^2)]^{1/2}\,. \tag{10.121}$$

For large damping constants ($\gamma \gg 2\omega_1$) we have $\mu_1 = \gamma$, $\mu_2 = \omega_1^2/\gamma$ and we obtain

$$\begin{aligned}\psi &= N\exp[-m(v-\omega_1^2 x/\gamma)^2/(2kT)]\,,\\ \psi^+ &= N\exp[-m(v+\gamma x)^2\omega_1^2/(2\gamma^2 kT)]\,.\end{aligned}$$

For small damping constants ($\gamma \ll 2\omega_1$) we have $\mu_1 = \mu_2 = \omega_1$, giving

$$\psi \;\; = N\exp[-m(v-\omega_1 x)^2\omega_1/(2\gamma kT)]\,,$$
$$\psi^+ = N\exp[-m(v+\omega_1 x)^2\omega_1/(2\gamma kT)]\,.$$

Free Brownian Motion

Without any force we can make the ansatz

$$\psi(x,v) = \mathrm{e}^{ikx} g_k(v)\exp[-(v/v_{\mathrm{th}})^2/4]\,. \tag{10.122}$$

Inserting (10.122) into $L_K\psi = -\lambda\psi$ with L_K given by (10.3 b) leads to

$$\left[\gamma v_{\mathrm{th}}^2\frac{\partial^2}{\partial v^2}+\frac{1}{2}\gamma-\frac{1}{4}\gamma\left(\frac{v}{v_{\mathrm{th}}}\right)^2-\mathrm{i}kv\right]g_k = -\lambda g_k\,, \tag{10.123}$$

which, by shifting the velocity according to

$$\hat{v} = v + 2\mathrm{i}kv_{\mathrm{th}}^2/\gamma\,, \tag{10.124}$$

is transformed to the harmonic oscillator equation

$$\left[\gamma v_{\mathrm{th}}^2\frac{\partial^2}{\partial \hat{v}^2}+\frac{1}{2}\gamma-\frac{1}{4}\gamma\left(\frac{\hat{v}}{v_{\mathrm{th}}}\right)^2+\lambda-\frac{k^2v_{\mathrm{th}}^2}{\gamma}\right]g_k = 0\,. \tag{10.125}$$

The eigensolutions and eigenvalues of (10.125) are [see (10.21, 38)]

$$g_{nk}(v) = \psi_n(\hat{v}) = \psi_n(v+2\mathrm{i}kv_{\mathrm{th}}^2/\gamma)\,, \tag{10.126}$$

$$\lambda_{nk}(v) = n\gamma + k^2v_{\mathrm{th}}^2/\gamma\,. \tag{10.127}$$

Without any boundary conditions in x, k is arbitrary and the eigenvalues (10.127) are continuous. For periodic boundary conditions in x with period 2π, $k = 0, \pm 1, \pm 2, \ldots$ and the eigenvalues are discrete. (See also Sect. 11.9 for a discussion of eigenvalues in a periodic potential.)

10.3 Matrix Continued-Fraction Solutions of the Kramers Equation

In this section we shall derive general solutions of the Kramers equation in terms of matrix continued fractions [9.19]. The matrix continued fractions are well suited for numerical calculations, as will be shown in Sects. 11.5–9, where we treat the Brownian motion problem in periodic potentials. The matrix continued fractions are, however, also useful to derive analytical results. In Sect. 10.4 the matrix continued fractions are expanded in the high-friction limit leading to the correction terms of the Smoluchowski equation.

The starting point of our consideration is the Kramers equation in the form of Brinkman's hierarchy (10.46 or 46a). This hierarchy of equations can be cast into the form of the tridiagonal recurrence relation discussed in Chap. 9 by further expanding the Hermite expansion coefficients $c_n(x,t)$ into a complete orthonormal set $\varphi^q(x)$,

$$\begin{aligned}&\int[\varphi^p(x)]^*\varphi^q(x)\,dx=\delta_{pq}\,,\\&\sum_p\varphi^p(x)[\varphi^p(x')]^*=\delta(x-x')\,,\end{aligned}\tag{10.128}$$

satisfying the boundary conditions in position space i.e.

$$c_n(x,t)=\sum_q c_n^q(t)\varphi^q(x)\,.\tag{10.129}$$

If we are looking for solutions of the Kramers equation which are periodic in x, as complete set a Fourier series will be used. If we are looking for solutions with natural boundary conditions in $x(-\infty<x<\infty)$, properly scaled Hermite functions seem to be a good choice. As the complete set we may also use the eigenfunctions of a Schrödinger equation corresponding to the Smoluchowski equation with the potential $f(x)$ (Sect. 5.4). Because these eigenfunctions must generally first be calculated numerically, determinaton of the matrix elements (10.131) with these eigenfunctions may thus become very complicated.

By inserting (10.129) into (10.46), multiplying the resulting equation with $(\varphi^p(x))^*$ and integrating, we obtain the one-sided tridiagonal vector recurrence relation (9.10) with

$$\boldsymbol{Q}_n^+=-\sqrt{n+1}\boldsymbol{D}\,,\quad \boldsymbol{Q}_n=-n\gamma\boldsymbol{I}\,,\quad \boldsymbol{Q}_n^-=-\sqrt{n}\hat{\boldsymbol{D}}\,.\tag{10.130}$$

The matrix elements of $\boldsymbol{D},\hat{\boldsymbol{D}}$ are defined by

$$A^{pq}=\int[\varphi^p(x)]^*A\,\varphi^q(x)\,dx\,,\tag{10.131}$$

where A has to be replaced by $D=v_{\text{th}}\partial/\partial x-\varepsilon f'/v_{\text{th}}$ and $\hat{D}=v_{\text{th}}\partial/\partial x+(1-\varepsilon)\times f'(x)/v_{\text{th}}$. The column vectors $\boldsymbol{c}_n$ have the components c_n^q, i.e.,

$$\boldsymbol{c}_n=(c_n^q)\,.\tag{10.132}$$

To deal with finite matrices of dimension $Q\times Q$, expansion (10.129) has to be truncated at $q=Q$.

The friction constant γ may depend on position (Sect. 10.4.4) but it should always be larger than zero $[\gamma(x)>0]$. The temperature T, i.e., v_{th}, is assumed to be independent of position. (If the temperature depends on position, the eigenfunctions $\psi_n(v)$ also depend on position. The operator $\partial/\partial x$ then also acts on the eigenfunctions ψ_n leading to a coupling of the $\boldsymbol{c}_n$ of higher order. By a suitable vector notation this higher-order recurrence relation can again be cast into a tridiagonal recurrence relation.)

10.3.1 Initial Value Problem

The general solution of (10.46) with the initial value $c_m(x,0)$ can be written as

$$c_n(x,t) = \sum_{m=0}^{\infty} \int G_{n,m}(x,x',t)\, c_m(x',0)\, dx' \tag{10.133}$$

where the Green's function has the initial value

$$G_{n,m}(x,x',0) = \delta(x-x')\,\delta_{n,m} \,. \tag{10.134}$$

The initial values for the coefficients of the expansion (10.43) are given by

$$c_m(x,0) = \int W(x,v,0)\, \psi_m(v) \exp[\varepsilon f(x)/v_{\mathrm{th}}^2]/\psi_0(v)\, dv \,. \tag{10.135}$$

In our vector notation (10.133, 134) take the form

$$\boldsymbol{c}_n = \sum_{m=0}^{\infty} \boldsymbol{G}_{n,m} \boldsymbol{c}_m(0) \,, \tag{10.133 a}$$

$$\boldsymbol{G}_{n,m}(0) = \boldsymbol{I}\,\delta_{n,m} \,, \tag{10.134 a}$$

where the matrix elements of $\boldsymbol{G}_{n,m}(t)$ are

$$G_{n,m}^{pq}(t) = \int\int [\varphi^p(x)]^* \, G_{n,m}(x,x',t)\, \varphi^q(x')\, dx\, dx' \,. \tag{10.136}$$

Green's function $G_{n,m}(x,x',t)$ in x representation expressed by the Green's function $G_{n,m}^{pq}(t)$ in φ^q representation reads

$$G_{n,m}(x,x',t) = \sum_{p,q} \varphi^p(x)\, G_{n,m}^{pq}(t)\, [\varphi^q(x')]^* \,. \tag{10.137}$$

According to (9.109) and because of (10.130), the Laplace transform of the Green's function matrix

$$\tilde{\boldsymbol{G}}_{n,m}(s) = \int_0^{\infty} \boldsymbol{G}_{n,m}(t)\, e^{-st}\, dt \tag{10.138}$$

can be expressed for $n = m$ by

$$\tilde{\boldsymbol{G}}_{m,m}(s) = [(s+m\gamma)\boldsymbol{I} - \tilde{\boldsymbol{K}}_m(s)]^{-1} \tag{10.139}$$

where the Laplace transform of the memory kernel $\boldsymbol{K}_m(t)$ is given by

$$\tilde{\boldsymbol{K}}_m(s) = -\sqrt{m+1}\,\boldsymbol{D}\tilde{\boldsymbol{S}}_m^+(s) - \sqrt{m}\,\hat{\boldsymbol{D}}\tilde{\boldsymbol{S}}_m^-(s) \,. \tag{10.140}$$

As mentioned in Chap. 9, the Green's function $\boldsymbol{G}_{m,m}(t)$ is a solution of the integral equation

$$\dot{G}_{m,m}(t) = -m\gamma G_{m,m}(t) + \int_0^t K_m(t-\tau)\, G_{m,m}(\tau)\,d\tau \tag{10.139a}$$

with the initial condition $G_{m,m}(0) = I$. Equation (10.139) is the Laplace transform of (10.139a).

The other solutions $n \neq m$ follow from

$$\begin{aligned}
\tilde{G}_{n,m}(s) &= \tilde{U}_{n,m}(s)\, \tilde{G}_{m,m}(s) \\
\tilde{U}_{n,m}(s) &= \tilde{S}^+_{n-1}(s)\tilde{S}^+_{n-2}(s)\ldots\tilde{S}^+_m(s) \quad \text{for} \quad n \geqq m+1 \\
\tilde{U}_{m,m}(s) &= I \\
\tilde{U}_{n,m}(s) &= \tilde{S}^-_{n+1}(s)\tilde{S}^-_{n+2}(s)\ldots\tilde{S}^-_m(s) \quad \text{for} \quad 0 \leqq n \leqq m-1\,.
\end{aligned} \tag{10.141}$$

Thus the Green's function $G_{n,m}(t)$ follows from $G_{m,m}(t)$ by convolution with $U_{n,m}(t)$ according to the first equation of (10.141).

The $\tilde{S}^+_n$ are given by the infinite continued fraction [cf. (9.104)]

$$\tilde{S}^+_n(s) = -\sqrt{n+1}\{sI + (n+1)\gamma I - (n+2)D[sI + (n+2)\gamma I - \ldots]^{-1}\hat{D}\}^{-1}\hat{D}\,, \tag{10.142}$$

whereas $\tilde{S}^-_n$ is given by the finite continued fraction [cf. (9.105) $n \geqq 3$]

$$\begin{aligned}
\tilde{S}^-_0(s) &= 0 \\
\tilde{S}^-_1(s) &= -s^{-1}D \\
\tilde{S}^-_2(s) &= -\sqrt{2}[sI + \gamma I - \hat{D}s^{-1}D]^{-1}D \\
\tilde{S}^-_n(s) &= -\sqrt{n}\{sI + (n-1)\gamma I - (n-1)\hat{D}[sI + (n-2)\gamma I \\
&\qquad \ldots - \hat{D}s^{-1}D\ldots]^{-1}D\}^{-1}D\,.
\end{aligned} \tag{10.143}$$

With the help of Green's function the transition probability $P(x,v,t\,|\,x',v',0)$ of the Kramers equation, i.e., the solution of (10.3) with the initial distribution

$$P(x,v,0\,|\,x',v',0) = \delta(x-x')\,\delta(v-v') \tag{10.144}$$

can be constructed. The successive insertion of (10.144) into (10.135) into (10.133) into (10.43) leads to

$$\begin{aligned}
&P(x,v,t\,|\,x',v',0) \\
&= \frac{\psi_0(v)\exp[-\varepsilon f(x)/v_{\text{th}}^2]}{\psi_0(v')\exp[-\varepsilon f(x')/v_{\text{th}}^2]} \sum_{n,m=0}^{\infty} G_{n,m}(x,x',t)\,\psi_n(v)\,\psi_m(v') \\
&= \frac{\psi_0(v)\exp[-\varepsilon f(x)/v_{\text{th}}^2]}{\psi_0(v')\exp[-\varepsilon f(x')/v_{\text{th}}^2]} \sum_{n,m=0}^{\infty} \sum_{p,q} \varphi^p(x)[\varphi^q(x')]^* G^{pq}_{n,m}(t)\,\psi_n(v)\,\psi_m(v')\,.
\end{aligned}$$

Thus the Laplace transform of the transition probability is expressed by the matrix elements $\tilde{G}^{pq}_{n,m}(s)$ of the matrix $\tilde{\boldsymbol{G}}_{n,m}(s)$, the solution of which is given by (10.139 – 143) in terms of matrix continued fractions.

The Laplace transform of the Green's function $\tilde{G}_{0,0}(x,x',s)$ is formally obtained from (10.139 – 143) by replacing the matrices $\boldsymbol{D}, \hat{\boldsymbol{D}}$ by $v_{\text{th}}\,\partial/\partial x - \varepsilon f'(x)/v_{\text{th}}$, $v_{\text{th}}\,\partial/\partial x + (1-\varepsilon)f'/v_{\text{th}}$. We then get a solution in terms of continued inverse operators. In (10.139) the inverse operators have to be applied to the δ function $\delta(x-x')$.

If a stationary solution of the Kramers equation exists and if the probability current is zero, the stationary solution is given by (10.49a). The stationary joint distribution is then given by ($t \geqq 0$)

$$\begin{aligned} W_2(x,v,t;x',v',0) &= P(x,v,t\,|\,x',v',0)\,W_{\text{st}}(x',v') \\ &= N\psi_0(v)\exp[-\varepsilon f(x)/v_{\text{th}}^2]\,\psi_0(v')\exp[-(1-\varepsilon)f(x')/v_{\text{th}}^2] \\ &\quad\times \sum_{n,m=0}^{\infty}\sum_{p,q} G^{pq}_{n,m}(t)\,\varphi^p(x)\,[\varphi^q(x')]^*\,\psi_n(v)\,\psi_m(v')\,. \end{aligned} \tag{10.146}$$

For $\varepsilon = \frac{1}{2}$ the expression takes a symmetric form in x and x'. The autocorrelation function of the velocity, $K_{vv}(t)$, is easily obtained from (10.146) by using $v_{\text{th}}\,\psi_1(v) = v\,\psi_0(v)$ and the normalization (10.41). The result for the half-sided Fourier transform $\tilde{K}_{vv}(\omega)$ is ($s = \mathrm{i}\omega$)

$$\begin{aligned} \tilde{K}_{vv}(\omega) &= \int_0^\infty \langle v(t)\,v(0)\rangle\,\mathrm{e}^{-\mathrm{i}\omega t}\,\mathrm{d}t \\ &= \int_0^\infty\!\!\iiiint v\,v'\,W_2(x,v,t;x',v',0)\,\mathrm{d}x\,\mathrm{d}x'\,\mathrm{d}v\,\mathrm{d}v'\,\mathrm{e}^{-\mathrm{i}\omega t}\,\mathrm{d}t \\ &= N v_{\text{th}}^2 \sum_{p,q}\left[\int \exp[-\varepsilon f(x)/v_{\text{th}}^2]\,\varphi^p(x)\,\mathrm{d}x\right] \\ &\quad\times\left[\int \exp[-(1-\varepsilon)f(x)/v_{\text{th}}^2]\,\varphi^q(x)\,\mathrm{d}x\right]^*\,\tilde{G}^{pq}_{1,1}(\mathrm{i}\omega)\,. \end{aligned} \tag{10.147}$$

The continued fraction needed for obtaining $\boldsymbol{G}_{1,1}$ is explicitly written down in (10.210, 211). As seen for the Brownian motion problem in periodic potentials, matrix continued fractions may be calculated numerically. The speed of convergence of the matrix continued fractions depends on the friction constant γ. For large friction only a few terms need be taken into account whereas for very small friction the number of iterations may be very large. The matrix continued fractions seem to converge even for very small damping constants. For very small friction one may use energy as a variable (Chap. 11, Sects. 4, 8.1, 9.1). Analytic expressions for matrix continued fractions are obtained only for special cases. In App. A3 we evaluate $\boldsymbol{G}_{0,0}(t)$ for the harmonic oscillator; in this chapter we derive an explicit result for $\boldsymbol{G}_{0,0}(t)$ for free Brownian motion from the general solution (10.139 – 143).

Free Brownian Motion

For free Brownian motion both D and $\hat{D}$ are equal to the differential operator $v_{\text{th}}\,\partial/\partial x$. For a position-independent damping force all matrices in the continued fractions commute. We may therefore formally treat the matrix continued fractions as ordinary continued fractions. For $\tilde{G}_{0,0}(s)$ we then have ($D = \hat{D} = v_{\text{th}}\,\partial/\partial x$) in x representation [see (10.180) for a general expression for $\tilde{K}_0(s)$]:

$$\tilde{G}_{0,0}(x,x',s) = [s - \tilde{K}_0(s)]^{-1}\delta(x-x')$$

with

$$[s-\tilde{K}_0(s)]^{-1} = \frac{1|}{|s} - \frac{D^2|}{|s+\gamma} - \frac{2D^2|}{|s+2\gamma} - \dots . \tag{10.148}$$

By introducing the notation

$$\alpha = \beta + s/\gamma\,, \quad \beta = -D^2/\gamma^2\,, \tag{10.149}$$

(10.148) may be written as

$$\frac{1}{s-\tilde{K}_0(s)} = \frac{1}{\gamma}\left\{\frac{1|}{|\alpha-\beta} + \frac{\beta|}{|\alpha-\beta+1} + \frac{2\beta|}{|\alpha-\beta+2} + \dots\right\}. \tag{10.150}$$

It follows from [Ref. 9.1, Sect. 48, Eqs. (23, 26)] that the continued fraction in the parenthesis is equal to

$$\begin{aligned}\{\,\} &= e^{\beta}\int_0^1 e^{-\beta u}u^{\alpha-1}du = \gamma\int_0^\infty \exp(\beta - \beta e^{-\gamma t} - \alpha\gamma t)dt \\ &= \gamma\int_0^\infty \exp[\beta(1-e^{-\gamma t}-\gamma t)]\,e^{-st}dt\,.\end{aligned} \tag{10.151}$$

In the second step we made the substitution $u = \exp(-\gamma t)$.

Thus we have

$$\tilde{G}_{0,0}(x,x',s) = \int_0^\infty \exp[\Omega(t)]\,e^{-st}dt\,\delta(x-x') \tag{10.152}$$

with [see (10.189) for definition of H_2)

$$\Omega(t) = D^2(\gamma t - 1 + e^{-\gamma t})/\gamma^2 = D^2 H_2(\gamma t)/\gamma^2\,, \tag{10.153}$$

and the inverse Laplace transform of $\tilde{G}_{0,0}(x,x',s)$ is given by

$$\begin{aligned}G_{0,0}(x,x',t) &= \exp[\Omega(t)]\,\delta(x-x') \\ &= \exp[v_{\text{th}}^2(\partial/\partial x)^2 H_2(\gamma t)/\gamma^2](2\pi)^{-1}\int_{-\infty}^{\infty}\exp[ik(x-x')]dk \\ &= (2\pi)^{-1}\int_{-\infty}^{\infty}\exp[-v_{\text{th}}^2 k^2 H_2(\gamma t)/\gamma^2 + ik(x-x')]dk\end{aligned}$$

$$= \frac{\gamma}{2 v_{\text{th}} \sqrt{\pi H_2(\gamma t)}} \exp\left(-\frac{\gamma^2 (x-x')^2}{4 H_2(\gamma t) v_{\text{th}}^2}\right). \tag{10.154}$$

This result may of course also be obtained in the usual way, i.e., by taking the average of the transition probability (10.55, 67, 68) for free Brownian motion over an initial Maxwell velocity distribution and by integration over the final velocity

$$G_{0,0}(x,x',t) = \iint P(x,v,t|x',v',0) \exp[-v'^2/(2v_{\text{th}}^2)]/(\sqrt{2\pi}\, v_{\text{th}})\, dv\, dv' . \tag{10.155}$$

From the first line of (10.154) it is easily seen that $G_{0,0}(x,x',t)$ is a solution of the time-dependent Smoluchowski equation

$$\begin{aligned} \dot{G}_{0,0} &= L_0(t) G_{0,0} \\ L_0(t) = \dot{\Omega} &= D^2 \gamma^{-2} dH_2(\gamma t)/dt = v_{\text{th}}^2 (\partial/\partial x)^2 (1 - e^{-\gamma t})/\gamma \end{aligned} \tag{10.156}$$

with the initial condition (10.134), a result also well known in the literature [10.3 – 5].

10.3.2 Eigenvalue Problem

As we have seen, Brinkman's hierarchy (10.46a), which is equivalent to the Kramers equation, can be cast into the tridiagonal vector recurrence relation by using expansion (10.129). We are now looking for eigensolutions of the Kramers equation, i.e., for solutions in which the expansion coefficients are of the form

$$c_n^q(t) = \hat{c}_n^q e^{-\lambda t} . \tag{10.157}$$

We now write $\hat{c}_n^q$ again as the column vector $\hat{\boldsymbol{c}}_n$. By eliminating all $\hat{\boldsymbol{c}}_n$ with $n \neq m$ we arrive at the homogeneous relation

$$[\lambda \boldsymbol{I} - m\gamma \boldsymbol{I} + \tilde{\boldsymbol{K}}_m(-\lambda)] \hat{\boldsymbol{c}}_m = \boldsymbol{0} \tag{10.158}$$

and the eigenvalues follow from, see (9.119),

$$D_m(\lambda) = \operatorname{Det}[\lambda \boldsymbol{I} - m\gamma \boldsymbol{I} + \tilde{\boldsymbol{K}}_m(-\lambda)] = 0 . \tag{10.159}$$

The eigenvectors $\hat{\boldsymbol{c}}_m$ follow from (10.158) and the other eigenvectors are then determined by

$$\hat{\boldsymbol{c}}_n = \tilde{\boldsymbol{U}}_{n,m}(-\lambda) \hat{\boldsymbol{c}}_m . \tag{10.160}$$

Because of (10.43, 129) the eigenfunctions ψ_λ of L_K then have the form

$$\psi_\lambda(x,v) = \psi_0(v) \exp[-\varepsilon f(x)/v_{\text{th}}^2] \sum_{n=0}^{\infty} \sum_q c_n^q \varphi^q(x) \psi_n(v) . \tag{10.161}$$

As mentioned in Sect. 9.3.2, any $D_m(\lambda)$ with $m=0,1,2,\ldots$ may be used to determine the eigenvalues λ. For the stationary eigenvalue $\lambda=0$ and its eigensolution of the Kramers equation, however, the index m must be equal to zero because $\tilde{S}_m^-(0)\,(m\geqq 0)$ does not exist. For large damping constants it may be advantageous to use $D_m(\lambda)$ for those eigenvalues where $\tilde{K}_m(-\lambda)$ is small (Sect. 10.4). Because the Fokker-Planck operator L_K or the transformed operator $\bar{L}_K$ are not Hermitian operators, the eigenvalues λ are generally complex. To find the eigenvalues λ, $D_m(\lambda)$ has to be calculated and the roots of $D_m(\lambda)=0$ have to be found with some complex root-finding technique. In Sect. 11.9 this method is applied to the Brownian motion of a pendulum.

Normalization

Because neither the Kramers operator L_K nor the transformed operator $\bar{L}_K = \bar{L}_{\text{rev}}+\bar{L}_{\text{ir}}$ are Hermitian operators, we must also find the eigenfunctions of L_K^+ or equivalently of $\bar{L}_K^+$ to form a biorthogonal set. For $\varepsilon=\frac{1}{2}$ we have (10.21, 26)

$$\bar{L}_K^+(x,v)=\bar{L}_K(x,-v)\,. \tag{10.162}$$

Thus the eigenfunctions $\bar{\psi}_\lambda^+(x,v)$ of $\bar{L}_K^+$ with eigenvalue λ are simply obtained from those of $\bar{L}_K$ with the same eigenvalue λ by replacing v by $-v$, i.e., $\bar{\psi}_\lambda^+(x,v)=\bar{\psi}_\lambda(x,-v)$. Because of the transformation (10.25) and because of (10.38), the connection between the eigenfunctions ψ_λ^+ of L_K^+

$$L_K^+\psi_\lambda^+(x,v)=-\lambda\,\psi_\lambda^+(x,v) \tag{10.163}$$

and those of L_K may be written as [N is the normalization in (10.49a)]

$$\psi_\lambda^+(x,v)=N^{-1}[\psi_0(v)]^{-2}\exp[f(x)/v_{\text{th}}^2]\,\psi_\lambda(x,-v)\,. \tag{10.164}$$

The function in front of ψ_λ is the inverse of the stationary distribution (10.49a) divided by the normalization constant in (10.49a). For $\varepsilon=\frac{1}{2}$ the normalization reads

$$\begin{aligned}1&=\iint\psi_\lambda^+(x,v)\,\psi_\lambda(x,v)\,dx\,dv\\&=N^{-1}\iint[\psi_0(v)]^{-2}\exp[f(x)/v_{\text{th}}^2]\,\psi_\lambda(x,v)\,\psi_\lambda(x,-v)\,dx\,dv\,.\end{aligned}$$

Inserting (10.161) and using (10.41) and $\psi_n(-v)=(-1)^n\psi_n(v)$ we finally have

$$N=\sum_{n=0}^{\infty}(-1)^n\sum_q\sum_p c_n^q c_n^p\int\varphi^q(x)\,\varphi^p(x)\,dx\,. \tag{10.165}$$

Symmetry Relations

If the potential is symmetric $[f(x)=f(-x)]$, i.e., if the force is antisymmetric $[f'(x)=-f'(-x)]$, and if the boundary conditions are also symmetric in x, we conclude form $L_K(x, v) = L_K(-x, -v)$ that the eigenfunctions must be either symmetric or antisymmetric

$$\psi_{\substack{s\\a}}(x, v) = \pm\, \psi_{\substack{s\\a}}(-x, -v)\,. \tag{10.166}$$

10.4 Inverse Friction Expansion

For large friction constants one may neglect in (10.1) the second order time derivative. The corresponding Fokker-Planck equation for the Langevin equation

$$\gamma\dot{x} + f'(x) = \Gamma(t) \tag{10.167}$$

with the Langevin force of (10.1) is called the Smoluchowski equation. This equation for the distribution function $W(x, t)$ for position only reads

$$\begin{aligned} &\partial W/\partial t = L_S W \\ &L_S = \frac{1}{\gamma}\frac{\partial}{\partial x} f'(x) + \frac{kT}{m\gamma}\frac{\partial^2}{\partial x^2}\,. \end{aligned} \tag{10.168}$$

Equation (10.168) may be derived from the Kramers equation (10.3) as follows. If we truncate system (10.46a) after the second $n = 1$ term (i.e., $c_2 = c_3 = \ldots = 0$ and omit the equations with $n \geqq 2$), (10.46a) reduces to

$$\begin{aligned} &\partial c_0/\partial t + Dc_1 = 0 \\ &\hat{D}c_0 + \partial c_1/\partial t + \gamma c_1 = 0\,. \end{aligned} \tag{10.169}$$

For large damping constants γ we may furthermore neglect the time derivative $\partial c_1/\partial t$ in the last equation, and by eliminating c_1 and putting $\varepsilon = 0$ obtain the Smoluchowski equation (10.168)

$$\begin{aligned} \partial c_0/\partial t &= -Dc_1 = \gamma^{-1}D\hat{D}c_0 \\ &= \frac{1}{\gamma}\frac{\partial}{\partial x}\left(f' + v_{\text{th}}^2\frac{\partial}{\partial x}\right)c_0\,, \end{aligned} \tag{10.170}$$

where

$$c_0(x, t) = \int W(x, v, t)\,dv \tag{10.171}$$

is the distribution in position only (10.44). If we do not neglect the time derivative $\partial c_1/\partial t$ in the last equation of (10.169), by eliminating c_1 we obtain an equation with a first and second order time derivative

$$\partial^2 c_0/\partial t^2 + \gamma\partial c_0/\partial t = v_{\text{th}}^2\partial^2 c_0/\partial x^2 + \partial(c_0 f')/\partial x \tag{10.172}$$

derived by *Brinkman* [10.1], as a Laplace transform of (10.172), and by *Sack* [10.6]. Without the force $-f'$ (10.172) reduces to the telegrapher's equation

$$\partial^2 c_0/\partial t^2 + \gamma \partial c_0/\partial t = v_{\text{th}}^2 \partial^2 c_0/\partial x^2 . \tag{10.173}$$

At first glance (10.172) seems to be preferable to (10.170) for the following two reasons. (i) In deriving (10.172) we did not neglect the time derivative $\partial c_1/\partial t$ as was done to obtain (10.170). (ii) Because (10.172) is essentially a wave equation (hyperbolic differential equation), a change in the particle density c_0 cannot travel faster than the thermal velocity $v_{\text{th}} = \sqrt{kT/m}$, whereas in the Smoluchowski equation (10.170) a change in particle density is present immediately afterwards at a finite distance (though it very strongly decreases with increasing distance). First we want to discuss that the last argument does not hold at second glance.

If one starts with a Maxwell distribution for the velocities it is not surprising that some change in particle density travels with a large speed because some particles have very large velocities according to the tail of the Maxwell distribution. Even had we started with the velocity distribution $\delta(v)$ (i.e., where the particles have zero velocity), the distribution function $W(x,v,t)$ would then have a finite (though very small) value for large velocities immediately afterwards. Thus a finite limit velocity (as in electrodynamics or in hydrodynamics) for the propagation of a disturbance does not exist in the case of Brownian motion because, due to the assumptions of the Langevin force, a momentum transfer of every size is possible.

The first argument does not hold either, as it is best seen by treating free Brownian motion, for which the telegrapher's equation can be solved exactly [10.7]. Assuming that initially the particles are at the position $x = x'$ and have a Maxwell velocity distribution, i.e.,

$$c_n(x,0) = \delta(x-x')\,\delta_{n0} , \tag{10.174}$$

the solution of the Kramers equation (10.3) integrated over the velocity or of (10.46a) is given by [see (10.154)]

$$\begin{aligned} c_0(x,t) &= (2v_{\text{th}}/\gamma)^{-1}[\pi H_2(\gamma t)]^{-1/2}\exp\{-\gamma^2(x-x')^2/[4H_2(\gamma t)\,v_{\text{th}}^2]\} ,\\ H_2(\gamma t) &= (\gamma t - 1 + \mathrm{e}^{-\gamma t}) . \end{aligned} \tag{10.175}$$

The solution of the telegrapher's equation (10.173) with $c_0(x,0) = \delta(x-x')$ and $\partial c_0(x,t)/\partial t|_{t=0} = -Dc_1(x,0) = 0$ is, according to *Hemmer* [10.7], given by

$$\begin{aligned} c_0(x,t) = \frac{1}{2}\mathrm{e}^{-\gamma t/2}\Bigg\{ &\delta(x-x'-v_{\text{th}}t) + \delta(x-x'+v_{\text{th}}t)\\ &+ \Theta(v_{\text{th}}t - |x-x'|)\left[\frac{\gamma}{2v_{\text{th}}}\mathrm{I}_0\left(\frac{\gamma}{2v_{\text{th}}}\sqrt{v_{\text{th}}^2t^2-(x-x')^2}\right)\right.\\ &\left.+\frac{\gamma t}{2\sqrt{v_{\text{th}}^2t^2-(x-x')^2}}\mathrm{I}_1\left(\frac{\gamma}{2v_{\text{th}}}\sqrt{v_{\text{th}}^2t^2-(x-x')^2}\right)\right]\Bigg\} . \end{aligned} \tag{10.176}$$

In (10.176) I_0 and I_1 are the modified Bessel functions and $\Theta(x)$ is the Heaveside jump function [$\Theta(x)=1$ for $x>0$, $\Theta(x)=0$ for $x<0$]. Using the asymptotic expansions of the modified Bessel functions

$$I_0(x) \approx I_1(x) \approx (2\pi x)^{-1/2} e^x , \tag{10.177}$$

one easily verifies that for large values of time ($\gamma t \gg 1$, $v_{\mathrm{th}} t \gg |x-x'|$) both (10.175, 176) approximately agree with the solution of the Smoluchowski equation (10.170) for $f'=0$ with the initial condition $c_0(x,0)=\delta(x-x')$,

$$c_0(x,t) = \sqrt{\gamma/(4\pi v_{\mathrm{th}}^2 t)} \exp[-\gamma(x-x')^2/(4 v_{\mathrm{th}}^2 t)] . \tag{10.178}$$

As seen in Fig. 10.2, the solution of the telegrapher's equation seems a little awkward for a distribution function, mainly because of the δ-function peaks. It should, however, be mentioned that both distributions (10.175, 176) have the same mean value $\langle x \rangle = 0$ and variance $\langle x^2 \rangle = 2 v_{\mathrm{th}}^2 H_2(\gamma t)/\gamma^2$. [The variance of (10.176) is best calculated by first deriving an equation for the variance directly from the telegrapher's equation (10.173) and then solving it.] Though the solution of the Smoluchowski equation does not have the correct variance, it resembles the exact distribution (because both are Gaussian) more than (10.176) resembles the latter.

Because of the δ-function peaks [which appear because we have not neglected the time derivative in the last equation of (10.169)] $Dc_1 = -\partial c_0/\partial t$ must have an infinite large value, where the δ-function peaks occur. As seen from the third equation of (10.46a), c_2 and c_3 then cannot both be considered small. Therefore not neglecting the time derivative in the last equation of (10.169) is inconsistent with the truncation of expansion (10.43) after the c_1 term, i.e., putting $c_2 = c_3 = \ldots = 0$. In the case of the Smoluchowski equation, $\partial c_0/\partial t$ is of the order γ^{-1}, c_2 is then of the order γ^{-2} and c_2 and higher coefficients may therefore be neglected for large γ.

We now want to derive (10.168) and its correction terms for the high-friction case in a more systematic way. Starting with the general solutions (10.139–143) we will show that the Smoluchowski equation and its correction terms can be obtained by expanding the matrix continued fractions in powers of γ^{-1}. The great advantage of this procedure is that no inverse matrices (in φ^q representation) or operators (in x representation) will occur. The disadvantage of this procedure is that through expanding the denominators, the inverse friction expansion converges only for friction constants which are sufficiently large. For a harmonic potential (10.52) the region of convergence is given by $\gamma > 2\omega_0$. For other potentials it may be valid only in the asymptotic limit $\gamma \to \infty$ (Sect. 11.9).

10.4.1 Inverse Friction Expansion for $K_0(t)$, $G_{0,0}(t)$ and $L_0(t)$

We first look for a solution of the Kramers equation where the initial distribution is the product of a Maxwell velocity distribution times the δ function $\delta(x-x')$. If we integrate this distribution function over the velocity, for $\varepsilon = 0$ this function is

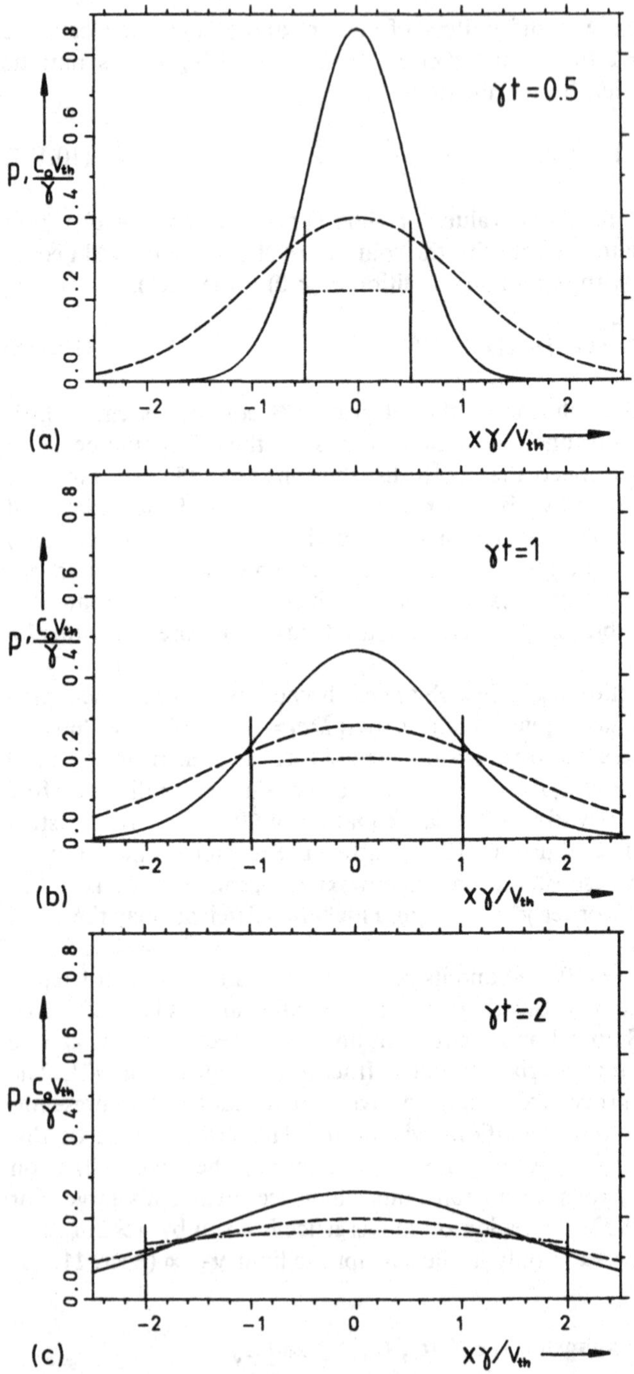

Fig. 10.2a–c. The exact solution (10.175) (*full line*), the solution (10.178) of the Smoluchowski equation (*broken line*) and the solution (10.176) of the telegrapher's equation (*broken line with dots*) as a function of x for three different times. The value of the probability p to find the particle at the δ-function peak is indicated by the length of the vertical line

identical to the Green's function $G_{0,0}(x,x',t)$, see (10.137). The solution of the Kramers equation, where the distribution function factorizes at time $t=0$ into a Maxwell velocity distribution and an arbitrary position-dependent distribution $c_0(x,0)$, can be expressed by the Green's function according to (10.43, 133). The distribution integrated over the velocity is then expressed by

$$\int W(x,v,t)\,\mathrm{d}v = c_0(x,t) = \int G_{0,0}(x,x',t)\,c_0(x',0)\,\mathrm{d}x' \,. \tag{10.179}$$

The Laplace transform of this Green's function is given in matrix notation by (10.139)

$$\tilde{G}_{0,0}(s) = [s\boldsymbol{I} - \tilde{\boldsymbol{K}}_0(s)]^{-1} \,, \tag{10.139a}$$

where according to (10.140, 142) $\tilde{\boldsymbol{K}}_0(s)$ is given by the infinite continued fraction

$$\tilde{\boldsymbol{K}}_0(s) = \boldsymbol{D}[(s+\gamma)\boldsymbol{I} - 2\boldsymbol{D}[(s+2\gamma)\boldsymbol{I} - 3\boldsymbol{D}[(s+3\gamma)\boldsymbol{I} - \dots]^{-1}\hat{\boldsymbol{D}}]^{-1}\hat{\boldsymbol{D}}]^{-1}\hat{\boldsymbol{D}} \,. \tag{10.180}$$

For large γ the continued fraction (10.180) is easily expanded into powers of γ^{-1}. (The variable s is also assumed to be of the order γ.) The result up to the order γ^{-5} reads

$$\begin{aligned}\tilde{\boldsymbol{K}}_0(s) = {} & (s+\gamma)^{-1}\boldsymbol{D}\hat{\boldsymbol{D}} + 2(s+\gamma)^{-2}(s+2\gamma)^{-1}\boldsymbol{D}^2\hat{\boldsymbol{D}}^2 \\ & + 6(s+\gamma)^{-2}(s+2\gamma)^{-2}(s+3\gamma)^{-1}\boldsymbol{D}^3\hat{\boldsymbol{D}}^3 \\ & + 4(s+\gamma)^{-3}(s+2\gamma)^{-2}\boldsymbol{D}(\boldsymbol{D}\hat{\boldsymbol{D}})^2\hat{\boldsymbol{D}} + O(\gamma^{-7}) \,.\end{aligned} \tag{10.181}$$

The inverse Laplace transform of (10.181) is the expansion of the memory function

$$\begin{aligned}\boldsymbol{K}_0(t) = {} & \mathrm{e}^{-\gamma t}\boldsymbol{D}\hat{\boldsymbol{D}} + \gamma^{-2}(2\gamma t\mathrm{e}^{-\gamma t} - 2\mathrm{e}^{-\gamma t} + 2\mathrm{e}^{-2\gamma t})\boldsymbol{D}^2\hat{\boldsymbol{D}}^2 \\ & + \gamma^{-4}\{[3\gamma t\mathrm{e}^{-\gamma t} - (15/2)\mathrm{e}^{-\gamma t} + 6\gamma t\mathrm{e}^{-2\gamma t} \\ & + 6\mathrm{e}^{-2\gamma t} + (3/2)\mathrm{e}^{-3\gamma t}]\boldsymbol{D}^3\hat{\boldsymbol{D}}^3 \\ & + [2(\gamma t)^2\mathrm{e}^{-\gamma t} - 8\gamma t\mathrm{e}^{-\gamma t} + 12\mathrm{e}^{-\gamma t} \\ & - 4\gamma t\mathrm{e}^{-2\gamma t} - 12\mathrm{e}^{-2\gamma t}]\boldsymbol{D}(\boldsymbol{D}\hat{\boldsymbol{D}})^2\hat{\boldsymbol{D}}\} + O(\gamma^{-6}) \,.\end{aligned} \tag{10.182}$$

With this memory function the solution $\boldsymbol{G}_{0,0}(t)$ may be determined by

$$\dot{\boldsymbol{G}}_{0,0}(t) = \int_0^t \boldsymbol{K}_0(t-\tau)\,\boldsymbol{G}_{0,0}(\tau)\,\mathrm{d}\tau \,. \tag{10.183}$$

For large γ the memory function rapidly decreases in time. Using proper partial integrations we may express (10.183) as

$$\dot{G}_{0,0}(t) = \int_0^\tau K_0(t-\tau')\,d\tau'\,G_{0,0}(\tau)\Big|_0^t - \int_0^t\int_0^\tau K_0(t-\tau')\,d\tau'\,\dot{G}_{0,0}(\tau)\,d\tau$$

$$= \int_0^t K_0(t-\tau)\,d\tau\,G_{0,0}(t)$$

$$-\int_0^t\left[\int_0^\tau K_0(t-\tau')\,d\tau'\int_0^\tau K_0(\tau-\tau'')\,G_{0,0}(\tau'')\,d\tau''\right]d\tau\,. \qquad (10.183\,a)$$

Because of the rapidly decreasing memory functions the last expression is much smaller than the first, and we may thus approximate (10.183) by

$$\begin{aligned}\dot{G}_{0,0}(t) &= L_0^{(1)}(t)\,G_{0,0}(t)\\ L_0^{(1)}(t) &= \int_0^t K_0(t-\tau)\,d\tau = \gamma^{-1}(1-e^{-\gamma t})D\hat{D}+\dots\,.\end{aligned} \qquad (10.184)$$

Thus we see that in the first step we simply replace $G_{0,0}(\tau)$ by $G_{0,0}(t)$ in the integral equation (10.183). This corresponds to the first Born approximation in the scattering theory of quantum mechanics. In this first step it is inconsistent to use expansion terms of (10.182) of the order γ^{-2} and higher. In the next approximation we replace $G_{0,0}(\tau'')$ in (10.183 a) by $G_{0,0}(t)$, so obtaining

$$L_0^{(2)}(t) = \int_0^t K_0(t-\tau)\,d\tau - \int_0^t\left[\int_0^\tau K_0(t-\tau')\,d\tau'\int_0^\tau K_0(\tau-\tau'')\,d\tau''\right]d\tau$$

$$= \gamma^{-1}(1-e^{-\gamma t})D\hat{D}+\gamma^{-3}(1-2\gamma t e^{-\gamma t}-e^{-2\gamma t})D[D,\hat{D}]\hat{D}+\dots, \qquad (10.184\,a)$$

which is already the beginning of the desired inverse friction expansion for the operator $L_0(t)$. In the first term on the right hande side of (10.184 a) the expansion of K_0 up to the term γ^{-2} must be used. The next iteration step is much more complicated because we have to perform a double partial integration to extract $G_{0,0}(t)$ from the last expression of (10.183 a). We therefore try to obtain first a formal expression for $G_{0,0}(t)$ directly from (10.139 a, 181). To do this we write $G_{0,0}(t)$ as

$$G_{0,0}(t) = \exp[\Omega(t)]\,, \qquad (10.185)$$

and try an expansion of $\Omega(t)$ in the form

$$\Omega(t) = \gamma^{-2}\Omega_2(\gamma t)+\gamma^{-4}\Omega_4(\gamma t)+\gamma^{-6}\Omega_6(\gamma t)+\dots\,. \qquad (10.186)$$

Note that Ω_4 and higher terms must contain only commutators of D and $\hat{D}$ because for commuting D and $\hat{D}$ the expansion (10.186) terminates after the first term, compare (10.153, 154) for $\hat{D}=D$. Because

$$\tilde{G}_{0,0}(s) = [sI-\tilde{K}_0(s)]^{-1} = s^{-1}I+s^{-2}\tilde{K}_0(s)+s^{-3}\tilde{K}_0^2(s)+s^{-4}\tilde{K}_0^3(s)+\dots \qquad (10.187)$$

the inverse friction expansion of $\tilde{G}_{0,0}(s)$ is also easily found. By an inverse Laplace transformation we then get $G_{0,0}(t)$. (To obtain the explicit expressions for the various inverse Laplace transforms, the reader should consult App. C and D of [7.7], where all the necessary transformations can be found.) Next we expand the exponential function (10.185) with $\boldsymbol{\Omega}$ given by (10.186) and compare the terms of equal powers of γ^{-1}. The final results of these lengthy calculations are

$$\begin{aligned}
\boldsymbol{\Omega}_2(\sigma) &= H_2(\sigma)\boldsymbol{D}\hat{\boldsymbol{D}} \\
\boldsymbol{\Omega}_4(\sigma) &= H_4(\sigma)\boldsymbol{D}[\boldsymbol{D},\hat{\boldsymbol{D}}]\hat{\boldsymbol{D}} \\
\boldsymbol{\Omega}_6(\sigma) &= 2[H_6^{(1)}(\sigma)+H_6^{(2)}(\sigma)]\boldsymbol{D}[\boldsymbol{D},\hat{\boldsymbol{D}}]^2\hat{\boldsymbol{D}} \\
&\quad + H_6^{(1)}(\sigma)\boldsymbol{D}\{[\boldsymbol{D},[\boldsymbol{D},\hat{\boldsymbol{D}}]]\hat{\boldsymbol{D}} - (1/2)[\boldsymbol{D}\hat{\boldsymbol{D}},[\boldsymbol{D},\hat{\boldsymbol{D}}]]\}\hat{\boldsymbol{D}}\,,
\end{aligned} \tag{10.188}$$

where the functions H_n are given by

$$\begin{aligned}
H_2(\sigma) &= \sigma - 1 + \mathrm{e}^{-\sigma} \\
H_4(\sigma) &= \sigma - (5/2) + 2\sigma\mathrm{e}^{-\sigma} + 2\mathrm{e}^{-\sigma} + (1/2)\mathrm{e}^{-2\sigma} \\
H_6^{(1)}(\sigma) &= (1/2)\sigma - (5/3) + 3\sigma\mathrm{e}^{-\sigma} - (3/2)\mathrm{e}^{-\sigma} + (3/2)\sigma\mathrm{e}^{-2\sigma} \\
&\quad + 3\mathrm{e}^{-2\sigma} + (1/6)\mathrm{e}^{-3\sigma} \\
H_6^{(2)}(\sigma) &= (1/2)\sigma - 2 + \sigma^2\mathrm{e}^{-\sigma} + 4\mathrm{e}^{-\sigma} - (1/2)\sigma\mathrm{e}^{-2\sigma} - 2\mathrm{e}^{-2\sigma}\,.
\end{aligned} \tag{10.189}$$

It is now easy to find the generalized Smoluchowski operator $\boldsymbol{L}_0(t)$, i.e., the operator of the equation

$$\dot{\boldsymbol{G}}_{0,0}(t) = \boldsymbol{L}_0(t)\,\boldsymbol{G}_{0,0}(t)\,. \tag{10.190}$$

Following *Weiss* and *Maradudin* [10.8], we first introduce the symbols $\{\}$ defined by

$$\begin{aligned}
\{u,x^0\} &= u\,, \\
\{u,x^n\} &= [\{u,x^{n-1}\},x]\,, \qquad n \geqq 1\,,
\end{aligned} \tag{10.191}$$

where $[A,B] = AB - BA$ is the commutator. Equation (3.20) of [10.8] reads

$$\begin{aligned}
\boldsymbol{L}_0(t) &= \left\{\dot{\boldsymbol{\Omega}}, \frac{1-\exp(-\boldsymbol{\Omega})}{\boldsymbol{\Omega}}\right\} \\
&= \dot{\boldsymbol{\Omega}} - (1/2)[\dot{\boldsymbol{\Omega}},\boldsymbol{\Omega}] + (1/6)[[\dot{\boldsymbol{\Omega}},\boldsymbol{\Omega}],\boldsymbol{\Omega}] \mp \dots\,.
\end{aligned} \tag{10.192}$$

Inserting (10.186, 188, 189) into (10.192) we finally obtain

$$\begin{aligned}L_0(t) = D\{&\gamma^{-1}h_1(\gamma t) + \gamma^{-3}h_3(\gamma t)[D,\hat{D}]\\&+\gamma^{-5}[2(h_5^{(1)}(\gamma t)+h_5^{(2)}(\gamma t))[D,\hat{D}]^2\\&+h_5^{(1)}(\gamma t)[D,[D,\hat{D}]]\hat{D} + h_5^{(2)}(\gamma t)[D\hat{D},[D,\hat{D}]]]\}\hat{D} + O(\gamma^{-7}).\end{aligned} \tag{10.193}$$

The functions h_n are given by $(\mathrm{d}/\mathrm{d}\sigma)H_{n+1}(\sigma)$ and read explicitly:

$$\begin{aligned}h_1(\sigma) &= 1-\mathrm{e}^{-\sigma} &&= \sigma+\dots\\h_3(\sigma) &= 1-2\sigma\mathrm{e}^{-\sigma}-\mathrm{e}^{-2\sigma} &&= 2\sigma^3/3!+\dots\\h_5^{(1)}(\sigma) &= (1/2)-3\sigma\mathrm{e}^{-\sigma}+(9/2)\mathrm{e}^{-\sigma}-3\sigma\mathrm{e}^{-2\sigma}\\&\quad-(9/2)\mathrm{e}^{-2\sigma}-(1/2)\mathrm{e}^{-3\sigma} &&= 6\sigma^5/5!+\dots\\h_5^{(2)}(\sigma) &= (1/2)-\sigma^2\mathrm{e}^{-\sigma}+2\sigma\mathrm{e}^{-\sigma}-4\mathrm{e}^{-\sigma}\\&\quad+\sigma\mathrm{e}^{-2\sigma}+(7/2)\mathrm{e}^{-2\sigma} &&= 2\sigma^5/5!+\dots\,.\end{aligned} \tag{10.194}$$

In deriving (10.193) we used the relationship

$$h_3H_2 - h_1H_4 = h_5^{(1)} + 2h_5^{(2)}\,. \tag{10.195}$$

Up to the term γ^{-3} (10.193, 194) agrees with (10.184a). In the x representation with $\varepsilon = 0$ the commutators are given by

$$\begin{aligned}&[D,\hat{D}] = f''\,; \quad [D,[D,\hat{D}]] = v_{\mathrm{th}}f'''\\&[D\hat{D},[D,\hat{D}]] = v_{\mathrm{th}}^2 f^{(\mathrm{IV})} + f'''(2v_{\mathrm{th}}^2\,\partial/\partial x + f')\,.\end{aligned} \tag{10.196}$$

It should be noted here that the distribution (10.179) also obeys (10.190). For large times $\gamma t \gg 1$, (10.193) reduces in the x representation for $\varepsilon = 0$ to

$$\begin{aligned}L_0(\infty) = \frac{\partial}{\partial x}\Bigg\{&\frac{1}{\gamma} + \frac{1}{\gamma^3}f'' + \frac{1}{\gamma^5}\left[2(f'')^2 + \frac{1}{2}v_{\mathrm{th}}^2 f^{(\mathrm{IV})} + \frac{3}{2}v_{\mathrm{th}}^2 f'''\frac{\partial}{\partial x} + f'''f'\right]\Bigg\}\\&\times\left(v_{\mathrm{th}}^2\frac{\partial}{\partial x} + f'\right) + O(\gamma^{-7})\,.\end{aligned} \tag{10.197}$$

One may object to the occurrence of the third-order x derivative in (10.197) because, due to Pawula's theorem (Sect. 4.3), the transition probability must then have negative values for sufficiently small times. As shown in Sect. 4.6 for a simple example, however, the negative values may be very small and the deviations from the exact distribution may be much larger for the distribution function following the truncation after the second derivative term than for the distribution function following the truncation after some higher derivative terms.

Equation (10.197) was derived in [10.2, 10.9] and up to the term γ^{-3} in [10.10–12] by different methods. Equations (10.193, 194) were obtained in [9.19] by using the expansion of the continued fraction, and up to the term γ^{-3} in [10.13] by using *van Kampen's* ordered cumulant expansion [10.14].

Application to the Parabolic Potential

To check the validity of expansion (10.193) we compare it to an exact $L_0(t)$. For the parabolic potential (10.52) the transition probability and therefore also $G_{0,0}(x,x',t)$ can be calculated exactly. [For free Brownian motion, the commutators in (10.193) vanish and the first term of the expansion already agrees with the exact result (10.156) for $\varepsilon = 0$.] By taking the average of the transition probability (10.55) over a Maxwell distribution for the initial velocities and by integration over the final velocity we obtain (see also [10.12])

$$G_{0,0}(x,x',t) = \frac{\omega_0}{v_{\text{th}}\sqrt{2\pi(1-y^2)}} \exp\left(-\frac{\omega_0^2(x-yx')^2}{2(1-y^2)v_{\text{th}}^2}\right),$$

$$y(t) = (\lambda_1 e^{-\lambda_2 t} - \lambda_2 e^{-\lambda_1 t})/(\lambda_1 - \lambda_2)\,, \tag{10.198}$$

$$\lambda_{\substack{1\\2}} = (\gamma \pm \sqrt{\gamma^2 - 4\omega_0^2})/2\,.$$

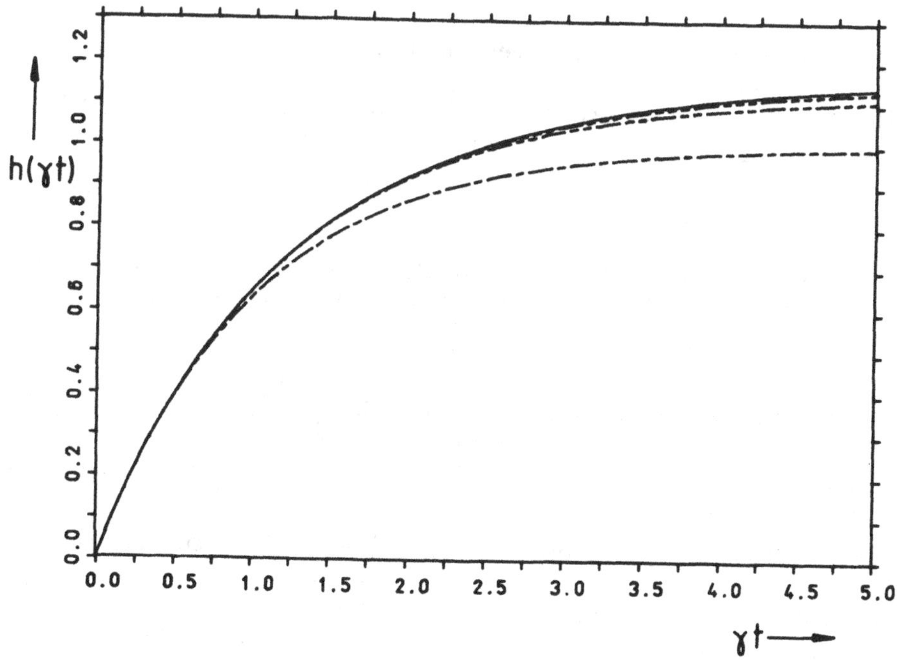

Fig. 10.3. The exact solution (10.199) (*solid line*) and the three different inverse friction expansions (*broken lines*) according to (10.200) as a function of γt for $2\omega_0/\gamma = 0.7$

As may be checked by insertion (10.198) satisfies (10.190) with $L_0(t)$ given by

$$L_0^{(\mathrm{ex})}(t) = h^{(\mathrm{ex})}(\gamma t)\frac{1}{\gamma}\frac{\partial}{\partial x}\left(v_{\mathrm{th}}^2\frac{\partial}{\partial x}+\omega_0^2 x\right)$$
$$h^{(\mathrm{ex})}(\gamma t) = -\gamma\dot{y}/(y\,\omega_0^2)\,. \tag{10.199}$$

This exact result has to be compared with (10.193) for the case of a harmonic potential with $[D,\hat{D}]=\omega_0^2$ (see 10.28) and where all the double commutators vanish, i.e., with

$$L_0(t) = h(\gamma t)\frac{1}{\gamma}\frac{\partial}{\partial x}\left(v_{\mathrm{th}}^2\frac{\partial}{\partial x}+\omega_0^2 x\right)$$
$$h(\gamma t) = h_1(\gamma t)+(\omega_0/\gamma)^2 h_3(\gamma t)+(\omega_0/\gamma)^4 2[h_5^{(1)}(\gamma t)+h_5^{(2)}(\gamma t)]+\dots\,. \tag{10.200}$$

A comparison between $h(\gamma t)$ and $h^{(\mathrm{ex})}(\gamma t)$ is shown in Fig. 10.3. It is seen that even for $2\omega_0 = 0.7\gamma$ the approximation is very good. A closer inspection shows that (10.200) is an expansion of $h^{(\mathrm{ex})}(\gamma t)$ into powers of the inverse friction constant. If should be mentioned that in the underdamped case $\gamma<2\omega_0$, $y(t)$ has zeros and the exact Liouville operator does not exist. For a discussion of cases where $L_0(t)$ does not exist, see [10.15]. In the underdamped case $\gamma<2\omega_0$, expansion (10.193) does not converge for any time $t>0$.

10.4.2 Determination of Eigenvalues and Eigenvectors

The eigenvalues and eigenvectors are determined by (10.158, 159). For large friction the eigenvalues are grouped together according to

$$\lambda_{m\nu} = m\gamma + \Delta_{m\nu}\,, \tag{10.201}$$

where $\Delta_{m\nu}$ is of the order γ^{-1}. To determine $\lambda_{m\nu}$ therefore (10.159) is most suitable. By expanding $\tilde{\boldsymbol{K}}_m(-\lambda_m)$ in powers of Δ_m we obtain (suppressing the index ν)

$$\left[\Delta_m\boldsymbol{I}+\tilde{\boldsymbol{K}}_m(-m\gamma)-\frac{d}{ds}\tilde{\boldsymbol{K}}_m(s)\bigg|_{s=-m\gamma}\Delta_m+\frac{1}{2}\left(\frac{d}{ds}\right)^2\tilde{\boldsymbol{K}}_m(s)\bigg|_{s=-m\gamma}\Delta_m^2+\dots\right]\hat{\boldsymbol{c}}_m$$
$$=0\,. \tag{10.202}$$

We now eliminate Δ_m in the third term and Δ_m^2 in the fourth term iteratively

$$\Delta_m^{(1)}\hat{\boldsymbol{c}}_m = -\tilde{\boldsymbol{K}}_m(-m\gamma)\hat{\boldsymbol{c}}_m$$
$$\Delta_m^{(2)}\hat{\boldsymbol{c}}_m = -\left[\tilde{\boldsymbol{K}}_m(-m\gamma)+\frac{d}{ds}\tilde{\boldsymbol{K}}_m(s)\bigg|_{s=-m\gamma}\tilde{\boldsymbol{K}}_m(-m\gamma)\right]\hat{\boldsymbol{c}}_m\,,$$

leading in the third approximation for $\lambda_m = m\gamma + \Delta_m^{(3)}$ to (not suppressing ν)

$$(\lambda_{m\nu}\boldsymbol{I} + \boldsymbol{L}_m)\hat{\boldsymbol{c}}_{m\nu} = \boldsymbol{0}\,, \tag{10.203}$$

where $\boldsymbol{L}_m$ is given by

$$\boldsymbol{L}_m = -m\gamma\boldsymbol{I} + \tilde{\boldsymbol{K}}_m(-m\gamma) + \frac{\mathrm{d}}{\mathrm{d}s}\tilde{\boldsymbol{K}}_m(s)\bigg|_{s=-m\gamma}\tilde{\boldsymbol{K}}_m(-m\gamma) + \left[\frac{\mathrm{d}}{\mathrm{d}s}\tilde{\boldsymbol{K}}_m(s)\bigg|_{s=-m\gamma}\right]^2$$
$$\times\tilde{\boldsymbol{K}}_m(-m\gamma) + \frac{1}{2}\frac{\mathrm{d}^2}{\mathrm{d}s^2}\tilde{\boldsymbol{K}}_m(s)\bigg|_{s=-m\gamma}[\tilde{\boldsymbol{K}}_m(-m\gamma)]^2 + \dots\,. \tag{10.204}$$

Using (10.204) for $m=0$ and the inverse friction expansion (10.180) for $\tilde{\boldsymbol{K}}_0(s)$ we obtain the operator (10.193) for large times $\gamma t \gg 1$. Thus in x representation with $\varepsilon = 0$, $\boldsymbol{L}_0 = \boldsymbol{L}_0(\infty)$ is given by (10.197). For the general case the expansion terms up to the order γ^{-3} read

$$\begin{aligned}\boldsymbol{L}_m = &-m\gamma\boldsymbol{I} + \gamma^{-1}((m+1)\boldsymbol{D}\hat{\boldsymbol{D}} - m\hat{\boldsymbol{D}}\boldsymbol{D}) + \gamma^{-3}[(1/2)(m+1)(m+2)\boldsymbol{D}^2\hat{\boldsymbol{D}}^2\\ &-(1/2)m(m-1)\hat{\boldsymbol{D}}^2\boldsymbol{D}^2 - (m+1)^2(\boldsymbol{D}\hat{\boldsymbol{D}})^2 + m(m+1)(\boldsymbol{D}\hat{\boldsymbol{D}}^2\boldsymbol{D} - \hat{\boldsymbol{D}}\boldsymbol{D}^2\hat{\boldsymbol{D}})\\ &+ m^2(\hat{\boldsymbol{D}}\boldsymbol{D})^2]\,.\end{aligned} \tag{10.205}$$

In the x representation with $\varepsilon = 0$, (10.205) agrees with the operator (4.2) of [Ref. 10.2]. (In (4.2) of [10.2] a factor 1/2 is missing in the term proportional to $m-1$.)

The other eigenvectors $\hat{\boldsymbol{c}}_n$, still having eigenvalues near $m\gamma$ and which are therefore termed $\hat{\boldsymbol{c}}_{n,m\nu}$, follow from

$$\hat{\boldsymbol{c}}_{n,m\nu} = \tilde{\boldsymbol{U}}_{n,m}(-\lambda_{m\nu})\hat{\boldsymbol{c}}_{m,m\nu}\,. \tag{10.206}$$

By performing a procedure similar to that used previously, we obtain

$$\begin{aligned}&\hat{\boldsymbol{c}}_{n,m\nu} = \boldsymbol{V}_{n,m}\hat{\boldsymbol{c}}_{m,m\nu}\\ &\boldsymbol{V}_{n,m} = \tilde{\boldsymbol{U}}_{n,m}(-m\gamma) + \frac{\mathrm{d}}{\mathrm{d}s}\tilde{\boldsymbol{U}}_{n,m}(s)\bigg|_{s=-m\gamma}(\boldsymbol{L}_m + m\gamma\boldsymbol{I}) + \dots\,.\end{aligned} \tag{10.207}$$

Evaluating the continued fractions (10.142, 143), we obtain the inverse friction expansion of $\tilde{\boldsymbol{U}}_{n,m}(s)$ (10.141) and of its derivative at $s = -m\gamma$. The final result is expressed in a more compact form if

$$\begin{aligned}\boldsymbol{Q}_n^+ &= \begin{cases}-\sqrt{n+1}\,\boldsymbol{D} & \text{for } n \geqq 0\\ 0 & \phantom{\text{for }} n \leqq -1\end{cases}\\ \boldsymbol{Q}_n^- &= \begin{cases}-\sqrt{n}\,\hat{\boldsymbol{D}} & \text{for } n \geqq 1\\ 0 & \phantom{\text{for }} n \leqq 0\end{cases}\end{aligned} \tag{10.208}$$

are used (10.130). The result at least correct up to γ^{-4} is $(n \geqq 0)$

$$\begin{aligned}
V_{n\mp4,n} &= \gamma^{-4}(4!)^{-1}Q^{\pm}_{n\mp4}Q^{\pm}_{n\mp3}Q^{\pm}_{n\mp2}Q^{\pm}_{n\mp1}\\
V_{n\mp3,n} &= \mp\gamma^{-3}(3!)^{-1}Q^{\pm}_{n\mp3}Q^{\pm}_{n\mp2}Q^{\pm}_{n\mp1}\\
V_{n\mp2,n} &= \gamma^{-2}(1/2)\,Q^{\pm}_{n\mp2}Q^{\pm}_{n\mp1}\\
&\quad+\gamma^{-4}(1/4)\,[Q^{\pm}_{n\mp2}Q^{\mp}_{n\mp1}Q^{\pm}_{n\mp2}Q^{\pm}_{n\mp1}\\
&\quad+(1/3)\,Q^{\mp}_{n\mp2}Q^{\pm}_{n\mp3}Q^{\pm}_{n\mp2}Q^{\pm}_{n\mp1}\\
&\quad+3\,Q^{\pm}_{n\mp2}Q^{\pm}_{n\mp1}(Q^{\pm}_{n}\,Q^{\mp}_{n\pm1}-Q^{\mp}_{n}\,Q^{\pm}_{n\mp1})]\\
V_{n\mp1,n} &= \mp\gamma^{-1}Q^{\pm}_{n\mp1}\mp\gamma^{-3}[(1/2)\,Q^{\mp}_{n\mp1}Q^{\pm}_{n\mp2}Q^{\pm}_{n\mp1}\\
&\quad+Q^{\pm}_{n\mp1}(Q^{\pm}_{n}\,Q^{\mp}_{n\pm1}-Q^{\mp}_{n}\,Q^{\pm}_{n\mp1})]\\
V_{n,n} &= I\,.
\end{aligned}\tag{10.209}$$

Up to a normalization constant, (4.1) of [10.2] agrees with (10.209) in the x representation for $\varepsilon=0$. [In (4.1) of [10.2] the last term in the third line from the bottom should read $(3/2)(2n-1)\beta$ instead of $(3/2)(n-1)\beta$.]

With the help of (10.208), (10.205) can be written as

$$\begin{aligned}
L_m &= -m\gamma I+A^{+}_{m}-A^{-}_{m}\\
A^{\pm}_{m} &= \gamma^{-1}Q^{\pm}_{m}Q^{\mp}_{m\pm1}+\gamma^{-3}[(1/2)\,Q^{\pm}_{m}\,Q^{\pm}_{m\pm1}Q^{\mp}_{m\pm2}\,Q^{\mp}_{m\pm1}\\
&\quad-(Q^{\pm}_{m}\,Q^{\mp}_{m\pm1})^{2}+Q^{\pm}_{m}\,Q^{\mp}_{m\pm1}Q^{\mp}_{m}\,Q^{\pm}_{m\mp1}]\,.
\end{aligned}\tag{10.205 a}$$

10.4.3 Expansion for the Green's Function $G_{n,m}(t)$

The expansion for $G_{0,0}(t)$ (10.185, 188, 189) cannot be used for other diagonal $G_{m,m}(t)$, as can best be seen by discussing $G_{1,1}(t)$. The exact solution for $\tilde{G}_{1,1}(s)$ reads:

$$\tilde{G}_{1,1}(s)=[(s+\gamma)I-\tilde{K}_1(s)]^{-1}\,,\tag{10.210}$$

$$\tilde{K}_1(s)\;=s^{-1}\hat{D}D+2D\{(s+2\gamma)I-3D[(s+3\gamma)I-\dots]^{-1}\hat{D}\}^{-1}\hat{D}\,.\tag{10.211}$$

The inverse friction expansion of (10.211) and its inverse Laplace transform is given by

$$\tilde{K}_1(s)=\frac{1}{s}\hat{D}D+2\frac{D\hat{D}}{s+2\gamma}+6\frac{D^2\hat{D}^2}{(s+2\gamma)^2(s+3\gamma)}+\dots\,,\tag{10.212}$$

$$K_1(t)=\hat{D}D+2\mathrm{e}^{-2\gamma t}D\hat{D}+6\gamma^{-2}(\gamma t\mathrm{e}^{-2\gamma t}-\mathrm{e}^{-2\gamma t}+\mathrm{e}^{-3\gamma t})D^2\hat{D}^2+\dots\,.\tag{10.213}$$

Notice that the memory kernel does not vanish for large times $[\boldsymbol{K}_1(\infty)=\hat{\boldsymbol{D}}\boldsymbol{D}]$. We therefore cannot use the Born approximation for solving the integral equation (10.139a) with $m=1$, as done for (10.183), and we cannot derive an operator $\boldsymbol{L}_1(t)$ for $G_{1,1}(t)$. Obviously the troublesome term in $\tilde{\boldsymbol{K}}_1(s)$ is $s^{-1}\hat{\boldsymbol{D}}\boldsymbol{D}$, which also occurs in other memory functions $\tilde{\boldsymbol{K}}_m(s)$ for $m \geqq 2$.

For very small s (s of the order $1/\gamma$) i.e., for large times (t of the order γ), this term becomes important and the inverse friction expansions for $\tilde{\boldsymbol{G}}_{m,m}(s)$ do not exist for $m \geqq 1$.

To find a different inverse friction expansion method we proceed as follows [9.19]. We first consider solutions of the tridiagonal recurrence relation

$$\dot{c}_n = \boldsymbol{Q}_n^- c_{n-1} - \gamma n c_n + \boldsymbol{Q}_n^+ c_{n+1}\,. \tag{10.214}$$

If the damping constants γ are large, the eigenvalues of (10.214) are grouped together according to (10.201). Let us consider solutions of (10.214) called $\boldsymbol{c}_{n,m}(t)$ which can be expanded into eigenfunctions of (10.214) belonging to the group of eigenvalues with the main part $m\gamma$, i.e.,

$$\boldsymbol{c}_{n,m}(t) = \sum_{\nu} b_{m\nu}\hat{\boldsymbol{c}}_{n,m\nu}\mathrm{e}^{-\lambda_{m\nu}t}\,, \tag{10.215}$$

where $b_{m\nu}$ are the expansion coefficients. In the x representation these solutions correspond to the solutions $P_{[m]}$ in [10.2], though no connection with the eigenvalues was made there. From (10.203, 207) it is seen that $\boldsymbol{c}_{n,m}(t)$ must satisfy

$$\boldsymbol{c}_{n,m} = \boldsymbol{V}_{n,m}\boldsymbol{c}_{m,m} \tag{10.216}$$

and

$$\dot{\boldsymbol{c}}_{m,m} = \boldsymbol{L}_m\boldsymbol{c}_{m,m}\,. \tag{10.217}$$

A formal solution of (10.216, 217) is given by

$$\boldsymbol{c}_{n,m}(t) = \boldsymbol{V}_{n,m}\boldsymbol{c}_{m,m}(t) = \boldsymbol{V}_{n,m}\mathrm{e}^{\boldsymbol{L}_m t}\boldsymbol{c}_{m,m}(0)\,. \tag{10.218}$$

A general solution of (10.214) is obtained by summation over all different m:

$$\boldsymbol{c}_n(t) = \sum_{m=0}^{\infty}\boldsymbol{c}_{n,m}(t) = \sum_{m=0}^{\infty}\boldsymbol{V}_{n,m}\mathrm{e}^{\boldsymbol{L}_m t}\boldsymbol{c}_{m,m}(0)\,. \tag{10.219}$$

The initial values $\boldsymbol{c}_n(0)$ of the general solution (10.219) and the initial values $\boldsymbol{c}_{m,m}(0)$ are connected by

$$\boldsymbol{c}_n(0) = \sum_{m=0}^{\infty}\boldsymbol{V}_{n,m}\boldsymbol{c}_{m,m}(0)\,. \tag{10.220}$$

Note that $\boldsymbol{c}_n(0)$ are column vectors and that $\boldsymbol{V}_{n,m}$ are matrices. We can, however, proceed in formal analogy to linear algebra and invert (10.220) by

$$c_{m,m}(0) = \sum_{r=0}^{\infty} \bar{V}_{m,r} c_r(0) \,, \tag{10.221}$$

where the "inverse" matrices $\bar{V}_{m,r}$ must fulfill the relation

$$\sum_{m=0}^{\infty} V_{n,m} \bar{V}_{m,r} = I \delta_{nr} \,. \tag{10.222}$$

Because $V_{m,m} = I$ and the other $V_{n,m}$ with $n \neq m$ are at least of the order γ^{-1}, $\bar{V}_{m,r}$ may be determined iteratively:

$$\begin{aligned} \bar{V}_{n,m} &= I\delta_{nm} - \bar{A}_{nm} + \sum_r \bar{A}_{nr}\bar{A}_{rm} \\ &\quad - \sum_{r,s} \bar{A}_{nr}\bar{A}_{rs}\bar{A}_{sm} + \sum_{r,s,t} \bar{A}_{nr}\bar{A}_{rs}\bar{A}_{st}\bar{A}_{tm} \mp \dots \,, \end{aligned} \tag{10.223}$$

$$\bar{A}_{nm} = V_{n,m} \quad \text{for} \quad n \neq m \,, \quad \bar{A}_{nm} = 0 \,.$$

The explicit calculation of the inverse friction expansion is a little cumbersome. The final result written in terms of $Q_n^{\pm}$ (10.208) reads

$$\begin{aligned} \bar{V}_{n\mp4,n} &= \gamma^{-4}(4!)^{-1} Q^{\pm}_{n\mp4} Q^{\pm}_{n\mp3} Q^{\pm}_{n\mp2} Q^{\pm}_{n\mp1} \\ \bar{V}_{n\mp3,n} &= \pm\gamma^{-3}(3!)^{-1} Q^{\pm}_{n\mp3} Q^{\pm}_{n\mp2} Q^{\pm}_{n\mp1} \\ \bar{V}_{n\mp2,n} &= \gamma^{-2}(1/2) Q^{\pm}_{n\mp2} Q^{\pm}_{n\mp1} \\ &\quad + \gamma^{-4}(1/4)[Q^{\pm}_{n\mp2} Q^{\pm}_{n\mp1} Q^{\mp}_{n} Q^{\pm}_{n\mp1} + (1/3) Q^{\pm}_{n\mp2} Q^{\pm}_{n\mp1} Q^{\pm}_{n} Q^{\mp}_{n\pm1} \\ &\quad + Q^{\mp}_{n\mp2} Q^{\pm}_{n\mp3} Q^{\pm}_{n\mp2} Q^{\pm}_{n\mp1} - 5 Q^{\pm}_{n\mp2} Q^{\mp}_{n\mp1} Q^{\pm}_{n\mp2} Q^{\pm}_{n\mp1}] \\ \bar{V}_{n\mp1,n} &= \pm\gamma^{-1} Q^{\pm}_{n\mp1} \pm \gamma^{-3}[(1/2) Q^{\pm}_{n\mp1} Q^{\pm}_{n} Q^{\mp}_{n\pm1} - 2 Q^{\pm}_{n\mp1} Q^{\mp}_{n} Q^{\pm}_{n\mp1}] \\ \bar{V}_{n,n} &= I - \gamma^{-2}(Q^{-}_{n} Q^{+}_{n-1} + Q^{+}_{n} Q^{-}_{n+1}) \\ &\quad - \gamma^{-4}\{Q^{-}_{n} Q^{+}_{n-1} Q^{+}_{n} Q^{-}_{n+1} + Q^{+}_{n} Q^{-}_{n+1} Q^{-}_{n} Q^{+}_{n-1} \\ &\quad + Q^{-}_{n}[(5/4) Q^{-}_{n-1} Q^{+}_{n-2} - 3 Q^{+}_{n-1} Q^{-}_{n}] Q^{+}_{n-1} \\ &\quad + Q^{+}_{n}[(5/4) Q^{+}_{n+1} Q^{-}_{n+2} - 3 Q^{-}_{n+1} Q^{+}_{n}] Q^{-}_{n+1}\} . \end{aligned} \tag{10.224}$$

This result is at least correct up to terms of the order γ^{-4}. Elements not written down are zero up to this order. As one may check, also the relation

$$\sum_{m=0}^{\infty} \bar{V}_{n,m} V_{m,r} = I \delta_{nr} \tag{10.222a}$$

holds up to this order.

By inserting (10.221) into (10.219) we obtain immediately the expression for the Green's function of system (10.214) (changing the notation of the indices)

$$c_n(t) = \sum_{m=0}^{\infty} G_{n,m}(t) c_m(0) \,, \qquad G_{n,m}(t) = \sum_{r=0}^{\infty} V_{n,r} e^{L_r t} \bar{V}_{r,m} \,. \tag{10.225}$$

This, together with inverse friction expansions (10.205, 209, 224), is the desired expansion of the Green's function. It has the formal analogy of an eigenfunction expansion of the Green's function though in x representation the "eigenvalues" L_r and "eigenfunctions" $V_{n,r}$, $\bar{V}_{r,m}$ are operators with respect to x. This analogy may be developed a little further. If we write (10.214) as

$$\dot{c}_n = \sum_m L_{n,m} c_m \tag{10.214a}$$

($L_{n,n} = -n\gamma I$; $L_{n,n+1} = Q_n^+$, $L_{n-1,n} = Q_n^-$, others are zero) the formal solution with the initial value $I\delta_{nm}$ is given by

$$G_{n,m}(t) = (e^{Lt})_{n,m} = I + L_{n,m} t + \frac{1}{2} \sum_s L_{n,s} L_{s,m} t^2 + \dots \,. \tag{10.226}$$

By comparison with (10.225) for small t we find

$$L_{n,m} = \sum_{r=0}^{\infty} V_{n,r} L_r \bar{V}_{r,m} \tag{10.227}$$

in analogy to the spectral decomposition of an operator. (Using (10.205a, 209, 224) the first terms of this relation may be checked.)

Equivalence of (10.225) *for* $n = m = 0$ *with* (10.185)

Before we discuss (10.225) and derive an expression for the Green's function in x representation, we want to show that for $n = m = 0$ (10.225) agrees with the Green's function $G_{0,0}(t)$ derived previously. Taking into account terms up to the order γ^{-4} only, we write (10.185, 186, 188, 189) in the form ($\sigma = \gamma t$)

$$\begin{aligned} G_{0,0}(t) &= \exp\{\gamma^{-2}(\sigma - 1 + e^{-\sigma}) D\hat{D} + \gamma^{-4}[\sigma - (5/2) + 2\sigma e^{-\sigma} + 2e^{-\sigma} \\ &\quad + (1/2) e^{-2\sigma}] D[D,\hat{D}]\hat{D}\} \\ &\approx \exp[(\gamma^{-1} D\hat{D} + \gamma^{-3} D[D,\hat{D}]\hat{D}) t] \exp\{\gamma^{-2}(-1 + e^{-\sigma}) D\hat{D} \\ &\quad + \gamma^{-4}[-(5/2) + 2\sigma e^{-\sigma} + 2e^{-\sigma} + (1/2) e^{-2\sigma}] D[D,\hat{D}]\hat{D}\} \,. \end{aligned} \tag{10.228}$$

In (10.228) we factorized the exponential function which is correct for terms up to the order γ^{-4}. In the first exponent we may write $L_0 t$, where L_0 is given by (10.205) for $m = 0$. The expansion of the second exponent up to the terms of the order γ^{-4} leads to

$$\begin{aligned} G_{0,0}(t) &= e^{L_0 t}\{I - \gamma^{-2} D\hat{D} + \gamma^{-2} e^{-\sigma} D\hat{D} \\ &\quad + \gamma^{-4}[-(5/2) + 2\sigma e^{-\sigma} + 2e^{-\sigma} + (1/2) e^{-2\sigma}] D[D,\hat{D}]\hat{D} \\ &\quad + (1/2)\gamma^{-4}(1 - 2e^{-\sigma} + e^{-2\sigma})(D\hat{D})^2\} \\ &= G_{0,0}^{(0)}(t) + G_{0,0}^{(1)}(t) + G_{0,0}^{(2)}(t) \end{aligned} \tag{10.229}$$

with

$$G_{0,0}^{(0)}(t) = e^{L_0 t}\{I - \gamma^{-2}D\hat{D} + \gamma^{-4}[-(5/2)D[D,\hat{D}]\hat{D} + (1/2)(D\hat{D})^2]\}$$
$$G_{0,0}^{(1)}(t) = e^{(-\gamma I + L_0)t}\{\gamma^{-2}D\hat{D} + \gamma^{-4}[2D^2\hat{D}^2 - 3(D\hat{D})^2] + 2\gamma^{-3}D[D,\hat{D}]\hat{D}\cdot t\}$$
$$G_{0,0}^{(2)}(t) = e^{(-2\gamma I + L_0)t}(1/2)\gamma^{-4}D^2\hat{D}^2 . \tag{10.230}$$

Equation (10.229) is the expansion (10.225) for $n = m = 0$. The first term in (10.230) is identical to

$$G_{0,0}^{(0)}(t) = I e^{L_0 t}\{I - \gamma^{-2}D\hat{D} + \gamma^{-4}[3(D\hat{D})^2 - (5/2)D^2\hat{D}^2]\} = V_{0,0}e^{L_0 t}\bar{V}_{0,0} ,$$

as is immediately seen. The third term may be approximately written in the form

$$G_{0,0}^{(2)}(t) \approx 2^{-1/2}\gamma^{-2}D^2 e^{-2\gamma I t}2^{-1/2}\gamma^{-2}\hat{D}^2 \approx V_{02}e^{L_2 t}\bar{V}_{20} .$$

In this small term $\sim\gamma^{-4}$ we need only the first term for L_2, V_{02}, $\bar{V}_{20}$. Because

$$e^{L_0 t}D \approx \exp(\gamma^{-1}D\hat{D}t)D = D\exp(\gamma^{-1}\hat{D}Dt)$$

we write the second term up to the order γ^{-4} as

$$G_{0,0}^{(1)}(t) = \gamma^{-1}D\exp[(-\gamma I + \gamma^{-1}\hat{D}D)t](I + \gamma^{-1}2[D,\hat{D}]t)$$
$$\times [I + \gamma^{-2}(2D\hat{D} - \hat{D}D)](\gamma^{-1}\hat{D} - 2\gamma^{-3}\hat{D}D\hat{D}) .$$

The first bracket is taken into the exponential function which gives the exponent, see (10.205)

$$(-\gamma I + \gamma^{-1}\hat{D}D + \gamma^{-1}2[D,\hat{D}])t = L_1 t .$$

The second bracket is put in front of the exponential function (the commutator would lead to terms of the order γ^{-5}) and we thus finally have

$$G_{0,0}^{(1)}(t) = V_{01}e^{L_1 t}\bar{V}_{10} .$$

Green's Function in x Representation

In the x representation, the Green's function of Brinkman's hierarchy (10.46a) is given by

$$G_{n,m}(x,x',t) = \sum_{r=0}^{\infty} V_{n,r}e^{L_r t}\bar{V}_{r,m}\delta(x-x') , \tag{10.231}$$

where the operators $V_{n,r}$, $\bar{V}_{r,m}$, L_r are obtained by inserting the differential operators (10.27) for D and $\hat{D}$. The transition probability of the Kramers equation then takes the form for $\varepsilon = 0$ (10.145)

$$P(x,v,t|x',v',0) = \frac{\psi_0(v)}{\psi_0(v')}\sum_{n,m,r=0}^{\infty}\psi_n(v)\psi_m(v')V_{n,r}e^{L_r t}\bar{V}_{r,m}\delta(x-x') . \tag{10.232}$$

The distribution in position only is obtained by integrating (10.232) over velocity. Because of (10.41) we have

$$P_0(x,t|x',v',0) = \int P(x,v,t|x',v',0)\,\mathrm{d}v$$
$$= \sum_{m,r=0}^{\infty} [\psi_m(v')/\psi_0(v')]\, V_{0,r} \mathrm{e}^{L_r t} \bar{V}_{r,m} \delta(x-x') \,. \tag{10.233}$$

For large times $\gamma t \gg 1$ only the term with $r=0$ survives in (10.232, 233). Equation (10.233) may then be written as ($V_{0,0}=1$)

$$P_0(x,t|x',v',0) = \mathrm{e}^{L_0 t} \sum_{m=0}^{\infty} [\psi_m(v')/\psi_0(v')]\, \bar{V}_{0,m} \delta(x-x') \,, \tag{10.234}$$

which shows that for large times P_0 is a solution of the Smoluchowski equation

$$\dot{P}_0 = L_0 P_0 \tag{10.235}$$

with the initial condition

$$P_0(x,0|x',v',0) = \sum_{m=0}^{\infty} [\psi_m(v')/\psi_0(v')]\, \bar{V}_{0,m} \delta(x-x') \,. \tag{10.236}$$

Therefore, for $\gamma t \gg 1$, $P(x,v,t|x',v',0)$ can be expressed by

$$P(x,v,t|x',v',0) = \psi_0(v) \sum_{n=0}^{\infty} \psi_n(v)\, V_{n,0} P_0(x,t|x',v',0) \,. \tag{10.237}$$

Thus, the distribution of the slow variable x completely determines the distribution function of the fast velocity variable for times much larger than the relaxation time $\tau = \gamma^{-1}$ of the velocity. If we start with a distribution function which factorizes into a Maxwell distribution and a position-dependent distribution

$$W(x,v,0) = \psi_0^2(v)\, g(x,0) \,, \tag{10.238}$$

the distribution $W(x,t)$ in position only may be obtained for large times as a solution of (10.235) with the initial distribution

$$W(x,0) = \bar{V}_{0,0}\, g(x,0) \,, \tag{10.239}$$

where $\bar{V}_{0,0}$ reads explicitly (10.208, 224)

$$\bar{V}_{0,0} = 1 - \gamma^{-2} D\hat{D} + \gamma^{-4} [3(D\hat{D})^2 - (5/2) D^2 \hat{D}^2] + \dots \,. \tag{10.240}$$

[As initial condition we may use $g(x,0) = \delta(x-x')$ in (10.238).] From (10.232) it is seen that for $\gamma t \gg 1$ the distribution in the velocity-position space can be expressed by the distribution $W(x,t)$ in position space only

$$W(x,v,t) = \sum_{n=0}^{\infty} \psi_0(v)\,\psi_n(v)\,V_{n,0}\,W(x,t)\,. \tag{10.241}$$

If fast variables are eliminated an initial slip of the distribution function of the slow variables similarly to (10.239) occurs [10.16, 17].

The joint distribution W_2 is the product of the transition probability (10.237)) and the stationary distribution (10.49a). For $\gamma t \gg 1$ we thus get

$$W_2(x,v,t;x',v',0)$$
$$= \sum_{n,m=0}^{\infty} \psi_0(v)\,\psi_n(v)\,\psi_0(v')\,\psi_m(v')\,V_{n,0}\mathrm{e}^{L_0 t}\,\bar{V}_{0,m}N\exp[-f(x')/v_{\mathrm{th}}^2]\,. \tag{10.242}$$

With this joint distribution any two time correlation functions can be calculated by integration (7.12). Using (10.41) and $v\,\psi_0(v) = v_{\mathrm{th}}\,\psi_1(v)$ the two-time correlation functions of position and velocity read for instance ($\gamma t \gg 1$)

$$\langle x(t)x(0)\rangle = N\int x\mathrm{e}^{L_0 t}\,\bar{V}_{0,0}x\exp[-f(x)/v_{\mathrm{th}}^2]\,\mathrm{d}x\,, \tag{10.243a}$$

$$\langle v(t)v(0)\rangle = v_{\mathrm{th}}^2 N\int V_{1,0}\mathrm{e}^{L_0 t}\,\bar{V}_{0,1}\exp[-f(x)/v_{\mathrm{th}}^2]\,\mathrm{d}x\,. \tag{10.243b}$$

The explicit results for the operators V and $\bar{V}$ are given by (10.240) and

$$V_{1,0} = \gamma^{-1}[1+\gamma^{-2}(D\hat{D}-\hat{D}D)]\hat{D}\,, \tag{10.244a}$$

$$\bar{V}_{0,1} = -\gamma^{-1}D[1+\gamma^{-2}(D\hat{D}-2\hat{D}D)]\,. \tag{10.244b}$$

Expressions $\exp(L_0 t)\cdot h(x)$ occurring in (10.243) may be obtained as solutions of the Smoluchowski equation (10.235) with the initial condition $h(x)$.

For the harmonic oscillator $f(x) = \omega_0^2 x^2/2$ the commutator (10.28) is constant i.e. $[D,\hat{D}] = \omega_0^2$. Because $\exp(D\hat{D}\alpha)D = D\exp(-\omega_0^2\alpha)\exp(D\hat{D}\alpha)$, $x = v_{\mathrm{th}}(D-\hat{D})/\omega_0^2$ and $\hat{D}\exp[-f(x)/v_{\mathrm{th}}^2] = 0$ we obtain up to the order γ^{-2}

$$\langle x(t)x(0)\rangle = \frac{v_{\mathrm{th}}^2}{\omega_0^2}(1+\omega_0^2/\gamma^2)\exp[-(\omega_0^2/\gamma)(1+\omega_0^2/\gamma^2)t]\,, \tag{10.245a}$$

$$\langle v(t)v(0)\rangle = -\frac{v_{\mathrm{th}}^2\omega_0^2}{\gamma^2}\exp[-(\omega_0^2/\gamma)(1+\omega_0^2/\gamma^2)t]\,. \tag{10.245b}$$

These correlation functions agree with $G_{xx}(t)\,\sigma_{xx}(\infty)$ (10.61, 64) and with (10.101) respectively for $\gamma \gg 2\omega_0$ and $\gamma t \gg 1$ [$\lambda_1 \approx \gamma - \omega_0^2/\gamma$, $\lambda_2 \approx (\omega_0^2/\gamma)\times(1+\omega_0^2/\gamma^2)$]. The correlation function $\langle x(t)x(0)\rangle$ of the slow variable x slips from its initial value $v_{\mathrm{th}}^2/\omega_0^2$ at $t=0$ to the slightly higher value $(v_{\mathrm{th}}^2/\omega_0^2)\times(1+\omega_0^2/\gamma^2)$ for times t in the range $1/\gamma \ll t \ll \gamma/\omega_0^2$. In contrast to this result the correlation function $\langle v(t)v(0)\rangle$ of the fast variable v shows a sharp transition from the value v_{th}^2 at $t=0$ to the value $-v_{\mathrm{th}}^2\omega_0^2/\gamma^2$ slightly below zero for times in the above range.

10.4.4 Position-Dependent Friction

For position-dependent friction $\gamma(x)$ the matrix continued fraction for $\tilde{\boldsymbol{K}}_0(s)$ corresponding to (10.180) has the form

$$\tilde{\boldsymbol{K}}_0(s) = \boldsymbol{D}[s\boldsymbol{I}+\boldsymbol{\gamma}-2\boldsymbol{D}(s\boldsymbol{I}+2\boldsymbol{\gamma}-\ldots)^{-1}\hat{\boldsymbol{D}}]^{-1}\hat{\boldsymbol{D}}\,, \tag{10.246}$$

where the matrix $\boldsymbol{\gamma}$ has to be calculated according to (10.131) with $A = \gamma(x)$. Because $\boldsymbol{\gamma}$ does not commute with $\boldsymbol{D}$ and $\hat{\boldsymbol{D}}$, some care must be taken in doing the inverse friction expansion. If $\boldsymbol{A}$ and $\boldsymbol{B}$ are matrices which do not commute, we have for small ξ

$$[\boldsymbol{A}-\xi\boldsymbol{B}]^{-1} = \boldsymbol{A}^{-1}+\boldsymbol{A}^{-1}\xi\boldsymbol{B}\boldsymbol{A}^{-1}+\ldots\,. \tag{10.247}$$

This may be checked by multiplying (10.247) with the matrix in the bracket. The expansion of (10.246) for the first two terms is $[\boldsymbol{A} = s\boldsymbol{I}+\boldsymbol{\gamma}$, $\xi\boldsymbol{B} = 2\boldsymbol{D}(s\boldsymbol{I}+2\boldsymbol{\gamma})^{-1}\hat{\boldsymbol{D}}]$

$$\tilde{\boldsymbol{K}}_0(s) = \boldsymbol{D}(s\boldsymbol{I}+\boldsymbol{\gamma})^{-1}\hat{\boldsymbol{D}}+\boldsymbol{D}(s\boldsymbol{I}+\boldsymbol{\gamma})^{-1}\boldsymbol{D}2(s\boldsymbol{I}+2\boldsymbol{\gamma})^{-1}\hat{\boldsymbol{D}}(s\boldsymbol{I}+\boldsymbol{\gamma})^{-1}\hat{\boldsymbol{D}}\,. \tag{10.248}$$

For the operator $\boldsymbol{L}_0$ of the Smoluchowski equation we get with the help of (10.204)

$$\boldsymbol{L}_0 = \boldsymbol{D}\boldsymbol{\gamma}^{-1}\hat{\boldsymbol{D}}+\boldsymbol{D}\boldsymbol{\gamma}^{-1}\boldsymbol{D}\boldsymbol{\gamma}^{-1}\hat{\boldsymbol{D}}\boldsymbol{\gamma}^{-1}\hat{\boldsymbol{D}}-\boldsymbol{D}\boldsymbol{\gamma}^{-2}\hat{\boldsymbol{D}}\boldsymbol{D}\boldsymbol{\gamma}^{-1}\hat{\boldsymbol{D}}\,, \tag{10.249}$$

which reads in x representation $[\varepsilon = 0; \gamma = \gamma(x)]$

$$L_0 = \frac{\partial}{\partial x}\left[\frac{1}{\gamma}+\frac{1}{\gamma^3}f''+\frac{1}{\gamma}\left(\frac{1}{\gamma}\right)'\left(v_{\rm th}^2\frac{\partial}{\partial x}+f'\right)\frac{1}{\gamma}\right]\left(v_{\rm th}^2\frac{\partial}{\partial x}+f'\right). \tag{10.250}$$

Notice that the third-order derivative $(\partial/\partial x)^3$ now already occurs in the second-order expansion term $\sim\gamma^{-3}$, whereas in (10.197) it first occurred in the third-order expansion term $\sim\gamma^{-5}$. In first order, (10.250) can still be brought to an Hermitian form

$$\begin{aligned}\bar{L} &= \exp[f(x)/(2v_{\rm th}^2)]\,L_0\exp[-f(x)/(2v_{\rm th}^2)]\\ &= \left(-v_{\rm th}\frac{\partial}{\partial x}+\frac{f'}{2v_{\rm th}}\right)\frac{1}{\gamma}\left(v_{\rm th}\frac{\partial}{\partial x}+\frac{f'}{2v_{\rm th}}\right) = \bar{L}^+\,.\end{aligned} \tag{10.251}$$

11. Brownian Motion in Periodic Potentials

In this chapter we apply some of the methods discussed in Chap. 10 for solving the Kramers equation for the problem of Brownian motion in a periodic potential. As discussed below, this problem arises in several fields of science, for instance in physics, chemical physics and communication theory. Restricting ourselves to the one-dimensional case, we deal with particles which are kicked around by the Langevin forces and move in a one-dimensional periodic potential (Fig. 11.1). Because of the excitation due to the Langevin forces the particles may leave the well and go either to the neighboring left or right well or they may move in the course of time to other wells which are further away. For long enough times the particles will thus diffuse in both directions of the x axis. As shown in Sect. 11.7 this diffusion can be described by a diffusion constant D, if we wait long enough. Thus the mean-square displacement is given by

$$\langle [x(t)-x(0)]^2\rangle = 2Dt \tag{11.1}$$

for large times t. (The particles are then distributed over many potential wells.)

If we apply an additional force F, which is independent of x, the particles will preferably diffuse in the direction of this force and in the average there is a drift velocity $\langle v\rangle$ which depends on the external force. For small forces the mobility μ defined by

$$\langle v\rangle = \mu F \tag{11.2}$$

will be independent of the force F (linear response), but for arbitrary forces F it will depend on the force F (nonlinear response). One problem is to calculate this linear and nonlinear mobility. As shown in Sect. 11.7, the diffusion constant is related to the mobility in linear response. For a time-varying force the Fourier

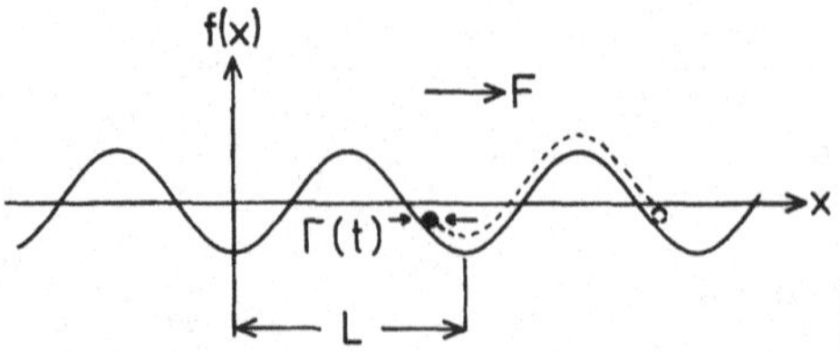

Fig. 11.1. Particle moving in the periodic potential $f(x)$

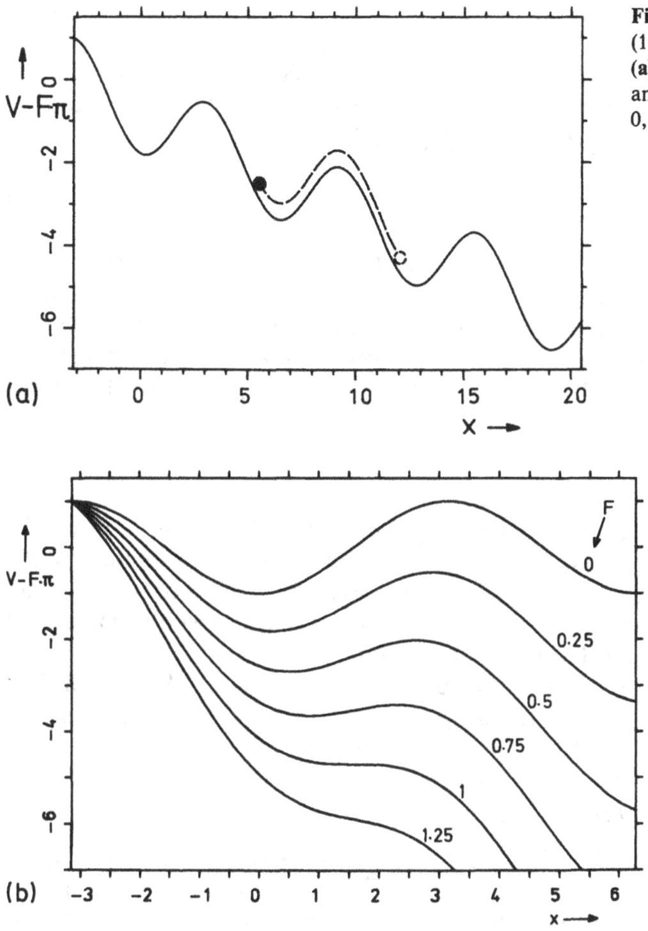

Fig. 11.2. The total potential (11.5) as a function of x for **(a)** $f(x) = -\cos x$, $F = 0.25$, and **(b)** $f(x) = -\cos x$, $F = 0, 0.25, \dots 1.25$

transform of $\langle v(t) \rangle$, denoted by $\langle \tilde{v}(\omega) \rangle$, in terms of the Fourier transform of $F(t)$, denoted by $\tilde{F}(\omega)$, takes the form in linear response

$$\langle \tilde{v}(\omega) \rangle = \chi(\omega) \tilde{F}(\omega) , \tag{11.2a}$$

where $\chi(\omega)$ is called the susceptibility.

As discussed in Chap. 10, the equation of motion for the coordinate $x(t)$ of the particle is the Langevin equation

$$\ddot{x} + \gamma \dot{x} + f'(x) = F + \Gamma(t) . \tag{11.3}$$

Here $f(x) = f(x+L)$ is the periodic potential with period L divided by the mass m of the particle, F is the external force divided by m, and γ is the damping constant. The Langevin force $\Gamma(t)$ describes white noise with zero mean and its correlation function is given by

$$\langle \Gamma(t)\,\Gamma(t')\rangle = 2\gamma(kT/m)\,\delta(t-t')\,, \tag{11.4}$$

where T is the temperature. For a qualitative discussion it is useful to plot the total potential $V(x)$ per mass, i.e., the sum of the periodic potential $f(x)$ and the potential $-Fx$ of the external force F (Fig. 11.2)

$$V(x) = f(x) - Fx\,. \tag{11.5}$$

This total potential V is a corrugated plane; the average slope is determined by the external force F. For large forces F, $V(x)$ has no minima, whereas for intermediate and small forces minima do occur. The Langevin equation (11.3) thus describes the Brownian motion of particles along such a corrugated plane. We now discuss qualitatively the motion of particles for large and small friction constants γ.

High-Friction Case

For large γ we can neglect inertial effects, i.e., we omit the $\ddot{x}$ term in (11.3). Without any noise $\Gamma(t)$ the particle performs a creeping motion. If minima of the total potential $V(x)$ exist, the particles finally reach them. This solution we call locked solution. If minima do not exist the particles move down the corrugated plane. This solution will be termed running solution. With noise, the particles do not stay permanently in the locked state but will sometimes be kicked out of their wells, moving to the lower neighboring well and so forth. The particles thus perform a hopping process from one well to the next lower one.

Intermediate- and Low-Friction Case

For smaller friction constants, inertial effects become important. Without noise a locked solution may occur if minima exist. We may, however, also have a running solution, even if the minima of the potential do occur. Because of their momentum the particles may overcome the next hill if the friction constant is small enough. This interesting bistability of (11.3) without noise is discussed in detail in Sect. 11.6. If we include noise, the particles may be kicked out of their well, i.e., out of the locked state. If the damping is small enough, they do not lose their energy very rapidly and therefore they may no longer be trapped in the neighboring lower well, as they are for large friction. The particles may thus get in the running state and may stay in this state for some time. Due to the Langevin forces, the energy of the particles fluctuates. The energy of the particles may thus decrease and they may again be trapped in one of the wells, now again belonging to the locked state. With the inclusion of noise we therefore get transitions between these two states.

It follows from these qualitative discussions that the low- and intermediate-friction case is more interesting than the high-friction case where no such bistability can occur.

Because of the nonlinearity $f'(x)$ in the Langevin equation (11.3), one uses rather the corresponding Fokker-Planck or Kramers equation (10.3) to calculate mobility or other expectation values. In the high-friction case the position x

becomes a slow variable and the velocity a fast variable (see Sect. 8.3 for a discussion of slow and fast variables). Therefore the Kramers equation then reduces to the Smoluchowski equation, i.e., a one-variable Fokker-Planck equation for the slow variable x. This one-variable Fokker-Planck equation can easily be solved in the stationary state, Sect. 5.2. For lower friction constants both variables x and v are relevant variables and the Fokker-Planck equation for both has to be solved. Solving this two-variable Fokker-Planck equation is much more difficult than solving the one-variable Smoluchowski equation. For very small friction constants (zero-friction limit), however, the energy becomes a slow variable, leading again to a one-variable equation.

The problem of Brownian motion in periodic potentials arises in quite different fields, Sect. 11.1, e.g. in solid-state physics, chemical physics and communication theory. By studying the synchronization of an oscillator, *Stratonovich* [11.1, 2] derived the stationary solution of the Smoluchowski equation, i.e., of the Fokker-Planck equation in the high-friction limit (see also [11.3 – 5]). *Haken* et al. [11.6] solved this Smoluchowski equation in connection with the locking of two laser modes, *Ivanchenko* and *Zil'berman* [11.7], and *Ambegaokar* and *Halperin* [11.8] solved it in connection with the Josephson tunneling junction. For a solution of the time-dependent Smoluchowski equation and its application to quantum noise in ring laser gyros, see [11.9, 10].

Solutions of the Fokker-Planck equation with two variables and with a small but time-varying field (linear response) or some expectation values of this Fokker-Planck equation have been obtained for not too low damping constants by different methods in connection with superionic conductors [11.11 – 15]. An approximate solution of this equation was proposed by *Das* and *Schwendimann* [11.16]. The one-dimensional rotation of dipoles in an external field also leads to the Langevin equation (11.3). Different methods have been developed in this case to obtain susceptibilities in linear response [7.7, 11.17 – 20].

For large but time-independent forces corrections to the Smoluchowski equation have been obtained by *Tikhonov* [11.21] and *Lee* [11.22]. *Kurkijärvi* and *Ambegaokar* [11.23], and *Schneider* et al. [11.24] have solved the two-variable Fokker-Planck equation by computer-simulation methods. A stochastic formulation for the motion out of the wells as given by *Nozières* and *Iche* [11.25] and a WKB-type expansion was developed by *Ben-Jacob* et al. [11.26].

In several papers *Vollmer* and the author [9.14 – 17, 11.27, 28] have applied the matrix continued-fraction method, which was developed for the Kramers equation in Chap. 10, to Brownian motion in periodic potentials. It turns out that this method works very well down to very low friction constants, so that even the connection to the zero-friction limit solution [11.29, 30] can be made. Results will be discussed for the stationary solution, susceptibilities, eigenvalues and their eigenfunctions in Sects. 11.5, 8, 9, respectively. The only limitation of the matrix continued-fraction method seems to be that the potential differences $f_{max} - f_{min}$ should not be too large compared to kT. Otherwise the dimension of the matrices which have to be inverted becomes too large and then the method ceases to be tractable.

11.1 Applications

Let us discuss some of the main applications of Brownian motion in periodic potentials.

11.1.1 Pendulum

The equation of motion for the angle of a mathematical pendulum with mass m in a viscous fluid and under the influence of an additional torque M (Fig. 11.3) is given by

$$ml\,\ddot{\varphi}+m\gamma l\dot{\varphi}+mg\sin\varphi=M+l\Gamma(t)\,. \tag{11.6}$$

The effects of the fluid are described by the damping constant $m\gamma$ and the fluctuating Langevin force $\Gamma(t)$. If we divide (11.6) by ml we recover the form (11.3) with $\varphi=x$ where the periodic potential $f(x)$ is given by the cosine potential

$$f(\varphi)=-(g/l)\cos\varphi\,. \tag{11.7}$$

An important quantity is the average angular velocity as a function of the torque M

$$\langle\dot{\varphi}\rangle=\mathrm{Fct}\,[M/(ml)]\,. \tag{11.8}$$

For a small torque M the pendulum will oscillate around its stable downside position, i.e., $\langle\dot{\varphi}\rangle$ will be zero. If the torque M is positive and large enough the pendulum will rotate clockwise, i.e., $\langle\dot{\varphi}\rangle$ will be different from zero. In the presence of the noise $\langle\dot{\varphi}\rangle$ will be a continuous function of the torque M.

11.1.2 Superionic Conductor

A superionic conductor consists of a nearly fixed ion lattice in which some other ions are highly movable. As an example we consider silver iodide (AgI). Here the

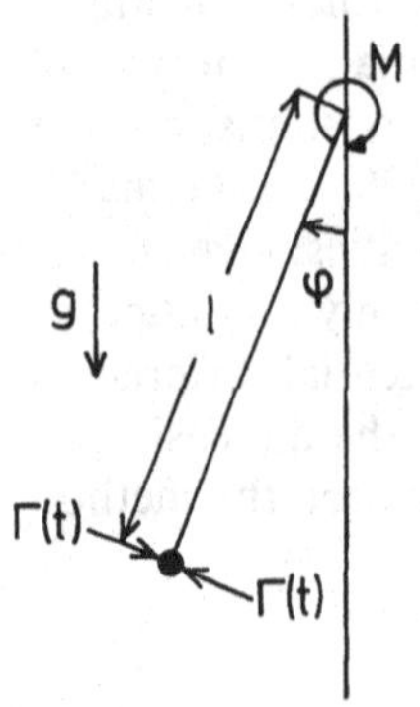

Fig. 11.3. Mathematical pendulum with an additional torque M and Langevin forces $\Gamma(t)$

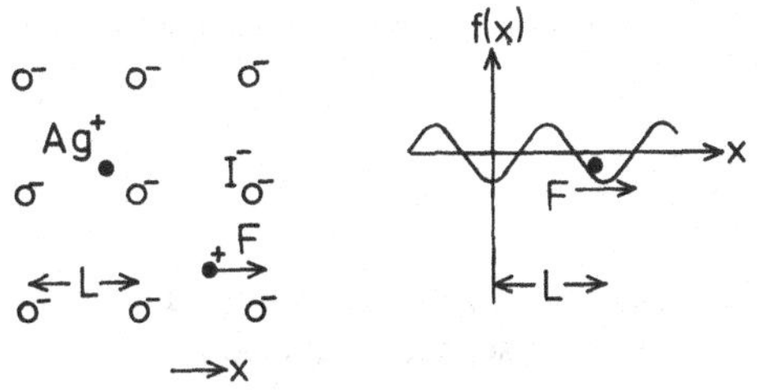

Fig. 11.4. Highly movable Ag^+ ions in the nearly fixed iodide lattice and the corresponding potential in one dimension

lattice consists of iodide (I^-) ions, while the silver ions (Ag^+) are highly mobile (Fig. 11.4). If an external field is applied to a one-dimensional model, neglecting interaction of different Ag^+ ions, then the equation of motion divided by the mass m in the periodic potential $mf(x)$ is [11.11 – 15]

$$\ddot{x}+\gamma\dot{x}+f'(x)=F+\Gamma(t)\,. \tag{11.9}$$

In (11.9) we added a damping force $\gamma\dot{x}$ and a Gaussian white-noise force $\Gamma(t)$ (per mass)

$$\langle\Gamma(t)\Gamma(t')\rangle=2\gamma(kT/m)\,\delta(t-t')\,. \tag{11.10}$$

By these two forces the effect of the small lattice vibrations on the motion of the Ag^+ ions is taken into account. If the motion of the Ag^+ ions is slow compared to the lattice vibrations the white-noise approximation in (11.10) is justified. In this application we are mainly interested in the current. This current can be expressed by the drift velocity or by the mobility (11.1), or in the dynamical case by the susceptibility $\chi(\omega)$ (11.2).

For more realistic treatment of superionic conduction this simple model (one dimension, one particle) should be generalized to three dimensions and the interaction between the mobile ions should be taken into account (Sects. 4.8.3, 4).

11.1.3 Josephson Tunneling Junction

A Josephson tunneling junction [11.31 – 33] consists of two superconductors which are separated by a thin oxide layer, Fig. 11.5. The phase difference

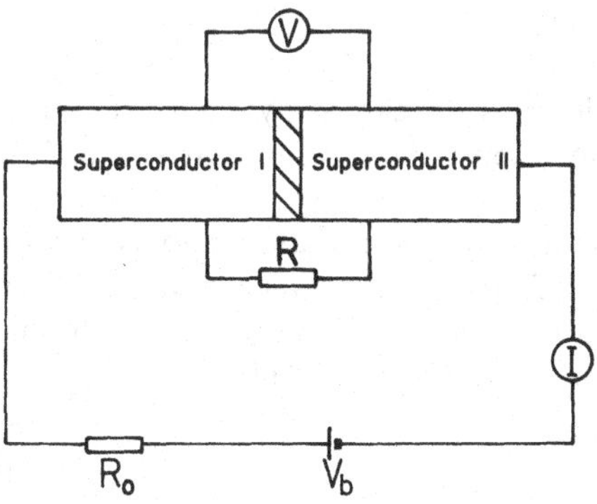

Fig. 11.5. Josephson tunneling junction

between the wave functions $\psi_{\rm I}$ and $\psi_{\rm II}$ of the Cooper pairs in the two superconductors is denoted by φ

$$\psi_{\rm II} = N\psi_{\rm I} e^{i\varphi} . \tag{11.11}$$

Furthermore the ratio $N = |\psi_{\rm II}|/|\psi_{\rm I}|$ is assumed to be constant. The time derivative of this phase difference is given by the Josephson equation

$$\dot{\varphi} = 2eV/\hbar , \tag{11.12}$$

where V is the potential difference across the oxide layer. If the resistance R_0 is very large, the current I is kept fixed. This total current can be written as

$$I = V/R - L(t) + C\dot{V} + I_{\rm max}\sin\varphi , \tag{11.13}$$

where the first two parts on the right-hand side stem from the resistor R. Here $-L(t)$ is a noise current, the correlation function of which is given by

$$\langle L(t)L(t')\rangle = (2/R)kT\delta(t-t') . \tag{11.14}$$

The term $C\dot{V}$ is the current due to the capacitance C of the junction. The last term is the current due to the Cooper pairs tunneling through the junction, where $I_{\rm max}$ is called the maximum Josephson current. Combining (11.12, 13), we obtain the form of the pendulum equation (11.6) [11.7, 8]

$$\left(\frac{\hbar}{2e}\right)^2 C\ddot{\varphi} + \left(\frac{\hbar}{2e}\right)^2 \frac{1}{R}\dot{\varphi} + \frac{\hbar}{2e} I_{\rm max}\sin\varphi = \frac{\hbar}{2e} I + \frac{\hbar}{2e} L(t) . \tag{11.15}$$

Here the capacitance C acts as a mass, the resistor R is responsible for damping and the current I is proportional to the torque. In this case we are interested mainly in the current voltage characteristic, i.e., the current I as a function of the voltage $\langle V\rangle = (\hbar/2e)\langle\dot{\varphi}\rangle$.

11.1.4 Rotation of Dipoles in a Constant Field

The model of Brownian rotation of dipoles first developed by *Debye* [11.34] is now widely used to explain infrared absorption by polar molecules, see [7.7, 11.17–20] for a review.

Here we consider one-dimensional Brownian rotation of dipoles in a constant field F_c, including inertial effects. The field F_c may for instance be a local field produced by some other immobile molecules. As is well known, the potential energy of a dipole of moment μ_0 in an electric field F_c is given by

$$V(\varphi) = -\mu_0 F_c \cos\varphi , \tag{11.16}$$

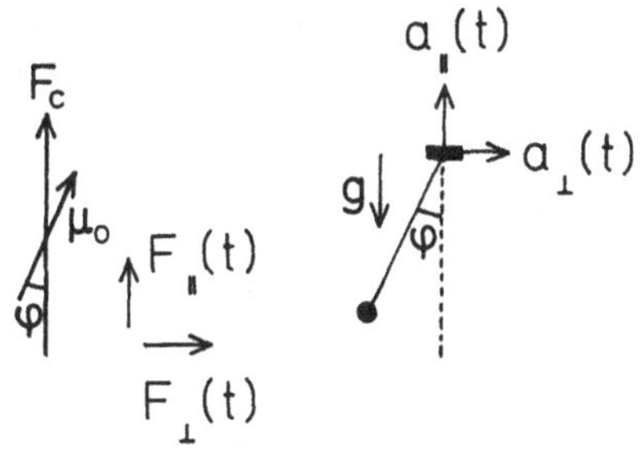

Fig. 11.6. Rotation of a dipole with moment μ_0 in a constant field F_c and additional time-varying field parallel and perpendicular to F_c. The equation of motion is equivalent to that of a pendulum where the support is accelerated parallel and perpendicular to the graviational force

where φ is the angle between the dipole and the field, Fig. 11.6. If we apply an additional time-varying field with components $F_\perp(t)$ and $F_\parallel(t)$ perpendicular and parallel to the constant field, the equation of motion for the phase φ of the dipoles with moment of inertia I_0 takes the form

$$I_0\,\ddot{\varphi} + \gamma I_0 \dot{\varphi} + \mu_0 F_c \sin\varphi = \mu_0[F_\perp(t)\cos\varphi - F_\parallel(t)\sin\varphi] + \Gamma(t)\,. \tag{11.17}$$

Here again the influence of the heat bath is taken into account by the damping torque $\gamma I_0 \dot{\varphi}$ and by the fluctuating torque Γ with

$$\langle \Gamma(t)\Gamma(t')\rangle = 2 I_0 \gamma k T\,\delta(t-t')\,. \tag{11.18}$$

Equation (11.17) is also the equation for the Brownian motion of a pendulum where the support of the pendulum is accelerated perpendicular and parallel to the gravitational field, Fig. 11.6.

Here we are interested mainly in the susceptibility, i.e., the linear response of the Fourier transform of the averaged dipole moment p divided by the Fourier transforms of the additional electric ac field

$$\chi_{\substack{\perp\\\parallel}}(\omega) = \tilde{p}_{\substack{\perp\\\parallel}}(\omega)/\tilde{F}_{\substack{\perp\\\parallel}}(\omega)\,. \tag{11.19}$$

Though the right-hand side of (11.17) is different to the right-hand side of (11.3), we shall see in Sect. 11.8 that $\chi_\perp(\omega)$ can be expressed by $\chi(\omega)$ from (11.2a).

11.1.5 Phase-Locked Loop

An ideal phase-locked loop (PLL) is a device where the phase of an oscillator signal follows exactly the phase of a reference signal. Phase-locked loops are used for instance in radio or TV sets to obtain stable tuning. [Here an oscillator is phase locked to an oscillation, the frequency of which is a rational multiple of the frequency of a built-in crystal oscillator (PLL synthesizers).] If there is some noise in the PLL, the phase of the oscillator and the signal are no longer locked exactly. If the phase difference is 2π it is said that a cycle slip has occurred. Obviously, the cycle slip rate/s is a quality measure for a real PLL.

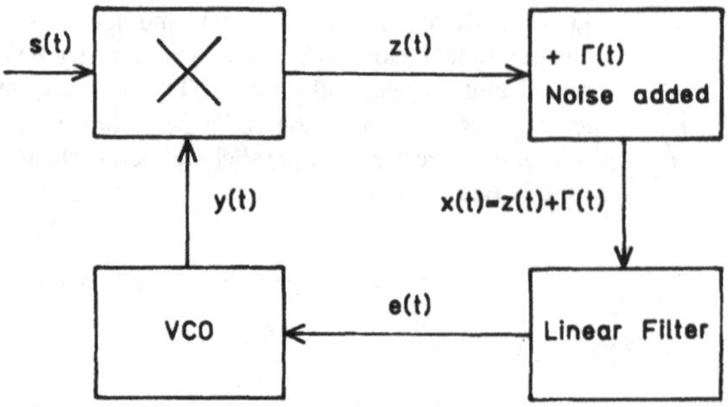

Fig. 11.7. A phase-locked loop (PLL). The multiplier is indicated by $\times$ and the voltage-controlled oscillator by VCO. For an ideal PLL the noise term Γ is missing

A type of PLL is shown in Fig. 11.7 [11.4, 5] whose equations follow. The phase ψ of the voltage-controlled oscillator (VCO) leads to the cosine oscillation

$$y(t) = \sqrt{2}K \cos\psi(t) , \tag{11.20}$$

where the derivative of the phase (frequency) is proportional to the input voltage $e(t)$ according to

$$\dot{\psi} = \omega_{\text{os}} = \omega_0 + \alpha \cdot e(t) . \tag{11.21}$$

In the multiplier the signal with frequency ω

$$s(t) = \sqrt{2}A \sin\theta ; \qquad \theta = \omega t \tag{11.22}$$

is mixed multiplicatively with the output of the VCO leading to

$$z(t) = sy = AK[\sin(\theta - \psi) + \underset{\sim\sim\sim\sim\sim}{\sin(\theta + \psi)}] . \tag{11.23}$$

The term with the wavy underline oscillates with approximately $2\omega_0$ and is assumed to be filtered out by a filter not shown in the diagram. As indicated in the diagram, a noise term $\Gamma(t)$ is added to $z(t)$, i.e., $x(t) = z(t) + \Gamma(t)$. Then $x(t)$ passes through a filter. We assume that this filter is linear and works so that the input voltage $x(t)$ is connected to the output voltage $e(t)$ by

$$\tau\dot{e} + e = x . \tag{11.24}$$

A filter where the input voltage x is applied to a resistor R and a capacitance C in series and where the output voltage e is the voltage of the capacitance leads to this relation with $\tau = RC$.

For the phase difference

$$\varphi = \theta - \psi \tag{11.25}$$

we then get the equation ($\dot{\varphi} = \dot{\theta} - \dot{\psi} = \omega - \omega_0 - \alpha e$)

$$\tau\ddot{\varphi} + \dot{\varphi} + \alpha AK \sin\varphi = (\omega - \omega_0) - \alpha\Gamma . \tag{11.26}$$

For white noise $\Gamma(t)$ we thus obtain, up to a negligible sign change in front of Γ, the same Langevin equation (11.3) as for a pendulum with a torque. The torque is here proportional to the detuning $\omega-\omega_0$. The corresponding total potential (11.5) reads in this case

$$\tau V(\varphi) = -\alpha A K \cos\varphi - (\omega-\omega_0)\varphi .$$

As may be easily derived from this expression, minima of $V(\varphi)$ occur if

$$|\alpha A K| > |\omega_0 - \omega| . \tag{11.27}$$

As discussed in the beginning of this chapter, then a locked solution with $\langle\dot\varphi\rangle = 0$ may occur if the noise is neglected. In other words, we then have an ideal synchronization to the signal frequency. If noise is taken into account $\langle\dot\varphi\rangle$ will no longer be zero. Therefore $\langle\dot\varphi\rangle$ as a function of the detuning $\omega-\omega_0$ will be an important relation.

A phase-locked loop with a linear filter leading to a first-order time derivative in the input-output relation is called a second-order loop, because the equation for φ is of second order. Without any filter or for $\tau=0$ then $e=x$ and the equation for φ reduces to a first-order equation and therefore the loop is called a first-order loop. As discussed in Sect. 11.6, the second-order loop may show hysteresis whereas the first-order loop never does.

11.1.6 Connection to the Sine-Gordon Equation

The Sine-Gordon equation reads

$$\frac{\partial^2\varphi}{\partial t^2} - c^2\frac{\partial\varphi^2}{\partial x^2} + d\sin\varphi = 0 . \tag{11.28}$$

It is used in quite a number of fields, for instance in solid-state physics for decribing dislocations [11.35] and long Josephson junctions [11.36]; in quantum optics for describing self-induced transparency [11.37]. As discussed in several textbooks [11.37 – 39], (11.27) possesses solutions in form of solitons, which propagate without being disturbed asymptotically by other solitons. One may generalize (11.28) by adding a damping term, a constant force term and a noise term [11.40], i.e.,

$$\begin{gathered}\frac{\partial^2\varphi}{\partial t^2} - c^2\frac{\partial^2\varphi}{\partial x^2} + \gamma\frac{\partial\varphi}{\partial t} + d\sin\varphi = F + \Gamma(x,t)\\ \langle\Gamma(x,t)\,\Gamma(x',t')\rangle = 2\gamma\Theta\,\delta(x-x')\,\delta(t-t') .\end{gathered} \tag{11.28a}$$

A chain of coupled pendula [11.41] also leads to (11.28a), if the phases of neighboring pendula differ only slightly. Without the space dependence $\partial^2\varphi/\partial x^2$, (11.28a) is of the form (11.3) in different notation. The position x in (11.3) is now denoted by φ. Equation (11.3) with $f'(x) = d\sin x$ may therefore be considered to be a Sine-Gordon equation in zero dimensions but with an additional noise and damping term.

11.2 Normalization of the Langevin and Fokker-Planck Equations

We have seen in the last section that several applications lead to the form of the Langevin equation (11.3) with a periodic potential $f(x) = f(x+L)$. In Applications 1, 3, 4, 5 we had a cosine potential with period 2π. In Application 2 the period was the lattice constant L. For further treatments it is convenient to transform the period to 2π and use the following variables, parameters and potentials

$$\begin{aligned} x_n &= \frac{2\pi}{L}x \;; \quad t_n = \frac{2\pi}{L}v_0 t\,; \quad v_n = \frac{v}{v_0}\,; \\ \gamma_n &= \frac{L}{2\pi}\frac{\gamma}{v_0}\,; \quad \Theta_n = \frac{kT}{mv_0^2} = \left(\frac{v_{\text{th}}}{v_0}\right)^2; \quad v_{\text{th}} = \sqrt{kT/m}\,; \\ F_n &= \frac{L}{2\pi}\frac{F}{v_0^2}\,; \quad f_n = \frac{f}{v_0^2}\,; \quad \frac{df_n}{dx_n} = \frac{L}{2\pi}\frac{1}{v_0^2}\frac{df}{dx}\,; \quad \Gamma_n(t_n) = \frac{L}{2\pi v_0^2}\Gamma(t)\,. \end{aligned} \tag{11.29}$$

The arbitrary quantity v_0 may be made either equal to one or it may be a properly chosen velocity. In the latter case all normalized quantities are dimensionless.

The Langevin equations (11.3, 4) read

$$\frac{d^2x_n}{dt_n^2} + \gamma_n\frac{dx_n}{dt_n} + \frac{df_n}{dx_n} = F_n + \Gamma_n\,; \quad f_n(x_n+2\pi) = f_n(x_n)\,, \tag{11.30}$$

$$\langle \Gamma_n(t_n)\Gamma_n(t_n')\rangle = 2\gamma_n\Theta_n\delta(t_n - t_n')\,. \tag{11.31}$$

For further applications it is worthwhile to note that the mobility times the damping constant is independent of the choice of the transformation

$$\gamma_n\mu_n = \gamma_n\langle v_n\rangle/F_n = \gamma\langle v\rangle/F = \gamma\mu\,. \tag{11.32}$$

More invariants like $\gamma_n t_n = \gamma t$ and $v_n t_n/x_n = vt/x$ may be found from (11.29).

If v_0 is chosen to be equal to the thermal velocity $v_{\text{th}} = \sqrt{kT/m}$, Θ_n is equal to one. This choice is very convenient for those calculations where the dependence of the drift velocity on F_n, the potential height or the damping constant γ_n is considered. If, however, the dependence on temperature is needed, this normalization is not appropriate because all the normalized quantities depend on temperature. Then we may normalize the height of the periodic potential equal to 1 by a properly chosen v_0. It is, of course, also possible to choose v_0 in such a way that either γ_n or F_n is equal to one.

In further treatment we omit the normalization index n, i.e., we write (11.30, 31) in the form

$$\ddot{x}+\gamma\dot{x}+f' = F+\Gamma(t)\,, \qquad f(x+2\pi)=f(x)\,, \tag{11.30a}$$

$$\langle\Gamma(t)\Gamma(t')\rangle = 2\gamma\Theta\delta(t-t')\,. \tag{11.31a}$$

Normalized Fokker-Planck Equation

The Fokker-Planck equation corresponding to the normalized Langevin equation (11.30a, 31a) takes the form

$$\frac{\partial W}{\partial t} = \left[-\frac{\partial}{\partial x}v+\frac{\partial}{\partial v}\left(\gamma v+f'-F+\gamma\Theta\frac{\partial}{\partial v}\right)\right]W\,. \tag{11.33}$$

With the exception of the additional constant force F this equation agrees with (10.3) after putting $\Theta = v_{\text{th}}^2$.

11.3 High-Friction Limit

In the high-friction limit we may omit the inertial term $\ddot{x}$ in (11.30a),

$$\gamma\dot{x} = F-f'(x)+\Gamma(t)\,. \tag{11.34}$$

In Sect. 10.4 an expansion of the Kramers equation in terms of inverse powers of γ is derived. This expansion can of course be applied to (11.33) [Replace f' by $f'-F$ in (10.197)]. The first approximation of this expansion is the Smoluchowski equation

$$\frac{\partial W}{\partial t} = \frac{1}{\gamma}\frac{\partial}{\partial x}\left(f'-F+\Theta\frac{\partial}{\partial x}\right)W = -\frac{\partial S}{\partial x}\,, \tag{11.35}$$

where $W = W(x,t) = \int W(x,v,t)\,\mathrm{d}v$ is the distribution function for x. Equation (11.35) is equivalent to the Fokker-Planck equation corresponding to (11.34, 31a). In (11.35) we introduced the probability current S.

11.3.1 Stationary Solution

Let us first look for the stationary solutions of (11.35). Because the probability current S is constant

$$\gamma S = (F-f')\,W-\Theta\partial W/\partial x\,, \tag{11.36}$$

we obtain immediately the solution (Sect. 5.2)

$$W(x) = \mathrm{e}^{-V(x)/\Theta}\left[N - \gamma(S/\Theta)\int_0^x \mathrm{e}^{V(x')/\Theta}\mathrm{d}x'\right], \tag{11.37}$$

where $V(x) = f(x) - Fx$ is the total potential (11.5). If x is an angle variable this distribution must be periodic in x with period 2π. If we require only that $W(x)$ is bounded for large enough x, it follows already from (11.37) that $W(x)$ must be periodic. To prove this, we first calculate the integral ($0 \leq x < 2\pi$)

$$\int_0^{2\pi n+x} \mathrm{e}^{V(x')/\Theta}\mathrm{d}x' = \int_0^{2\pi} \mathrm{e}^{V(x')/\Theta}\mathrm{d}x' + \ldots + \int_{2\pi(n-1)}^{2\pi n} \mathrm{e}^{V(x')/\Theta}\mathrm{d}x' + \int_{2\pi n}^{2\pi n+x} \mathrm{e}^{V(x')/\Theta}\mathrm{d}x'.$$

Because of $V(x+2\pi n) = V(x) - 2\pi nF$ we write (after using a proper shift in the integration variables)

$$\begin{aligned}\int_0^{2\pi n+x} \mathrm{e}^{V(x')/\Theta}\mathrm{d}x &= I + I\mathrm{e}^{-2\pi F/\Theta} + \ldots + I\mathrm{e}^{-2\pi(n-1)F/\Theta} + \int_0^x \mathrm{e}^{V(x')/\Theta}\mathrm{d}x'\,\mathrm{e}^{-2\pi nF/\Theta} \\ &= I\frac{1-\mathrm{e}^{-2\pi nF/\Theta}}{1-\mathrm{e}^{-2\pi F/\Theta}} + \mathrm{e}^{-2\pi nF/\Theta}\int_0^x \mathrm{e}^{V(x')/\Theta}\mathrm{d}x',\end{aligned}$$

where I is defined by

$$I = \int_0^{2\pi} \mathrm{e}^{V(x)/\Theta}\mathrm{d}x. \tag{11.38}$$

Thus we have

$$\begin{aligned}W(x+2\pi n) &= \mathrm{e}^{-V(x)/\Theta}\left[N - \frac{\gamma SI}{\Theta(1-\mathrm{e}^{-2\pi F/\Theta})}\right]\mathrm{e}^{2\pi nF/\Theta} \\ &\quad + \mathrm{e}^{-V(x)/\Theta}\left[\frac{\gamma SI}{\Theta(1-\mathrm{e}^{-2\pi F/\Theta})} - \gamma\frac{S}{\Theta}\int_0^x \mathrm{e}^{V(x')/\Theta}\mathrm{d}x'\right].\end{aligned} \tag{11.39}$$

For $F>0$ ($F<0$), this expression can be bounded only in the limit $n \to +\infty$ ($n \to -\infty$) if the first bracket on the right-hand side vanishes, i.e.,

$$\gamma SI = \Theta N(1-\mathrm{e}^{-2\pi F/\Theta}). \tag{11.40}$$

Hence, we obtain from (11.39, 37)

$$W(x+2\pi) = W(x) \tag{11.41}$$

which proves our statement. Because of the periodicity we normalize the distribution in the periodicity interval

$$\int_0^{2\pi} W(x)\,dx = N\int_0^{2\pi} e^{-V(x)/\Theta}dx - \gamma(S/\Theta)\int_0^{2\pi} e^{-V(x)/\Theta}\left(\int_0^{x} e^{V(x')/\Theta}dx'\right)dx.$$
$$= 1\,. \tag{11.42}$$

The mean drift velocity $\langle v\rangle$ is given by the constant probability current times 2π

$$\begin{aligned}\langle v\rangle &= \langle \dot{x}\rangle = \gamma^{-1}\langle F - f'(x) + \Gamma(t)\rangle \\ &= \gamma^{-1}\langle F - f'(x)\rangle = \gamma^{-1}\int_0^{2\pi}[F - f'(x)]\,W dx \\ &= \gamma^{-1}\int_0^{2\pi}(\gamma S + \Theta\,\partial W/\partial x)\,dx = 2\pi S\,.\end{aligned} \tag{11.43}$$

In deriving (11.43) we used (11.34, 36, 41). The two equations (11.40, 42) determine the integration constants N and S. By eliminating N we obtain the drift velocity

$$\gamma\langle v\rangle = \frac{2\pi\Theta(1-e^{-2\pi F/\Theta})}{\int_0^{2\pi} e^{V(x)/\Theta}dx \int_0^{2\pi} e^{-V(x)/\Theta}dx - (1-e^{-2\pi F/\Theta})\int_0^{2\pi} e^{-V(x)/\Theta}\int_0^{x} e^{V(x')/\Theta}dx'\,dx}\,. \tag{11.44}$$

If we consider only the mobility in linear response i.e. for $F\to 0$, the double integral term vanishes and in the other integrals we can replace $V(x)$ by $f(x)$:

$$\gamma\mu(0) = \lim_{F\to 0}\frac{\gamma\langle v\rangle}{F} = \frac{2\pi}{\int_0^{2\pi} e^{f(x)/\Theta}dx}\;\frac{2\pi}{\int_0^{2\pi} e^{-f(x)/\Theta}dx}\,. \tag{11.45}$$

For the cosine potential

$$f(x) = -d\cos x \tag{11.46}$$

(11.45) reduces to

$$\gamma\mu(0) = [\mathrm{I}_0(d/\Theta)]^{-2}\,, \tag{11.47}$$

where I_0 is the modified Bessel function. For model potentials, which are piece-by-piece linear, the integrals in (11.44, 45) can be evaluated analytically. For the cosine potential (11.46) one may express $\gamma\mu$ by Bessel functions of complex order [11.2, 6].

Continued-Fraction Expansion

For the cosine potential (11.46) we can obtain a very useful expansion in terms of continued fractions as follows [11.7]. Inserting the Fourier expansion

$$W(x)=\sum_{n=-\infty}^{\infty} c_n \mathrm{e}^{\mathrm{i}nx}, \qquad c_n=c^*_{-n} \tag{11.48}$$

in (11.36) we obtain for the cosine potential the tridiagonal recurrence relation

$$(F-\mathrm{i}n\Theta)c_n+\tfrac{1}{2}\mathrm{i}d(c_{n-1}-c_{n+1})=\gamma S\delta_{n0}\,. \tag{11.49}$$

For $n\geqq 1$, these equations give an infinite continued fraction for $c_1/c_0=c^*_{-1}/c_0$, Sect. 9.2.1. If we then use (11.49) for $n=0$ and the normalization condition $c_0=(2\pi)^{-1}$ we can express γS by a continued fraction. The final result for the mobility times the damping constant reads

$$\gamma\mu=\gamma 2\pi S/F=1-\langle\sin x\rangle d/F\,, \tag{11.50}$$

where $-\langle\sin x\rangle$ is the imaginary part of the continued fraction [see (9.31) for this notation]

$$2\pi c_1=\langle \mathrm{e}^{-\mathrm{i}x}\rangle=\langle\cos x\rangle-\mathrm{i}\langle\sin x\rangle$$

$$=2\cdot\frac{0.25|}{|\Theta/d+\mathrm{i}F/d}+\frac{0.25|}{|2\Theta/d+\mathrm{i}F/d}+\frac{0.25|}{|3\Theta/d+\mathrm{i}F/d}+\dots\,. \tag{11.51}$$

This expression can easily be evaluated on a programmable pocket calculator even for very low Θ/d (e.g, $\Theta/d=10^{-4}$). The results are shown in Fig. 11.8. The zero temperature limit $\Theta\to 0$ is obtained as follows. For $F>d$ the deterministic equation of motion is given by (11.34) without the noise term. If we write this relation for the cosine potential as

$$\frac{1}{T}\,\frac{\gamma\,\mathrm{d}x}{F-d\sin x}=W_{\Theta\to 0}(x)\,\mathrm{d}x=\frac{\mathrm{d}t}{T}\,,$$

where T is the time which a particle needs to travel the distance 2π

$$T=\int_{-\pi}^{\pi}\frac{\gamma\,\mathrm{d}x}{F-d\sin x}=\frac{\gamma 2\pi}{\sqrt{F^2-d^2}}\,, \tag{11.52}$$

we get the following expression for the stationary distribution function in the zero-temperature limit

$$W_{\Theta\to 0}(x)=\frac{1}{2\pi}\,\frac{\sqrt{F^2-d^2}}{F-d\sin x}\,, \qquad |F|>d\,. \tag{11.53}$$

For $|F|<d$ we have $\dot{x}=0$ and therefore the zero-temperature distribution is the δ function $\delta(x-\arcsin(F/d))$. Thus for the mobility times the damping constant we obtain

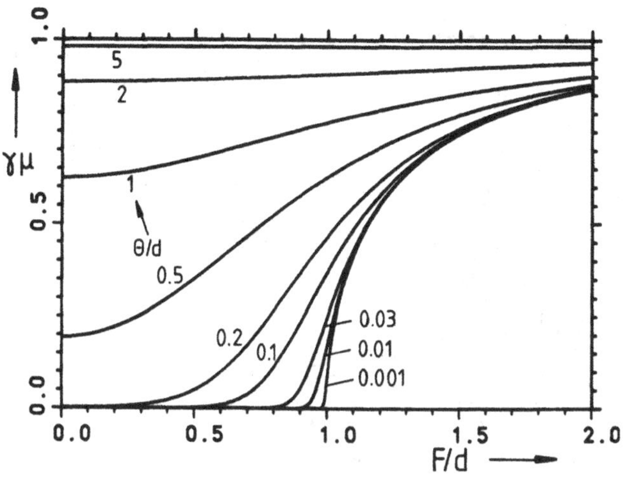

Fig. 11.8. The mobility times the damping constant as a function of F/d for various temperatures Θ/d. The zero-temperature limit (11.54) practically agrees with the $\Theta/d = 0.001$ curve

$$\gamma\mu_{\Theta\to 0} = \frac{1}{F}\int_0^{2\pi}(F - d\sin x)\,W_{\Theta\to 0}(x)\,dx$$

$$= \begin{cases} \sqrt{F^2 - d^2}/F & \text{for} \quad |F| \geqq d \\ 0 & \quad\quad\; |F| \leqq d \end{cases} . \tag{11.54}$$

This result can also be derived from the continued fraction (11.51). For $\Theta \to 0$ (11.51) becomes a periodic continued fraction having the value

$$\begin{aligned} \langle e^{-ix}\rangle_{\Theta\to 0} &= -iF/d + \sqrt{1-(F/d)^2}\,, \\ \langle \cos x\rangle_{\Theta\to 0} &= \begin{cases} \sqrt{1-(F/d)^2} & \text{for} \quad |F| \leqq d \\ 0 & \quad\quad\; F \geqq d\,, \end{cases} \\ \langle \sin x\rangle_{\Theta\to 0} &= \begin{cases} F/d & \text{for} \quad |F| \leqq d \\ F/d - \sqrt{(F/d)^2 - 1} & \quad\quad\; F \geqq d \end{cases} . \end{aligned} \tag{11.55}$$

The distribution function for finite temperatures can be obtained from (11.49) by up-iteration. Because we now know c_0 and c_1, the other coefficients c_n with $n \geqq 2$ and therefore the distribution function follow by up-iteration. As mentioned in Sect. 9.2.1 this up-iteration is numerically unstable. Therefore higher accuracy is now required if we use this procedure. As also mentioned in Sect. 9.2.1 the up-iteration according to (9.28a), however, is numerically stable. The results for the stationary distribution function are shown in Fig. 11.9. The asymmetry for small Θ in Fig. 11.9b is explained as follows. At $x = \pi/2$ the slope of the potential V is zero, compare Fig. 11.2b for $F/d = 1$. The particles which are kicked to the right of $x = \pi/2$ move further down the potential, whereas the particles kicked to the left are reflected at the high potential barrier and thus accumulate at $x \lesssim \pi/2$.

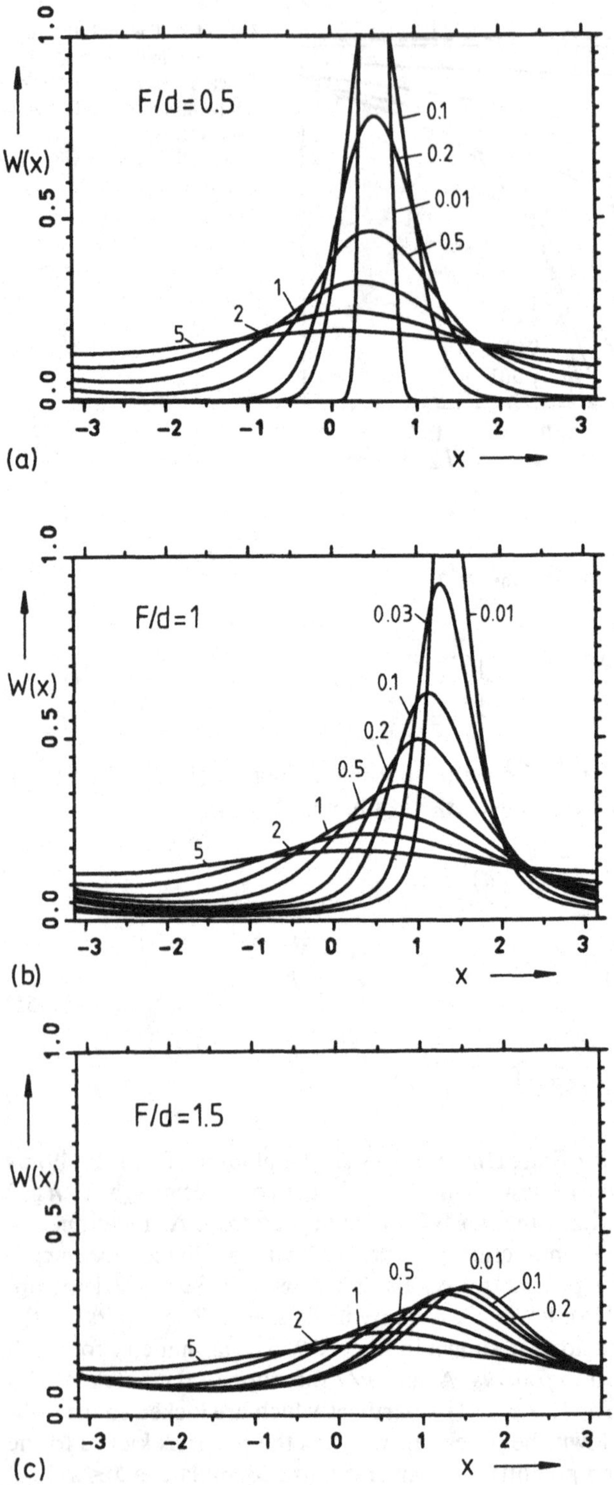

Fig. 11.9. The stationary distribution function $W(x)$ for various temperatures Θ/d and for $F/d = 0.5$ (**a**), $F/d = 1$ (**b**), $F/d = 1.5$ (**c**). In (**c**) the curve for $\Theta/d = 0.01$ practically agrees with the zero-temperature distribution (11.53)

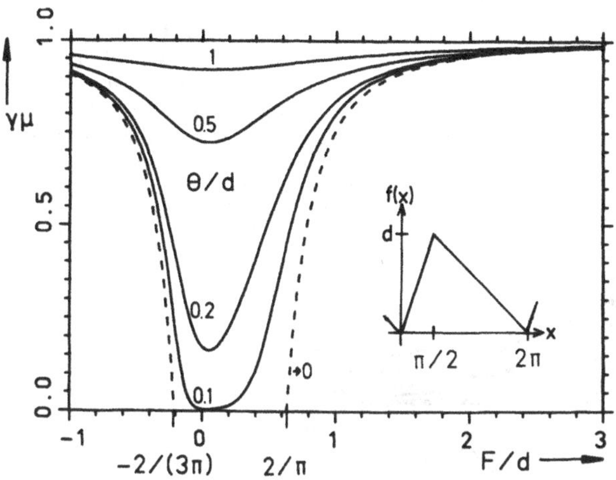

Fig. 11.10. The mobility times the damping constant for the periodic saw-tooth potential $f(x) = (2d/\pi)x$ if $0 \leqq x \leqq \pi/2$; $f(x) = [2d/(3\pi)](2\pi - x)$ if $\pi/2 \leqq x \leqq 2\pi$ as a function of F/d for various temperatures Θ/d

For symmetric potentials the mobility $\mu(F)$ is an even function in F; for asymmetric potentials this is no longer true. In Fig. 11.10 the result is shown for a saw-tooth potential [11.42].

Inverse Friction Expansion

The next correction term to the Smoluchowski equation (11.35) reads [(10.250), $\partial\gamma/\partial x = 0$, $f' \to f' - F$, $v_{\text{th}}^2 = \Theta$]

$$\frac{\partial W}{\partial t} = -\frac{\partial S}{\partial x}, \quad S = -\frac{1}{\gamma}\left(1 + \frac{f''}{\gamma^2}\right)\left(f' - F + \Theta\frac{\partial}{\partial x}\right)W. \tag{11.56}$$

In the stationary state the probability current S is again constant. Because the correction term f''/γ^2 is small we may write

$$\gamma S(1 - f''/\gamma^2) = (F - f')W - \Theta\partial W/\partial x. \tag{11.57}$$

The drift velocity and the constant probability current S are still connected by

$$\langle v \rangle = 2\pi S, \tag{11.58}$$

which may be proved in analogy to (11.43) by using (11.3, 57) and the periodicity of $f(x)$ and $W(x)$. Equation (11.57) can be solved similarly to (11.36). For the cosine potential (11.46) we can again use expansion (11.48). We then obtain (11.49) with the additional term $-Sd(\delta_{n,1} + \delta_{n,-1})/(2\gamma)$ on the right-hand side of (11.49). The same method used previously to solve (11.49) now leads to

$$\gamma\mu = [1 - \langle \sin x \rangle d/F][1 - \langle \cos x \rangle d/\gamma^2] + O(\gamma^{-4}). \tag{11.59}$$

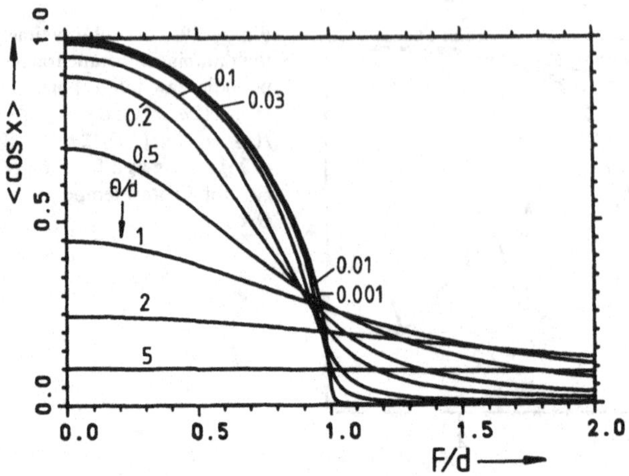

Fig. 11.11. The average $\langle \cos x \rangle$ as a function of F/d for various temperatures Θ/d. The zero-temperature limit, see (11.55), practically agrees with the $\Theta = 0.001$ curve

Here, $\langle \cos x \rangle$ and $-\langle \sin x \rangle$ are the real and imaginary parts of the continued fraction (11.51). The result for $\langle \cos x \rangle$ as a function of the field F/d is shown in Fig. 11.11. The zero-temperature result follows from (11.55).

11.3.2 Time-Dependent Solution

Because the time does not explicitly occur in the Smoluchowski equation (11.35) we can make the separation 'ansatz'

$$W(x,t) = \varphi(x)\mathrm{e}^{-\lambda t} \tag{11.60}$$

which leads to the eigenvalue problem

$$L_\mathrm{S}\varphi_n = -\lambda_n \varphi_n\,, \tag{11.61}$$

where L_S is the operator of the Smoluchowski equation

$$L_\mathrm{S} = \frac{1}{\gamma}\,\frac{\partial}{\partial x}\left[(f' - F) + \Theta\frac{\partial}{\partial x}\right]. \tag{11.62}$$

In general the solution $W(x,t)$ need not be periodic in x. It follows from Floquet's theorem [11.43–45] that the eigenfunction may be chosen in the form of Bloch waves

$$\varphi_n(x) = \mathrm{e}^{\mathrm{i}kx} v_n(k,x)\,. \tag{11.63}$$

Here $v_n(k,x)$ is a periodic function in x

$$v_n(k, x+2\pi) = v_n(k,x)\,. \tag{11.64}$$

If we require the solutions to be bounded for $x \to \pm\infty$, k must be real. The k values can be restricted to the first Brillouin zone

$$-\tfrac{1}{2} \leqq k \leqq \tfrac{1}{2}. \tag{11.65}$$

By the transformation $\psi(x) = \exp[V(x)/(2\Theta)]\,\varphi(x)$, where $V(x)$ is the total potential (11.5), one formally obtains a Schrödinger equation for $\psi(x)$ with the periodic potential $\gamma V_S = V'^2/(4\Theta) - V''/2 = (f' - F)^2/(4\Theta) - f''/2$ (Sect. 5.4). The boundary condition (11.63, 64) for $\varphi(x)$ transforms to the boundary condition

$$\psi_n(x) = \exp(\mathrm{i}k_S)\,\tilde{v}_n(k,x) = \exp\{[-F/(2\Theta) + \mathrm{i}k]x\}\,\tilde{v}_n(k,x)\,, \tag{11.63a}$$

$$\tilde{v}_n(k,x+2\pi) = \tilde{v}_n(k,x) = \exp[f(x)/(2\Theta)]\,v_n(k,x) \tag{11.64a}$$

for the wave function $\psi(x)$. Usually the boundary condition (11.63a, 64a) with real k_S is used for the Schrödinger equation. If we know the eigenvalues $\tilde{\lambda}_n(k_S)$ of the Schrödinger equation we may obtain the eigenvalues $\lambda_n(k)$ of the Smoluchowski equation by analytic continuation (by replacing k_S by $k + \mathrm{i}F/(2\Theta)$), i.e. $\lambda_n(k) = \tilde{\lambda}_n(k + \mathrm{i}F/(2\Theta))$.

In some applications only periodic solutions need be considered. If x is an angle variable and if we do not distinguish between full rotations, the solution $W(x,t)$ must be periodic in x and therefore k can take only the value 0. To calculate expectation values of periodic functions we also need only periodic solutions. If, however, the diffusion over the infinite x axis is investigated, the k dependence must be taken into account. The eigenvalues λ_n then depend on k. For vanishing F but arbitrary k the eigenvalues λ_n are real, because the problem is then essentially equivalent to a Schrödinger equation problem. For nonvanishing F the eigenvalues λ_n become complex.

Generally, eigenvalues and eigenfunctions may be obtained by numerical integration or by some other numerical methods. For simple models, however, the eigenvalues and eigenfunctions may be obtained analytically.

Periodic Potential Model

For the potential shown in Fig. 11.12, using the jump condition in Sect. 5.6 we obtain for a vanishing external field F the eigenvalues [5.12]:

$$\lambda_n(k) = (\Theta/\gamma)(n + \nu)^2; \qquad n = 0, \pm 1, \pm 2, \ldots \tag{11.66}$$

where ν is given by $(-\tfrac{1}{2} < \nu \leqq \tfrac{1}{2})$

$$\nu = \pi^{-1} \arcsin\{\sin \pi k / \cosh[f_0/(2\Theta)]\}. \tag{11.67}$$

Some of the lowest eigenvalues are shown in Fig. 11.13. The eigenvalues are arranged in bands, one for every $|n|$ satisfying

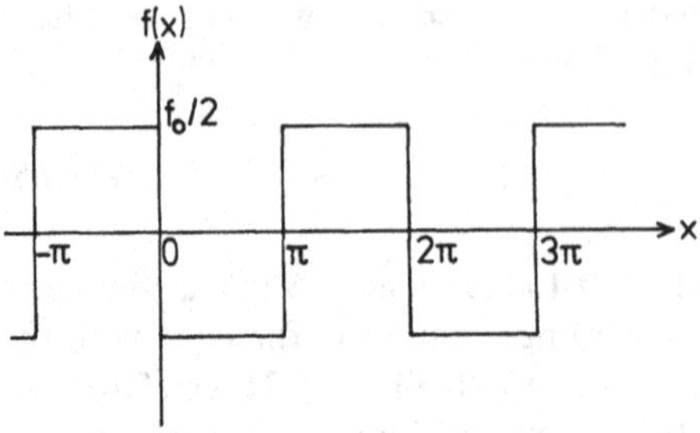

Fig. 11.12. Periodic potential model

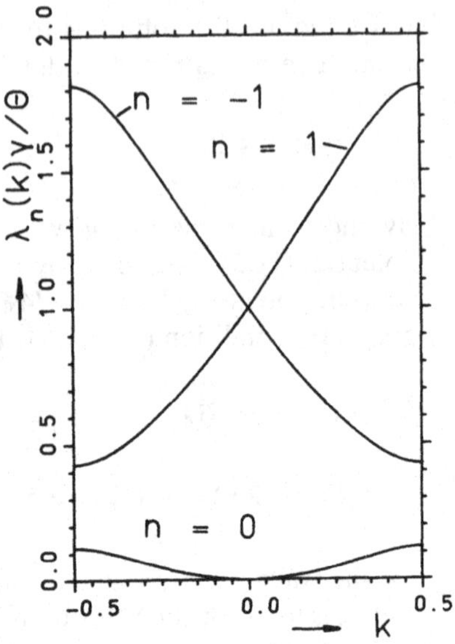

Fig. 11.13. Eigenvalue bands for the periodic potential model in Fig. 11.12 for $f_0 = \Theta$

$$\lambda_n(-\tfrac{1}{2}) = \lambda_{-n}(\tfrac{1}{2}) \leq \begin{matrix}\lambda_n(k) \\ \lambda_{-n}(k)\end{matrix} \leq \lambda_n(\tfrac{1}{2}) = \lambda_{-n}(-\tfrac{1}{2}) \, .$$

For different bands the following relations hold:

$$\begin{aligned} &\lambda_0(k) < \lambda_{-1}(k) < \lambda_1(k) < \lambda_{-2}(k) < \ldots \quad \text{for} \quad k > 0 \, , \\ &\lambda_0(k) < \lambda_1(k) < \lambda_{-1}(k) < \lambda_2(k) < \ldots \quad \text{for} \quad k < 0 \, . \end{aligned}$$

The eigenfunctions φ_n and φ_n^+ of L and L^+ are expressed by the eigenfunctions ψ_n of the Hermitian operator

$$L = \mathrm{e}^{f(x)/(2\Theta)} \gamma L_{\mathrm{S}} \mathrm{e}^{-f(x)/(2\Theta)} = \Theta \frac{\partial^2}{\partial x^2} - \frac{1}{4} \frac{f'^2}{\Theta} + \frac{1}{2} f'' \tag{11.68}$$

(5.39, 54, 55) in the form

$$\begin{aligned} \varphi_n &= \mathrm{e}^{-f(x)/(2\Theta)} \psi_n = \mathrm{e}^{-f(x)/(2\Theta)} \mathrm{e}^{ikx} u_n(k,x) \, , \\ \varphi_n^+ &= \mathrm{e}^{f(x)/(2\Theta)} \psi_n \; = \mathrm{e}^{f(x)/(2\Theta)} \mathrm{e}^{ikx} u_n(k,x) \, . \end{aligned} \tag{11.69}$$

The periodic function $u_n(k,x)$ normalized according to

$$\int_0^{2\pi} u_n^*(k,x) u_n(k,x) \, \mathrm{d}x = 1 \tag{11.70}$$

is given by

$$u_n(k,x) = \begin{cases} A_n(k)[\mathrm{e}^{\mathrm{i}(n+\nu-k)x} - a(k)\mathrm{e}^{\mathrm{i}\nu\pi}\mathrm{e}^{-\mathrm{i}(n+\nu+k)x}] & \text{for} \quad 0<x<\pi \\ A_n(k)\mathrm{e}^{\mathrm{i}n\pi}[\mathrm{e}^{\mathrm{i}(n+\nu-k)(x-\pi)} + a(k)\mathrm{e}^{\mathrm{i}\nu\pi}\mathrm{e}^{-\mathrm{i}(n+\nu-k)(x-\pi)}] & \\ & \text{for} \quad \pi<x<2\pi, \end{cases} \tag{11.71}$$

with

$$\begin{aligned} a(k) &= (\cos k\pi - \sqrt{\cosh^2[f_0/(2\Theta)] - \sin^2 k\pi})/\sin[f_0/(2\Theta)] \\ A_n(k) &= \{2\pi[1+a(k)^2]\}^{-1/2}. \end{aligned} \tag{11.72}$$

The function $u_n(k,x)$ for the stationary distribution reads

$$u_0(0,x) = \exp[-f(x)/(2\Theta)]\{2\pi\cosh[f_0/(2\Theta)]\}^{-1/2}$$

and the transition probability in terms of eigenvalues and eigenfunctions finally takes the form

$$\begin{aligned} P(x,t|x',0) = &\exp\left[-\frac{f(x)}{2\Theta} + \frac{f(x')}{2\Theta}\right] \\ &\times \sum_{n=-\infty}^{\infty} \int_{-1/2}^{1/2} \mathrm{e}^{\mathrm{i}k(x-x')} u_n(k,x) u_n^*(k,x') \mathrm{e}^{-\lambda_n(k)t} dk. \end{aligned} \tag{11.73}$$

Cosine Potential

The general time-dependent solution can be written as

$$W(x,t) = \int_{-1/2}^{1/2} \mathrm{e}^{\mathrm{i}kx} \sum_{n=-\infty}^{\infty} c_n(k,t)\mathrm{e}^{\mathrm{i}nx} dk. \tag{11.74}$$

For periodic W the integral has to be omitted and we have to put $k=0$, i.e.,

$$W(x,t) = \sum_{n=-\infty}^{\infty} c_n(t)\mathrm{e}^{\mathrm{i}nx}; \qquad c_n = c_{-n}^*. \tag{11.74a}$$

We obtain the following tridiagonal recurrence relation for the cosine potential (11.46) by inserting (11.74) into (11.35):

$$\gamma\dot{c}_n = [-\mathrm{i}(n+k)F - \Theta(n+k)^2]c_n + \tfrac{1}{2}(n+k)d(c_{n-1} - c_{n+1}). \tag{11.75}$$

As discussed in Sect. 9.2, the Laplace transform of the Green's function as well as the equation for determining the eigenvalues can be given in terms of ordinary continued fractions. For instance, the ansatz $c_n = \exp(-\lambda t)\hat{c}_n$ leads for $k=0$ to

$$\tfrac{1}{2}nd\hat{c}_{n+1} + (\Theta n^2 + \mathrm{i}nF - \gamma\lambda)\hat{c}_n - \tfrac{1}{2}nd\hat{c}_{n-1} = 0. \tag{11.76}$$

For $\lambda \neq 0$ (the stationary problem $\lambda = 0$ is treated in Sect. 11.3.1), (11.76) leads to $\hat{c}_0 = 0$. Thus (11.76) splits into two systems, one for $n > 0$ and the other for $n < 0$. For real λ (11.74a) must be real and therefore we have $\hat{c}_n = \hat{c}^*_{-n}$. Both systems are thus equivalent and either one can be used to find the real eigenvalues. For complex eigenvalues λ one system is the complex conjugate of the other one. Thus one determines λ the other λ^*.

For real and complex eigenvalues $\lambda \neq 0$ we can eliminate all $\hat{c}_n$ with $n > 1$ finally leading to

$$[\gamma\lambda - \mathrm{i}F - \Theta + \tilde{K}_1(-\lambda)]\,\hat{c}_1 = 0 \tag{11.77}$$

where $\tilde{K}_1(s)$ is given by the ordinary continued fraction

$$\tilde{K}_1(s) = -\frac{\frac{1}{4}\cdot 1\cdot 2\cdot d^2|}{|\gamma s + 2\mathrm{i}F + 2^2\Theta} + \frac{\frac{1}{4}\cdot 2\cdot 3\cdot d^2|}{|\gamma s + 3\mathrm{i}F + 3^2\Theta} + \dots\,. \tag{11.78}$$

Solutions of (11.77) with $\hat{c}_1 \neq 0$ are only possible if

$$D_1(\lambda) = \gamma\lambda - \mathrm{i}F - \Theta + \tilde{K}_1(-\lambda) = 0\,. \tag{11.79}$$

From this equation the eigenvalues can be determined. In Fig. 11.14a, b the eigenvalues are shown as a function of the external force for various d/Θ ratios. As is seen, the eigenvalues $\lambda \neq 0$ are always complex for $F \neq 0$ though the imaginary parts have extremely low values for small Θ/d and $F/d < 1$. In Fig. 11.14b the fundamental frequencies ω and 2ω of the noiseless overdamped motion are also shown. This fundamental frequency is 2π divided by the time (11.52), which a particle needs to travel the distance 2π, i.e.

$$\omega = \sqrt{F^2 - d^2}/\gamma\,. \tag{11.80}$$

Eigenvalues for Vanishing Force

If $F = 0$ the eigenvalues are real. To obtain these eigenvalues for small d/Θ ratios it is not advisable to use (11.77) for higher eigenvalues λ_r with $r \geqq 2$. In this case we eliminate all $\hat{c}_n$ with $n \neq r$ for the system with $n > 0$ leading to

$$[\gamma\lambda - r^2\Theta + \tilde{K}_r(-\lambda)]\,\hat{c}_r = 0\,, \tag{11.77a}$$

where $\tilde{K}_r(s)$ is given by a sum of an infinite and a finite continued fraction

$$\begin{aligned}\tilde{K}_r(s) = &-\frac{\frac{1}{4}r(r+1)d^2|}{|\gamma s + (r+1)^2\Theta} + \frac{\frac{1}{4}(r+1)(r+2)d^2|}{|\gamma s + (r+2)^2\Theta} + \dots\\ &-\frac{\frac{1}{4}r(r-1)d^2|}{|\gamma s + (r-1)^2\Theta} + \frac{\frac{1}{4}(r-1)(r-2)d^2|}{|\gamma s + (r-2)^2\Theta} + \dots\,.\end{aligned} \tag{11.78a}$$

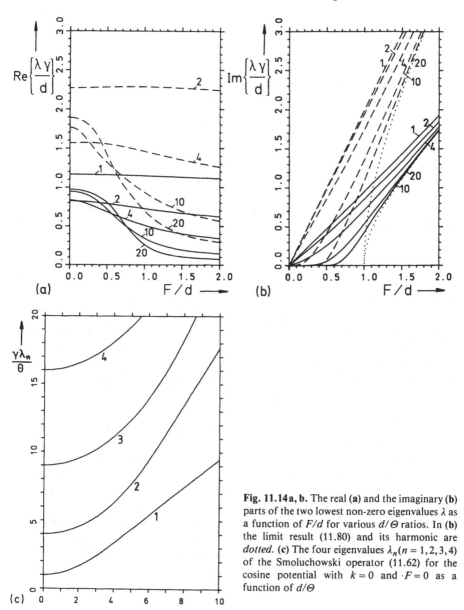

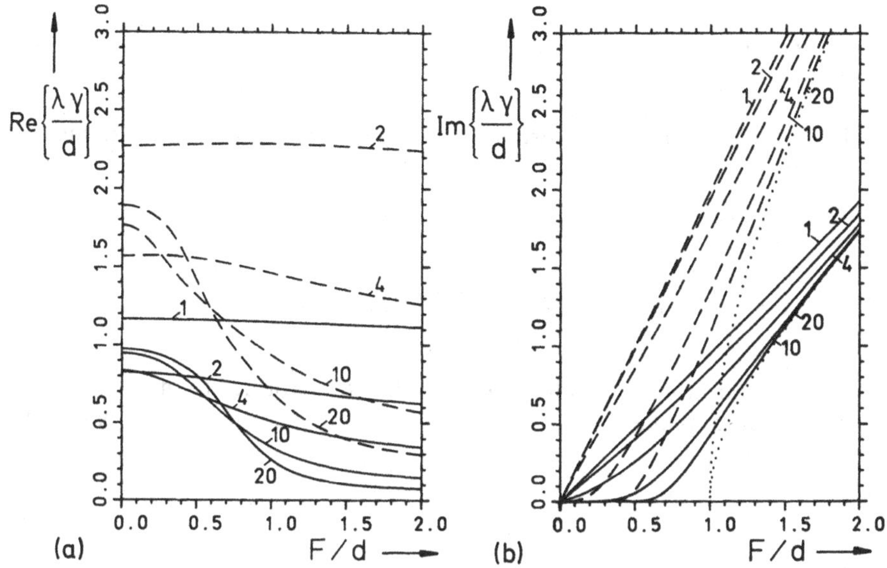

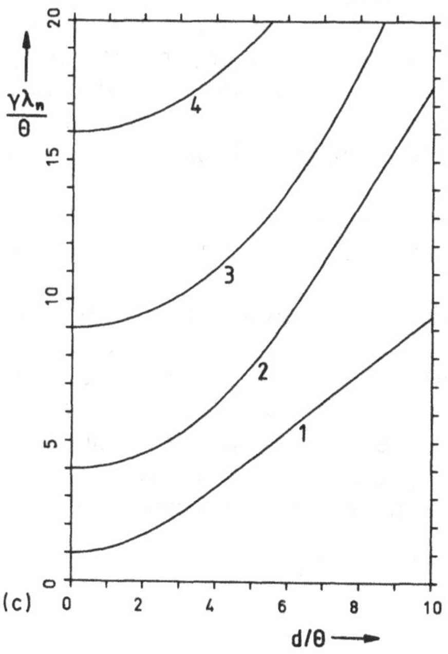

Fig. 11.14a, b. The real **(a)** and the imaginary **(b)** parts of the two lowest non-zero eigenvalues λ as a function of F/d for various d/Θ ratios. In **(b)** the limit result (11.80) and its harmonic are *dotted*. **(c)** The four eigenvalues $\lambda_n (n = 1, 2, 3, 4)$ of the Smoluchowski operator (11.62) for the cosine potential with $k = 0$ and $F = 0$ as a function of d/Θ

The eigenvalues follow from the equation

$$\gamma\lambda - r^2\Theta + \tilde{K}_r(-\lambda) = 0 \,. \tag{11.79a}$$

Equation (11.79a) is now preferable because $\tilde{K}_r$ is small for small d/Θ and $\gamma\lambda_r \approx r^2\Theta$. For small d/Θ we can expand the continued fraction (11.78a) in powers of d/Θ leading to

$$\gamma\lambda_n/\Theta = n^2 + \frac{1}{2}\,\frac{n^2}{4n^2-1}\left(\frac{d}{\Theta}\right)^2 \quad \text{for} \quad d/\Theta \ll 1\,. \tag{11.81}$$

For large d/Θ we expand the cosine potential at its minimum and maximum

$$-\cos x \approx -1 + \frac{1}{2}x^2 - \frac{1}{24}x^4 \quad \text{for} \quad x \approx 0\,,$$

$$-\cos x \approx 1 - \frac{1}{2}(x-\pi)^2 - \frac{1}{24}(x-\pi)^4 \quad \text{for} \quad x \approx \pi\,.$$

If the Smoluchowski equation (11.35) is then transformed to the Schrödinger equation (Sect. 5.4) and if the x^4 terms are taken into account by first-order perturbation theory, we get

$$\gamma\lambda_n/\Theta = nd/\Theta - n^2/2 \quad \text{for} \quad d/\Theta \gg 1\,. \tag{11.82}$$

In Fig. 11.14c the ratio $\gamma\lambda_n/\Theta$ is plotted as a function of d/Θ.

The coefficients are determined from (11.76) by upward and downward iteration. Because the coefficients can be chosen to be real or purely imaginary the eigenfunctions are even or odd. Thus all except the stationary eigenvalues are twofold degenerate. (If $F \neq 0$ we have pairs of complex conjugate eigenvalues, which are not degenerate for $\lambda \neq 0$. For the stationary solution $c_0 \neq 0$ must be real, and no degeneracy occurs for $F = 0$.)

For an application of (11.74a, 75) for $k = 0$ to the quantum noise problem in ring laser gyros and for the determination of the spectrum by a continued fraction, see [11.10].

As discussed at the beginning of this chapter, we may look for the diffusion constant of a particle without an applied force F. This problem was treated in [11.46–48] for the high-friction case where the Smoluchowski equation is applicable. The result is that the diffusion constant D is given by the mobility in linear response times the noise strength Θ. In Sect. 11.7 we shall derive this result for arbitrary friction γ. For the periodic potential model of Fig. 11.12, for instance, it follows from (11.234, 66) that the diffusion constant is given by $D = (\Theta/\gamma)\,[\cosh(f_0/(2\Theta))]^{-2}$.

11.4 Low-Friction Limit

If the damping constant and the external force are zero, the energy

$$E = \tfrac{1}{2}v^2 + f(x) \tag{11.83}$$

will be a constant of motion. (The Langevin force then also vanishes.) For very small friction the energy will slowly vary in the course of time. In the low-friction limit the energy will therefore become the relevant or slow variable and position x or velocity v will become the irrelevant or fast variable (Sect. 8.3). (In the high-friction limit the position variable x is the slow variable, whereas the velocity v is the fast variable.) If the friction is small the external force F must also be small, because otherwise the energy gain of the particles due to the external force F cannot be compensated by energy dissipation and no stationary solution would exist in the low-friction limit for finite forces F. If

$$F = \gamma F_0 \tag{11.84}$$

it turns out that in the low-friction limit a stationary solution does exist for finite F_0.

A transformation to an energy variable for particles moving in an arbitrary potential was already made in [Ref. 1.10, Vol. I, p. 115ff.]. However, in order to obtain velocity expectation values, the method of [1.10] has to be modified so that two separate energy distribution functions, one for each sign of the velocity, have to be taken into account.

11.4.1 Transformation to E and x Variables

We now express the distribution functions $W(x, v, t)$ by the space coordinate x and the energy E. The lines of constant energy for a cosine potential are shown in Fig. 11.15. To retain full information of the distribution function, we have to introduce the two energy and space distribution functions, W_+ for positive and W_- for negative velocities

$$\begin{aligned} W_+(x,E,t) &= W(x, v(x,E), t)\,, \\ W_-(x,E,t) &= W(x, -v(x,E), t)\,, \end{aligned} \tag{11.85}$$

$$v(x,E) = +\sqrt{2[E-f(x)]}\,. \tag{11.86}$$

For further calculations the sum (S) and the difference (D) of W_+ and W_- are sometimes more suitable:

$$W_{\substack{S\\D}}(x,E,t) = W_+(x,E,t) \pm W_-(x,E,t)\,. \tag{11.87}$$

In this section we are looking for solutions of the Fokker-Planck equation (11.33) which are periodic in x with period 2π, i.e., $W(x, v, t) = W(x+2\pi, v, t)$. As shown in Sect. 11.5, the stationary solution must always be periodic in x. For simplicity we assume that the potential has only one maximum and therefore only one minimum in the period length 2π. Assuming further that the maxima E_0 of the periodic potential are located at $-\pi$ and π and that $x_1 = x_1(E)$ and $x_2 = x_2(E)$ are the minimum and maximum values for the space coordinate for $E < E_0$ in the region $-\pi \leqq x \leqq \pi$ (Fig. 11.16), we require the following continuity conditions for the distribution functions

$$\begin{aligned} W_\pm(-\pi,E,t) &= W_\pm(\pi,E,t) && \text{for} \quad E>E_0 \\ W_+(x_{\substack{1\\2}},E,t) &= W_-(x_{\substack{1\\2}},E,t) && \text{for} \quad E<E_0 \quad \text{or} \end{aligned} \tag{11.88}$$

$$\begin{aligned} W_{\substack{S\\D}}(-\pi,E,t) &= W_{\substack{S\\D}}(\pi,E,t) && \text{for} \quad E>E_0 \\ W_D(x_1,E,t) &= W_D(x_2,E,t) = 0 && \text{for} \quad E<E_0\,. \end{aligned} \tag{11.89}$$

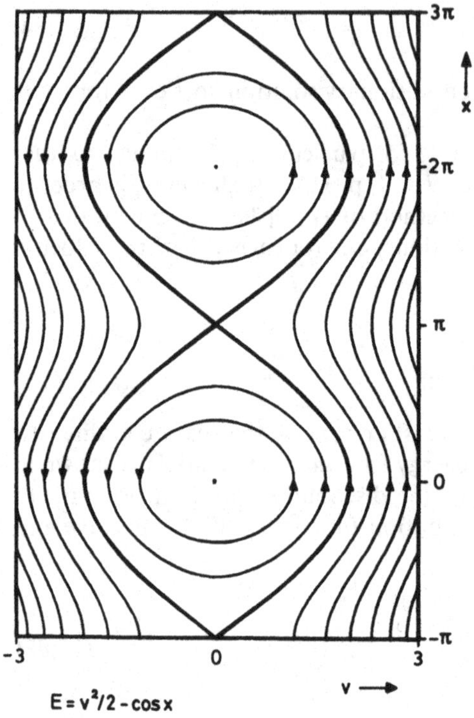

Fig. 11.15. The lines $E=$ const in the phase space shown for the cosine potential $f(x) = -\cos x$. The *bold curve* corresponds to the maximum $E=E_0=1$ of the periodic potential. It separates the running from the oscillating solutions

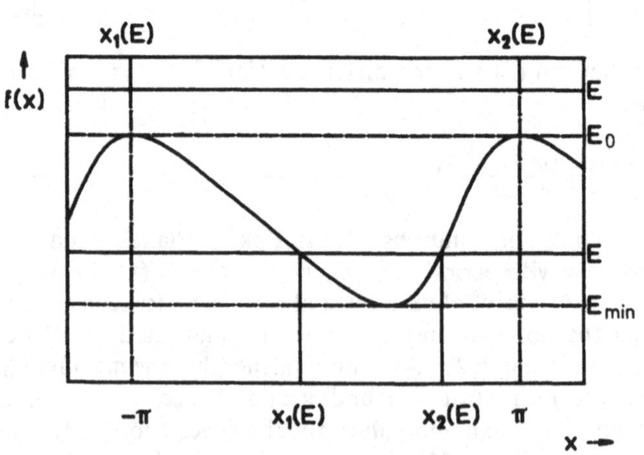

Fig. 11.16. The periodic potential $f(x)$, E_0, $E_{\min}$ and $x_1(E)$, $x_2(E)$ for $E \leqq E_0$. In the following $x_1(E) = -\pi$, $x_2(E)=\pi$ is used for $E \geqq E_0$

The Fokker-Planck equation (11.33) now reads for the functions (11.85)

$$\frac{1}{v(x,E)}\frac{\partial}{\partial t}W_{\pm} = \mp\frac{\partial}{\partial x}W_{\pm} + \gamma\frac{\partial}{\partial E}\left[v(x,E)\left(1+\Theta\frac{\partial}{\partial E}\right)\mp F_0\right]W_{\pm} \tag{11.90}$$

and for (11.87)

$$\frac{1}{v(x,E)}\frac{\partial}{\partial t}W_{\substack{\mathrm{D}\\ \mathrm{S}}} = -\frac{\partial}{\partial x}W_{\substack{\mathrm{S}\\ \mathrm{D}}} + \gamma\frac{\partial}{\partial E}\left[v(x,E)\left(1+\Theta\frac{\partial}{\partial E}\right)W_{\substack{\mathrm{D}\\ \mathrm{S}}} - F_0 W_{\substack{\mathrm{S}\\ \mathrm{D}}}\right], \tag{11.91}$$

where F_0 is defined by (11.84). It should be emphasized that $W_{\pm}$ are the distribution functions in (x, v) space. Those in (x, E) space would be obtained by multiplying $W_{\pm}$ with the Jacobian

$$\frac{\partial(x,v)}{\partial(x,E)} = \frac{\mathrm{d}v}{\mathrm{d}E} = \frac{1}{v(x,E)} = \frac{1}{\sqrt{2[E-f(x)]}}. \tag{11.92}$$

The distribution in (x, E) space is not introduced because the equations for $W_{\pm}$ and $W_{\substack{\mathrm{S}\\ \mathrm{D}}}$ have a simpler form.

Expectation Values

The expectation value of an arbitrary function $h(x, v)$ is given by

$$\begin{aligned}\langle h(x,v)\rangle &= \int_{-\infty}^{\infty}\int_{-\pi}^{\pi} h(x,v)\,W(x,v,t)\,\mathrm{d}v\,\mathrm{d}x\\
&= \int_{E_{\min}}^{\infty}\int_{x_1(E)}^{x_2(E)} [h(x,v(x,E))\,W_+(x,E,t) + h(x,-v(x,E))\\
&\quad\times W_-(x,E,t)]\,\frac{\mathrm{d}E\,\mathrm{d}x}{v(x,E)}\\
&= \int_{E_{\min}}^{\infty}\int_{x_1(E)}^{x_2(E)} \frac{1}{2}[h(x,v(x,E)) + h(x,-v(x,E))]\,W_\mathrm{S}(x,E,t)\,\frac{\mathrm{d}E\,\mathrm{d}x}{v(x,E)}\\
&\quad+ \int_{E_{\min}}^{\infty}\int_{x_1(E)}^{x_2(E)} \frac{1}{2}[h(x,v(x,E)) - h(x,-v(x,E))]\,W_\mathrm{D}(x,E,t)\,\frac{\mathrm{d}E\,\mathrm{d}x}{v(x,E)}.\end{aligned} \tag{11.93}$$

Here $E_{\min}$ is the minimum energy of the potential. The minimum and maximum values $x_1(E)$ and $x_2(E)$ are $-\pi$ and π for $E > E_0$. For $h(x, v) = 1$ and $h(x, v) = v$ (11.93) specializes to

$$1 = \int_{E_{\min}}^{\infty}\int_{x_1(E)}^{x_2(E)} W_\mathrm{S}(x,E,t)\,\frac{\mathrm{d}E\,\mathrm{d}x}{v(x,E)} \quad \text{(normalization)}, \tag{11.94}$$

$$\langle v\rangle = \int_{E_{\min}}^{\infty}\int_{x_1(E)}^{x_2(E)} W_\mathrm{D}(x,E,t)\,\mathrm{d}E\,\mathrm{d}x \quad \text{(drift velocity)}. \tag{11.95}$$

11.4.2 'Ansatz' for the Stationary Distribution Functions

In the stationary state (11.90) reduces to

$$\pm\frac{\partial W_\pm}{\partial x}=\gamma\frac{\partial}{\partial E}\left[v(x,E)\left(1+\Theta\frac{\partial}{\partial E}\right)\mp F_0\right]W_\pm\,. \tag{11.96}$$

If the distribution function does not change very rapidly in E, $W_\pm$ is thus nearly independent of x in the low-friction limit $\gamma\to 0$. In this limit there is a motion mainly along the lines $E=$ const, Fig. 11.15. The dependence of $W_\pm$ on energy is determined by the small change of energy. The Langevin equation (11.3) transforms for the energy variable (11.83) to

$$\dot{E}=-2\gamma[E-f(x)]\pm\gamma v(x,E)F_0\pm v(x,E)\Gamma \quad \text{for} \quad \begin{cases} v>0 \\ v<0 \end{cases}, \tag{11.97}$$

where $v(x,E)$ is defined by (11.86). For the distribution function the energy change is described by the right-hand side of (11.96). As best seen from (11.97), the small energy gain $\gamma v(x,E)F_0$ due to the external field for $v>0$ cancels the small energy loss $-\gamma v(x,E)F_0$ for $v<0$ if the motion is closed, i.e., for $E<E_0$. Therefore the distribution functions $W_+(E)$ and $W_-(E)$ are identical (and W_D is zero), in agreement with the second boundary condition in (11.88 or 89).

For $E>E_0$ the motion of the particles for $v>0$ and $v<0$ is separated. Therefore the particles may gain (lose) energy by the external field F_0 for $v>v(x,E_0)$ ($v<-v(x,E_0)$) and the distribution functions $W_\pm$ will be different for $E>E_0$, compatible with the first boundary condition in (11.88). The distribution functions $W_\pm$ and their first order derivative with respect to E should be continuous at $E=E_0$. As shown below this continuity condition of $W_\pm$ and their first order derivatives cannot be fulfilled by taking into account only x-independent functions. Therefore an x dependence must be considered near $E=E_0$. Because $\partial^2 W/\partial E^2$ must be large if an x dependence is present (11.96), there is strong diffusion perpendicular to the $E=E_0$ trajectory. This strong diffusion is expected also from the following consideration. Particles moving along closed trajectories near $E\lesseqgtr E_0$ are confronted at each turn with two different groups of particles; in one group the particles move upwards ($v>0$) along trajectories near $E\gtrsim E_0$, and in the other they move downwards ($v<0$) along trajectories near $E\gtrsim E_0$, Fig. 11.15. Therefore strong diffusion perpendicular to $E=E_0$ is also expected in a boundary layer around $E=E_0$ from this consideration. Outside this boundary layer near $E=E_0$ one expects that the distribution functions no longer depend on x as discussed before.

We therefore make the following 'ansatz' for $W_S(x,E)$ and $W_D(x,E)$:

$$W_{\substack{S\\D}}(x,E)=\tilde{W}_{\substack{S\\D}}(E)+w_{\substack{S\\D}}(x,E)\,. \tag{11.98}$$

Here $\tilde{W}_{\substack{S\\D}}(E)$ are slowly varying functions in E only. The $w_{\substack{S\\D}}(x,E)$ are rapidly varying functions in E and slowly varying functions in x that contribute only in a thin boundary layer (skin) around $E = E_0$. As shown below, the thickness of the skin is of the order of magnitude of $\sqrt{\gamma}$. As it turns out, the amplitudes of w_S are also of the order $\sqrt{\gamma}$. For small γ values the distribution shifts in x direction near the maxima at $v = 0$ proportional to γF_0, Sect. 11.5. This shift cannot be treated by 'ansatz' (11.98) by which, however, terms up to the order $\sqrt{\gamma}$ can be calculated. For $w_{\substack{S\\D}}$ only terms in lowest order will be taken into account. In accordance with this approximation we put $x_{\substack{1\\2}}(E) = \mp\pi$ in $w_\pm(x_{\substack{1\\2}}, E)$. Thus the continuity conditions (11.89) simplify for small γ values to

$$
\begin{aligned}
w_{\substack{S\\D}}(\pi,E) &= w_{\substack{S\\D}}(-\pi,E) \qquad \text{for} \quad E > E_0\,,\\
w_D(\pm\pi,E) &= \tilde{W}_D(E) = 0 \qquad \text{for} \quad E < E_0\,.
\end{aligned}
\tag{11.99}
$$

Because of the assumptions made for $w_{\substack{S\\D}}$, the first derivatives in E are neglected compared with the second derivatives in E. Furthermore the velocity need be considered for $E = E_0$ only. Therefore the stationary equations (11.91) for x-dependent solutions reduce to

$$\partial w_{\substack{S\\D}}/\partial x = \gamma\Theta v(x,E_0)\,\partial^2 w_{\substack{D\\S}}/\partial E^2\,. \tag{11.100}$$

Instead of the space coordinate x we introduce the variable u defined by

$$u = u(x) = -\pi + \int_{-\pi}^{x} v(\xi,E_0)\,d\xi/\bar{v}(E_0)\,, \tag{11.101}$$

$$\bar{v}(E) = \frac{1}{2\pi}\int_{-\pi}^{\pi} v(x,E)\,dx = \frac{1}{2\pi}\int_{-\pi}^{\pi}\sqrt{2[E-f(x)]}\,dx\,, \qquad E \geqq E_0\,. \tag{11.102}$$

For the cosine potential

$$f(x) = -d\cos x \tag{11.103}$$

we have ($E_0 = d$)

$$u = \pi\sin(x/2)\,, \tag{11.104}$$

$$\bar{v}(E_0) = 4\sqrt{d}/\pi\,. \tag{11.105}$$

Using (11.101, 102), (11.100) simplifies to

$$\partial w_{\substack{S\\D}}/\partial u = \gamma\Theta\bar{v}(E_0)\,\partial^2 w_{\substack{D\\S}}/\partial E^2\,. \tag{11.106}$$

To get an equation for the slowly varying functions $\tilde{W}_{\substack{S\\D}}$, we take an x average of the stationary equations (11.91), which is equivalent to a time average of the Fokker-Planck equation along the trajectories $E = \text{const}$, i.e. a time average of (11.91) multiplied by $v(x,E)$. As discussed before, F_0 does not change the energy of the particles if the motion is closed, i.e. F_0 must be omitted and $\tilde{W}_D$ is zero for $E < E_0$. Because the constant probability current in E direction must vanish, we finally have

$E > E_0$:

$$\bar{v}(E)\left(1+\Theta\frac{\partial}{\partial E}\right)\tilde{W}_{\substack{D\\S}}(E)-F_0\tilde{W}_{\substack{S\\D}}(E)=0\,, \tag{11.107}$$

$E < E_0$:

$$\left(1+\Theta\frac{\partial}{\partial E}\right)\tilde{W}_S(E)=0\,,\qquad \tilde{W}_D(E)=0\,. \tag{11.108}$$

In (11.107) $\bar{v}(E)$ is the x-averaged velocity (11.102) which is given by (11.141) for the cosine potential (11.103).

11.4.3 *x*-Independent Functions

Equations (11.107, 108) are easily solved. The final result is [11.29, 30]

$E < E_0$:

$$\tilde{W}_D(E)=0\,,\qquad \tilde{W}_S(E)=2N\mathrm{e}^{-E/\Theta}\,, \tag{11.109}$$

$E > E_0$:

$$\begin{aligned}\tilde{W}_D(E)&=\mathrm{e}^{-E/\Theta}\{B\cosh[F_0g(E)/\Theta]+C\sinh[F_0g(E)/\Theta]\}\,,\\ \tilde{W}_S(E)&=\mathrm{e}^{-E/\Theta}\{B\sinh[F_0g(E)/\Theta]+C\cosh[F_0g(E)/\Theta]\}\,.\end{aligned} \tag{11.110}$$

Here $g(E)$ is defined by ($E \geqq E_0$)

$$g(E)=\int_{E_0}^{E}\mathrm{d}E'/\bar{v}(E') \tag{11.111}$$

and N, B and C are integration constants.

If the x-dependent functions w_S and w_D are not taken into account, we can require only that the functions are continuous at $E = E_0$. Because of $g(E_0) = 0$ this immediately leads to $B = 0$ and $C = 2N$, i.e.,

$$\begin{aligned}\tilde{W}_\pm&=N\exp(-E/\Theta) &&\text{for}\quad E<E_0\,,\\ \tilde{W}_\pm&=N\exp\{-[E\mp F_0g(E)]/\Theta\} &&\text{for}\quad E>E_0\,.\end{aligned} \tag{11.112}$$

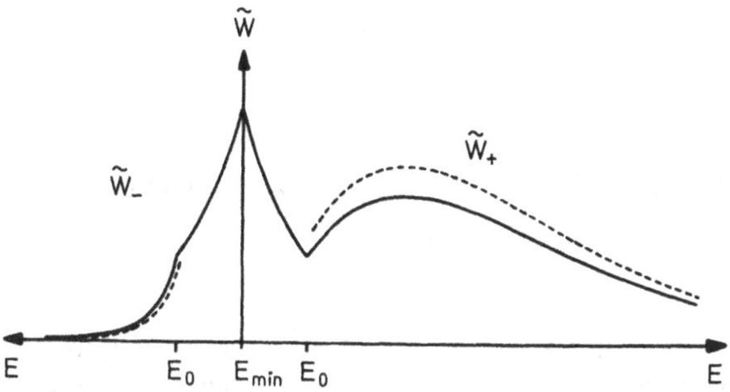

Fig. 11.17. Plot of the distributions $\tilde{W}_+(E) = (\tilde{W}_S + \tilde{W}_D)/2$ for positive velocities (*right*) and $\tilde{W}_-(E) = (\tilde{W}_S - \tilde{W}_D)/2$ for negative velocities (*left*) for the cosine potential (11.103) with $d/\Theta = 0.5$, $F_0/\sqrt{\Theta} = 2$. The *full curves* show the distributions in the limit of vanishing friction [according to (11.112)], the *broken curves* hold for finite friction [according to (11.126, 127), $\varepsilon\kappa = 0.2$]. To simplify the plot the same normalization constant N was chosen for both cases. Therefore both distributions here are the same for $E < E_0$

$$N^{-1} = \sqrt{2\pi\Theta} \int_{-\pi}^{\pi} \mathrm{e}^{-f(x)/\Theta}\mathrm{d}x + 4\pi \int_{E_0}^{\infty} \frac{\mathrm{d}\bar{v}(E)}{\mathrm{d}E}\, \mathrm{e}^{-E/\Theta}\{\cosh[F_0 g(E)/\Theta] - 1\}\mathrm{d}E\,. \tag{11.112a}$$

A plot of the distribution function (11.112) is shown in Fig. 11.17. In lowest order $W_\pm$ are continuous, their derivatives are discontinuous at $E = E_0$. (The derivative of W_S is continuous but the derivative of W_D is discontinuous at $E = E_0$.) In next order the x-independent parts $\tilde{W}_\pm$ are not continuous, however, the full distribution $W_\pm(x, E)$ and their derivatives are continuous at $E = E_0$ (Sect. 11.4.4).

11.4.4 *x*-Dependent Functions

Solutions of (11.106) consistent with (11.99) and different from zero only in the vicinity of E_0 are $(-\pi \leqq u \leqq \pi)$

$E > E_0$:

$$w_D = w_0 \sum_{n=1}^{\infty} \mathrm{Im}\{a_n \exp[-(1+\mathrm{i})\,\alpha\sqrt{n}(E-E_0)/\Theta]\}\cos nu\,, \tag{11.113}$$

$$w_S = w_0 \sum_{n=1}^{\infty} \mathrm{Re}\{a_n \exp[-(1+\mathrm{i})\,\alpha\sqrt{n}(E-E_0)/\Theta]\}\sin nu\,, \tag{11.114}$$

$E < E_0$:

$$w_D = w_0 \sum_{n=0}^{\infty} \mathrm{Im}\{b_n \exp[(1+\mathrm{i})\,\alpha\sqrt{n+1/2}(E-E_0)/\Theta]\}\cos(n+1/2)u\,, \tag{11.115}$$

$$w_S = w_0 \sum_{n=0}^{\infty} \mathrm{Re}\{b_n \exp[(1+\mathrm{i})\,\alpha\sqrt{n+1/2}(E-E_0)/\Theta]\}\sin(n+1/2)u\,. \tag{11.116}$$

Here α is given by

$$\alpha = \sqrt{\Theta/[2\gamma \bar{v}(E_0)]} \tag{11.117}$$

and the imaginary and real parts are indicated by Im{} and Re{}. The prefactor

$$w_0 = \Theta \tilde{W}'_D(E_0+0)/\alpha \tag{11.118}$$

is chosen so that the complex amplitudes a_n and b_n are of the order of magnitude one. In (11.113, 114) the terms with $n=0$ must be excluded, because they would give x-independent functions, which are already included in $\tilde{W}_{\substack{S\\D}}(E)$.

Continuity Conditions at $E = E_0$

We now require that at $E=E_0$, $W_{\substack{S\\D}}(x,E)$ and their derivatives with respect to E are continuous

$$W_{\substack{S\\D}}(x,E_0-0) = W_{\substack{S\\D}}(x,E_0+0)\,, \tag{11.119}$$

$$\partial W_{\substack{S\\D}}(x,E)/\partial E\Big|_{E=E_0-0} = \partial W_{\substack{S\\D}}(x,E)/\partial E\Big|_{E=E_0+0}\,. \tag{11.120}$$

Because $w_S(x,E)$ and $\partial w_S(x,E)/\partial E$ are antisymmetric in $u(x)$ we obtain by inserting (11.98) and using $\tilde{W}_D(E)=0$ for $E<E_0$

$$w_S(x,E_0-0) = w_S(x,E_0+0)\,, \tag{11.119a}$$

$$w_D(x,E_0-0) = \tilde{W}_D(E_0+0) + w_D(x,E_0+0)\,, \tag{11.119b}$$

$$\partial w_S(x,E)/\partial E\,|_{E=E_0-0} = \partial w_S(x,E)/\partial E\,|_{E=E_0+0}\,, \tag{11.120a}$$

$$\partial w_D(x,E)/\partial E\,|_{E=E_0-0} = \tilde{W}'_D(E_0+0) + \partial w_D(x,E)/\partial E\,|_{E=E_0+0}\,. \tag{11.120b}$$

The derivative of $\tilde{W}_D(E)$ with respect to E is denoted by a prime. The x-independent part of (11.119) for $W_S(x,E)$ leads to

$$\tilde{W}_S(E_0-0) = \tilde{W}_S(E_0+0) \equiv \tilde{W}_S(E_0)\,, \tag{11.121}$$

whereas the x-independent part of (11.120) for $W_S(x,E)$ gives

$$\tilde{W}'_S(E_0-0) = \tilde{W}'_S(E_0+0) + O(\sqrt{\gamma})\,. \tag{11.121a}$$

Because $\tilde{W}_S$ is slowly varying in E and w_S is rapidly varying in E (11.121 a) is only correct in the lowest order term $\sqrt{\gamma}^0$ [(11.121) is correct up to terms of the order $\sqrt{\gamma}$]. One concludes from the continuity of the probability current for $\tilde{W}_S$ and from (11.121) that the jump condition

$$\tilde{W}'_S(E_0-0) = \tilde{W}'_S(E_0+0) - \tilde{W}_D(E_0+0)F_0/[\bar{v}(E_0)\,\Theta] \tag{11.121b}$$

holds even in terms of the order $\sqrt{\gamma}$. Introducing the constant $\varkappa$ by

$$\tilde{W}_D(E_0+0) = \varkappa w_0 = \varkappa \Theta \tilde{W}'_D(E_0+0)/\alpha \tag{11.122}$$

and inserting (11.113 – 116) into the continuity conditions (11.119a – 120b) and using (11.118), we have

$$\sum_{n=0}^{\infty} b_n^{(r)} \sin(n+1/2)u = \sum_{n=1}^{\infty} a_n^{(r)} \sin nu\,, \tag{11.119a'}$$

$$\sum_{n=0}^{\infty} b_n^{(i)} \cos(n+1/2)u = \varkappa + \sum_{n=1}^{\infty} a_n^{(i)} \cos nu\,, \tag{11.119b'}$$

$$\sum_{n=0}^{\infty} (b_n^{(r)} - b_n^{(i)}) \sqrt{n+1/2}\, \sin(n+1/2)u = \sum_{n=1}^{\infty} (-a_n^{(r)} + a_n^{(i)}) \sqrt{n}\, \sin nu\,, \tag{11.120a'}$$

$$\sum_{n=0}^{\infty} (b_n^{(r)} + b_n^{(i)}) \sqrt{n+1/2}\, \cos(n+1/2)u = 1 + \sum_{n=1}^{\infty} (-a_n^{(r)} - a_n^{(i)}) \sqrt{n}\, \cos nu\,. \tag{11.120b'}$$

Here the amplitudes with an upper index r (i) are the real (imaginary) parts of the corresponding amplitudes. From (11.119a' – 120b') the amplitudes $a_n^{(r)}$, $a_n^{(i)}$, $b_n^{(r)}$, $b_n^{(i)}$ and the constant $\varkappa$ have to be determined.

Determination of the Constant $\varkappa$

Equation (11.119a' – 120b') must be fulfilled for all u values. To derive equations for the expansion coefficients we may expand $\cos(n+1/2)u$ $[\sin(n+1/2)u]$ into $\cos ru$ $[\sin ru]$ in the range $-\pi \leqq u \leqq \pi$ (or vice versa). In [11.30] the following expansions were used:

$$\cos(n+1/2)u = C_n + \sum_{r=1}^{\infty} C_{nr} \cos ru\,,$$

$$C_n = (-1)^n/[\pi(n+1/2)]\,,$$

$$C_{nr} = [(-1)^{n+r}/(n+r+1/2) + (-1)^{n-r}/(n-r+1/2)]/\pi\,,$$

$$\sin nu = \sum_{r=0}^{\infty} S_{nr} \sin(r+1/2)u\,,$$

$$S_{nr} = [(-1)^{r-n}/(r-n+1/2) - (-1)^{r+n}/(r+n+1/2)]/\pi\,.$$

Inserting $\cos(n+1/2)u$ and $\sin nu$ into (11.119a' – 120b') and comparing terms in $\cos ru$ and $\sin(r+1/2)u$, the following set of linear equations is obtained with $N = \infty$:

$$b_n^{(r)} - \sum_{m=1}^{N} S_{mn} a_m^{(r)} = 0 \quad (0 \leqq n \leqq N),$$

$$\varkappa - \sum_{m=0}^{N} C_m b_m^{(i)} = 0,$$

$$a_n^{(i)} - \sum_{m=0}^{N} C_{mn} b_m^{(i)} = 0 \quad (1 \leqq n \leqq N)$$

$$(b_n^{(r)} - b_n^{(i)})\sqrt{n+1/2} + \sum_{m=1}^{N} S_{mn}(a_m^{(r)} - a_m^{(i)})\sqrt{m} = 0 \quad (0 \leqq n \leqq N-1),$$

$$\sum_{m=0}^{N} (b_m^{(r)} + b_m^{(i)})\sqrt{m+1/2}\, C_m = 1,$$

$$(a_n^{(r)} + a_n^{(i)})\sqrt{n} + \sum_{m=0}^{N} C_{mn}(b_m^{(r)} + b_m^{(i)})\sqrt{m+1/2} = 0 \quad (1 \leqq n \leqq N).$$

To solve this system a finite N was used in [11.30]. The number of unknown coefficients is $4N+3$, i.e., $N+1$ coefficients $b_n^{(r)}$ and $b_n^{(i)}$ and N coefficients $a_n^{(r)}$ and $a_n^{(i)}$ and one $\varkappa$. The number of equations must be the same. Therefore, the index n runs only to $N-1$ in the fourth equation. By using a difference scheme, expression $\varkappa_N = 0.859 + 0.476/(N+2)$ was found for N values in the range $10 \leqq N \leqq 18$ in [11.30]. Extrapolation of this result to $N \to \infty$ gives the value $\varkappa = 0.859$. More accurate calculations lead to the slightly lower value [11.49]

$$\varkappa = 0.855(4). \tag{11.123}$$

11.4.5 Corrected x-Independent Functions and Mobility

Because of the boundary conditions (11.121, 122) the coefficients N, B, C of the x-independent functions (11.109, 110) are now connected by

$$C = 2N,\ B = \varkappa C F_0 g'(E_0)/\alpha[1 + O(\sqrt{\gamma})] = 2N\varepsilon\varkappa[1 + O(\sqrt{\gamma})] \tag{11.124}$$

where ε is an abbreviation for

$$\varepsilon = F_0 g'(E_0)/\alpha = F_0/(\bar{v}(E_0)\,\alpha) = [\sqrt{2}F_0/\sqrt{\bar{v}(E_0)\,\Theta}]\sqrt{\gamma}. \tag{11.125}$$

Thus the x-independent functions correct up to the order $\sqrt{\gamma}$ read:

$$\begin{aligned} \tilde{W}_S &= 2N e^{-E/\Theta} \quad \text{for} \quad E < E_0 \\ \tilde{W}_S &= 2N e^{-E/\Theta} + 2N e^{-E/\Theta}\{\cosh[F_0 g(E)/\Theta] - 1\} \\ &\quad + 2N\varepsilon\varkappa e^{-E/\Theta}\sinh[F_0 g(E)/\Theta] \quad \text{for} \quad E > E_0; \end{aligned} \tag{11.126}$$

$$\tilde{W}_D = 0 \quad \text{for} \quad E < E_0$$

$$\tilde{W}_D = 2N e^{-E/\Theta} \sinh[F_0 g(E)/\Theta] + 2N\varepsilon\varkappa e^{-E/\Theta} \cosh[F_0 g(E)/\Theta] \quad \text{for} \quad E > E_0 . \tag{11.127}$$

Because the x-dependent solutions w_S and w_D are of the order $\sqrt{\gamma}$ and because they are different from zero only in a region of width $\sqrt{\gamma}$ around $E = E_0$, the contribution to the normalization and to the drift velocity is of the order γ. Here we consider only terms up to the order $\sqrt{\gamma}$. Therefore, w_S and w_D are neglected in calculating the normalization and the drift velocity. The x-dependent solutions enter via the continuity conditions (11.119, 120) in such a way that the corrected x-independent solutions deviate from (11.112) by terms of the order $\sqrt{\gamma}$. Inserting (11.126, 127) into (11.94, 95), we have [$x_1(E)$ and $x_2(E)$ are defined in Fig. 11.16]

$$\begin{aligned} N^{-1} &= 2 \int_{E_{\min}}^{\infty} \int_{x_1(E)}^{x_2(E)} e^{-E/\Theta} dE\, dx/v(x,E) \\ &\quad + 2 \int_{E_0}^{\infty} \int_{-\pi}^{\pi} e^{-E/\Theta} \{\cosh[F_0 g(E)/\Theta] - 1\} dE\, dx/v(x,E) \\ &\quad + 2\varepsilon\varkappa \int_{E_0}^{\infty} \int_{-\pi}^{\pi} e^{-E/\Theta} \sinh[F_0 g(E)/\Theta]\, dE\, dx \\ &= 4\pi(A_0 + A_1 + \varepsilon\varkappa A_2) , \end{aligned} \tag{11.128}$$

$$\begin{aligned} \langle v \rangle &= 2N \int_{E_0}^{\infty} \int_{-\pi}^{\pi} e^{-E/\Theta} \sinh[F_0 g(E)/\Theta]\, dE\, dx \\ &\quad + 2N\varepsilon\varkappa \int_{E_0}^{\infty} \int_{-\pi}^{\pi} e^{-E/\Theta} \cosh[F_0 g(E)/\Theta]\, dE\, dx \\ &= 4\pi N(A_3 F_0 + \varepsilon\varkappa A_4) . \end{aligned} \tag{11.129}$$

The first integral A_0 on the right-hand side of (11.128) is equal to

$$A_0 = \frac{1}{4\pi} \int_{-\infty}^{\infty} \int_{-\pi}^{\pi} e^{-v^2/(2\Theta) - f(x)/\Theta} dx\, dv = \frac{\sqrt{2\pi\Theta}}{4\pi} \int_{-\pi}^{\pi} e^{-f(x)/\Theta} dx . \tag{11.130}$$

Because

$$\bar{v}'(E) = \frac{d\bar{v}(E)}{dE} = \frac{d}{dE} \frac{1}{2\pi} \int_{-\pi}^{\pi} v(x,E)\, dx = \frac{1}{2\pi} \int_{-\pi}^{\pi} \frac{dx}{v(x,E)} ,$$

we may write the other two integrals in (11.128) as

$$A_1 = \int_{E_0}^{\infty} \bar{v}'(E) e^{-E/\Theta} \{\cosh[F_0 g(E)/\Theta] - 1\} dE \,, \tag{11.131}$$

$$A_2 = \int_{E_0}^{\infty} \bar{v}'(E) e^{-E/\Theta} \sinh[F_0 g(E)/\Theta] \, dE \,. \tag{11.132}$$

After performing a partial integration of the first integral on the right-hand side in (11.129), we obtain for A_3 and A_4

$$A_3 = \int_{E_0}^{\infty} \frac{e^{-E/\Theta}}{\bar{v}(E)} \cosh[F_0 g(E)/\Theta] \, dE \,, \tag{11.133}$$

$$A_4 = \int_{E_0}^{\infty} e^{-E/\Theta} \cosh[F_0 g(E)/\Theta] \, dE \,. \tag{11.134}$$

Taking into account only terms up to the order $\sqrt{\gamma}$, we finally get for the mobility times the damping constant

$$\gamma\mu = \gamma\langle v\rangle/F = \langle v\rangle/F_0 = C + D\sqrt{\gamma/\sqrt{\Theta}} \,, \tag{11.135}$$

$$C = \frac{A_3}{A_0 + A_1} \,, \tag{11.136}$$

$$D = \frac{\varkappa\sqrt{2}}{\sqrt{\bar{v}(E_0)\sqrt{\Theta}}} \left(\frac{A_4}{A_0 + A_1} - F_0 \frac{A_2 A_3}{(A_0 + A_1)^2} \right) . \tag{11.137}$$

In the linear response ($F_0 \to 0$) (11.136 – 137) simplify to

$$C = \int_{E_0}^{\infty} [e^{-E/\Theta}/\bar{v}(E)] \, dE/A_0 \,, \tag{11.138}$$

$$D = \sqrt{2}\varkappa\Theta e^{-E_0/\Theta} / [\sqrt{\bar{v}(E_0)\sqrt{\Theta}} A_0] \,. \tag{11.139}$$

For the *cosine potential* (11.103) we have ($E \geqq d$)

$$E_0 = d \,, \quad \bar{v}(E_0) = 4\sqrt{d}/\pi \,, \tag{11.140}$$

$$\bar{v}(E) = \sqrt{2} \cdot 2\mathbf{E}[2d/(d+E)] \sqrt{E+d}/\pi \,, \tag{11.141}$$

$$\bar{v}'(E) = \sqrt{2}\mathbf{K}[2d/(d+E)]/(\sqrt{E+d}\,\pi) \,, \tag{11.142}$$

$$A_0 = \sqrt{\pi\Theta/2}\, \mathrm{I}_0(d/\Theta) \,. \tag{11.143}$$

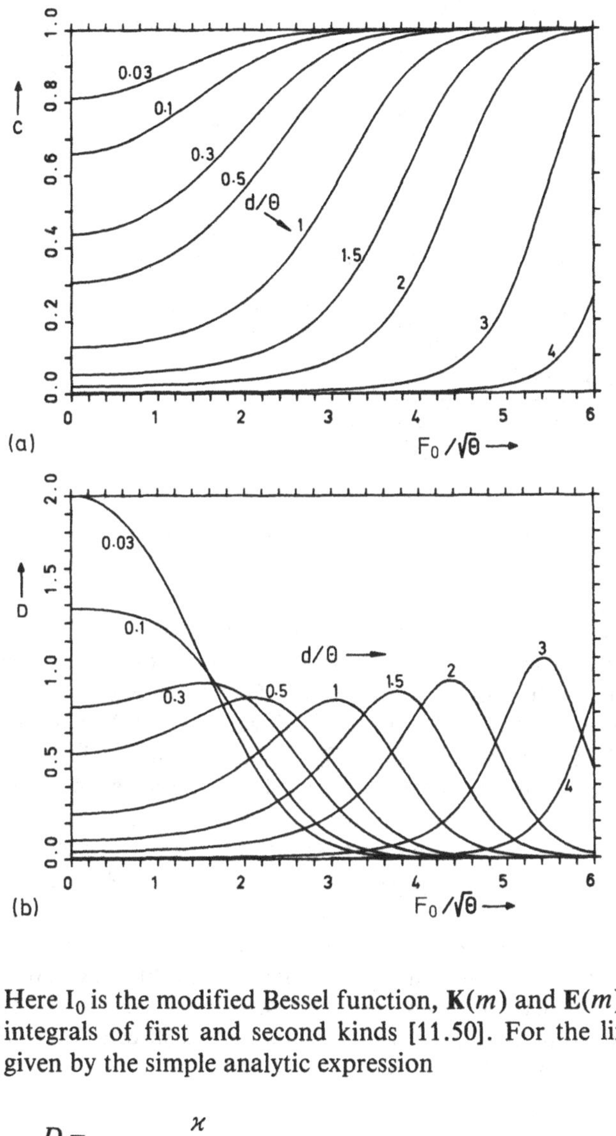

Fig. 11.18. The first expansion coefficients C (**a**) and D (**b**) of the mobility times damping constant as a function of the external force $F_0/\sqrt{\Theta} = F/(\gamma\sqrt{\Theta})$ for different amplitudes of the cosine potential (11.103)

Here I_0 is the modified Bessel function, $\mathbf{K}(m)$ and $\mathbf{E}(m)$ are the complete elliptic integrals of first and second kinds [11.50]. For the linear response, D is thus given by the simple analytic expression

$$D = \frac{\varkappa}{\sqrt[4]{d/\Theta}\, e^{d/\Theta} \mathrm{I}_0(d/\Theta)} . \tag{11.144}$$

Because of the elliptic function in (11.138) we give only an analytic expression for C in the limit $d/\Theta \to \infty$ which reads [$\bar{v}(E)$ is approximated by $\bar{v}(E_0) = 4\sqrt{d}/\pi$ and the modified Bessel function by (10.177)]

$$\gamma\mu = C = (\pi/2)\exp(-2d/\Theta) \quad \text{for} \quad \gamma \to 0 , \quad d/\Theta \to \infty , \quad F_0 \to 0 . \tag{11.145}$$

It is interesting to compare it with the large friction result $\gamma \to \infty$ (11.47) which in the limit $d/\Theta \to \infty$ is given by

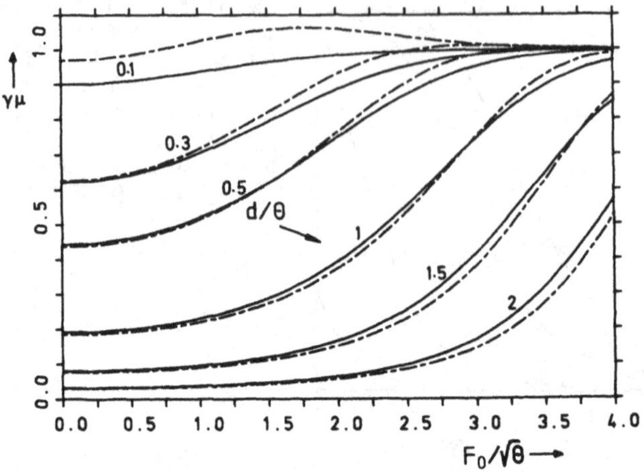

Fig. 11.19. The mobility times the damping constant according to the expansion (11.135) (*broken line*) compared with the exact result obtained by the matrix continued-fraction method of Sect. 11.5 (*full line*), as functions of the external force $F_0/\sqrt{\Theta} = F/(\gamma\sqrt{\Theta})$ for different amplitudes d/Θ of the cosine potential (11.103) and for $\gamma/\sqrt{\Theta} = 0.05$

$$\gamma\mu = I_0(d/\Theta)^{-2} = 2\pi(d/\Theta)\exp(-2d/\Theta) \qquad \text{for} \qquad \gamma \to \infty, d/\Theta \to \infty, F \to 0\,. \tag{11.146}$$

It is seen that both expressions have the same Boltzmann factor $\exp(-2d/\Theta)$ and differ by a prefactor of $4d/\Theta$. As may be seen from the expressions, C and D (11.130–137) depend only on the combination d/Θ and $F_0/\sqrt{\Theta}$. In Fig. 11.18, C and D are shown as a function of $F_0/\sqrt{\Theta}$ for fixed d/Θ values [11.30]. Another choice would be to plot C and D as a function of $F_0/\sqrt{d}$ for fixed d/Θ values. As discussed in Sect. 11.6, in the limit of zero noise strength C then shows a jump at a critical force. In Fig. 11.19 the approximate result (11.135) is compared with the exact result obtained in Sect. 11.5 by the matrix continued-fraction method. It is seen that the fit is quite good for $\gamma/\sqrt{\Theta} = 0.05$. For small d/Θ, one has appreciable deviations. These deviations, however, are expected from our derivation because the thickness $\Theta/\alpha \sim \sqrt{\gamma}$ of the skin around E_0 should be small compared to $E_0 - E_{min} = 2d$. This is not fulfilled for small d/Θ values. For large forces $F_0/\sqrt{\Theta}$ there are considerable deviations, best seen from the curves for large d/Θ values, resulting from the shift of the distribution in x-direction which becomes larger than the thickness of the skin.

11.5 Stationary Solutions for Arbitrary Friction

For intermediate friction constants γ the position and velocity variables or the energy and position variables cannot be associated with different time scales, as in Sects. 11.3, 4. Therefore both variables x and v (or E and x) are now relevant variables and the Fokker-Planck equation (11.33) for two variables must be solved. As discussed in Sect. 10.1.4, we may first expand the distribution function in Hermite functions $\psi_n(v)$ (10.39, 40), see (10.43). Because (10.3) is formally identical to (11.33) with $v_{th}^2 = \Theta$ and $f' \to f' - F$, we use the Hermite functions (10.39, 40) with $v_{th} = \sqrt{\Theta}$. For $\varepsilon = 0$ this expansion then reads

$$W(x,v) = \psi_0(v) \sum_{n=0}^{\infty} c_n(x)\, \psi_n(v)\,. \tag{11.147}$$

In the stationary state the expansion coefficients do not depend on time t. By inserting this expansion into the Fokker-Planck equation (11.33) we obtain the following special form of the Brinkman hierarchy (10.46a)

$$\begin{aligned}
&\sqrt{1}Dc_1 &= 0\\
&\sqrt{1}\hat{D}c_0 + 1\,\gamma c_1 + \sqrt{2}Dc_2 &= 0\\
&\sqrt{2}\hat{D}c_1 + 2\,\gamma c_2 + \sqrt{3}Dc_3 &= 0\\
&\sqrt{3}\hat{D}c_2 + 3\,\gamma c_3 + \sqrt{4}Dc_4 &= 0\\
&\ldots \quad \ldots \quad \ldots &= 0\,.
\end{aligned} \tag{11.148}$$

For the cosine potential *Tikhonov* [11.21] already derived this coupled system of differential equations (it appeared, however, in a different form). The operators D and $\hat{D}$ are defined in (10.27). They read for the present case [$\varepsilon = 0$; $v_{\text{th}} = \sqrt{\Theta}$, the derivative of the potential f' in Chap. 10 has to be replaced by the derivative $V' = f'(x) - F$ of the total potential (11.5)]

$$D = \sqrt{\Theta}\,\partial/\partial x\,, \qquad \hat{D} = \sqrt{\Theta}[\partial/\partial x + (f' - F)/\Theta]\,. \tag{11.149}$$

It is immediately seen from the first equation of (11.148) that the coefficient c_1 must be independent of x

$$c_1(x) = c = \text{const}\,. \tag{11.150}$$

11.5.1 Periodicity of the Stationary Distribution Function

Because the function $f'(x)$ in the Fokker-Planck equation (11.33) or in (11.148) is periodic with period 2π, the Fokker-Planck operator in (11.33) commutes with the translation operator T defined by

$$TW(x,v) = W(x + 2\pi, v)\,. \tag{11.151}$$

Therefore the solutions of the Fokker-Planck equation can be chosen in such a way that they are also eigenfunctions of this translation operator. Denoting the eigenvalues of the translation operator by $\exp(\mathrm{i}k2\pi)$, we may therefore write $-1/2 < \mathrm{Re}\{k\} \leqq 1/2$ (k may be complex)

$$W(x,v) = \mathrm{e}^{\mathrm{i}kx} u(k,x,v)\,, \qquad u(k,x,v) = u(k, x + 2\pi, v)\,, \tag{11.152}$$

i.e., u is a periodic function in x. The expansion coefficients c_n must then also have the form

$$c_n(x) = \mathrm{e}^{\mathrm{i}kx} u_n(k,x)\,, \qquad u_n(k,x) = u_n(k, x + 2\pi)\,. \tag{11.153}$$

It immediately follows from (11.150, 153) that k must be zero if $c_1 \neq 0$. Hence, stationary solutions with $k \neq 0$ can occur only if $c_1 = 0$. By expanding the periodic functions in (11.153) into a Fourier series and by using the down-iteration procedure for solving (11.148), described below, we can express the Fourier coefficients c_2^p in terms of those of c_1^p. If c_1 is zero, $c_2(x)$ must therefore also be zero and it then follows from the second equation of (11.148) that

$$c_0(x) = N e^{-V(x)/\Theta} = N e^{-f(x)/\Theta} e^{Fx/\Theta},$$

$$c_n(x) = 0 \quad \text{for} \quad n \geqq 1 .$$

Thus we have either the nonperiodic solution ($F \neq 0$)

$$\hat{W}(x,v) = (2\pi\Theta)^{-1/2} N \exp\left(-\frac{v^2}{2\Theta} - \frac{f(x)}{\Theta} + \frac{Fx}{\Theta}\right) \tag{11.154}$$

or the periodic solution

$$W(x,v) = W(x+2\pi, v) \tag{11.155}$$

or a combination of both, i.e., $W_c(x,v) = A\, W(x,v) + B\, \hat{W}(x,v)$. Because

$$\hat{W}(x+2\pi n, v) = \exp(2\pi n F/\Theta)\, \hat{W}(x,v),$$

we then recover the form (11.39) obtained for the Smoluchowski equation. If we require that the solutions are bounded, we must discard solution (11.154). Thus the solution must be periodic in x in the stationary state and we therefore normalize the distribution function in one period, i.e.,

$$1 = \int_0^{2\pi} \int_{-\infty}^{\infty} W(x,v)\, dx\, dv = \int_0^{2\pi} c_0(x)\, dx .$$

The drift velocity is given by

$$\begin{aligned}\langle v \rangle &= \int_0^{2\pi} \int_{-\infty}^{\infty} v\, W(x,v)\, dx\, dv = \int_0^{2\pi} \int_{-\infty}^{\infty} S_x(x,v)\, dx\, dv \\ &= \int_0^{2\pi} \int_{-\infty}^{\infty} v\, \psi_0(v) \sum_{n=0}^{\infty} c_n(x)\, \psi_n(v)\, dx\, dv \\ &= \sqrt{\Theta}\, 2\pi c .\end{aligned} \tag{11.156}$$

In deriving this result we used

$$v\, \psi_0(v) = \sqrt{\Theta}(b + b^+)\, \psi_0(v) = \sqrt{\Theta}\, \psi_1(v) \tag{11.157}$$

[see (10.22, 38)], the normalization (10.41) and (11.150). It is seen from (11.33, 10.4b) that the probability current density in x direction is given by $S_x = v\, W$.

We conclude from (11.148, 149) that the coefficient $c_1 = c$ depends only on the combination $\gamma/\sqrt{\Theta}$, F/Θ and f'/Θ. The mobility times the damping constant may be written as

$$\gamma\mu = \frac{\gamma\langle v\rangle}{F} = \frac{\gamma}{\sqrt{\Theta}}\,\frac{\Theta}{F}\,2\pi c\,. \tag{11.158}$$

11.5.2 Matrix Continued-Fraction Method

To solve (11.148) we expand the periodic coefficients c_n into a truncated Fourier series

$$c_n(x) = (2\pi)^{-1/2}\sum_{p=-Q}^{Q} c_n^p \mathrm{e}^{\mathrm{i}px}\,. \tag{11.159}$$

By introducing the vector notation

$$\boldsymbol{c}_n = \begin{pmatrix} c_n^{-Q} \\ \vdots \\ c_n^{Q} \end{pmatrix} \tag{11.160}$$

the system (11.148) of coupled differential equations may then be cast into the stationary form of the tridiagonal vector recurrence relation (9.10), where the matrix elements of the matrices $\boldsymbol{Q}_n$ and $\boldsymbol{Q}_n^{\pm}$ are given by

$$Q_n^{pq} = -n\gamma\delta_{pq},\; (Q_n^{+})^{pq} = -\sqrt{n+1}\,D^{pq},\; (Q_n^{-})^{pq} = -\sqrt{n}\,\hat{D}^{pq} \tag{11.161}$$

with

$$D^{pq} = \frac{\sqrt{\Theta}}{2\pi}\int_0^{2\pi}\mathrm{e}^{-\mathrm{i}px}\frac{\partial}{\partial x}\mathrm{e}^{\mathrm{i}qx}\mathrm{d}x = \mathrm{i}\sqrt{\Theta}q\,\delta_{pq} \tag{11.162}$$

and

$$\hat{D}^{pq} = \frac{\sqrt{\Theta}}{2\pi}\int_0^{2\pi}\mathrm{e}^{-\mathrm{i}px}\left(\frac{\partial}{\partial x} + \frac{f'}{\Theta} - \frac{F}{\Theta}\right)\mathrm{e}^{\mathrm{i}qx}\mathrm{d}x$$
$$= \sqrt{\Theta}[(\mathrm{i}q - F/\Theta)\,\delta_{pq} + f'_{p-q}/\Theta]\,. \tag{11.163}$$

Here f'_r are the expansion coefficients of the negative periodic force f'

$$f'(x) = \sum_r f'_r \mathrm{e}^{\mathrm{i}rx}\,. \tag{11.164}$$

For the cosine potential $f(x) = -d\cos x$ (11.163) simplifies to

$$\hat{D}^{pq} = \sqrt{\Theta}[(\mathrm{i}q - F/\Theta)\,\delta_{pq} - \mathrm{i}(\delta_{p,q+1} - \delta_{p,q-1})d/(2\Theta)]\,. \tag{11.165}$$

The elimination of $\boldsymbol{c}_n$ with $n \geqq 2$ leads to

$$c_1 = \tilde{S}_0^+ (s=0) c_0 , \tag{11.166}$$

where $\tilde{S}_0^+ (s)$ is given by (10.142). Because we treat the stationary problem here, $s = 0$ has to be used. Expressing c_0 in terms of c_1 gives

$$c_0 = H c_1 , \quad H = [\tilde{S}_0^+ (s=0)]^{-1} . \tag{11.166a}$$

By slight manipulations of the continued fractions (10.142) it may be seen that H can be written as [9.14]

$$H = -\gamma \hat{D}^{-1} \left\{ I - \frac{1}{\gamma^2} D \left[I - \frac{1}{2\gamma^2} D \left[I - \frac{1}{3\gamma^2} D [I \ldots]^{-1} \hat{D} \right]^{-1} \hat{D} \right]^{-1} \hat{D} \right\} , \tag{11.167}$$

where D and $\hat{D}$ are the matrices with elements given by (11.162, 163). From (11.150, 166a)

$$c_0^p = \sum_q H^{pq} c_1^q = \sum_q H^{pq} c \sqrt{2\pi}\, \delta_{q0} = H^{p0} c \sqrt{2\pi} . \tag{11.168}$$

The normalization condition requires $c_0^0 = (2\pi)^{-1/2}$, so that the drift velocity (11.156) is finally given by

$$\langle v \rangle = \sqrt{\Theta}\, 2\pi c = \sqrt{\Theta}/H^{00} . \tag{11.169}$$

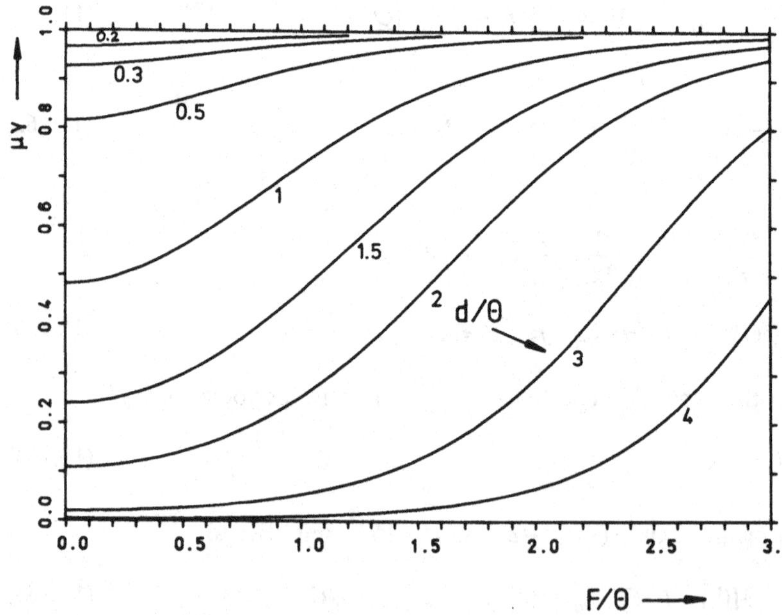

Fig. 11.20. The mobility times the damping constant as a function of F/Θ for various amplitudes of d/Θ of the cosine potential $f(x) = -d \cos x$ and for $\gamma/\sqrt{\Theta} = 1$

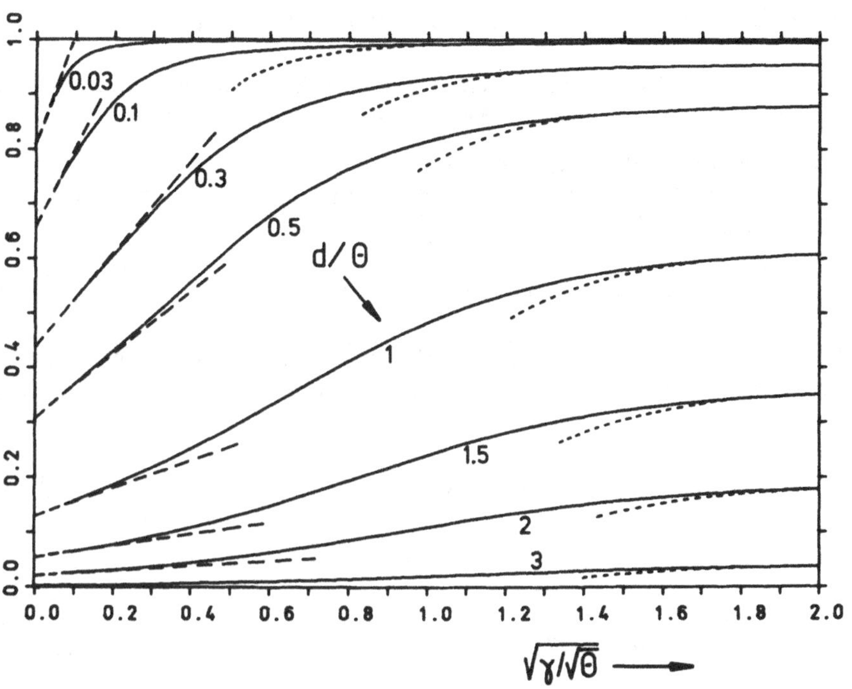

Fig. 11.21. The mobility times the damping constant as a function of $\sqrt{\gamma/\sqrt{\Theta}}$ for various amplitudes d/Θ of the cosine potential in linear response $F \to 0$. The *full curves* were obtained by the matrix continued-fraction method [9.14], the *broken lines*, from the low-friction expression (11.135), and the *dotted lines* are the results of the inverse friction expansion (11.59)

Thus, the main task in determining drift velocity or mobility (11.158) is calculating the matrix continued fraction (11.167). To evaluate this infinite continued fraction, it was approximated in [9.14] by its Nth approximant. The number N and the truncation number Q of the Fourier series (11.159) were determined in [9.14] so that a further increase of N and Q did not alter the result of $\langle v \rangle$ beyond a given accuracy. It turned out that the number N for giving results accurate to three decimals was of the order $N_0 = 20\sqrt{\Theta}/\gamma$, and $Q = 12$ was sufficient for d values up to $d/\Theta = 4$. In Fig. 11.20 the mobility times the damping constant is shown as a function of F/Θ for various amplitudes d/Θ of the cosine potential. For large forces the effect of the periodic potential becomes negligible and therefore one obtains the result $\gamma\mu = 1$ which is also valid for vanishing force $f' = 0$. (This result is immediately obtained by averaging (11.3) in the stationary state, i.e., $\langle \ddot{x} \rangle = 0$, $\langle \dot{x} \rangle = \langle v \rangle = F/\gamma$.) In Fig. 11.19 the mobility times damping constant was shown for very low damping constants. The result for a small external force F (linear response) is shown in Fig. 11.21 as a function of $\sqrt{\gamma/\sqrt{\Theta}}$. [To evaluate (11.167) for $F = 0$ a small F must be used, because the expression will become infinite for zero force.] For small damping constants the plots fit very accurately with the low-friction approximation (11.135) and for large damping constants with the high-friction approximation (11.59).

11.5.3 Calculation of the Stationary Distribution Function

To obtain the distribution function (11.147) higher expansion coefficients $\boldsymbol{c}_n = (c_n^p)$ with $n \geqq 2$ are needed. We now discuss two ways to calculate $\boldsymbol{c}_n$:

(i) Iteration of the Recurrence Relation
According to (11.168, 150) the Fourier expansion coefficients of $c_0(x)$ and $c_1(x) = c$ are given by $[c_0^0 = (2\pi)^{-1/2}]$

$$c_0^p = H^{p0}/(\sqrt{2\pi}\,H^{00})\,; \qquad c_1^p = \delta_{p0}/(\sqrt{2\pi}\,H^{00})\,. \tag{11.168a}$$

By up-iteration of (11.148) one then obtains higher expansion coefficients. For the Fourier coefficients c_n^p, the system (11.148) takes the form ($c_n^q = 0$ for $n \leqq -1$)

$$\sqrt{n}\sum_q \hat{D}^{pq} c_{n-1}^q + n\gamma c_n^p + \sqrt{(n+1)\Theta}\,\mathrm{i}p c_{n+1}^p = 0\,. \tag{11.170}$$

Thus, this up-iteration now reads explicitly for $p \neq 0$; $n \geqq 2$

$$c_n^p = \frac{\mathrm{i}}{p\sqrt{n\Theta}}\left[(n-1)\gamma c_{n-1}^p + \sqrt{n-1}\sum_q \hat{D}^{pq} c_{n-2}^q\right] \tag{11.171a}$$

and for $p = 0$; $n \geqq 2$

$$c_n^0 = -\frac{1}{\gamma\sqrt{n}}\sum_q \hat{D}^{0q} c_{n-1}^q\,. \tag{11.171b}$$

By starting with (11.168a) it is thus in principle easy to obtain both higher expansion coefficients and also, with the help of (11.159) the distribution function (11.147). This up-iteration, however, is numerically unstable. (Down-iteration leading to the matrix continued fraction (11.167) is stable.) Therefore, up-iteration can be performed only up to a $n = N_0$. As revealed in [9.15], the magnitude of the coefficients c_n^q decreases (after an eventual initial increase) and after reaching a minimum at N_0 again increases due to the numerical instability. By using a higher numerical accuracy, the number N_0, at which the numerical instability becomes dominant, can be shifted to higher values. It was found in [9.15] that a good check for convergence is the positivity of the distribution function.

(ii) Iteration by the Matrices $\tilde{S}_n^+$
As already mentioned in Sect. 9.2.1 the up-iteration of the tridiagonal scalar recurrence relation is numerically unstable for the minimal solution, but the up-iteration according to (9.28a) is numerically stable. The same seems to be true for the tridiagonal vector recurrence relation. As just discussed, the up-iteration of (11.170) is unstable but the up-iteration according to the first two equations of (10.141) with $m = 0$ and (10.142) seems to be numerically stable. If we use instead of the matrices $\tilde{S}_n^+$ $(s = 0)$ the matrices $\boldsymbol{A}_n$ defined by

$$\begin{aligned}A_n &= -\gamma\sqrt{n+1}\,\tilde{S}_n^+(s=0)\\ &= [I-\gamma^{-2}(n+1)^{-1}DA_{n+1}]^{-1}\hat{D}\\ &= [I-\gamma^{-2}(n+1)^{-1}D[I-\gamma^{-2}(n+2)^{-1}D[I-\ldots]^{-1}\hat{D}]^{-1}\hat{D}]^{-1}\hat{D}\end{aligned} \tag{11.172}$$

this up-iteration reads ($n \geqq 1$):

$$c_n = (-\gamma)^{-n}(n!)^{-1/2}A_{n-1}\ldots A_1 A_0 c_0 . \tag{11.173}$$

To obtain c_n all matrices $A_{n-1},\ldots,A_0$ are needed. They do not have to be calculated separately, however, because for calculating $H = -\gamma A_0^{-1}$ with the downward iteration (Sect. 9.5.2) all matrices A_n with $1 \leqq n \leqq N$ are obtained in intermediate steps. In contrast to the calculation in Sect. 11.5.2 all A_n with $1 \leqq n \leqq N$ must now be stored.

In Fig. 11.22, 22a perspective plots and in Fig. 11.23 altitude charts of the distribution function obtained by the first method in [9.15] are shown for the cosine potential $f(x) = -d\cos x$. To simplify the labeling in the plots, the normalization $v_0 = v_{\mathrm{th}}$ in (11.29) was used, i.e., $\Theta = 1$. If they exist, trajectories of the running solution of the noiseless equation (11.3) as well as the position of its locked solution are also indicated in the altitude charts. For the free-field case ($F=0$), the distribution function is given by the Boltzmann distribution $W(x,v) = \exp(-v^2/2 + d\cos x)/[(2\pi)^{3/2}\mathrm{I}_0(d)]$, which is independent of the damping constant γ. For very large forces the distribution function is the shifted Maxwell distribution $W(x,v) = (2\pi)^{-3/2}\exp[-(v-F/\gamma)^2/2]$. This distribution is also the exact distribution for a zero potential ($d=0$). Going from the bottom to the top plots in Fig. 11.22, the transition from the Boltzmann distribution towards the shifted Maxwell distribution is seen. For intermediate forces the distribution functions depend strongly on the damping constant γ. For γ values of the order one or less, the distributions have a complicated shape. Let us now discuss the plots by going from large to small friction constants.

For the large friction constant $\gamma = 5$ the distribution is not very structured. All plots have features similar to those of the distribution for $F=0$, the maxima of the distribution are shifted in the positive v and x directions, the amount of the shift increases with increasing external force. Furthermore, the x variation of the distribution becomes flatter with increasing field, whereas the v variation does not change its shape appreciably. The trajectory of the running solution for large friction constants γ is approximately given by $v = (F - d\sin x)/\gamma$. The ridge of the distribution agrees with the trajectory of the running solution (Fig. 11.23).

For $\gamma = 1$, significant deviations from the Boltzmann distribution occur at lower forces F. The variation in x direction at the ridge of the distribution is no longer symmetric with respect to the maxima. The tails towards the positive x and v directions are due to an appreciable mean free path [see (10.10) for a definition of the mean free path]. Because of the noise, the ridge deviates appreciably from the trajectory of the running noiseless solution.

For $\gamma = 0.25$, the mean free path is about the period of the potential. In this case, the distribution has a complicated shape for intermediate forces. For

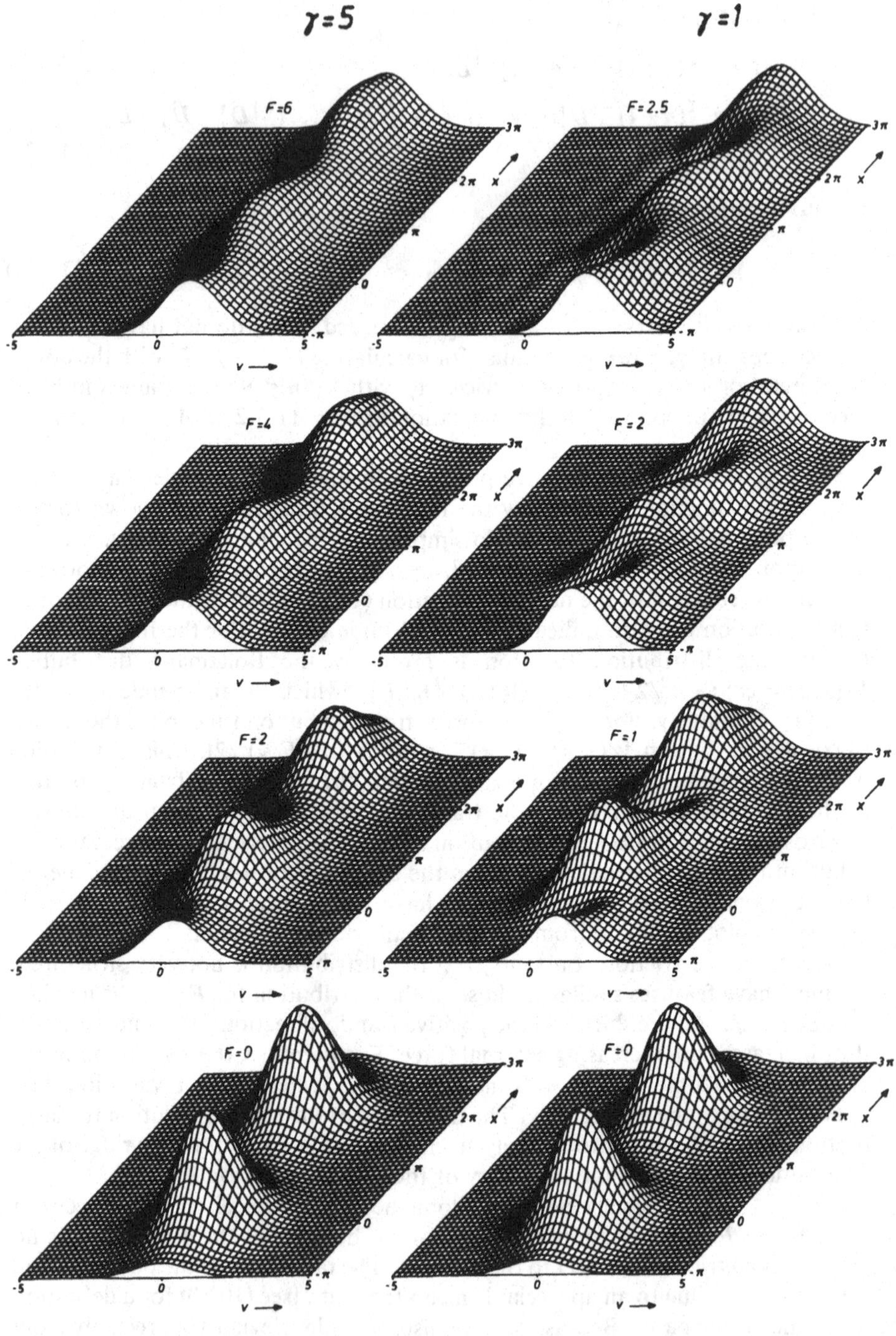

Fig. 11.22. Perspective plots of the probability distribution $W(x, v)$ for different forces F (increasing from bottom to top) and different friction constants γ (decreasing from left to right) for a potential

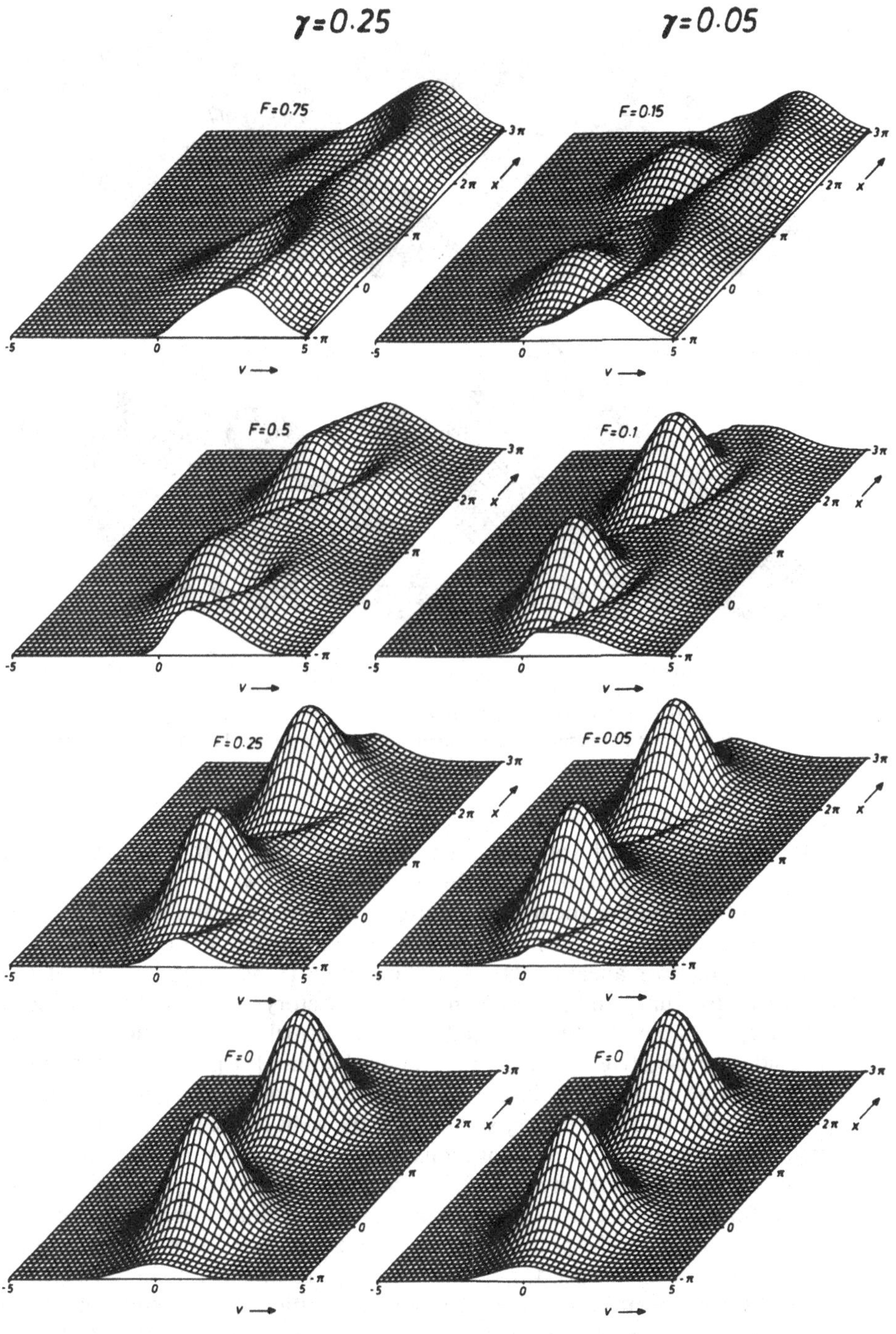

Fig. 11.22 (continued)
amplitude $d = 1$. All plots are shown in the range of $v = -5 \ldots +5$ and $x = -\pi \ldots 3\pi$ (2 periods) using the normalization $\Theta = 1$

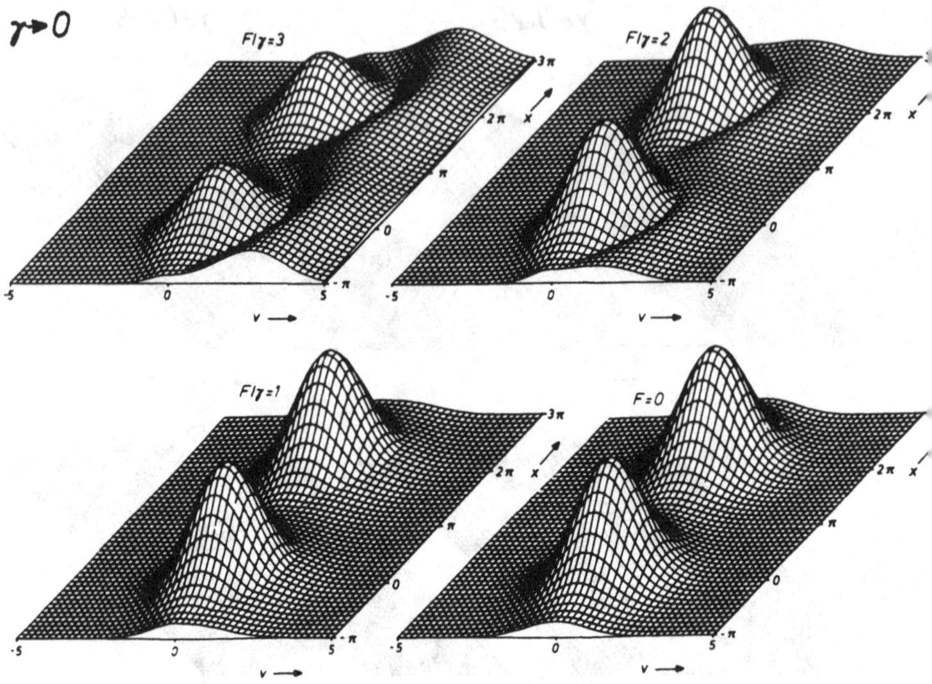

Fig. 11.22a. Perspective plots of the probability distribution $W(x, v)$ according to (11.112) for different forces F/γ for $d = 1$ in the small-friction limit $\gamma \to 0$ (similar to Fig. 11.22) using the normalization $\Theta = 1$

$F = 0.25$, only a locked solution exists, whereas for $F = 0.5$ and 0.75 both running and locked solutions exist. Whereas for $F = 0.5$ we still find a maximum near the locked solution, the locked solution has a low probability for $F = 0.75$. This is explained by the flat minimum of the total potential $V(x)$, Fig. 11.2b. (Because of the normalization $\Theta = 1$, the potential is measured in units of Θ.) The distribution function has a ridge near the trajectory of the running solution. At the steepest decents of the potential (11.5) the distribution function decreases very rapidly for negative velocities, i.e., the probability for particles being at the downward slope of the potential (11.5) and going in backward direction with an appreciable velocity becomes very small. Near the wells of the potential (11.5) the particles oscillate, thus having approximately the same probability for the same positive and negative velocity ($F = 0.25$ and 0.5). The saddle point of the distribution for $F = 0.5$ is crucial in so far as particles with an appreciably smaller velocity are trapped in the potential wells, whereas particles with an appreciably larger velocity are not trapped, i.e., they move on to the next potential well.

For the very small value $\gamma = 0.05$ the distribution has maxima at $v \approx 0$, $x \approx 2n\pi$. For the forces $F = 0.1$ and $F = 0.15$ a ridge exists which agrees well with the trajectory of the running solution. Except for $E \approx d$, the lines of constant altitude agree well with the lines of constant energy. Whereas in the vicinity of $E = d$ [i.e., $v = \pm 2\cos(x/2)$] the distribution increases slowly in x for positive velocities, the distribution at $E = d$ decreases rapidly for negative velocities, see especially the altitude chart for $F = 0.15$. The low increase is explained by the low diffusion constant γ, the rapid decrease by the fact that the probability of particles going backwards at the tops of the potential $V(x)$ is very small.

Finally, in Fig. 11.22a the low-friction limit distribution (11.112) is shown. It agrees well with the distribution for $\gamma = 0.05$ for the same ratios of F/γ. This distribution depends only on the energy. As discussed in Sect. 11.4, x dependence must be taken into account in a skin around $E = d$ for finite γ.

11.5.4 Alternative Matrix Continued Fraction for the Cosine Potential

For the cosine potential $f(x) = -d\cos x$ the matrices $\hat{\boldsymbol{D}}$ and $\boldsymbol{D}$ (11.162, 165) also lead to tridiagonal coupling of the coefficients in the upper index. Introducing the vector

$$\boldsymbol{\alpha}_p = \begin{pmatrix} \alpha_p^0 = c_0^p \\ \alpha_p^1 = c_1^p \\ \vdots \\ \alpha_p^N = c_N^p \end{pmatrix} \tag{11.174}$$

(11.170) can be cast into the two-sided tridiagonal recurrence relation

$$\boldsymbol{Q}_p^- \boldsymbol{\alpha}_{p-1} + \boldsymbol{Q}_p \boldsymbol{\alpha}_p + \boldsymbol{Q}_p^+ \boldsymbol{\alpha}_{p+1} = 0\,, \tag{11.175}$$

where the matrix elements of the matrices $\boldsymbol{Q}_p^\pm$ and $\boldsymbol{Q}_p$ are given by

$$\begin{aligned} (\boldsymbol{Q}_p^\pm)^{nm} &= \pm \tfrac{1}{2}\mathrm{i}\sqrt{n}\,\delta_{n-1,m} \\ (\boldsymbol{Q}_p)^{nm} &= \sqrt{n}\left(\mathrm{i}p\frac{\Theta}{d} - \frac{F}{d}\right)\delta_{n-1,m} + \sqrt{n+1}\,\mathrm{i}p\frac{\Theta}{d}\,\delta_{n+1,m} + n\frac{\gamma}{\sqrt{\Theta}}\,\frac{\Theta}{d}\,\delta_{nm}\,. \end{aligned} \tag{11.175a}$$

For a two-sided recurrence relation $\boldsymbol{\alpha}_p$ with $p \neq 0$ can also be eliminated. It leads to

$$\boldsymbol{M}\boldsymbol{\alpha}_0 = 0\,, \quad \boldsymbol{M} = \boldsymbol{Q}_0 + \tilde{\boldsymbol{K}}_0^+(0) + \tilde{\boldsymbol{K}}_0^-(0)\,, \tag{11.176}$$

where $\tilde{\boldsymbol{K}}_0^+(s)$ is the matrix continued fraction (9.112) and $\tilde{\boldsymbol{K}}_0^-(s)$ is the corresponding infinite matrix continued fraction for the elimination of $\boldsymbol{\alpha}_p$ with negative p. Because $\boldsymbol{\alpha}_p = \boldsymbol{\alpha}_{-p}^*$ the final expression for $\boldsymbol{M}$ can be written as

$$\boldsymbol{M} = \boldsymbol{Q}_0 + 2\,\mathrm{Re}\{\tilde{\boldsymbol{K}}_0^+(0)\}\,.$$

Using an index notation, (11.176) then becomes

$$\sum_{m=1}^{N} M^{nm}\alpha_0^m = -M^{n0}\alpha_0^0\,; \quad n = 1,\ldots,N\,. \tag{11.177}$$

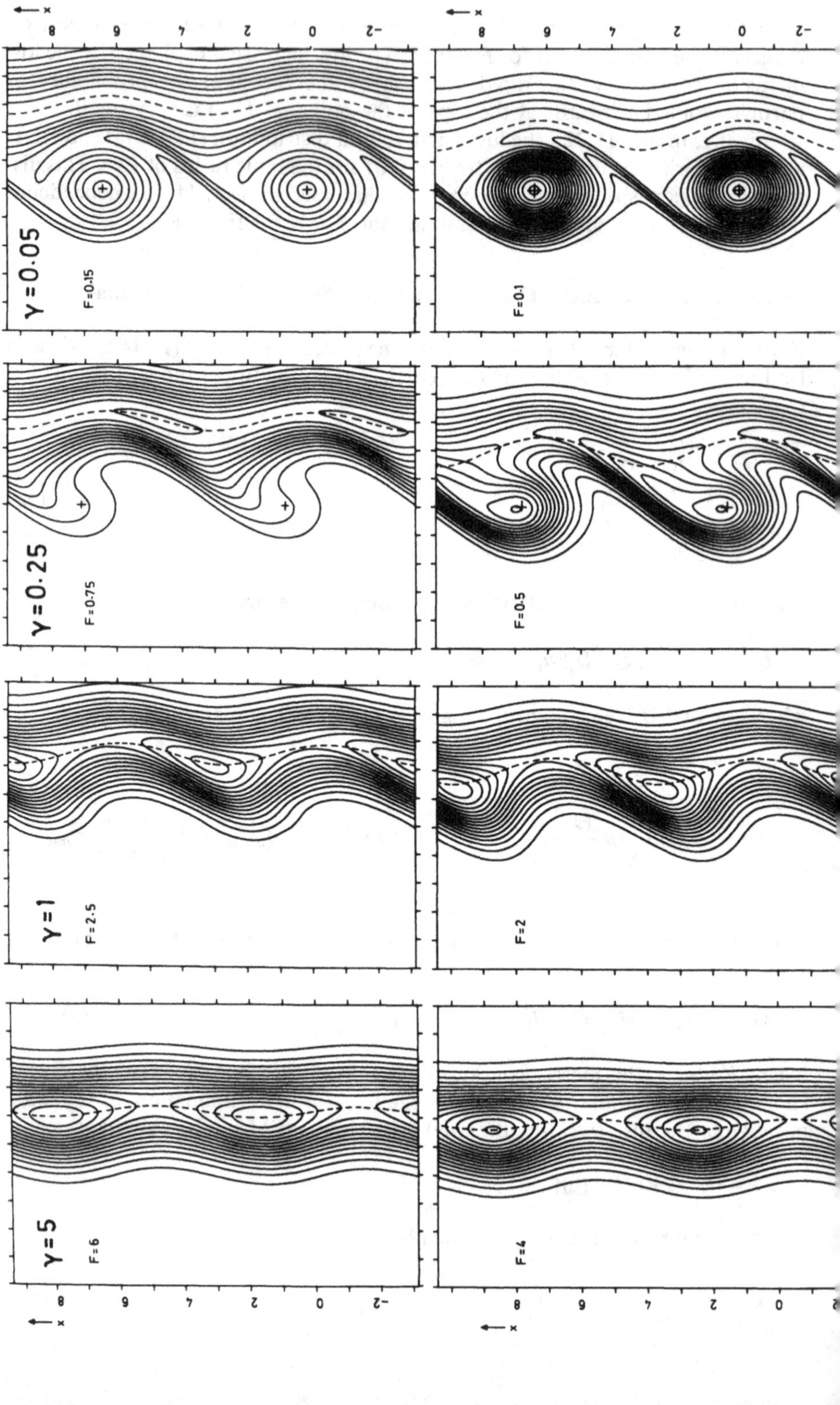
γ = 0.05
F=0.15
F=0.1
γ = 0.25
F=0.75
F=0.5
γ = 1
F=2.5
F=2
γ = 5
F=6
F=4
x
8
6
4
2
0
-2

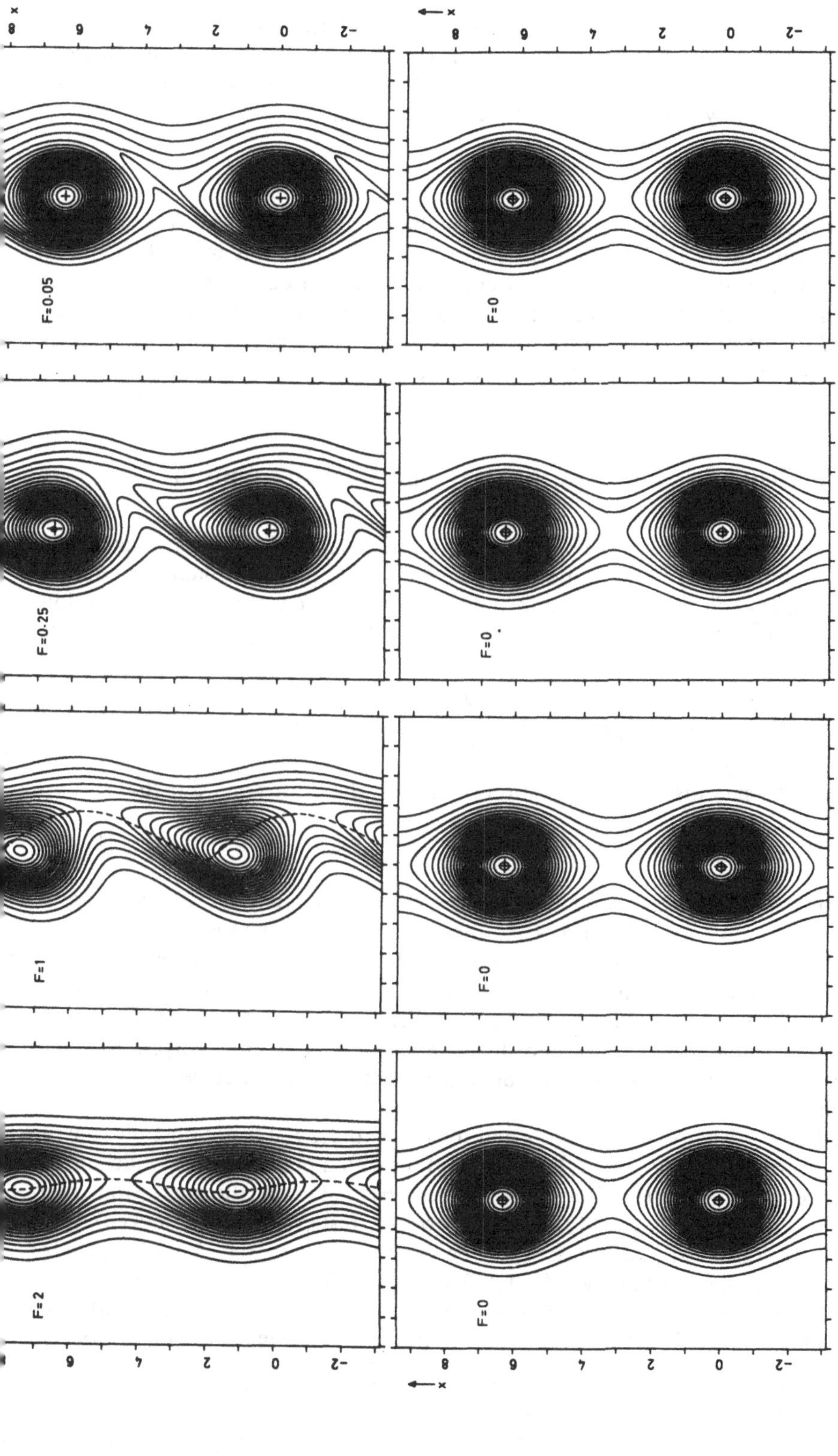

Fig. 11.23. Altitude charts of the probability distribution $W(x, v)$ for several forces F and friction constants γ for $d = 1$ using the normalization $\Theta = 1$. The lines start at $W = 0.005$ (*outermost lines*) and differ by 0.005. If they exist, the positions of the locked solution are indicated by a *cross*, the stationary trajectory of the running solution is shown by a *broken line*

Because M^{0m} is zero we have discarded the equation with $n=0$. The term $\alpha_0^0=(2\pi)^{-1/2}$ follows from normalization. Therefore, (11.177) is an inhomogeneous equation for the N unknown coefficients $\alpha_0^1 \dots \alpha_0^N$. The drift velocity is finally given by [see (11.156) with $\sqrt{2\pi}\,c = c_1^0 = \alpha_0^1$]

$$\langle v\rangle = \sqrt{2\pi\Theta}\,\alpha_0^1 . \tag{11.178}$$

The similar procedure in the high-friction limit suggests that the drift velocity can be evaluated by this method for lower Θ/d values, if γ is large. For smaller friction constants γ, however, the continued fraction described in Sect. 11.5.2 seems to be preferable, because for small friction the number N of expansion coefficients of the Hermite functions is large and therefore the dimension of the matrices which have to be inverted becomes large, too.

11.6 Bistability between Running and Locked Solution

As already discussed at the beginning of this chapter, two solutions, one locked, one running, exist in the stationary state for zero noise, if the damping constant is small enough. When noise is included, transitions between these two solutions occur. In the low-friction limit it is even possible to define an effective potential

$$V_{\mathrm{LF}}^{\pm}(E) = \begin{cases} E & \text{for} \quad E \leqq E_0 \\ E \mp F_0 g(E) & \phantom{\text{for}} \quad E \geqq E_0 . \end{cases} \tag{11.179}$$

The stationary distribution function (11.112) is the Boltzmann distribution

$$\tilde{W}_{\pm}(E) = N \exp[-V_{\mathrm{LF}}^{\pm}(E)/\Theta] . \tag{11.180}$$

A plot of the potential (11.179) for the cosine potential (11.103) is shown in Fig. 11.24. As seen for forces F_0 above the first critical force F_{01}, a second minimum corresponding to the running solution occurs. For forces F_0 in the range $F_{01} < F_0 < F_{02}$, this minimum is higher than the minimum at $E+d=0$ corresponding to the locked solution and therefore the running solution is not globally stable in this region. For forces $F_0 > F_{02}$ the minimum corresponding to the running solution is the lowest and the running solution then becomes globally stable.

For finite damping constants and for finite forces F it is not possible to write the distribution function in the form (11.180) with a potential being independent of the temperature Θ. [This is already seen by looking at the expression for x-dependent solutions (11.113 – 116).] However, bistability of the locked versus running solutions, just discussed for the low-friction case, seems to exist also for finite damping constants $\gamma\sqrt{d} \lesssim 1.193$, Sect. 11.6.2.

Let us now investigate solutions without noise and then discuss the effect of an added noise. With the exception of a special model, we restrict ourselves here to the cosine potential.

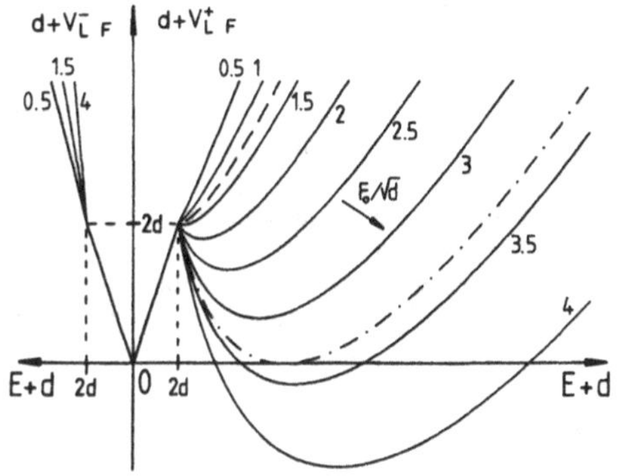

Fig. 11.24. The effective potential (11.179) as a function of $E+d$ for various external forces $F_0/\sqrt{d}$ and for the cosine potential $f(x) = -d\cos x$. The effective potential $V_{\mathrm{LF}}^{+}(V_{\mathrm{LF}}^{-})$ for positive (negative) velocities is plotted to the *right* (*left*). The effective potential for the critical force $F_{01}/\sqrt{d} = 4/\pi$ is shown by a *broken line* and for the critical force $F_{02}/\sqrt{d} = 3.3576$ by a *broken line with dots*

11.6.1 Solutions Without Noise

Without the noise term, (11.3) becomes a deterministic equation, which reads for the cosine potential $f(x) = -d\cos x$

$$\ddot{x} + \gamma\dot{x} + d\sin x = F. \tag{11.181}$$

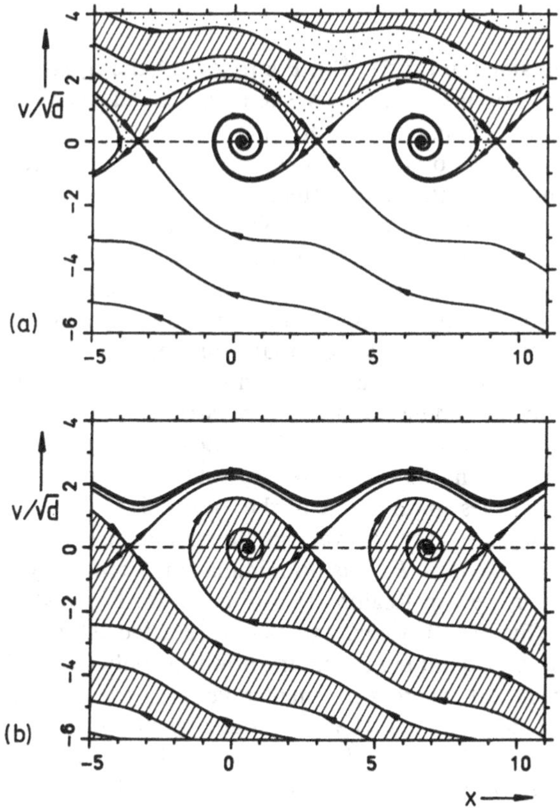

Fig. 11.25. The trajectories of the equation (11.181) or of the normalized equation (11.181a) ($v/\sqrt{d} = \mathrm{d}x/\mathrm{d}\tau$) going through the saddle points for $\gamma/\sqrt{d} = 0.25$ and for $F/d = 0.25$ **(a)** and $F/d = 0.5$ **(b)**

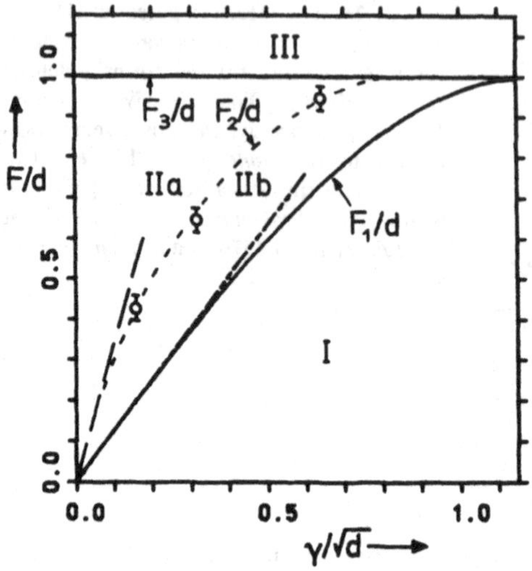

Fig. 11.26. The three critical forces F_1/d, F_2/d and F_3/d as a function of the friction constant $\gamma/\sqrt{d}$ for the cosine potential (11.103). The *straight lines* ---- and – – – are the low-friction approximations $F_1 = (4/\pi)\sqrt{d}\,\gamma$ and $F_2 = F_{02}\,\gamma$, with F_{02} given by (11.196). The *circles* indicate extrapolations for $\Theta \to 0$ of the critical force F_2/d calculated with the matrix continued-fraction method. For the meaning of Regions I, II a, II b and III, see text

By dividing (11.181) by d and using $\tau = \sqrt{d}\,t$ instead of t we obtain the normalized version of (11.181)

$$d^2x/d\tau^2 + (\gamma/\sqrt{d})\,dx/d\tau + \sin x = F/d\,. \tag{11.181 a}$$

Numerical results have been obtained in [11.51, 52]. For $|F|/d < 1$ the equation (11.181) has two kinds of singular points, i.e., points where both $\dot{x}$ and $\dot{v}$ vanish. For zero velocity and for positions belonging to the maxima of the total potential $V(x) = -d\cos x - Fx$ we have saddle points whereas for zero velocity and for positions belonging to the minima of the total potential $V(x)$ we have focal points. In Fig. 11.25 a, b the trajectories of (11.181) going through one of the saddle points are shown. In Fig. 11.25 a only a locked stationary solution exists whereas in Fig. 11.25 b a locked and a running stationary solution exist. Note that trajectories not going through singular points cannot cross each other, because the solution of the second order differential equation (11.181) is uniquely defined by $v = \dot{x}$ and x outside singular points. In Fig. 11.25 b, the trajectory going in the direction to the saddle point is a separatrix [11.26] which separates the trajectories which finally spiral in one of the focal points (shaded regions) from the trajectories which end up at the running solution (nonshaded regions). In Fig. 11.25 a all trajectories spiral in one of the focal points.

In Fig. 11.26 the "phase diagram" for the various possibilities is shown. Because the total potential $V = -d\cos x - Fx$ has no minima for $F > d$, only a running solution exists for $F > d$, i.e., in Region III in Fig. 11.26. For $F < d$, minima of the total potential (Fig. 11.2) occur, and therefore a locked and a running solution may exist. For large damping constants, however, only a locked solution exists (Region I in Fig. 11.26), whereas for smaller damping constants in Region II both solutions coexist. Region I and II are separated by a critical force F_1/d, which depends only on $\gamma/\sqrt{d}$ as seen from the normalized equation (11.181 a). The critical force F_1/d as a function of d/γ^2 was obtained in

[11.51, 52] by numerical integration of (11.181 a), see also [11.24, 27]. As seen, the region where both solutions coexist vanishes for damping constants $\gamma/\sqrt{d} \gtrsim 1.193$.

For the model potential

$$f(x) = 2d(x/\pi)^2 \quad \text{for} \quad |x| \leqq \pi\,, \quad f(x) = f(x+2\pi) \tag{11.182}$$

the critical force F_1 can be obtained analytically. Integrating the equation of motion $\ddot{x} + \gamma\dot{x} + f'(x) = F$ for an initial condition where the particle starts from rest at $x = -\pi$, the particle reaches $x = +\pi$ only if $F \geqq F_1$. The explicit result for F_1 reads

$$\frac{F_1}{d} = \frac{4}{\pi}\tanh\frac{\pi\gamma/\sqrt{d}}{2\sqrt{4/\pi^2 - (\gamma/\sqrt{d})^2}} \tag{11.183}$$

for $\gamma/\sqrt{d} \leqq 2/\pi$. The critical force F_3, i.e., the limit force above which the total potential $V(x) = f(x) - Fx$ has no minima and maxima, is given by

$$F_3/d = 4/\pi\,. \tag{11.184}$$

For $\gamma/\sqrt{d} = 2/\pi$ the critical force F_1 coincides with this critical force F_3 and no bistability region is found for $\gamma/\sqrt{d} \geqq 2/\pi$. The model potential's critical forces F_1 and F_3 show a similar dependence on $\gamma/\sqrt{d}$ to F_1 and F_3 in Fig. 11.26 for the cosine potential. Because the particles in the potential (11.182) reach higher velocities for a larger time interval than those in the cosine potential with the same energy difference $2d$, the damping force has larger influence in the model potential. Therefore the bistability region of the model potential (11.182) already vanishes for a $\gamma/\sqrt{d}$ value smaller than that for the cosine potential.

The drift velocity of the running solution may be defined by (T is the period of the stationary running solution)

$$\langle v\rangle_r = \frac{1}{T}\int_0^T v\,dt = \frac{1}{T}\int_{-\pi}^{\pi} dx = \frac{2\pi}{T} \tag{11.185}$$

and again can be obtained by numerical integration of (11.181).

Low-Friction Mobility Without Noise

For the low-friction limit the drift velocity as well as the critical force F_1 can be obtained exactly. This case will therefore be discussed in some detail. Introducing the energy $E = v^2/2 - d\cos x$ and the force to friction ratio $F_0 = F/\gamma$ as done in Sect. 11.4, (11.181) is transformed into

$$\frac{dE}{dt} = \gamma(F_0 - v)\,v\,. \tag{11.186}$$

In the first approximation with respect to γ,

$$v = v(x,E) = \sqrt{2(E+d\cos x)}\,, \qquad E = \text{const} = \bar{E} \tag{11.187}$$

can be used on the right-hand side of (11.186). The mobility, defined by (11.1, 185) reads

$$\mu_r = \frac{\langle v\rangle_r}{F} = \frac{1}{FT}\int_{-\pi}^{\pi} dx = \frac{2\pi}{FT}\,,$$

where the period T is defined by

$$T = \int_0^T dt = \int_{-\pi}^{\pi}\frac{dx}{v} = \int_{-\pi}^{\pi}\frac{dx}{\sqrt{2(\bar{E}+d\cos x)}} = 2\pi\frac{d}{d\bar{E}}\bar{v}(\bar{E})\,,$$
$$\bar{v}(\bar{E}) = \frac{1}{2\pi}\int_{-\pi}^{\pi}\sqrt{2(\bar{E}+d\cos x)}\,dx\,. \tag{11.188}$$

[The averaged velocity $\bar{v}(E)$ and its derivative are given by (11.141, 142).] Therefore

$$\gamma\mu_r = \gamma\langle v\rangle_r/F = [\bar{v}'(\bar{E})F_0]^{-1}\,. \tag{11.189}$$

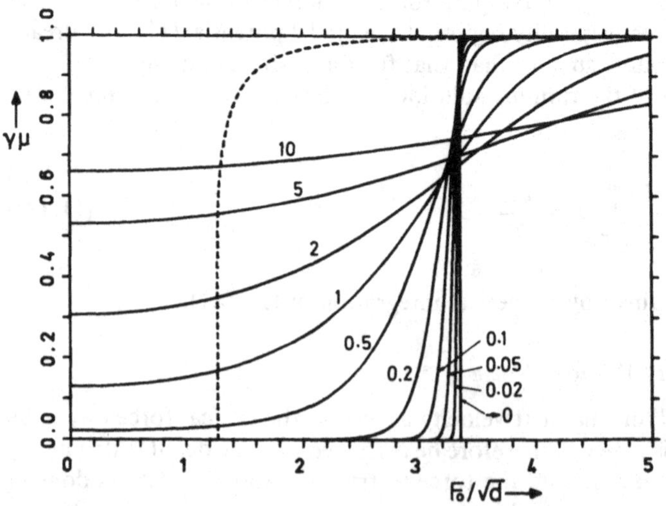

Fig. 11.27. The mobility times the damping constant of the running solution for the noiseless case (*dotted line*) and the mobility μ times the damping constant γ as a function of the external force $F_0/\sqrt{d}$ in the low-friction limit including noise for various temperatures Θ/d (*solid lines*). For $\Theta/d \geqq 0.1$ (11.136) was used and for $\Theta/d \leqq 0.1$, approximation (11.194)

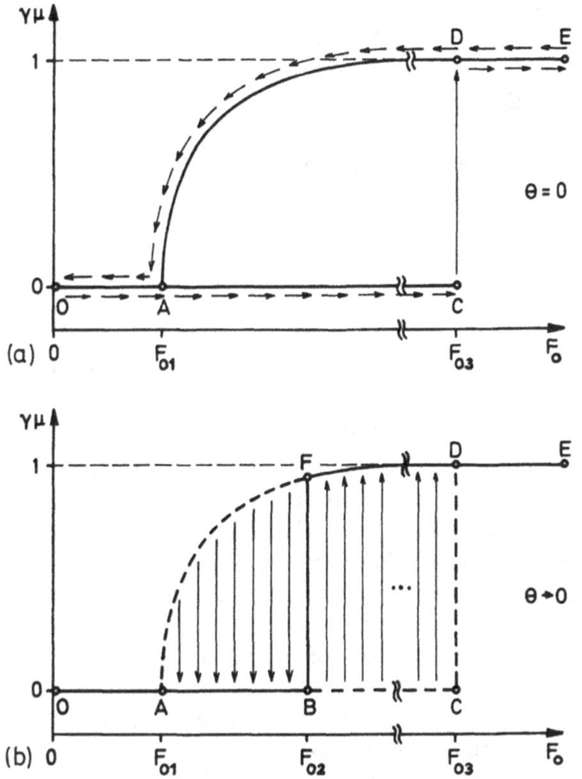

Fig. 11.28. (**a**) Qualitative plot of mobility without noise as a function of external force. The *arrows* indicate the behavior of the mobility on increasing and decreasing the external force F_0. In the region $F_{01} < F_0 < F_{03}$ bistability occurs. (**b**) Qualitative plot of mobility with noise as a function of the external force in the low-temperature limit. At the critical force F_{02} the mobility jumps from zero mobility to the mobility of the running solution without noise and vice versa. For further discussion, see text

The averaged velocity $\bar{v}(\bar{E})$ in the stationary state $\bar{E}$ is determined by balancing the energy gain due to the field, and the energy loss due to friction, over one period according to (11.186)

$$\gamma F_0 \int_0^T v\,\mathrm{d}t = \gamma F_0 \int_{-\pi}^{\pi} \mathrm{d}x = \gamma \int_0^T v^2 \mathrm{d}t = \gamma \int_{-\pi}^{\pi} v\,\mathrm{d}x\,,$$

i.e., it is determined by

$$F_0 = \frac{1}{2\pi} \int_{-\pi}^{\pi} v\,\mathrm{d}x = \bar{v}(\bar{E})\,. \tag{11.190}$$

Equation (11.190) for given F_0 determines $\bar{E}$ which may be used to calculate $\gamma\mu_r$ from (11.189). The result is shown in Fig. 11.27 by the dotted line. This line starts at $F_0 = (4/\pi)\sqrt{d}$ because (11.190) has solutions only for $F_0 \geqq \bar{v}(d) = (4/\pi)\sqrt{d}$.

A qualitative plot of this situation is given in Fig. 11.28a, where $\gamma\mu$ as a function of the external force F_0 is shown. Starting at zero force (point 0) and switching on adiabatically the force F_0, the mobility remains zero until the critical force $F_{03} = d/\gamma$ is reached, where the minimum of the effective potential disappears.

At that point C the mobility jumps to the value of the running solution (point D) and follows it for increasing force (point E). Decreasing the force adiabatically, the running solutions remain running ones, so that the mobility decreases according to the curve DA and becomes zero for $F_{01} = (4/\pi)\sqrt{d}$ (point A). It is seen that for adiabatical switching hysteresis between the forces F_{01} and F_{03} occurs. All the other points in the region ACDA may occur if an ensemble of independent particles is considered with a suitable mixture of particles in both running and locked states.

11.6.2 Solutions With Noise

For a nonvanishing Θ the Fokker-Planck equation (11.33) has to be solved. In the stationary state this leads to a unique distribution function and therefore also the mobility is a unique smooth function of the external force showing no bistability. If the strength of the noise force is lowered, the mobility versus force plots get steeper. In the limit of vanishing noise force ($\Theta \to 0$), it seems that there is a sharp jump at a critical force F_2, where the locked solution jumps to the running solution of the noiseless equation. Thus an arbitrary small noise force seems to change the double-valued behavior of the deterministic equation to single-valued. The critical force must lie in the bistability region of the deterministic equation, i.e., $F_1 \leqq F_2 \leqq F_3$. With the help of the matrix continued-fraction method (Sect. 11.5.2) we can calculate the mobility as a function of F/d. In Fig. 11.29 the mobility times the damping constant is shown as a function of F/d. It is clearly seen in this plot that for smaller Θ the curve gets steeper, and a critical force F_2/d of about 0.8 seems to exist for the parameters of the plot. Unfortunately, the matrix continued-fraction method of Sect. 11.5.2 does not

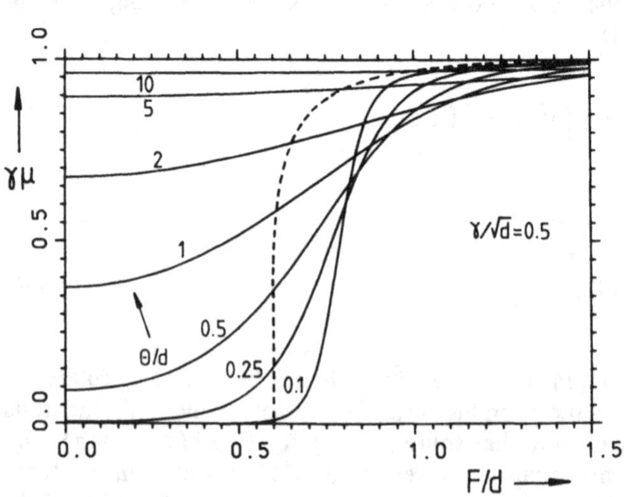

Fig. 11.29. Mobility μ times damping constant γ as a function of the external force F/d for $\gamma = \sqrt{d}/2$ and for various temperatures Θ/d (*full line*). The corresponding expression for the running solution without noise $\gamma\mu_r = \gamma\langle v\rangle_r/F$ with $\langle v\rangle_r$ given by (11.185) is shown by the *broken line*

work very well in the limit of small Θ, so that the critical force F_2 cannot be obtained very accurately. To see this steepening, it is essential to plot $\gamma\mu$ as a function of F/d. If we plot $\gamma\mu$ as a function of F/Θ as in Fig. 11.20, neither steepening nor critical force occurs. For further conclusions see the end of Sect. 11.6.3.

11.6.3 Low-Friction Mobility with Noise

If this section we shall discuss the low-friction limit in some detail. Here the distribution function can be obtained analytically [(11.112, 179, 180) and Fig. 11.24] and therefore also the limit $\Theta \to 0$ can be easily performed. As seen by comparison with (11.126) the distribution function (11.112) is only valid for small $\varepsilon \sim \sqrt{\gamma/\Theta}$, see (11.125). If we perform the limit $\Theta \to 0$, we still have to require $\varepsilon \ll 1$, i.e., γ/Θ must still be small. If we do not take into account the terms proportional to $\sqrt{\gamma}$ (11.135), the mobility times damping constant is now given by (11.136), where A_0, A_1 and A_3 are defined by (11.143, 131, 133) and where $\bar{v}(E)$ and $\bar{v}'(E)$ occurring in A_1 and A_3 can be expressed by the complete elliptic integrals according to (11.141, 142). By numerical integration of the remaining integrals, the mobility times the damping constant was obtained in [11.29, 30] as a function of $F_0/\sqrt{\Theta}$ for Θ values down to $0.03\,d$. In Fig. 11.27 $\gamma\mu$ is shown as a function of $F_0/\sqrt{d}$.

For lower temperatures Θ the exponents in the integrals A_1 and A_3 lead to high and narrow maxima, making numerical integration methods inadequate. Then an expansion of the effective potential (11.179) near the minima is more suitable. Neglecting lower-order terms like 1 and $\exp[-F_0 g/\Theta]$, the integrands in A_1 and A_3 are proportional to $\exp[-V_{\mathrm{LF}}^{+}(E)/\Theta]$. Because $\mathrm{d}g/\mathrm{d}E = 1/\bar{v}(E)$, the maximum of the distribution function (minimum of the effective potential) for $E > d$ occurs at $\bar{E}$, where $\bar{E}$ is determined by $\bar{v}(\bar{E}) = F_0$, i.e., by (11.190) for the noiseless case. Near this minimum, expansion of the effective potential up to second order

$$V_{\mathrm{LF}}^{+}(E) = E - F_0 g(E) = \bar{E} - F_0 g(\bar{E}) + \frac{1}{2} F_0 \frac{\bar{v}'(\bar{E})}{\bar{v}^2(\bar{E})}(E-\bar{E})^2 \tag{11.191}$$

and using $\bar{v}'(E) \approx \bar{v}'(\bar{E})$, $\bar{v}(E) \approx \bar{v}(\bar{E}) = F_0$ in the integrand leads to

$$\begin{aligned} A_1/\bar{v}'(\bar{E}) = F_0 A_3 &= \frac{1}{2}\exp\{[-\bar{E} + g(\bar{E})F_0]/\Theta\} \int_{-\infty}^{\infty} \exp\left(-\frac{\bar{v}'(\bar{E})}{2F_0\Theta}x^2\right)\mathrm{d}x \\ &= \sqrt{\frac{\pi F_0 \Theta}{2\bar{v}'(\bar{E})}}\exp[-V_{\mathrm{LF}}^{+}(\bar{E})/\Theta]\,. \end{aligned} \tag{11.192}$$

Similarly, using $\mathrm{I}_0(x) \sim \mathrm{e}^x/\sqrt{2\pi x}$ for $x \to \infty$, we get

$$A_0 = [\Theta/(2\sqrt{d})]\exp(d/\Theta) \tag{11.193}$$

and finally ($F_0 \geqq F_{01}$) [11.27]

$$\gamma\mu = \frac{\gamma\mu_r}{1+\sqrt{\gamma\mu_r\Theta/(2\pi d)}\,\exp\{[d+V_{LF}^{+}(\bar{E})]/\Theta\}}\,. \tag{11.194}$$

Here $\mu_r = \langle v\rangle_r/(\gamma F_0)$ is the averaged mobility of the deterministic motion with energy $\bar{E}$ (11.189).

Let us now discuss the low-temperature approximation (11.194). For $\Theta \to 0$, $\gamma\mu$ remains 0 for increasing F_0 until the sign of the exponent changes from positive to negative values. At the force $F_0 = F_{02}$ determined by

$$d+V_{LF}^{+}(\bar{E}) = d+\bar{E}-F_0 g(\bar{E}) = d+V_{LF}^{+}(-d) = 0 \tag{11.195}$$

the mobility jumps to the value of the noiseless mobility of the running solution and is equal to it for $F_0 > F_{02}$ in this limit. Obviously (11.195) is the condition that the two minima in Fig. 11.24 have the same value. The value of F_{02} and the values of $\bar{E}$ and $\gamma\mu_r$ at this critical force have been calculated numerically from (11.190, 195) [11.27]

$$F_{02} = 3.3576\sqrt{d}\,, \quad \bar{E} = 5.6591\,d\,, \quad \gamma\mu_r = \gamma\mu_{\Theta\to 0} = 0.9960\,. \tag{11.196}$$

These parameters have to be compared with the critical parameters of the noiseless equation, which read in the low-friction limit

$$\begin{aligned} &F_{01} = (4/\pi)\sqrt{d}\,, && \bar{E} = d\,, && \gamma\mu_r = 0 \\ &F_{03} = d/\gamma\,, && \bar{E} = \tfrac{1}{2}F_{03}^2\,, && \gamma\mu_r = 1\,. \end{aligned} \tag{11.197}$$

For $\Theta = 0.1\,d$ (11.194) agrees very well with the exact expression (11.136). For $\Theta < 0.1\,d$, $\gamma\mu$ in Fig. 11.27 was obtained by using (11.194).

In analogy with the discussion of a force cycle in Fig. 11.28a in the noiseless case, we make a similar consideration here in the limit $\Theta \to 0$ (Fig. 11.28b). An adiabatic increase of the force from point 0 to A and B does not change the mobility until B is reached. At B the mobility jumps to its noiseless value (point F) and, on increasing the force further, increases further to D and E on the noiseless curve. The decrease of the force F_0 leads to the same curve EDFBAO. Thus the ambiguity in the mobility of Fig. 11.28a is removed in the stationary state. If one starts with a distribution sharply peaked around the locked (running) solution of the deterministic equation with forces $F_0 > F_{02}$ ($F_0 < F_{02}$), one obtains transitions of the mobility from the lower to the upper (upper to lower) curves as indicated by the arrows for small but finite temperatures Θ. By changing the force in a nonadiabatic way, hysteresis effects should be found. To calculate the transition rate and hysteresis effects, the time-dependent equation (11.33) has to be solved (Sect. 11.9).

In Fig. 11.30 mobility is plotted as a function of temperature Θ for different values for the force F_0. Below the critical force $F_0 < F_{02}$ the mobility finally drops to zero for $\Theta \to 0$, approaching $\Theta = 0$ with horizontal tangent. The critical curve, given by (11.194) with $F_0 = F_{02}$, approaches the mobility of the running solution

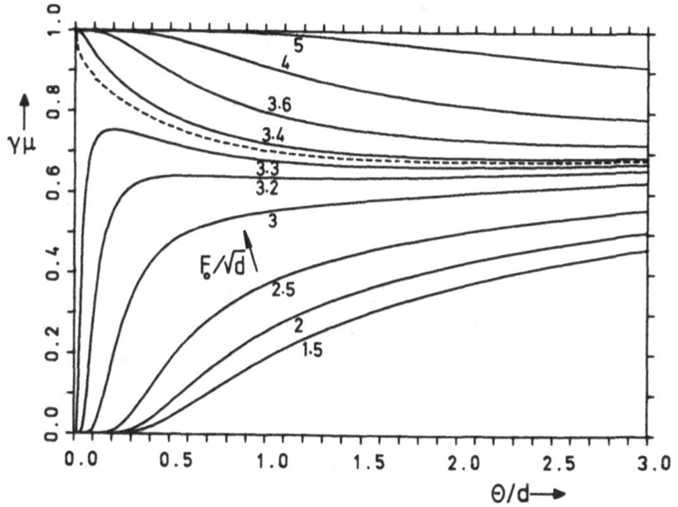

Fig. 11.30. Mobility μ times damping constant γ as a function of temperature Θ for various external forces $F_0/\sqrt{d}$. The *dotted line* indicates the critical forces $F_0 = F_{02}$. It should be noted that the curves for $F_0 \geqq F_{02}$ do not reach exactly the value $\gamma\mu = 1$ for $\Theta = 0$, but end at the corresponding values of the mobility of the running solution for the noiseless case. For the critical force F_{02} this value is 0.996 (11.196)

without noise with vertical tangent, but the curves for higher external forces approach the values of this noiseless mobility for $\Theta \to 0$ again with horizontal tangent.

Conclusion

With the inclusion of noise the solution is unique even in the limit $\Theta \to 0$ and no coexistence occurs. Region II is therefore separated into a part IIa (only running solution) and a part IIb (only locked solution). Hence, in the limit $\Theta \to 0$, the only relevant curve in the stationary state is the line F_2/d separating IIa and IIb. Whereas for small γ the critical force F_2 was calculated exactly [the slope $F_2/(\gamma\sqrt{d}) = F_{02}/\sqrt{d}$ is given by (11.196)], only approximate values have been obtained in Sect. 11.6.2 with the matrix continued-fraction method for larger damping constants.

11.7 Instationary Solutions

Though some of the instationary solutions of the Fokker-Planck equation (11.33) may be periodic in x, as always the case for the bounded stationary solution (Sect. 11.5), the instationary solution generally does not need to be periodic. For calculating expectation values of expressions periodic in x, only the periodic solutions are needed. If x is an angle variable and if we do not distinguish whether a full rotation was made or not, we also require that the instationary solution must be periodic in x. If, however, we want to calculate the diffusion of particles in an infinite periodic potential (Fig. 11.1), the nonperiodic solutions must be considered.

Because of Floquet's theorem [11.43–45] we can make the ansatz for the nonperiodic solution (k real)

$$W(x, v, t) = \int_{-1/2}^{1/2} \tilde{W}(k, x, v, t)\, e^{ikx} dk \,, \tag{11.198}$$

where $\tilde{W}$ is periodic in x with period 2π. In (11.198) we assumed that k is restricted to the first Brillouin zone. The periodic function $\tilde{W}$ can be expanded in a truncated Fourier series with respect to x and into Hermite functions with respect to v as done before,

$$\tilde{W}(k, x, v, t) = \psi_0(v)\, e^{-\varepsilon f(x)/\Theta} \sum_{n=0}^{\infty} \sum_{p=-Q}^{Q} c_n^p(k, t)\, e^{ipx}\, \psi_n(v) \,. \tag{11.199}$$

As in (10.43) we split off the factor $\exp[-\varepsilon f(x)/\Theta]$ because some expectation values for $\varepsilon = 1/2$ can then be written symmetrically. In the notation of Sect. 10.3, the complete set $\varphi^p(k, x)$, in which the expansion coefficients $c_n(x, t)$ of $W(x, v, t)$ are expanded with respect to x [compare (10.129)], is then given by

$$\varphi^p(k, x) = (2\pi)^{-1/2} e^{i(p+k)x} \,, \tag{11.200}$$

satisfying the orthonormality condition

$$\int_{-\infty}^{\infty} [\varphi^p(k, x)]^* \varphi^q(k', x)\, dx = \delta_{p,q}\, \delta(k - k') \tag{11.201}$$

and the completeness relation

$$\sum_p \int_{-1/2}^{1/2} [\varphi^p(k, x)]^* \varphi^p(k, x')\, dk = \delta(x - x') \,. \tag{11.202}$$

Using the vector notation (11.160), where c_n^p are now the expansion coefficients $c_n^p(k, t)$, we again obtain the tridiagonal form (9.10) with $Q_n^{\pm}$ and Q_n given by (11.161) and where the matrix elements of the matrices D and $\hat{D}$ read

$$\begin{aligned} D^{p,q}(k) &= \sqrt{\Theta}\,(2\pi)^{-1} \int_0^{2\pi} e^{-i(p+k)x} \left(\frac{\partial}{\partial x} - \frac{\varepsilon f'}{\Theta} \right) e^{i(q+k)x} dx \\ &= \sqrt{\Theta}\,[i(q+k)\,\delta_{pq} - \varepsilon f'_{p-q}/\Theta] \,, \end{aligned} \tag{11.203}$$

$$\begin{aligned} \hat{D}^{p,q}(k) &= \sqrt{\Theta}\,(2\pi)^{-1} \int_0^{2\pi} e^{-i(p+k)x} \left[\frac{\partial}{\partial x} + (1-\varepsilon) \frac{f'}{\Theta} - \frac{F}{\Theta} \right] e^{i(q+k)x} dx \\ &= \sqrt{\Theta}\,\{[i(q+k) - F/\Theta]\,\delta_{pq} + (1-\varepsilon) f'_{p-q}/\Theta\} \,, \end{aligned} \tag{11.204}$$

where f_r' are the expansion coefficients of $f'(x)$ as in (11.164). The transition probability $P(x, v, t\,|\,x', v', 0)$ is the solution of (11.33) with the initial condition

$$P(x, v, 0\,|\,x', v', 0) = \delta(x-x')\,\delta(v-v')\,. \tag{11.205}$$

It may be written as

$$P(x, v, t\,|\,x', v', 0) = \frac{\psi_0(v)}{\psi_0(v')}\,\frac{\exp[-\varepsilon f(x)/\Theta]}{\exp[-\varepsilon f(x')/\Theta]}\sum_{n,m}\psi_n(v)\,\psi_m(v') \times \int_{-1/2}^{1/2}\sum_{p,q}(2\pi)^{-1}\mathrm{e}^{\mathrm{i}px-\mathrm{i}qx'}\mathrm{e}^{\mathrm{i}k(x-x')}G_{n,m}^{p,q}(k,t)\,\mathrm{d}k\,, \tag{11.206}$$

where $G_{n,m}^{p,q}(k, t)$ satisfies the initial condition

$$G_{n,m}^{p,q}(k, 0) = \delta_{nm}\,\delta_{pq}\,. \tag{11.207}$$

Eigenfunction Expansion

The operator of the Fokker-Planck equation (11.33) L_{FP} and its adjoint operator L_{FP}^+

$$\begin{aligned} L_{\mathrm{FP}} &= -\frac{\partial}{\partial x}v + \frac{\partial}{\partial v}\left(\gamma v + f' - F + \gamma\Theta\frac{\partial}{\partial v}\right), \\ L_{\mathrm{FP}}^+ &= v\frac{\partial}{\partial x} + \left(-\gamma v - f' + F + \gamma\Theta\frac{\partial}{\partial v}\right)\frac{\partial}{\partial v} \end{aligned} \tag{11.208}$$

are invariant by the replacement $x \to x + 2\pi n$ for periodic potentials. Therefore the eigenfunctions $\varphi_n(k,x,v)$ and $\varphi_n^+(k,x,v)$ of these operators, i.e.,

$$\begin{aligned} L_{\mathrm{FP}}\varphi_n(k,x,v) &= -\lambda_n(k)\,\varphi_n(k,x,v) \\ L_{\mathrm{FP}}^+\varphi_n^+(k,x,v) &= -\lambda_n(k)\,\varphi_n^+(k,x,v)\,, \end{aligned} \tag{11.209}$$

can also be written in form of Bloch waves, i.e.,

$$\begin{aligned} \varphi_n(k,x,v) &= u_n(k,x,v)\,\mathrm{e}^{\mathrm{i}kx} \\ \varphi_n^+(k,x,v) &= u_n^+(k,x,v)\,\mathrm{e}^{-\mathrm{i}kx}\,, \end{aligned} \tag{11.210}$$

where u_n and u_n^+ are periodic functions in x.

Inserting (11.210) into (11.209) leads to

$$\begin{aligned} (L_{\mathrm{FP}} - \mathrm{i}kv)\,u_n &= -\lambda_n u_n \\ (L_{\mathrm{FP}}^+ - \mathrm{i}kv)\,u_n^+ &= -\lambda_n u_n^+\,. \end{aligned} \tag{11.209a}$$

The orthonormality relation reads

$$\int_0^{2\pi}\int_{-\infty}^{\infty} u_n^+(k,x,v)\,u_m(k,x,v)\,\mathrm{d}x\,\mathrm{d}v = \delta_{nm}\,. \tag{11.211}$$

With the help of the completeness relation

$$\int_{-1/2}^{1/2}\sum_n u_n^+(k,x',v')\,u_n(k,x,v)\,\mathrm{e}^{\mathrm{i}k(x-x')}\mathrm{d}k = \delta(x-x')\,\delta(v-v')\,, \tag{11.212}$$

which is assumed to be valid, it is simple to express the transition probability by the eigenfunctions

$$\begin{aligned} &P(x,v,t\,|\,x',v',0)\\ &\quad= \int_{-1/2}^{1/2}\sum_n u_n^+(k,x',v')\,u_n(k,x,v)\,\mathrm{e}^{\mathrm{i}k(x-x')}\,\mathrm{e}^{-\lambda_n(k)t}\mathrm{d}k\,. \end{aligned} \tag{11.213}$$

The eigenvalues are arranged in bands, as for the Smoluchowski equation. The eigenvalues and eigenfunctions for $k=0$ are determined in Sect. 11.9.

Calculation of the Generalized Dynamic Structure Factor

We now want to find an expression for the dynamic structure factor defined by

$$\tilde{S}(k_1,k_2,\omega) = (2\pi)^{-1}\int_0^{\infty}\mathrm{e}^{-\mathrm{i}\omega t}S(k_1,k_2,t)\,\mathrm{d}t\,, \tag{11.214}$$

$$S(k_1,k_2,t) = \langle \mathrm{e}^{-\mathrm{i}k_1x(t)+\mathrm{i}k_2x(0)}\rangle\,. \tag{11.215}$$

This structure factor plays an essential role in light and neutron scattering experiments [11.15, 53, 54].

In the following we restrict ourselves to the case where the external force F vanishes. In this case the stationary distribution is the Boltzmann distribution

$$W_{\mathrm{st}}(x,v) = N\,\psi_0^2(v)\,\mathrm{e}^{-f(x)/\Theta}\,,\qquad N^{-1} = \int_0^{2\pi}\mathrm{e}^{-f(x)/\Theta}\mathrm{d}x\,. \tag{11.216}$$

The expectation value (11.215) can then be calculated by

$$\begin{aligned} S(k_1,k_2,t) = \lim_{M\to\infty}(2M)^{-1}&\int_{-\infty}^{\infty}\mathrm{d}x\int_{-\infty}^{\infty}\mathrm{d}v\int_{-2\pi M}^{2\pi M}\mathrm{d}x'\int_{-\infty}^{\infty}\mathrm{d}v'\,\mathrm{e}^{-\mathrm{i}k_1x}\mathrm{e}^{\mathrm{i}k_2x'}\\ &\times P(x,v,t\,|\,x',v',0)\,W_{\mathrm{st}}(x',v')\,. \end{aligned} \tag{11.217}$$

Because the stationary solution (11.216) was normalized in the interval $[0, 2\pi]$, for the x' integration we have to divide by the number of these intervals. We now insert (11.206, 216) into (11.217). Because for periodic functions $g(x) = g(x+2\pi)$ we have $(-1/2 < k \leq 1/2)$

$$\int_{-\infty}^{\infty} e^{ikx} g(x)\,dx = \sum_{\nu=-\infty}^{\infty} e^{ik2\pi\nu} \int_0^{2\pi} e^{ikx} g(x)\,dx = \delta(k) \int_0^{2\pi} e^{ikx} g(x)\,dx$$

$$= \delta(k) \int_0^{2\pi} g(x)\,dx$$

and

$$\lim_{M\to\infty} \frac{1}{2M} \int_{-2\pi M}^{2\pi M} e^{ikx} g(x)\,dx \begin{cases} = \int_0^{2\pi} g(x)\,dx & \\ & \text{for} \\ = 0 & \end{cases} \begin{matrix} k = 0 \\ \\ k \neq 0 . \end{matrix}$$

Thus (11.217) can be different from zero only if k_1 and k_2 differ by an integer. (In the language of a solid-state physicist, the difference between k_1 and k_2 must be multiples of the reciprocal lattice vector, here equal to 1.) We therefore write

$$k_1 = l_1 + k , \quad k_2 = l_2 + k ; \quad -1/2 < k \leq 1/2 , \tag{11.218}$$

where k is restricted to the first Brillouin zone, and where l_1, l_2 are integers.

By using the orthonormality condition (10.41) for $\psi_n(v)$, we finally obtain by inserting (11.206, 216) into (11.217)

$$S(l_1+k, l_2+k, t) = 2\pi N \sum_{p,q} G_{0,0}^{p,q}(k,t) M_{p-l_1} \hat{M}^*_{q-l_2} , \tag{11.219}$$

where M_r and $\hat{M}_r$ are defined by

$$\begin{aligned} M_r &= (2\pi)^{-1} \int_0^{2\pi} e^{-\varepsilon f(x)/\Theta + irx} dx , \\ \hat{M}_r &= (2\pi)^{-1} \int_0^{2\pi} e^{-(1-\varepsilon) f(x)/\Theta + irx} dx . \end{aligned} \tag{11.220}$$

For $\varepsilon = 1/2$, M_r and $\hat{M}_r$ are equal. The dynamic structure factor (11.214) can be expressed by

$$\tilde{S}(l_1+k, l_2+k, \omega) = N \sum_{p,q} \tilde{G}_{0,0}^{p,q}(k, i\omega) M_{p-l_1} \hat{M}^*_{q-l_2} , \tag{11.221}$$

where $\tilde{G}_{0,0}^{p,q}(k,s)$ are the Laplace transforms of the matrix elements $G_{0,0}^{p,q}(k,t)$, i.e.,

$$\tilde{G}_{0,0}^{p,q}(k,s) = \int_0^{\infty} G_{0,0}^{p,q}(k,t)\, e^{-st} dt . \tag{11.222}$$

These matrix elements form a matrix $\tilde{\boldsymbol{G}}_{0,0}(k,s)$ which can be obtained by the matrix continued fraction [see (10.139, 140, 142) for $m = 0$], i.e., by

$$(\tilde{G}^{p,q}_{0,0}(k,\mathrm{i}\,\omega)) \equiv \tilde{\boldsymbol{G}}_{0,0}(k,\mathrm{i}\,\omega) = [\mathrm{i}\,\omega\boldsymbol{I} - \tilde{\boldsymbol{K}}_0(\mathrm{i}\,\omega)]^{-1}, \tag{11.223}$$

$$\tilde{\boldsymbol{K}}_0(\mathrm{i}\,\omega) = \boldsymbol{D}[(\mathrm{i}\,\omega+\gamma)\boldsymbol{I} - 2\boldsymbol{D}[(\mathrm{i}\,\omega+2\gamma)\boldsymbol{I} \\ -3\boldsymbol{D}[(\mathrm{i}\,\omega+3\gamma)\boldsymbol{I} - \ldots]^{-1}\hat{\boldsymbol{D}}]^{-1}\hat{\boldsymbol{D}}]^{-1}\hat{\boldsymbol{D}}\,. \tag{11.224}$$

The matrix elements of the matrices $\boldsymbol{D}$ and $\hat{\boldsymbol{D}}$ are given by (11.203, 204) with $F=0$. Thus the problem of calculating the dynamic structure factor is essentially to evaluate the matrix continued fraction (11.224). The dynamic structure factor (11.221) was obtained by *Dieterich* et al. [11.53] by integrating the truncated coupled system of differential equations for the expansion coefficients c^p_n, as done for $k=0$ in [11.13]. The structure factor for $k=0$ is needed in Sect. 11.8 for the calculation of various susceptibilities. As will be shown, the matrix continued fraction (11.224) can even be evaluated for such small friction constants so that the connection to the zero friction limit can be made.

11.7.1 Diffusion Constant

We now want to calculate the diffusion of particles moving in the infinite periodic potential of Fig. 11.1 without an additional external force. (With an additional force an additional drift due to the external force would be present, see the remark at the end of this subsection.) If initially the particles are near $x=0$ they will diffuse to adjacent wells. For very large times they are distributed over many wells, as shown in Fig. 11.31. As discussed in the beginning of this chapter, the particles hop from one well to an adjacent well for large damping constants, whereas for smaller γ they move over quite a number of wells before they get trapped in one and are then excited again to move to other wells. The diffusion constant defined by either of the two relations

$$D = \frac{1}{2}\lim_{t\to\infty}\frac{\mathrm{d}}{\mathrm{d}t}\langle[x(t)-x(0)]^2\rangle\,, \tag{11.225a}$$

$$D = \frac{1}{2}\lim_{t\to\infty}\frac{1}{t}\langle[x(t)-x(0)]^2\rangle \tag{11.225b}$$

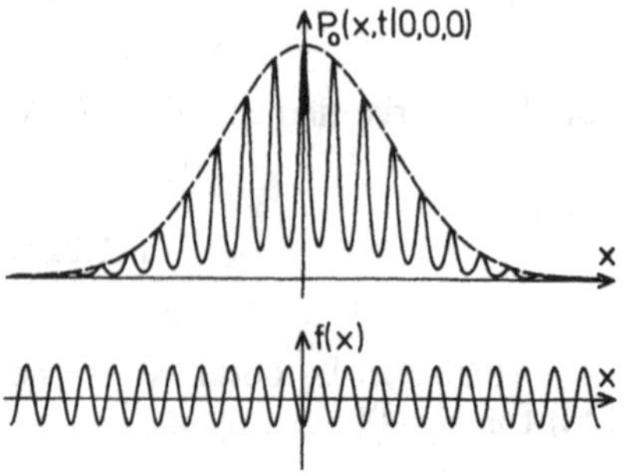

Fig. 11.31. The transition probability (11.236b) for the cosine potential (11.46) with $d/\Theta=1$ for $Dt=4$

describes the diffusion in the infinite periodic potential for large times (without any drift velocity). It is easy to see that (11.225 b) follows from (11.225 a). The reverse is also true if one assumes that $\langle [x(t)-x(0)]^2\rangle/t$ can be expanded in powers of $1/t$ (apply l'Hôpital's rule).

The diffusion constant is connected to the mobility. This was proved for the Smoluchowski equation (i.e. for the high-friction limit of the Kramers equation) in [11.46–48]. The simplest proof for Kramers equation seems to be the following: Starting with (11.225 a) and differentiating inside the expectation value, we have

$$\begin{aligned} D &= \lim_{t\to\infty} \langle \dot{x}(t)[x(t)-x(0)]\rangle = \lim_{t\to\infty} \langle v(t) \int_0^t v(t')\,dt'\rangle \\ &= \lim_{t\to\infty} \int_0^t \langle v(t)v(t')\rangle\,dt' = \lim_{t\to\infty} \int_0^t \langle v(t)v(t-\tau)\rangle\,d\tau\,. \end{aligned}$$

The velocity correlation function $\langle v(t)v(t-\tau)\rangle$ is different from zero only for finite τ. For very large t the correlation function becomes independent of t (stationary process). Therefore, we may write (replace t by $t_0+\tau$)

$$\langle v(t)v(t-\tau)\rangle = \langle v(t_0+\tau)v(t_0)\rangle\,.$$

Because the time integral of the stationary velocity correlation function is connected to the dc mobility in linear response [compare (7.11 b) for $t\to\infty$ and (7.23)], we thus have in normalized and unnormalized units (indicated by an index un)

$$D = \Theta \int_0^\infty R_{v,L}(t)\,dt = \Theta\mu = \gamma\mu\cdot\Theta/\gamma\,, \tag{11.226a}$$

$$D_{\text{un}} = kT\mu_{\text{un}} = (\gamma_{\text{un}}\mu_{\text{un}})kT/(m\gamma_{\text{un}})\,. \tag{11.226b}$$

The diffusion constant in a periodic potential is thus given by the diffusion constant for free Brownian motion (3.19) multiplied by the factor $\gamma\mu = \gamma_{\text{un}}\mu_{\text{un}}$. This factor was already obtained in Sect. 11.5 and plotted in Fig. 11.21.

11.7.2 Transition Probability for Large Times

The eigenfunction expansion (11.213) is very suitable to determine the transition probability for large times t. For $k=0$ and $n=0$ the eigenvalue $\lambda_n(k)$ is zero, for $k\neq 0$ or $n\geqq 1$ or both the eigenvalue will be larger than zero. Assuming that eigenvalues $\lambda_n(0)$ with $n>0$ are separated from the zero eigenvalue $\lambda_0(0)=0$ by a finite value, only the term with $n=0$ of expansion (11.213) will survive for large times:

$$P(x,v,t\,|\,x',v',0) \approx \int_{-1/2}^{1/2} u_0^+(k,x',v')\,u_0(k,x,v)\,e^{ik(x-x')}\,e^{-\lambda_0(k)t}\,dk\,. \tag{11.227}$$

For $k=0$ and $F=0$ we have the stationary solution

$$\lambda_0(0)=0\,,\qquad u_0(0,x,v)=W_{\rm st}\,,\qquad u_0^+(0,x,v)=1\,,$$

with $W_{\rm st}$ given by (11.216). For large times we need only the integrand for small k. We therefore expand the eigenvalue $\lambda_0(k)$ and eigenfunctions u_0 and u_0^+ into a Taylor series

$$\begin{aligned}\lambda_0(k) &= 0+\lambda_0^{(1)}k+\lambda_0^{(2)}k^2+\dots\\ u_0(k,x,v) &= W_{\rm st}+u_0^{(1)}k+u_0^{(2)}k^2+\dots\\ u_0^+(k,x,v) &= 1+u_0^{+(1)}k+u_0^{+(2)}k^2+\dots\,.\end{aligned}\tag{11.228}$$

Inserting (11.228) into (11.209a), we get by comparing terms proportional to k

$$L_{\rm FP}u_0^{(1)}=(\mathrm{i}v-\lambda_0^{(1)})\,W_{\rm st}\,,\tag{11.229a}$$

and by comparing terms proportional to k^2

$$L_{\rm FP}u_0^{(2)}=(\mathrm{i}v-\lambda_0^{(1)})\,u_0^{(1)}-\lambda_0^{(2)}\,W_{\rm st}\,.\tag{11.229b}$$

By integrating (11.229a) over the interval $0\leqq x<2\pi$, $-\infty<v<\infty$, the left-hand side vanishes, because of the periodic boundary condition for $u_0^{(1)}$. Since $W_{\rm st}$ is symmetric in v we immediately conclude that $\lambda_0^{(1)}$ must vanish, i.e.,

$$\lambda_0^{(1)}=0\,,\qquad L_{\rm FP}u_0^{(1)}=\mathrm{i}v\,W_{\rm st}\,.\tag{11.230}$$

Similarly, by integrating (11.229b) over the same interval, the left-hand side again vanishes and we get

$$\lambda_0^{(2)}=\mathrm{i}\int_0^{2\pi}\int_{-\infty}^{\infty}v\,u_0^{(1)}\,\mathrm{d}x\,\mathrm{d}v\,.\tag{11.230a}$$

From (11.230, 230a) $\lambda_0^{(2)}$ and $u_0^{(1)}$ can be calculated.

Connection to the Stationary Distribution Function in Linear Response

With an additional force F the stationary solution $W(F,x,v)$ has to satisfy

$$(L_{\rm FP}-F\partial/\partial v)\,W=0\,,$$

where $L_{\rm FP}$ is given by (11.208) with $F=0$. For small F we make a Taylor series expansion

$$W(F)=W_{\rm st}+Fw+\dots$$

and we obtain in linear response for w

$$L_{\rm FP} w = \frac{\partial}{\partial v} W_{\rm st} = -\frac{v}{\Theta} W_{\rm st} . \tag{11.231}$$

By comparing (11.231) with the last equation in (11.230), we get the relation

$$u_0^{(1)} = -\mathrm{i}\Theta w . \tag{11.232}$$

The mobility is defined by

$$\mu = \lim_{F\to 0} \frac{\langle v \rangle}{F} = \lim_{F\to 0} \frac{1}{F} \int_0^{2\pi} \int_{-\infty}^{\infty} v\, W(F, x, v)\, \mathrm{d}x\, \mathrm{d}v = \int_0^{2\pi} \int_{-\infty}^{\infty} v\, w\, \mathrm{d}x\, \mathrm{d}v . \tag{11.233}$$

By comparing this expression with (11.230a) and by using (11.232, 226) we thus conclude that the second derivative of $\lambda_0(k)$ at $k = 0$ can be expressed by

$$\lambda_0^{(2)} = \frac{1}{2} \left. \frac{\mathrm{d}^2 \lambda_0(k)}{\mathrm{d}k^2} \right|_{k=0} = \Theta\mu = D . \tag{11.234}$$

This relation was derived for the Smoluchowski equation (in the three-dimensional case) by *Festa* and *d'Agliano* [11.46].

If we insert (11.228) into (11.227) and extend the integration to $\pm\infty$, we obtain ($\lambda_0^{(1)} = 0$, $\lambda_0^{(2)} = D$)

$$\begin{aligned} &P(x, v, t | x', v', 0) \\ &= \int_{-\infty}^{\infty} \{W_{\rm st}(x, v) + k[u_0^{(1)}(x, v) + u_0^{+(1)}(x', v')\, W_{\rm st}(x, v)] + \ldots\} \mathrm{e}^{\mathrm{i}k(x-x') - k^2 D t} \mathrm{d}k \\ &= \{W_{\rm st}(x, v) + \mathrm{i}[u_0^{(1)}(x, v) + u_0^{+(1)}(x', v')\, W_{\rm st}(x, v)](x - x')/(2Dt) + \ldots\} \\ &\quad \times \exp[-(x - x')^2/(4Dt)]/\sqrt{\pi D t} . \end{aligned} \tag{11.235}$$

Let us now consider symmetric potentials $f(x) = f(-x)$. It then follows from $L_{\rm FP}(x, v) = L_{\rm FP}(-x, -v)$ and from (11.229a) with $\lambda_0^{(1)} = 0$ that $u_0^{(1)}(x, v)$ must be an antisymmetric function $u_0^{(1)}(x, v) = -u_0^{(1)}(-x, -v)$. Similarly, one may prove that also $u_0^{+(1)}(x, v) = -u_0^{+(1)}(-x, -v)$ is an antisymmetric function. For $x' = v' = 0$ [$u_0^{+(1)}(0, 0) = 0$] (11.235) then reduces to

$$\begin{aligned} P(x, v, t | 0, 0, 0) &= [W_{\rm st}(x, v) + \Theta w(x, v) x/(2Dt) + \ldots] \\ &\quad \times \exp[-x^2/(4Dt)]/\sqrt{\pi D t} . \end{aligned} \tag{11.235a}$$

Here we expressed $u_0^{(1)}$ by w (11.232).

Thus this transition probability is completely determined by the stationary distribution $W_{\rm st}$, the stationary distribution in linear response w and the diffusion constant $D = \Theta\mu$. If we integrate (11.235a) over v we obtain the transition probability for the position coordinate

$$P(x,t|0,0,0) = \int_{-\infty}^{\infty} P(x,v,t|0,0,0)\,dv = P_0(x,t|0,0,0) + O(t^{-3/2}), \quad (11.236\text{a})$$

$$P_0(x,t|0,0,0) = N e^{-f(x)/\Theta} \exp[-x^2/(4Dt)]/\sqrt{\pi D t}\,. \quad (11.236\text{b})$$

For large t, P_0 is the leading term of this transition probability. As seen from Fig. 11.31, it consists of a smooth envelope and varies in the periodicity interval according to the stationary distribution $\sim \exp[-f(x)/\Theta]$. If we integrate (11.236b) over the periodicity interval $(\bar{x}-\pi, \bar{x}+\pi)$ we obtain after replacing the variable x by $\bar{x}$ in the slowly varying envelope

$$\bar{P}(\bar{x},t) = \exp[-\bar{x}^2/(4Dt)]/\sqrt{\pi D t}\,. \quad (11.236\text{c})$$

If we multiply (11.235a) by v and integrate the resulting expression over v and the periodicity interval $(\bar{x}-\pi, \bar{x}+\pi)$, we obtain the leading term for large t of the probability current:

$$\begin{aligned}\bar{S}(\bar{x},t) &= \int_{\bar{x}-\pi}^{\bar{x}+\pi} \int_{-\infty}^{\infty} v P(x,v,t|0,0,0)\,dx\,dv \\ &= \bar{x} \exp[-\bar{x}^2/(4Dt)]/\sqrt{4\pi D t^3}\,. \end{aligned} \quad (11.237)$$

In deriving this relation we used (11.233, 234) and replaced x by $\bar{x}$ in the slowly varying envelope $x\exp[-x^2/(4Dt)]$. As may be checked by insertion, (11.236c, 237) are connected by the continuity equation

$$\partial \bar{P}/\partial t + \partial \bar{S}/\partial \bar{x} = 0\,.$$

By inserting the leading term of (11.235) into (11.217) we obtain an expression for the dynamic structure factor (11.214) for small $k_1 = k_2 = k$ and for small ω, i.e.,

$$\tilde{S}(k,k,\omega) = \frac{1}{2\pi}\int_0^{\infty} e^{-i\omega t} e^{-Dk^2 t}\,dt = \frac{1}{2\pi}\,\frac{1}{i\omega + Dk^2}\,. \quad (11.238)$$

Finite External Force

For a finite external force we may define a diffusion constant by (starting at $t=0$ with the sharp value $x(0)$)

$$D = \frac{1}{2}\lim_{t\to\infty}\frac{d}{dt}\langle [x(t) - \langle x(t)\rangle]^2\rangle\,. \quad (11.239)$$

For large t the expansion (11.228) can still be made, where W_{st} is now the stationary solution with an external force. The first coefficient of the eigenvalue expansion can then be expressed by the mean drift velocity

$$\lambda_0^{(1)} = i\langle v\rangle$$

leading to a $\langle v \rangle t$ shift of the x-dependent envelope of the distribution. Because $\partial W_{st}/\partial v$ is now no longer equal to $-v W_{st}/\Theta$ (11.231), the diffusion constant is no longer connected to the change of the stationary drift velocity due to an additional small force ΔF, as it is the case for $F = 0$.

11.8 Susceptibilities

As discussed in Chap. 7, the linear response function and its Fourier transform (i.e., the susceptibility) describes the response of the system to a small time-dependent external force. In this section we mainly consider two cases.

In the first case we investigate the linear response of the velocity of particles moving in a periodic potential to an external force. After making a Fourier transform, this response is described by the susceptibility $\chi_v(\omega)$ in (11.2a). If applied to a superionic conductor, the frequency-dependent conductivity $\sigma(\omega)$ is proportional to this susceptibility

$$\sigma(\omega) = (e^2 n/m)\,\chi_v(\omega)\,, \tag{11.240}$$

where e is the elementary charge, n the density of particles and m their mass. (The force per mass is given by eE/m, where E is the electric field.) The real part of $\chi_v(\omega)$ describes the absorption.

In the second case we deal with the rotation of dipoles in a constant external field, Sect. 11.1.4. Here, one is interested in the polarization, i.e., in the averaged dipole moment times the density of dipole moments. If we apply a small additional time-dependent external field either perpendicular or parallel to the external field (Fig. 11.6), in the stationary state there is only an averaged dipole moment $p_\perp(t)$ perpendicular to or an averaged dipole moment $p_\parallel(t)$ parallel to the external field given by

$$\begin{aligned} p_\perp(t) &= \mu_0 \langle \sin\varphi(t) \rangle\,, \\ p_\parallel(t) &= \mu_0 [\langle \cos\varphi(t) \rangle - \langle \cos\varphi \rangle_{st}]\,. \end{aligned} \tag{11.241}$$

Here, $\langle \cos\varphi \rangle_{st}$ is the averaged value without the additional time-dependent external field. The response of the dipoles to the time-dependent external field is thus described by the susceptibility $\chi_{\substack{\perp \\ \parallel}}(\omega)$ (11.19). Because the energy dissipation is given by the time average of $\dot{P}(t)F(t)$, the absorption is now proportional to the imaginary part of $\omega\chi_{\substack{\perp \\ \parallel}}(\omega)$. Though the equation of motion, see (11.17), is somewhat different to the first case, both can be treated in the same manner. Furthermore we show that $\chi_\perp(\omega)$ can be expressed by $\chi_v(\omega)$.

Response Functions

We restrict ourselves to a cosine potential in case I. Applying the normalization (11.29), the Langevin equation reads

Case I

$$\ddot{x}+\gamma\dot{x}+d\sin x = F(t)+\Gamma(t)\,, \tag{11.242}$$

$$\langle \Gamma(t)\Gamma(t')\rangle = 2\gamma\Theta\delta(t-t')\,. \tag{11.243}$$

If we divide (11.17) by the angular momentum of inertia, we have with $\varphi = x$ for

Case II

$$\ddot{x}+\gamma\dot{x}+d\sin x = F_{\perp}(t)\cos x - F_{\parallel}(t)\sin x + \Gamma(t)\,, \tag{11.244}$$

where the correlation function of $\Gamma(t)$ is the same as in Case I. The constants d and Θ are given by

$$d = \mu_0 F_0/I_0\,, \qquad \Theta = kT/I_0 \tag{11.245}$$

and $F_{\perp, \parallel}$ are normalized fields

$$F_{\perp, \parallel}(t) = \mu_0 F^{(\mathrm{un})}_{\perp, \parallel}(t)/I_0\,, \tag{11.246}$$

where the index 'un' indicates the unnormalized fields in (11.17).

The Fokker-Planck equation corresponding to (11.242, 244) with (11.243) reads

$$\partial W/\partial t = [L_{\mathrm{FP}} + L_{\mathrm{ext}}(t)]\,W\,. \tag{11.247}$$

Here, the time-independent Kramers operator L_{FP} is given by

$$L_{\mathrm{FP}} = -v\frac{\partial}{\partial x} + d\sin x\frac{\partial}{\partial v} + \gamma\frac{\partial}{\partial v}\left(v + \Theta\frac{\partial}{\partial v}\right), \tag{11.248}$$

and the time-dependent operator $L_{\mathrm{ext}}(t)$ has the form

Case I

$$L_{\mathrm{ext}}(t) = -F(t)\,\partial/\partial v\,, \tag{11.249}$$

Case II

$$L_{\mathrm{ext}}(t) = [-F_{\perp}(t)\cos x + F_{\parallel}(t)\sin x]\,\partial/\partial v\,. \tag{11.250}$$

The stationary solution of (11.247) with $L_{\mathrm{ext}}(t) = 0$ is the Boltzmann distribution (11.216) with $f(x) = -d\cos x$.

The general expression for the linear response function was derived in Chap. 7 (7.9, 10, 13), where it was further shown that the velocity response function $R_{v,L}(t)$ in Case I can be expressed by the velocity autocorrelation function $K_{vv}(t)$ (7.23). For the normalization (11.29) we have

$$R_{v,L}(t) = K_{vv}(t)/\Theta \,. \tag{11.251}$$

By using (7.10, 13) and

$$\begin{aligned} L_{\mathrm{FP}} \cos x \, W_{\mathrm{st}} &= -\sin x \frac{\partial}{\partial v} W_{\mathrm{st}} \Theta \,, \\ L_{\mathrm{FP}} \sin x \, W_{\mathrm{st}} &= \cos x \frac{\partial}{\partial v} W_{\mathrm{st}} \Theta \,, \end{aligned} \tag{11.252}$$

it is seen that in Case II the response function $R_{\perp}(t)$ ($R_{\parallel}(t)$) of the normalized dipole moment $p_{\substack{\perp \\ \parallel}}(t) = p^{(\mathrm{un})}(t)/\mu_0$ to the normalized fields (11.246) can be expressed by the time derivative of the sine (cosine) autocorrelation function

$$R_{\perp}(t) = -\frac{\mathrm{d}}{\mathrm{d}t} K_{ss}(t)/\Theta \,, \quad K_{ss}(t) = \langle \sin x(t) \sin x(0) \rangle \,, \tag{11.253}$$

$$R_{\parallel}(t) = -\frac{\mathrm{d}}{\mathrm{d}t} K_{cc}(t)/\Theta \,, \quad K_{cc}(t) = \langle \cos x(t) \cos x(0) \rangle \,. \tag{11.254}$$

It follows from (7.24) and (11.251, 253) that the response function $R_{\perp}(t)$ in Case II and $R_{v,L}(t)$ in Case I are connected by

$$d^2 R_{\perp}(t) = \frac{\mathrm{d}}{\mathrm{d}t} \left(\frac{\mathrm{d}}{\mathrm{d}t} + \gamma \right)^2 R_{v,L}(t) \,. \tag{11.255}$$

Susceptibilities

As shown in Sect. 7.3 the susceptibilities are the Fourier transforms of the response functions. From (11.251, 253, 254) we thus obtain for the normalized susceptibilities ($\chi_{\substack{\perp \\ \parallel}} = \mu_0^2 \chi_{\substack{\perp \\ \parallel}}^{(\mathrm{un})}$)

$$\begin{aligned} \Theta \chi_v(\omega) &= \tilde{K}_{vv}(\omega) \,, \\ \Theta \chi_{\perp}(\omega) &= \langle \sin^2 x \rangle_{\mathrm{st}} - \mathrm{i} \omega \tilde{K}_{ss}(\omega) \,, \\ \Theta \chi_{\parallel}(\omega) &= \langle \cos^2 x \rangle_{\mathrm{st}} - \mathrm{i} \omega \tilde{K}_{cc}(\omega) \,, \end{aligned} \tag{11.256}$$

where $\tilde{K}_{vv}$, $\tilde{K}_{ss}$, $\tilde{K}_{cc}$ are the half-sided Fourier transforms of the corresponding correlation functions. The response to time-independent fields is described by the static susceptibilities $\chi_v(0)$, $\chi_{\perp}(0)$ and $\chi_{\parallel}(0)$. The static susceptibilities $\chi_{\perp}(0)$ and $\chi_{\parallel}(0)$ can be expressed by the modified Bessel functions I_n according to

$$\Theta\chi_{\perp}(0) = K_{ss}(0) - K_{ss}(\infty) = \langle \sin^2 x\rangle_{st} = \frac{\Theta \mathrm{I}_1(d/\Theta)}{d\mathrm{I}_0(d/\Theta)}, \tag{11.257}$$

$$\begin{aligned}\Theta\chi_{\parallel}(0) &= K_{cc}(0) - K_{cc}(\infty) = \langle \cos^2 x\rangle_{st} - (\langle \cos x\rangle_{st})^2 \\ &= 1 - \frac{\Theta \mathrm{I}_1(d/\Theta)}{d\mathrm{I}_0(d/\Theta)} - \left(\frac{\mathrm{I}_1(d/\Theta)}{\mathrm{I}_0(d/\Theta)}\right)^2 .\end{aligned} \tag{11.258}$$

Here we have used

$$K_{ss}(\infty) = (\langle \sin \rangle_{st})^2 = 0\,, \quad K_{cc}(\infty) = (\langle \cos \rangle_{st})^2\,. \tag{11.259}$$

The static susceptibility $\chi_v(0)$ is identical to the mobility μ in linear response (Sect. 11.5). Because

$$R_{v,L}(0) = 1\,, \quad \dot{R}_{v,L}(0) = -\gamma\,, \quad \left(\frac{\mathrm{d}}{\mathrm{d}t} + \gamma\right)^2 R_{v,L}\bigg|_{t=0} = -d^2\langle \sin^2 x\rangle_{st}/\Theta\,,$$

the connection (11.255) between the response functions $R_{v,L}(t)$ and $R_{\perp}(t)$ leads to the following connection between the susceptibilities $\chi_{\perp}(\omega)$ and $\chi_v(\omega)$:

$$d^2(\chi_{\perp}(\omega) - \langle \sin^2 x\rangle_{st}/\Theta) = \mathrm{i}\omega(\mathrm{i}\omega + \gamma)^2\chi_v(\omega) - \mathrm{i}\omega(\mathrm{i}\omega + \gamma)\,. \tag{11.260}$$

Matrix Continued-Fraction Expansion

The half-sided Fourier transform of the correlation functions and therefore also the susceptibilities can be calculated by matrix continued fractions. Because $\chi_v(\omega)$ can be expressed by $\chi_{\perp}(\omega)$, we need to calculate only $\tilde{K}_{ss}(\omega)$ and $\tilde{K}_{cc}(\omega)$. These correlation functions can be reduced to special values of the dynamic structure factor (11.214)

$$\begin{aligned}\tilde{K}_{\substack{ss\\cc}}(\omega) &= \int_0^\infty e^{-\mathrm{i}\omega t}\begin{bmatrix}\langle \sin x(t) \sin x(0)\rangle \\ \langle \cos x(t) \cos x(0)\rangle\end{bmatrix}\mathrm{d}t \\ &= (\pi/2)[\tilde{S}(1,1,\omega) + \tilde{S}(-1,-1,\omega) \mp \tilde{S}(1,-1,\omega) \mp \tilde{S}(-1,1,\omega)]\,.\end{aligned} \tag{11.261}$$

For the symmetric potential one may show [9.17] that $\tilde{S}(1,1,\omega)$ is equal to $\tilde{S}(-1,-1,\omega)$ and that $\tilde{S}(1,-1,\omega)$ is equal to $\tilde{S}(-1,1,\omega)$. In Sect. 11.7 it was already shown that this dynamic structure factor can be expressed in terms of the matrix continued fraction (11.224), see (11.221, 223). Thus the only problem in obtaining the susceptibilities is the evaluation of the matrix continued fraction for the cosine potential, where one has to put $k = 0$ and $F = 0$ in the matrix elements (11.203, 204).

Results

We now discuss the results of the numerical calculation done in [9.17]. As seen from the expressions, $\sqrt{\Theta}\chi_v$, $\Theta\chi_\perp$ and $\Theta\chi_\parallel$ depend only on the combination $\omega/\sqrt{\Theta}$ and d/Θ. We first discuss the susceptibility χ_v for Case I, shown in Figs. 11.32–35. As already discussed, the absorption of energy is proportional to the real part χ_v' of this susceptibility which is shown in Fig. 11.32a for $d/\Theta = 1$ and in Figs. 11.34, 35 for $d/\Theta = 2$. For small damping constants the $\chi_v'(\omega)$ curve has two maxima. One maximum occurs approximately at the frequency $\omega_0 = \sqrt{d}$: it stems from the oscillations of the particles in the well of the cosine potential. The other maximum occurs at $\omega = 0$: it stems from those particles which go over the hills of the cosine potential. For large damping constants there is only one broad absorption maximum at a finite frequency. This transition from a low value at $\omega = 0$ to a broad plateau is typical for a hopping process [11.15].

The eigenvalues of the Fokker-Planck operator (11.248) are discussed in Sect. 11.9. For damping constants in the region $0.1 < \gamma/\sqrt{\Theta} < 1$, the frequency $\omega_{max} \approx \omega_0$, where one of the maxima occurs, is given quite accurately by the imaginary part of the complex eigenvalue belonging to an antisymmetric eigenfunction and with lowest real part, whereas the width $\Delta\omega$ is given quite accurately by the real part of this eigenvalue. The width $\Delta\omega$ of the maximum at $\omega = 0$ is given quite accurately by the lowest real eigenvalue which belongs to an antisymmetric eigenfunction.

For very small damping constants the maximum at $\omega = 0$ will become very large ($\sim 1/\gamma$), whereas the other maximum is finite. In Fig. 11.35 this transition to the zero-friction limit, obtained in Sect. 11.8.1, is clearly seen.

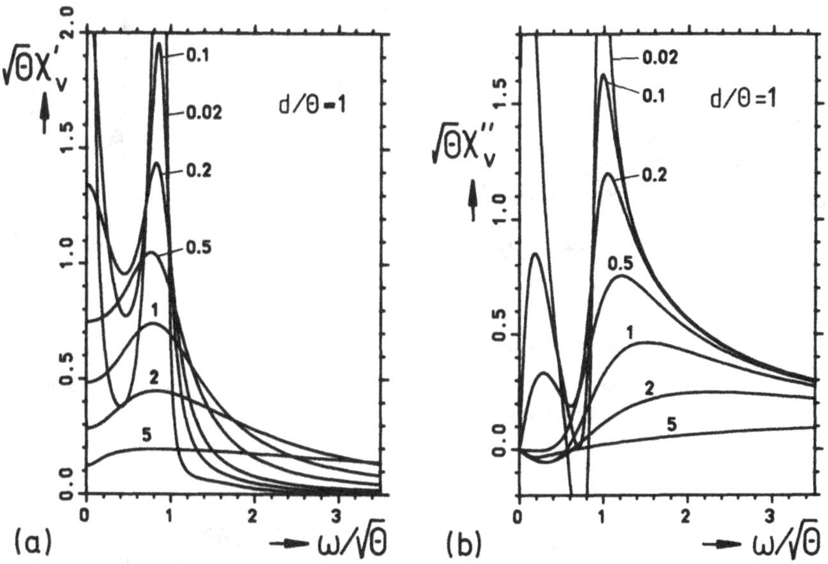

Fig. 11.32. Real (a) and negative imaginary (b) parts of the susceptibility (Fourier transform of the velocity autocorrelation function) multiplied by $\sqrt{\Theta}$ for $d/\Theta = 1$ as a function of the frequency $\omega/\sqrt{\Theta}$ for various friction constants $\gamma/\sqrt{\Theta}$

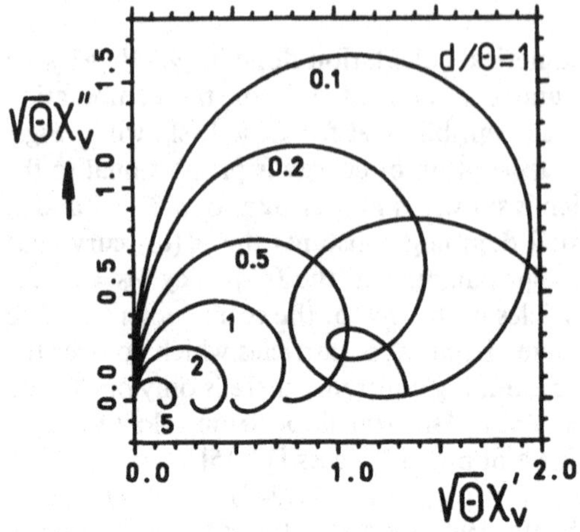

Fig. 11.33. The negative imaginary part times $\sqrt{\Theta}$ as a function of the real part of the susceptibilities for the parameters of Fig. 11.32. The frequency varies along the curve and is infinite at the origin

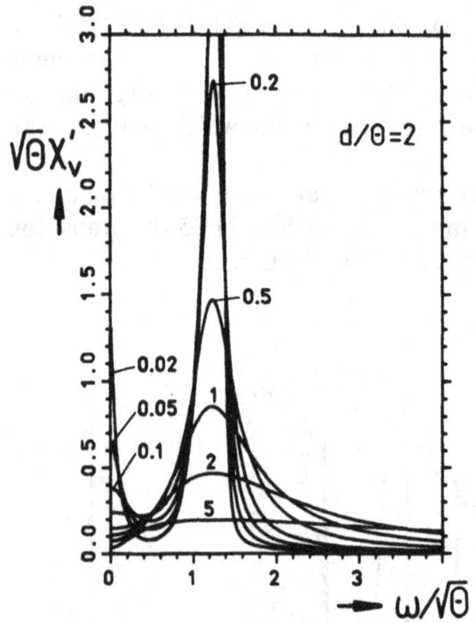

Fig. 11.34

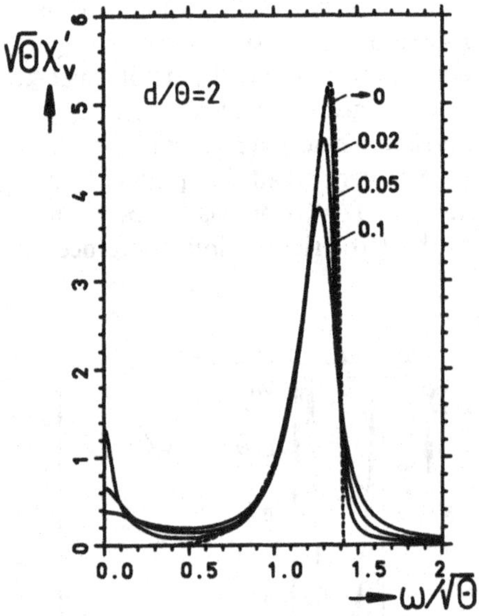

Fig. 11.35

Fig. 11.34. Same as Fig. 11.32a, but for the potential $d/\Theta = 2$

Fig. 11.35. The real part of the susceptibility $\sqrt{\Theta}\chi_v$ as a function of the frequency for some low friction constants (*solid lines*). The leading term $\sigma_1^{(1)}(\omega)$ of the zero-friction limit (11.279) is shown by the *dotted curve*

In Fig. 11.32b the imaginary part $\sqrt{\Theta}/\chi_v''$ is shown for $d/\Theta = 1$. The real part of the Fourier transform of the averaged position $\langle x \rangle$ is given by $\mathrm{Re}\{\mathrm{i}\langle v \rangle/\omega\} = [-\chi_v''(\omega)/\omega]F(\omega)$ for real $F(\omega)$. For large frequencies $\omega \gtrsim \omega_0$, $\langle x \rangle$ oscillates in opposite direction to the field, whereas for smaller frequencies and large damping forces $\langle x \rangle$ oscillates in the direction of the field, similar to the motion of a forced oscillator. For lower frequencies and damping constants $\langle x \rangle$ oscillates again opposite to the field. This is explained as follows. The particles which go over the hill of the cosine potential behave like free particles for small friction. Because free particles show a displacement opposite to the harmonic driving field, one therefore expects a positive sign of χ_v'' for small frequencies and small damping constants, if an appreciable amount of particles is excited enough to go over the hill.

The imaginary part of $\chi_v(\omega)$ versus its real part (Cole-Cole plot) is plotted in Fig. 11.33 for frequencies from $\omega = 0$ to $\omega = \infty$. For large damping constants the Cole-Cole plot is approximately a semicircle whereas for small damping constants it is approximately a full circle plus a semicircle. The impedance of an electric circuit which consists of two parallel branches, one having an inductance and a resistance in series, the other an inductance, a capacitance and a resistance in series, reproduces approximately the same loci for varying frequencies. The particles oscillating in the well of the cosine potential are described by the resonant circuit in series, whereas the particles going over the hill may be described by the other branch having an inductance and a resistance in series.

Next we discuss the susceptibilities $\chi_\perp(\omega)$ and $\chi_\parallel(\omega)$ for the Brownian rotation of a dipole in a constant field. Here the absorption of energy is proportional to $\omega\chi_\perp''(\omega)$ or $\omega\chi_\parallel''(\omega)$, respectively. The negative imaginary part $\chi_\perp''(\omega)$ is shown in Fig. 11.36b for $d/\Theta = 1$ and in Fig. 11.37b for $d/\Theta = 2$. For small damping constants there is a pronounced maximum at $\omega_{\max} \approx \omega_0 = \sqrt{d}$ similar to χ_v'. For very small damping constants and $d/\Theta = 1$ there another maximum occurs at a frequency $\omega_{\max}'$, which should be considered as the third harmonic of the frequency $\omega_{\max}$ of the main maximum. (Because of the damping constant and because the frequency decreases with increasing energy, $\omega_{\max}'$ is less than $3\,\omega_{\max}$. This third harmonic is hardly visible in $\chi_v'(\omega)$ curve.) For $d/\Theta = 2$ the maximum at $\omega_{\max} \approx \omega_0 = \sqrt{d}$ is more peaked for small damping constants because it is now more closely related to a harmonic oscillator, Fig. 11.38. The real part $\chi_\perp'(\omega)$ plotted in Fig. 11.36a shows the typical behavior of the susceptibility of an oscillator. The static value for $\omega = 0$ given by (11.257) is independent of the damping force because the stationary solution is also independent of it.

The negative imaginary part $\chi_\parallel''(\omega)$ of the susceptibility $\chi_\parallel(\omega)$ is shown in Fig. 11.37a for $d/\Theta = 1$ and Fig. 11.39 for $d/\Theta = 2$. For small damping constants a sharp peak occurs approximately at the second harmonic frequency $\omega_{\max} \approx 2\,\omega_0 = 2\sqrt{d}$, as one may expect from the parametric oscillation of the pendulum. A more precise location of this frequency for $d/\Theta = 2$ is given by the imaginary part of the lowest eigenvalue with lowest real part which belongs to the symmetric eigenfunction (Sect. 11.9), whereas the width is approximately given by the real part of this eigenvalue. The absolute amount of the susceptibilities for fields parallel to the static field is much smaller than that for fields perpendicular to the static field, especially for large d/Θ. For $d/\Theta = 1$ the real part $\chi_\parallel'$ of $\chi_\parallel$ is plotted in Fig. 11.37a, again showing the susceptibility of an oscillator with frequency $\omega_{\max} \approx 2\,\omega_0 = 2\sqrt{d}$.

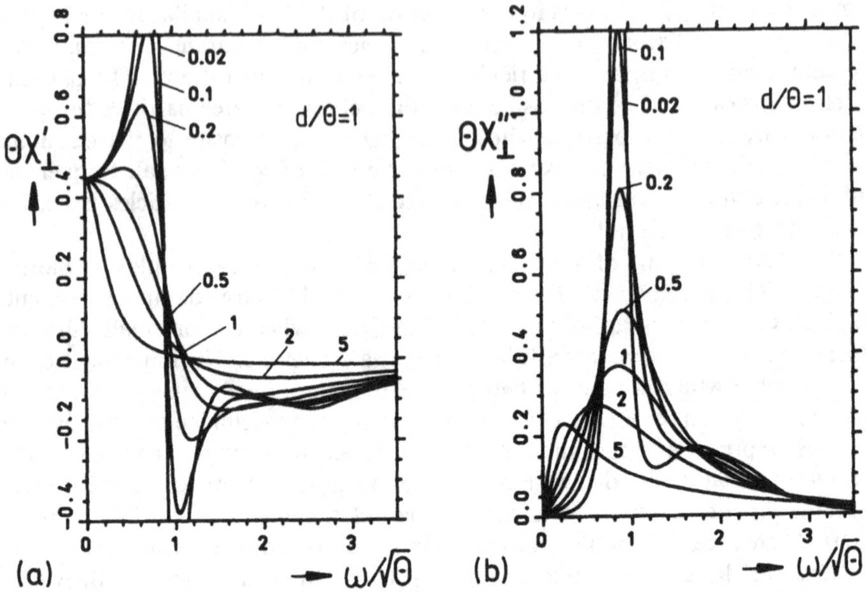

Fig. 11.36. Real **(a)** and negative imaginary **(b)** parts of the susceptibility $\chi_\perp$ (Fourier transform of the time derivative of the sine autocorrelation function) multiplied by Θ for $d/\Theta = 1$ as a function of the frequency $\omega/\sqrt{\Theta}$ for various friction constants $\gamma/\sqrt{\Theta}$

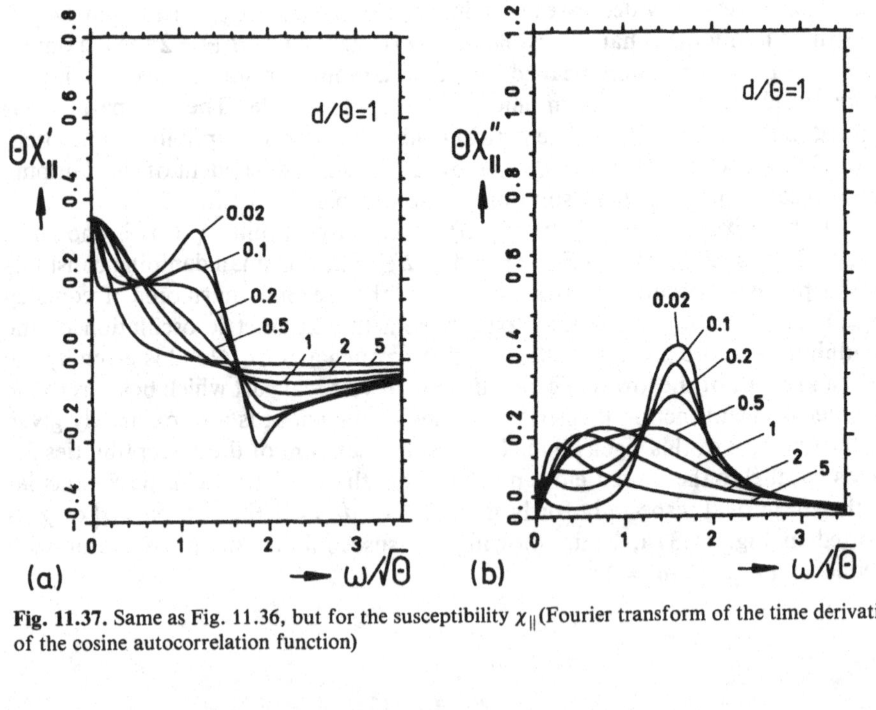

Fig. 11.37. Same as Fig. 11.36, but for the susceptibility $\chi_\parallel$ (Fourier transform of the time derivative of the cosine autocorrelation function)

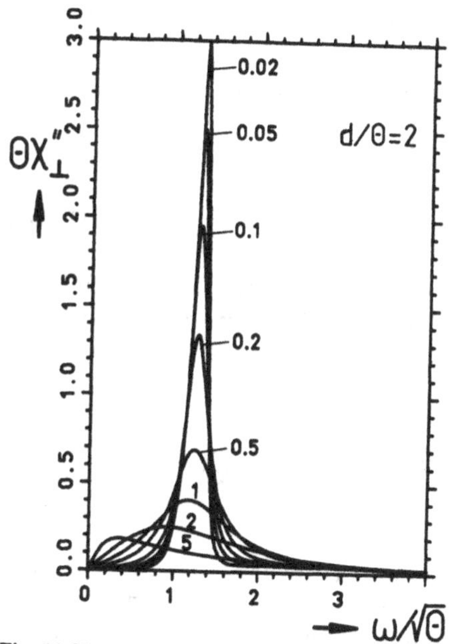

Fig. 11.38. Same as Fig. 11.36b, but for $d/\Theta = 2$

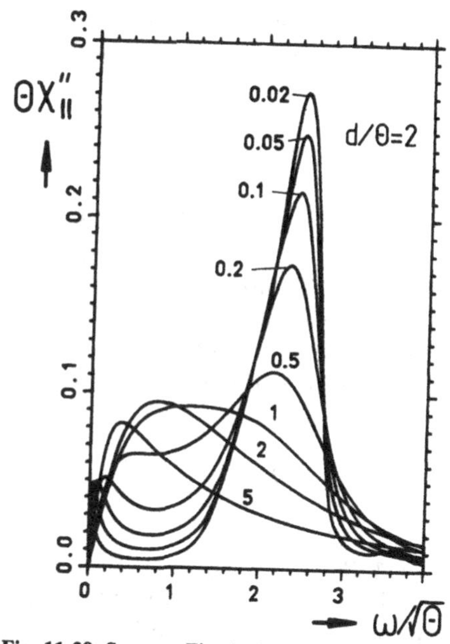

Fig. 11.39. Same as Fig. 11.37b, but for $d/\Theta = 2$

11.8.1 Zero-Friction Limit

Because the stationary solution is independent of γ, in the zero-friction limit it is still given by (11.216) with $f(x) = -d\cos x$. (For small γ it will of course take a long time till this stationary solution is finally established.) In the zero-friction limit a stationary autocorrelation function of some expression $f(x, v)$ for finite times t can be obtained by averaging the initial values of the deterministic motion with $\gamma = 0$ with respect to this stationary distribution, i.e.,

$$\begin{aligned}&\langle f(x(t), v(t)) f(x(0), v(0))\rangle\\&\quad = \iint f(x(x', v', t), v(x', v', t)) f(x', v')\, W_{\mathrm{st}}(x', v')\,\mathrm{d}x'\,\mathrm{d}v' \,. \end{aligned}\tag{11.262}$$

Here $x(x', v', t), v(x', v', t)$ is a solution of the deterministic equation of motion with $\gamma = 0$ and $F = 0$ in the cosine potential, i.e.,

$$\dot{x} = v\,, \qquad \dot{v} = -d\sin x \tag{11.263}$$

with the initial condition

$$x(x', v', 0) = x'\,, \qquad v(x', v', 0) = v'\,. \tag{11.264}$$

Here we restrict the calculations to the velocity autocorrelation function and the susceptibility $\chi_v(\omega)$. The susceptibility $\chi_\perp(\omega)$ follows from $\chi_v(\omega)$ by (11.260) with $\gamma = 0$. The susceptibility $\chi_{\|}(\omega)$ can be obtained with the same procedure; the expressions, however, are more complicated.

Inserting in (11.216) $f(x) = -d\cos x$ and $\psi_0(v)$ we thus have to calculate

$$K_{vv}(t) = \langle v(t)\,v(0)\rangle = [\sqrt{2\pi\Theta}\,2\pi \mathrm{I}_0(d/\Theta)]^{-1}\int_0^{2\pi}\int_{-\infty}^{\infty} v(x',v',t)\,v' \times \exp[-v'^2/(2\Theta)+(d/\Theta)\cos x']\,\mathrm{d}x'\,\mathrm{d}v' , \tag{11.265}$$

where I_0 is the modified Bessel function of zeroth order. For system (11.263) the energy $E = v^2/2 - d\cos x$ is a constant of motion. Therefore the energy or a function of it should now be used as one variable. Here it is convenient to introduce the variable

$$\varepsilon = \varepsilon(x',v') = \frac{E+d}{2d} = \left(\frac{v}{2\omega_0}\right)^2 + \sin^2\frac{x}{2} = \left(\frac{v'}{2\omega_0}\right)^2 + \sin^2\frac{x'}{2} . \tag{11.266}$$

In (11.266) the frequency for small amplitudes is denoted by $\omega_0 = \sqrt{d}$. The solutions of system (11.263) are oscillatory for $\varepsilon < 1$ (i.e., librations in the case of a dipole), whereas for $\varepsilon > 1$ they are of a running type (i.e., rotations in a case of a dipole).

Instead of the variables x' and v' we use in (11.265) the energy (11.266) and the angle φ defined by the integral

$$\varphi = \varphi(x',v') = \int_0^{x'} \frac{\mathrm{d}\xi}{2\sqrt{\varepsilon(x',v') - \sin^2\xi/2}} . \tag{11.267}$$

The solutions of system (11.263) with the initial condition (11.264) may then be expressed in terms of the Jacobian elliptic functions ([11.55]; for the elliptic functions we use the notation of [11.50]) for

$$\begin{aligned} &\varepsilon < 1 \quad \text{by} \\ &x(x',v',t) = 2\arcsin[\sqrt{\varepsilon}\,\mathrm{sn}(\omega_0 t + \varphi, \varepsilon)] \\ &v(x',v',t) = 2\omega_0\sqrt{\varepsilon}\,\mathrm{cn}(\omega_0 t + \varphi, \varepsilon) \end{aligned} \tag{11.268}$$

and similarly for

$$\begin{aligned} &\varepsilon > 1 \quad \text{by} \\ &x(x',v',t) = 2\arcsin\{\mathrm{sn}[\sqrt{\varepsilon}(\omega_0 t + \varphi), 1/\varepsilon]\} \\ &v(x',v',t) = \pm 2\omega_0\sqrt{\varepsilon}\,\mathrm{dn}[\sqrt{\varepsilon}(\omega_0 t + \varphi), 1/\varepsilon] . \end{aligned} \tag{11.269}$$

For $t = 0$ (11.268, 269) express $x' = x(x',v',0)$ and $v' = v(x',v',0)$ as a function of ε and φ. The period of the functions $\mathrm{sn}(\varphi,\varepsilon)$ and $\mathrm{cn}(\varphi,\varepsilon)$ is $4\mathbf{K}(\varepsilon)$, whereas the period of the function $\mathrm{dn}(\varphi,\varepsilon)$ is $2\mathbf{K}(\varepsilon)$, where $\mathbf{K}(\varepsilon)$ is the complete elliptic integral of the first kind [11.50]. If we use the variables φ and ε in (11.265) instead of the variables x', v', the double integral is transformed to

$$K_{vv}(t) = [\sqrt{2\pi\Theta}\,2\pi\,\mathrm{I}_0(d/\Theta)]^{-1}(2\omega_0)^3\left[\int_0^1 \varepsilon\,\mathrm{e}^{-\omega_0^2(2\varepsilon-1)/\Theta} I_1(\varepsilon,t)\,\mathrm{d}\varepsilon + 2\int_1^\infty \varepsilon\,\mathrm{e}^{-\omega_0^2(2\varepsilon-1)/\Theta} I_2(\varepsilon,t)\,\mathrm{d}\varepsilon\right], \tag{11.270}$$

with

$$I_1(\varepsilon,t) = \int_0^{4\mathbf{K}(\varepsilon)} \mathrm{cn}(\omega_0 t+\varphi,\varepsilon)\,\mathrm{cn}(\varphi,\varepsilon)\,\mathrm{d}\varphi\,, \tag{11.271}$$

and

$$I_2(\varepsilon,t) = \int_0^{2\mathbf{K}(1/\varepsilon)} \mathrm{dn}[\sqrt{\varepsilon}(\omega_0 t+\varphi),1/\varepsilon]\,\mathrm{dn}[\sqrt{\varepsilon}\varphi,1/\varepsilon]\,\mathrm{d}\varphi\,. \tag{11.272}$$

In deriving (11.270) we have used the Jacobian of the transformation from x' and v' to the ε and φ variables

$$J = (\mathrm{d}x'\,\mathrm{d}v')/(\mathrm{d}\varepsilon\,\mathrm{d}\varphi) = 2\omega_0\,. \tag{11.273}$$

The factor 2 in front of the last integral in (11.270) stems from the fact that the velocity can have positive and negative values for $\varepsilon > 1$, see the last expression in (11.269).

To evaluate the integrals (11.271, 272) we use the following series expansion of the elliptic functions cn and dn (see 16.23.2–3 [Ref. 11.50]):

$$\mathrm{cn}(u,m) = \frac{2\pi}{\sqrt{m}\,\mathbf{K}(m)}\sum_{n=0}^{\infty}\frac{\sqrt{q}\,q^n}{1+q^{2n+1}}\cos\left[(2n+1)\frac{\pi u}{2\mathbf{K}(m)}\right], \tag{11.274}$$

$$\mathrm{dn}(u,m) = \frac{\pi}{2\mathbf{K}(m)} + \frac{2\pi}{\mathbf{K}(m)}\sum_{n=1}^{\infty}\frac{q^n}{1+q^{2n}}\cos\left(2n\frac{\pi u}{2\mathbf{K}(m)}\right), \tag{11.275}$$

where the nome q is given by

$$q(m) = \exp[-\pi\mathbf{K}(1-m)/\mathbf{K}(m)]\,. \tag{11.276}$$

By inserting the appropriate expressions into (11.271, 272) we can perform the integration over φ, leading to

$$I_1(\varepsilon,t) = \frac{8\pi^2 q(\varepsilon)}{\varepsilon\mathbf{K}(\varepsilon)}\sum_{n=0}^{\infty}\frac{q^{2n}(\varepsilon)}{[1+q^{2n+1}(\varepsilon)]^2}\cos\left[(2n+1)\frac{\pi\omega_0 t}{2\mathbf{K}(\varepsilon)}\right], \tag{11.277}$$

$$I_2(\varepsilon,t) = \frac{\pi^2}{2\mathbf{K}(1/\varepsilon)}\left[1+8\sum_{n=1}^{\infty}\frac{q^{2n}(1/\varepsilon)}{[1+q^{2n}(1/\varepsilon)]^2}\cos\left(2n\frac{\pi\sqrt{\varepsilon}\,\omega_0 t}{2\mathbf{K}(1/\varepsilon)}\right)\right]. \tag{11.278}$$

The main frequency of the $I_1(\varepsilon, t)$ term is given by

$$\omega(\varepsilon) = \pi\omega_0/[2\mathbf{K}(\varepsilon)] \approx \omega_0(1-\varepsilon/4 \pm \ldots) .$$

Because we are interested mainly in the susceptibilities we make a half-sided Fourier transform of the correlation function (11.270). Further treatment is restricted to the calculation of the real part $\chi_v'(\omega)$. The imaginary part $\chi_v''(\omega)$ follows from the Kramers-Kronig relations (7.42). Because

$$\int_0^\infty \cos\alpha t \cos\omega t \, dt = \frac{1}{2}\int_{-\infty}^{\infty} \cos\alpha t \, e^{-i\omega t} dt = \frac{\pi}{2}[\delta(\alpha-\omega)+\delta(\alpha+\omega)] ,$$

we can immediately perform the cos-transformation of $I_1(\varepsilon, t)$ and $I_2(\varepsilon, t)$. Due to the δ function we can then easily perform the integration over ε and finally arrive at

$$\chi_v'(\omega) = \sigma(\omega) = \sum_{n=0}^{\infty} \sigma_1^{(n)}(\omega) + \sigma_2^{(0)}(\omega) + \sum_{n=1}^{\infty} \sigma_2^{(n)}(\omega) . \tag{11.279}$$

Here the terms $\sigma_1^{(n)}(\omega)$ are given by

$$\sigma_1^{(n)}(\omega) = \frac{32\pi\omega_0^2 \mathbf{K}(\varepsilon) q(\varepsilon)^{2n+1} e^{-\omega_0^2(2\varepsilon-1)/\Theta}}{\sqrt{2\pi\Theta}\,\Theta I_0(d/\Theta)(2n+1) d\mathbf{K}(\varepsilon)/d\varepsilon \{1+[q(\varepsilon)]^{2n+1}\}^2} , \tag{11.280}$$

where the positive variable ε ($0 \leqq \varepsilon < 1$) is related to the frequency ω by

$$\omega = \pm\omega_0(2n+1)\pi/[2\mathbf{K}(\varepsilon)] ; \quad n = 0, 1, 2, \ldots . \tag{11.281}$$

The derivative of the complete elliptic integral of first kind can be expressed by the elliptic integrals of first and second kind, i.e., by $\mathbf{K}(\varepsilon)$ and $\mathbf{E}(\varepsilon)$ according to

$$\frac{d\mathbf{K}(\varepsilon)}{d\varepsilon} = \frac{1}{2}\left(\frac{\mathbf{E}(\varepsilon)}{\varepsilon(1-\varepsilon)} - \frac{\mathbf{K}(\varepsilon)}{\varepsilon}\right). \tag{11.282}$$

The term $\sigma_2^{(0)}(\omega)$ reads

$$\sigma_2^{(0)}(\omega) = \frac{8\pi\omega_0^3}{\sqrt{2\pi\Theta}\,\Theta I_0(d/\Theta)} \int_1^\infty \frac{\pi\varepsilon}{2\mathbf{K}(1/\varepsilon)} e^{-\omega_0^2(2\varepsilon-1)/\Theta} d\varepsilon \, \delta(\omega) . \tag{11.283}$$

It has a δ-function singularity at $\omega = 0$. For an external dc field the mobility is proportional to $1/\gamma$ for small damping constants, Sect. 11.5. Therefore the response to a dc field must diverge in the zero-friction limit as shown in (11.283).

Finally, the terms $\sigma_2^{(n)}(\omega)$ read

$$\sigma_2^{(n)}(\omega) = \frac{32\pi\omega_0^2 \mathbf{K}(1/\varepsilon)}{n\sqrt{2\pi\Theta}\,\Theta I_0(d/\Theta)\mathbf{E}(1/\varepsilon)} (\varepsilon-1)\sqrt{\varepsilon} \frac{q(1/\varepsilon)^{2n} e^{-\omega_0^2(2\varepsilon-1)/\Theta}}{\{1+[q(1/\varepsilon)]^{2n}\}^2} . \tag{11.284}$$

Here the variable $\varepsilon > 1$ is related to the frequency ω by

$$\omega = \pm 2n\pi\sqrt{\varepsilon}[2\mathbf{K}(1/\varepsilon)]\,, \qquad n = 1, 2, \ldots\,. \tag{11.285}$$

The explicit values for $\sigma_1^{(n)}(\omega)$ and $\sigma_2^{(n)}(\omega)$ $(n = 1, 2, \ldots)$ can easily be obtained by applying the polynomial expansions for the complete elliptic integrals of first and second kind [11.50]. For large $d/\Theta = \omega_0^2/\Theta$ values, $\sigma_1^{(1)}(\omega)$ is the leading term. In Fig. 11.35 this leading term is plotted as a function of $\omega/\sqrt{\Theta}$ for $\omega_0^2/\Theta = 2$. For large ω_0^2/Θ we can simplify the expression still further. Making a series expansion of $q(\varepsilon)$, $\mathbf{K}(\varepsilon)$, $\mathbf{E}(\varepsilon)$ and an asymptotic expansion of $\mathrm{I}_0(d/\Theta)$, we obtain

$$\begin{aligned} \sigma_1^{(1)}(\omega) &= 32\pi\Theta^{-2}\omega_0^2(\omega_0 - \omega)\,\mathrm{e}^{-8\omega_0(\omega_0-\omega)/\Theta} && \text{for} \quad 0 < \omega \leqq \omega_0\,, \\ \sigma_1^{(1)}(\omega) &= 0 && \text{for} \quad \omega \geqq \omega_0\,, \end{aligned} \tag{11.286}$$

which has a sharp maximum $\sigma_{1\,\mathrm{max}}^{(1)}$ at ω_{max} given by

$$\omega_{\mathrm{max}} = \omega_0 - \Theta/(8\omega_0)\,, \qquad \sigma_{1\,\mathrm{max}}^{(1)} = (4\pi/\mathrm{e})\,\omega_0/\Theta\,. \tag{11.286a}$$

This result can be derived directly from (11.265) by using the harmonic approximation for x and v with energy-dependent frequency $\omega(E) = \omega_0 - (E - E_{\mathrm{min}})/(8\omega_0)$ and by averaging over the energy.

11.9 Eigenvalues and Eigenfunctions

As discussed in Sect. 10.3.2, the eigenvalues and eigenfunctions of the Kramers equation follow from (10.159) and the eigenfunctions from (10.158, 160). First we want to apply this method to calculate the eigenvalues of the Fokker-Planck operator (11.248) without the force F. Then we determine the eigenvalues of (11.208) including F.

Eigenvalues for $F = 0$

To calculate the eigenvalues with lowest real parts we need consider only (10.159) for $m = 0$. Thus the final equation to determine the general complex eigenvalues λ is given by

$$\mathrm{Det}\,[\lambda\boldsymbol{I} + \tilde{\boldsymbol{K}}_0(-\lambda)] = 0\,, \tag{11.287}$$

where $\tilde{\boldsymbol{K}}_0(s)$ is the matrix continued fraction (11.224). Obviously the zeros of (11.287) are the poles $\mathrm{i}\omega_n = -\lambda_n$ of the Green's function (11.223). We now present the results obtained in [9.16] for the periodic eigenfunctions, i.e., for $k = 0$. In Fig. 11.40 the eigenvalues divided by $\sqrt{\Theta}$ are shown as a function of $\gamma/\sqrt{\Theta}$ for $d/\Theta = 2$ ($\lambda/\sqrt{\Theta}$ depends only on $\gamma/\sqrt{\Theta}$ and d/Θ). We may roughly distinguish four regions of the damping constant. For large damping constants (Region IV with $\gamma/\sqrt{\Theta} \gtrsim 6$) the eigenvalues are always real. They are grouped together according to (10.201), the lowest ones $(m = 0)$ follow from the

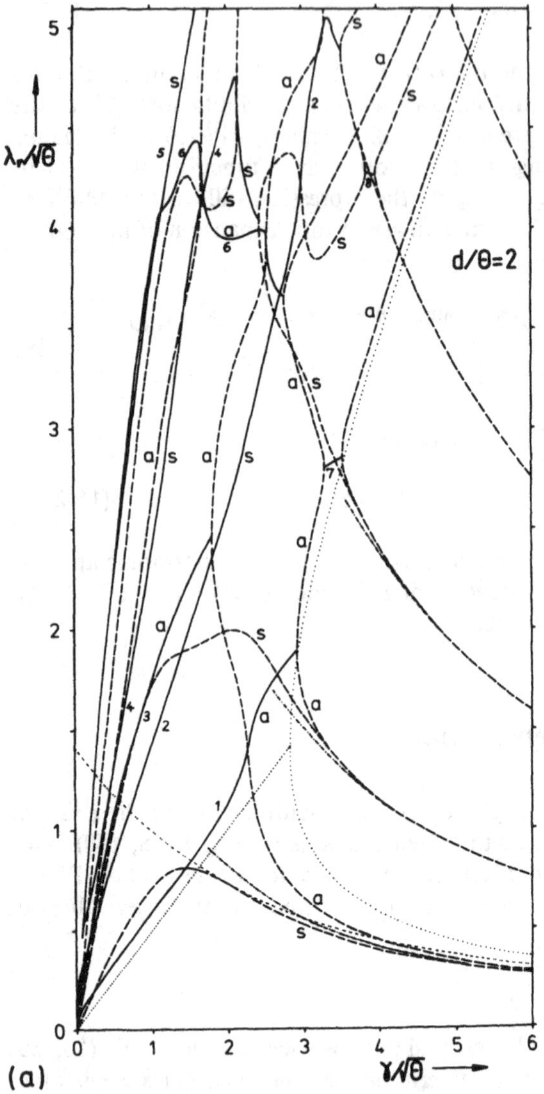

Fig. 11.40. The eigenvalues $\lambda/\sqrt{\Theta}$ of the Kramers equation for the cosine potential with $d/\Theta = 2$ as a function of the friction constant $\gamma/\sqrt{\Theta}$. The *broken lines* show real eigenvalues, the *solid lines* in **(a)**, the real part $\lambda_r/\sqrt{\Theta}$, and in **(b)**, imaginary part $\lambda_i/\sqrt{\Theta}$ of the complex eigenvalues. For the complex eigenvalues the complex conjugate is also a solution. The *numbers* indicate corresponding real and imaginary parts. The symbols "s" denote symmetric, "a" antisymmetric eigenfunctions belonging to the eigenvalue. No symbols are given if two eigenfunctions with different symmetry are nearly degenerate. For the lines, ···· denotes the lowest oscillator eigenvalues, see (10.83); .. .. .. the inverted parabolic potential eigenvalue, see (10.108); —·—·—·— the Smoluchowski equation eigenvalues according to Fig. 11.14c; and – – – – in **(b)** are the extrapolated eigenvalues according to Fig. 11.44

Smoluchowski equation (Sect. 11.3 and Fig. 11.14c). With the exception of the zeroth eigenvalue they are twofold degenerate [see the discussion following (11.82)]. For smaller $\gamma/\sqrt{\Theta} \gtrsim 3$ (Region III) this degeneracy is removed, the deviation from the results of the Smoluchowski equation may be calculated by correction terms to the Smoluchowski equation discussed in Sect. 10.4.2. For the inverse friction expansion to be valid, it is essential that the different m groups of eigenvalues (10.201) do not mix. As seen from Fig. 11.40, this essential feature breaks down for those eigenvalues which become complex. The points where complex eigenvalues occur are shifted to higher damping constants for higher eigenvalues, therefore the validity of the inverse friction expansion is also shifted to larger damping constants if higher eigenvalues are considered. For smaller

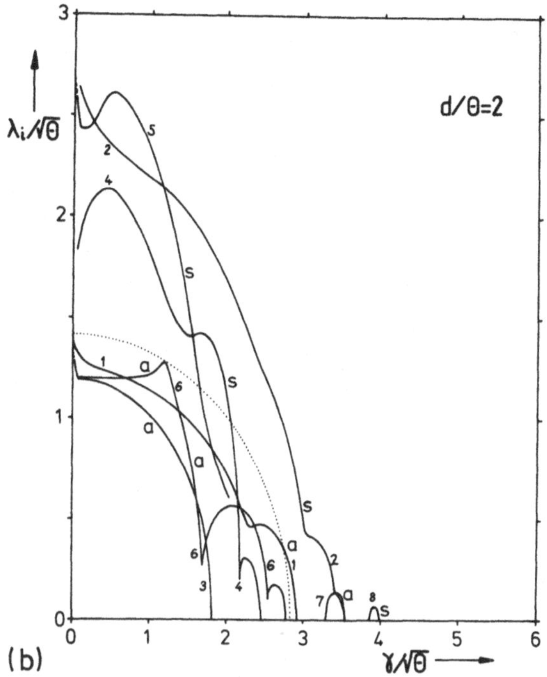

Fig. 11.40 b

γ ($\gamma/\sqrt{\Theta} \lesssim 3$) we distinguish two regions for the complex eigenvalues, i.e., $\gamma/\sqrt{\Theta} \lesssim 0.05$ (Region I) and $0.05 \lesssim \gamma/\sqrt{\Theta} \lesssim 3$ (Region II). In Region I the real and imaginary parts of the eigenvalues show square-root dependence according to (11.301 a) and Fig. 11.44. In Region II the real parts and roughly also the imaginary parts behave like those of the harmonic oscillator according to (10.83, 84). The real parts, however, have an additional constant term which may

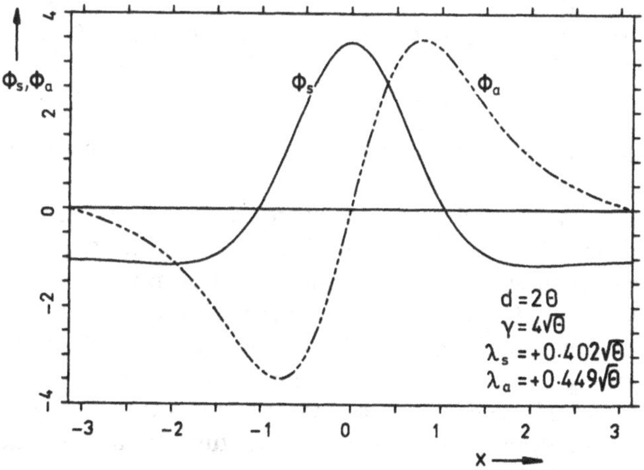

Fig. 11.41. The velocity integrated eigenfunctions (11.288) for large friction ($\gamma/\sqrt{\Theta} = 4$) for the lowest two eigenvalues. The eigenfunctions are nearly degenerate and symmetric (s) or antisymmetric (a). (Ordinate in arbitrary units)

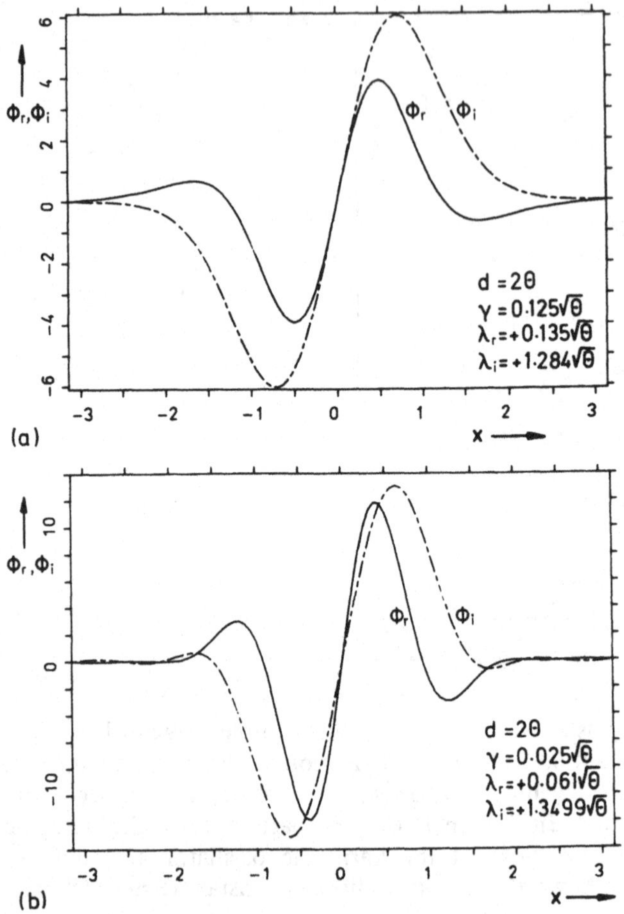

Fig. 11.42. Real Φ_r and imaginary Φ_i parts of the velocity integrated eigenfunction (11.288) in arbitrary units. **(a)** For intermediate friction $\gamma/\sqrt{\Theta} = 0.125$. **(b)** For low friction $\gamma/\sqrt{\Theta} = 0.025$ the eigenfunction contracts to small x values

be interpreted as the additional damping term due to the different frequencies $\omega(E)$ (11.292).

The imaginary parts of Curves 1, 3, 6 in Fig. 11.40 are appreciably lower than the corresponding frequencies of the harmonic oscillator because the frequency decreases with increasing amplitude. In Region II a complicated mixture of real and complex eigenvalues is found. Consider for instance Curve 7. Two real eigenvalues become two complex eigenvalues and again become two real eigenvalues for decreasing damping constant. Some of the eigenvalues remain complex down to $\gamma = 0$. In Region II the different asymptotic eigenvalues for large and small damping constants mix together in a netlike structure. Because of the symmetry of the problem the netlike structure decomposes into eigenvalues with symmetric, respectively, antisymmetric eigenfunctions (10.166). Some of the eigenfunctions (10.161) integrated over the velocity

$$\Phi(x) = \int_{-\infty}^{+\infty} \Phi(x, v)\,dv \tag{11.288}$$

are shown for the cosine potential for different values of $\gamma/\sqrt{\Theta}$ in Figs. 11.41, 42. For large $\gamma/\sqrt{\Theta}$ values the eigenfunctions are real. They belong to nearly degenerate eigenvalues, Fig. 11.41. For smaller $\gamma/\sqrt{\Theta}$ the eigenfunctions are complex if the eigenvalues are complex, Fig. 11.42a, b. For very small $\gamma/\sqrt{\Theta}$ the contraction of the eigenfunctions (Sect. 11.9.1) is clearly seen in Fig. 11.42b.

Eigenvalues for $F \neq 0$

The eigenvalues for $F \neq 0$ follow by the same procedure as used for $F = 0$, i.e., by (11.287) with $\tilde{\boldsymbol{K}}_0$ given by (11.224). For the matrices $\boldsymbol{D}$ and $\hat{\boldsymbol{D}}$ in (11.224) the matrix elements (11.203, 204) with $F \neq 0$ must be used. In Figs. 11.43a, b, c the eigenvalues are shown as a function of the external force for three typical values of the damping constant and for various noise powers Θ. Because we are mainly interested in the low temperature $\Theta \to 0$ limit, we normalize the eigenvalue by the frequency $\sqrt{d}$. (As may be easily seen from (11.208, 209a) for $k = 0$, by dividing (11.208) by $\sqrt{d}$ and by using the normalized velocity $\tilde{v} = v/\sqrt{d}$, the eigenvalue divided by $\sqrt{d}$ is a function of $\gamma/\sqrt{d}$, Θ/d and F/d only. We now discuss the results obtained in [11.28] by going from small to large friction constants. In Fig. 11.43a the two lowest non-zero real eigenvalues are shown for the friction constant $\gamma/\sqrt{d} = 0.5$, which lies in the middle of the bistability region $0 \leqq \gamma/\sqrt{d} < 1.193$, see Fig. 11.26. The most remarkable feature of Fig. 11.43a is that the lowest non-zero eigenvalue tends to zero for decreasing noise power Θ/d for forces F/d in the bistability region ($F_1 < F < F_3$). The explanation for this behaviour runs as follows. Without any Langevin force (i.e., $\Theta/d = 0$) (11.30a) has two stable solutions, a running and a locked one. For finite noise strength Θ/d one gets transitions between these two solutions. For $\Theta/d \to 0$ the transition rates and therefore also the eigenvalue must vanish. (For a further discussion of transition rates see [11.28].) For finite but small noise power Θ the stationary solution W_{st} of the Fokker-Planck equation and some stationary expectation values show a sharp transition at a critical force $F_2 (F_1 < F_2 < F_3)$ as already discussed in Sect. 11.6.2. As seen in Fig. 11.43a the lowest non-zero eigenvalue is smallest at approximately $F = F_2$. Thus a very long time is needed to establish the stationary solution at this critical force. This is similar to the critical slowing down at a second order phase transition which we find here for a transition reminiscent of a first order phase transition. In the limit $\Theta/d \to 0$ one expects that the $\lambda/\sqrt{d}$ curve as a function of F makes a sharp transition to zero if F reaches the first critical force F_1 and stays zero till F reaches the third critical force F_3 where $\lambda/\sqrt{d}$ will jump to a finite value which is approximately given by $\gamma/\sqrt{d}$. This value is obtained as follows: For large forces $F \gg d$ one can neglect the periodic force in (11.30a) and thus obtains the damping constant $\lambda = \gamma$ of the Brownian motion without a potential. (The constant force can be absorbed in the shifted velocity $v_s = v - F$, i.e., one obtains for v_s the equation $\dot{v}_s + \gamma v_s = \Gamma(t)$ of free Brownian motion.) For low forces F and low noise powers Θ/d the

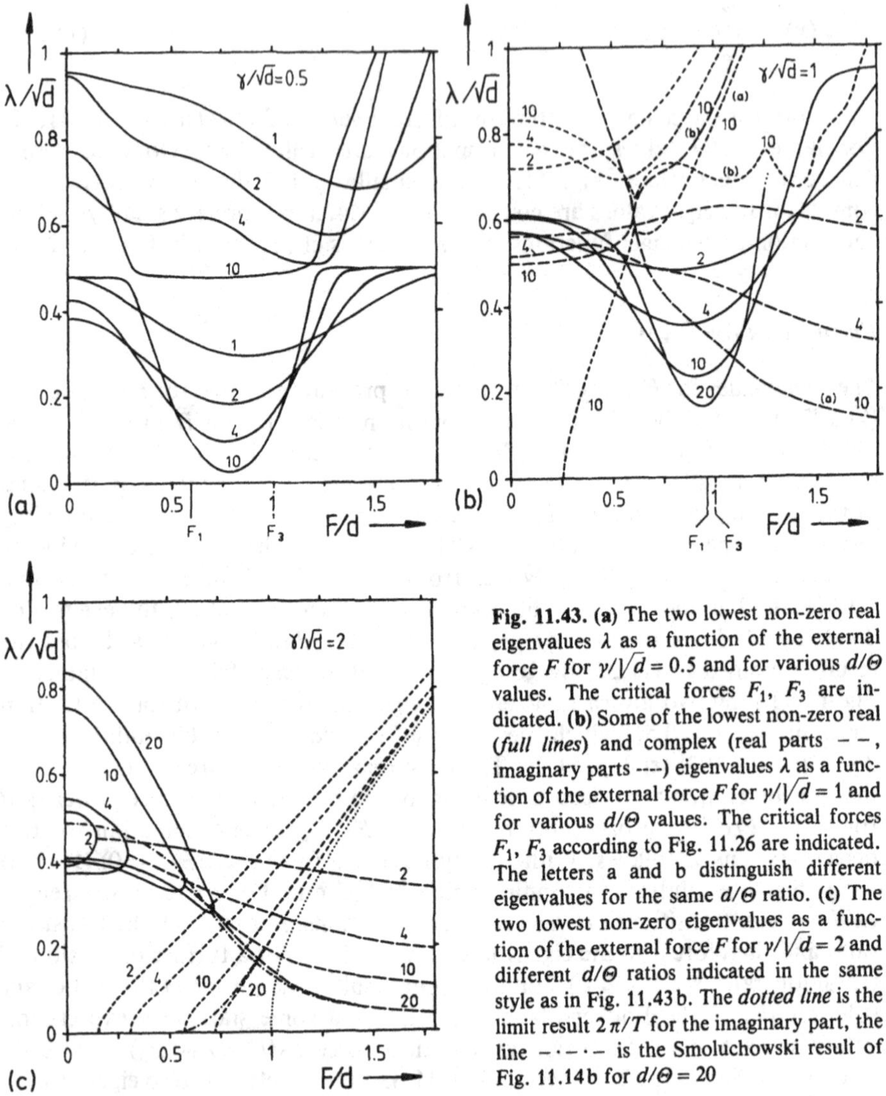

Fig. 11.43. (**a**) The two lowest non-zero real eigenvalues λ as a function of the external force F for $\gamma/\sqrt{d} = 0.5$ and for various d/Θ values. The critical forces F_1, F_3 are indicated. (**b**) Some of the lowest non-zero real (*full lines*) and complex (real parts – –, imaginary parts ----) eigenvalues λ as a function of the external force F for $\gamma/\sqrt{d} = 1$ and for various d/Θ values. The critical forces F_1, F_3 according to Fig. 11.26 are indicated. The letters a and b distinguish different eigenvalues for the same d/Θ ratio. (**c**) The two lowest non-zero eigenvalues as a function of the external force F for $\gamma/\sqrt{d} = 2$ and different d/Θ ratios indicated in the same style as in Fig. 11.43b. The *dotted line* is the limit result $2\pi/T$ for the imaginary part, the line –·–·– is the Smoluchowski result of Fig. 11.14b for $d/\Theta = 20$

particles oscillate in a well which is approximately parabolic. The first non-zero real eigenvalue in such a parabolic potential is given by γ, see (10.83). (The deviations from this value for finite Θ/d stem from the deviations from the parabolic form.) As seen from Fig. 11.43a in the bistability region the next real eigenvalue takes over the value $\lambda \approx \gamma$ of the lowest non-zero eigenvalue outside the bistability region for very low noise power Θ. Complex eigenvalues are also obtained. They are not plotted in Fig. 11.43a in order not to overload this figure. The real parts of these complex eigenvalues did not show signs of critical slowing down in the bistability region as was the case for the real eigenvalues.

In Fig. 11.43 b real and complex eigenvalues are shown for $\gamma/\sqrt{d}=1$, i.e. just near the end of the bistability region at $\gamma/\sqrt{d}=1.193\ldots$, see Fig. 11.26. The lowest non-zero real eigenvalue still shows the critical slowing down in the bistability region; however, it does not reach the low values as in Fig. 11.43 a even for the lower $\Theta/d=0.05$. For large forces the real parts of the complex eigenvalues decrease for increasing F whereas the imaginary parts increase with increasing F as in Fig. 11.43 c. Thus at $\gamma/\sqrt{d}=1$ the eigenvalue dependence shows features of the eigenvalues for small and large $\gamma/\sqrt{d}$. Therefore the eigenvalues show a complicated structure in this intermediate region.

In Fig. 11.43 c at $\gamma/\sqrt{d}=2$ we are well outside the bistability region and the dependence of the eigenvalues on the force simplifies again. For $F>d$ only a running solution occurs for zero noise ($\Theta/d=0$). This running solution shows oscillation in time with frequency components being multiples of a fundamental frequency $\omega=2\pi/T$. If we neglect the second time derivative in (11.181) the time to travel the distance 2π is given by (11.52). The imaginary parts of the eigenvalues with low real parts agree approximately with $\omega=2\pi/T$ for $F/d>1$ and low Θ/d. For smaller forces F/d the imaginary parts of the eigenvalues disappear and two real eigenvalues appear instead of the two complex conjugate ones. (Because L_{FP} is real the complex conjugate of a complex eigenvalue is also an eigenvalue of L_{FP}.) The bend in the real part of the eigenvalue at $F\approx d$ for low Θ may be considered as a rudiment of the critical slowing down of the lowest non-zero real eigenvalue in the bistability region.

In the Smoluchowski limit in Fig. 11.14 a, b the eigenvalues behave similarly as in Fig. 11.43 c. In contrast to Fig. 11.14 b, however, the imaginary parts disappear suddenly for decreasing F in Fig. 11.43 c. In Fig. 11.14 b the imaginary parts are zero for $F=0$ only, though they have extremly low values for small Θ/d and $F/d<1$. The real parts in Fig. 11.14 a behave similar as those in Fig. 11.43 c, however, for smaller F/d a bifurcation is observed in Fig. 11.43 c.

11.9.1 Eigenvalues and Eigenfunctions in the Low-Friction Limit

In the low-friction limit the energy $E=v^2/2+f(x)$ becomes a slow variable, see Sect. 8.3 for a discussion of slow and fast variables. Therefore one should use E or a function of E like the action integral [1.17] as one variable. For the other variable one may use x or the angle variable ψ. The x variable has the advantage that no second derivatives in x occur (no diffusion in x), whereas for the angle variable a second derivative in ψ does occur (diffusion in ψ). As it turns out, for the real eigenvalues one should use E and x variables, i.e., (11.90) or (11.91) and average over x as in Sect. 11.4, whereas for the complex eigenvalues it seems to be more suitable to use E and ψ variables. Because the complex eigenvalues (at least those with vanishing real parts for $\gamma\to 0$) can be calculated analytically, we first determine the complex eigenvalues.

Complex Eigenvalues

As the numerical solutions in Fig. 11.42 b show, for very small damping constants the eigenfunctions contract to the E region near $E=E_{\mathrm{min}}$, because the frequency depends on energy and therefore, due to diffusion in E, only the oscillations with $E \approx E_{\min}$ will survive in the ensemble for $\gamma \to 0$. This effect may be treated analytically as follows. For E and ψ variables the Fokker-Planck equation (11.33) with $F = 0$ transforms to ([9.16])

$$\dot{W} = \left[-\omega(E) \frac{\partial}{\partial \psi} + L_{\text{ir}} \right] W, \tag{11.289}$$

where $\omega(E)$ is the energy-dependent frequency for $\gamma = 0$ and L_{ir} is the collision operator (10.13) with $v_{\text{th}}^2 = \Theta$. The reversible part in (11.289) follows from the transformation of the reversible equations $\dot{x} = v$, $\dot{v} = -f'(x)$ to the angle variable equations $\dot{\psi} = \omega(E)$, $\dot{E} = 0$ [11.57]. The angle variable ψ and the variable φ defined in (11.267) are connected by

$$\psi = [\omega(E)/\omega_0]\,\varphi\,. \tag{11.290}$$

For small damping constants γ, the 2π periodic eigenfunctions Φ are assumed to have the form which will be shown to be consistent with later results ($\nu = 1, 2, 3, \ldots$)

$$W = \Phi(E, \psi)\mathrm{e}^{-\lambda t} = \mathrm{e}^{\mathrm{i}\nu\psi} h_\nu(E)\mathrm{e}^{-\lambda t} + O(\sqrt{\gamma})\,. \tag{11.291}$$

The E-dependent function $h_\nu(E)$ and the eigenvalue are determined by the collision operator L_{ir}. If we further assume that the eigenfunction $h_\nu(E)$ is different from zero only near $E \approx E_{\min}$, we may approximate $\omega(E)$ by

$$\omega(E) = \omega_0 + a(E - E_{\min})\,, \tag{11.292}$$

where ω_0^2 is the second derivative of the potential $f(x)$ at its minimum. The following calculations are applicable for an arbitrary potential with a single minimum. We therefore give an expression for the constant a for an arbitrary potential with a minimum at $x = 0$. Assuming that the negative force is

$$f'(x) = \omega_0^2 x + c_2 x^2 + c_3 x^3\,, \tag{11.293}$$

the constant a has the value [11.56]

$$a = \left.\frac{\mathrm{d}\omega(E)}{\mathrm{d}E}\right|_{E=E_{\min}} = \frac{3}{4}\frac{c_3}{\omega_0^3} - \frac{5}{6}\frac{c_2^2}{\omega_0^5}\,, \tag{11.292a}$$

and for $f' = d \sin x$

$$a = -1/(8\sqrt{d})\,. \tag{11.292b}$$

[The negative force $f'(x)$ may for instance be the negative force $V'(x)$ of the total potential (11.5).]

For the small collision operator we approximate the transformation to E and ψ variables by the transformation for the harmonic oscillator, i.e., by ($E \approx E_{\min}$)

$$x = (\sqrt{2\Delta}/\omega_0)\sin\psi\,, \quad v = \sqrt{2\Delta}\cos\psi \quad \text{with} \quad \Delta = E - E_{\min}\,, \tag{11.294}$$

which leads to ([9.16])

$$L_{\text{ir}} = \gamma\left\{\Theta(1+\cos 2\psi)\Delta\frac{\partial^2}{\partial\Delta^2} + [\Theta + \Delta(1+\cos 2\psi)]\frac{\partial}{\partial\Delta} - \Theta\sin 2\psi\frac{\partial^2}{\partial\psi\,\partial\Delta}\right.$$
$$\left. + \frac{1}{2}\left(\frac{\Theta}{\Delta} - 1\right)\sin 2\psi\frac{\partial}{\partial\psi} + \frac{\Theta}{4\Delta}(1-\cos 2\psi)\frac{\partial^2}{\partial\psi^2} + 1\right\}. \tag{11.295}$$

Writing

$$\Delta = E - E_{\min} = \hat{\varepsilon}\alpha\sqrt{\Theta}\,, \tag{11.296}$$

where the complex constant α is assumed to be proportional to $\sqrt{\gamma}$, we get by inserting (11.291) in (11.289), using (11.292, 295) and neglecting terms of the order $\sqrt{\gamma}$ (higher harmonics are also of the order $\sqrt{\gamma}$):

$$\left[\hat{\varepsilon}\frac{\partial^2}{\partial\hat{\varepsilon}^2} + \frac{\partial}{\partial\hat{\varepsilon}} - \frac{\nu^2}{4\hat{\varepsilon}} + \frac{\alpha}{\gamma\sqrt{\Theta}}(\lambda - \mathrm{i}\nu\omega_0) - \mathrm{i}\nu a\frac{\alpha^2}{\gamma}\hat{\varepsilon}\right]h_\nu = 0\,. \tag{11.297}$$

This equation is solved by

$$h_\nu = \hat{\varepsilon}^{\nu/2}\mathrm{e}^{-\hat{\varepsilon}/2}\mathrm{L}_n^{(\nu)}(\hat{\varepsilon}) \tag{11.298}$$

provided that

$$\mathrm{i}\nu a\,\alpha^2/\gamma = \tfrac{1}{4}\,; \quad (\lambda - \mathrm{i}\nu\omega_0)\,\alpha/(\gamma\sqrt{\Theta}) = n + \tfrac{1}{2}(\nu+1)\,. \tag{11.299}$$

For a definition of the Laguerre polynomials $\mathrm{L}_n^{(\nu)}$ and for their differential equations see [9.26]. Thus, we obtain ($\nu = 1, 2, 3, \ldots,\ n = 0, 1, 2, \ldots$)

$$\alpha = \frac{1}{2}\sqrt{\frac{\gamma}{2\nu|a|}}\left(1 - \frac{a}{|a|}\mathrm{i}\right), \tag{11.300}$$

$$\lambda_{\nu n} = \mathrm{i}\nu\omega_0 + (2n+\nu+1)\sqrt{\frac{1}{2}|a|\Theta\nu\gamma}\left(1 + \frac{a}{|a|}\mathrm{i}\right). \tag{11.301}$$

[The sign of α has to be chosen in such a way that the eigenfunctions (11.298) decrease with increasing energy.] All the assumptions made in the beginning are fulfilled by (11.298, 300). Therefore, we have found consistent solutions of

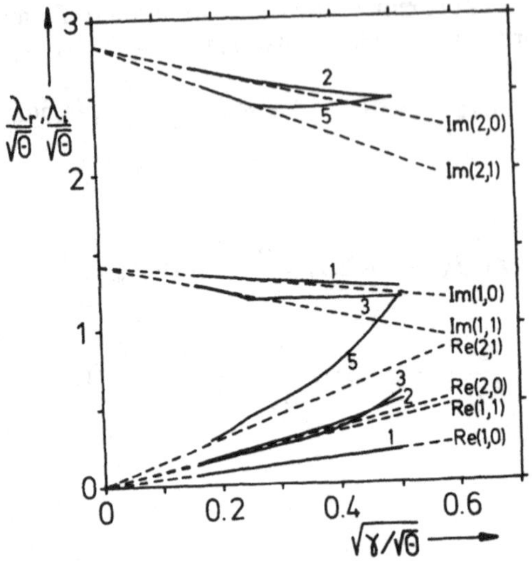

Fig. 11.44. A plot of real and imaginary parts of some complex eigenvalues divided by $\sqrt{\Theta}$ as a function of the square root of the friction constant (*solid lines*) compared to the low-friction results (11.301 a) for the indicated pairs (ν, n) (*broken lines*) for $d/\Theta = \omega_0^2/\Theta = 2$ and $F = 0$. The numbers 1, 2, 3, 5 indicate the same curves as in Fig. 11.40

(11.289) in the limit $\gamma \to 0$. Other solutions not fulfilling our assumptions may also occur.

The present derivation holds only for $\nu \neq 0$ and $a \neq 0$. For real eigenvalues or for a strict parabolic potential the eigenfunctions do not show this contraction to the minimal energy with width $\sim\sqrt{\gamma}$. It may, of course, happen that $a = 0$ and some other values of $\mathrm{d}^n\omega(E)/(\mathrm{d}E)^n$ for $E = E_{\min}$ are different from zero for some special potentials $f(x)$. In this case, one also gets a contraction to $E \approx E_{\min}$, but the dependence of λ on γ will then have a different form.

For the cosine potential, where a is given by (11.292 b), (11.301) specializes to

$$\lambda_{\nu n} = \mathrm{i}\,\nu\omega_0 + \tfrac{1}{4}(2n+\nu+1)(1-\mathrm{i})\sqrt{\nu\Theta\gamma/\omega_0}\,, \tag{11.301 a}$$

$$\omega_0 = \sqrt{d}\,; \quad \nu = 1,2,3,\dots; \quad n = 0,1,2,\dots\,.$$

The complex conjugate of (11.301 a) is also an eigenvalue. In Fig. 11.44 the real and imaginary parts of (11.301 a) are compared with the results of the matrix continued-fraction method, showing a very good fit for small damping constants.

The eigenvalues and eigenfunctions can be used to obtain expressions for the susceptibilities. As already mentioned in Sect. 11.8, the maximum at the finite frequency agrees quite well with the imaginary part and the width with the real part of one of the eigenvalues, if the damping constant is not too low. Because of the contraction of eigenfunctions, more and more eigenvalues and eigenfunctions enter for very small γ, and in the limit $\gamma \to 0$ an infinite number of eigenvalues and eigenfunctions must be used.

Real Eigenvalues

For the determination of the real eigenvalues, where the eigenfunctions depend on the energy only, we have to average (11.90) or (11.91) over a trajectory

$E = \text{const}$ of the noiseless equation (11.263). For this purpose we first introduce the action integral $I(E)$ and its derivative defined by

$$I(E) = \int_{x_1(E)}^{x_2(E)} v(x,E)\,\mathrm{d}x\,, \quad I'(E) = T(E) = \int_{x_1(E)}^{x_2(E)} [1/v(x,E)]\,\mathrm{d}x\,. \tag{11.302}$$

Here $x_1(E)$ and $x_2(E)$ are the minimal and maximal x-values in the potential, see Fig. 11.16. For $E > E_0$ we have $x_1 = -\pi$, $x_2 = +\pi$ and the averaged velocity (11.102) is then given by the first part of (11.302) divided by 2π i.e.

$$\bar{v}(E) = I(E)/(2\pi) \quad \text{for} \quad E > E_0\,. \tag{11.303}$$

It should be noted that the usual definition $\hat{I}$ of an action integral for $E < E_0$ is twice the value (11.302)

$$\hat{I}(E) = \oint v(x,E)\,\mathrm{d}x = 2I(E)\,. \tag{11.304}$$

We do not use the definition (11.304) because the action integral would then be discontinuous at $E = E_0$. The time for one cycle for $E < E_0$ is twice the derivative of the action integral i.e., $2I'(E)$. For the potential $-d\cos x$, $I(E)$ and $T(E)$ are given by:

$$\begin{aligned}
&E_{\min} = -d \leqq E \leqq E_0 = d\\
&\qquad I(E) = 8\sqrt{d}\{\mathbf{E}[(E+d)/(2d)] - [1-(E+d)/(2d)]\,\mathbf{K}[(d+E)/(2d)]\}\\
&\qquad\quad \approx (\pi/\sqrt{d})(E+d)[1+(E+d)/(16d)] \quad \text{for} \quad E \gtrsim E_{\min}\,,\\
&I'(E) = T(E) = (2/\sqrt{d})\,\mathbf{K}[(E+d)/(2d)]\\
&\qquad\quad \approx (\pi/\sqrt{d})[1+(E+d)/(8d)] \quad \text{for} \quad E \gtrsim E_{\min}\,,
\end{aligned} \tag{11.305a}$$

$$\begin{aligned}
&E \geqq E_0 = d\\
&\qquad I(E) = 4\sqrt{2(E+d)}\,\mathbf{E}[2d/(E+d)]\\
&\qquad\quad \approx 2\pi\sqrt{2(E+d)} - 2\pi d/\sqrt{2(E+d)} \quad \text{for} \quad E \gg d\\
&I'(E) = T(E) = (4\sqrt{2(E+d)})\,\mathbf{K}[2d/(E+d)]\\
&\qquad\quad \approx 2\pi/\sqrt{2(E+d)} + 2\pi d/\sqrt{2(E+d)}^{\,3} \quad \text{for} \quad E \gg d\,.
\end{aligned} \tag{11.305b}$$

Here $\mathbf{K}(m)$ and $\mathbf{E}(m)$ are the complete elliptic integrals of first and second kind [11.50]. The action integral (11.305) and its derivative are plotted in Fig. 11.45. At $E = E_0$, $T(E)$ has a weak logarithmic singularity of the form

$$I'(E) = T(E) \approx (2/\sqrt{d})\ln(32d/|E-d|) \quad \text{for} \quad E \approx d\,. \tag{11.306}$$

If we average (11.91) over x or equivalently if we take a time-average of (11.91) multiplied by $v(x,E)$ we obtain

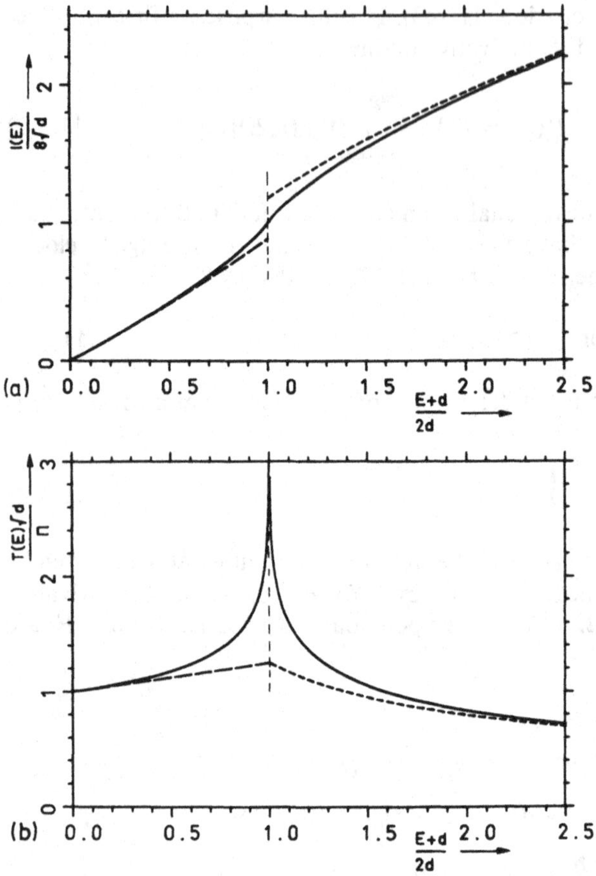

Fig. 11.45. The action integral $I(E)$ (**a**) and its derivative $T(E) = I'(E)$ (**b**) for the potential $-d\cos x$ (*solid line*). The expressions of (11.305) for $E+d \ll d$ and $E \gg d$ are shown by the *broken line*

$$E < E_0$$

$$I'(E)\frac{\partial \tilde{W}_S}{\partial t} = \gamma \frac{\partial}{\partial E} I(E)\left(1 + \Theta\frac{\partial}{\partial E}\right)\tilde{W}_S$$

$$\tilde{W}_D = 0\,, \tag{11.307a}$$

$$E > E_0$$

$$I'(E)\frac{\partial \tilde{W}_{\substack{D\\S}}}{\partial t} = \gamma \frac{\partial}{\partial E}\left[I(E)\left(1 + \Theta\frac{\partial}{\partial E}\right)\tilde{W}_{\substack{D\\S}} - 2\pi F_0 \tilde{W}_{\substack{S\\D}}\right]. \tag{11.307b}$$

As discussed in Sect. 11.4.2 the force $F_0 = F/\gamma$ drops out in (11.307a) because of the closed motion. The separation ansatz

$$\tilde{W}_{\substack{D\\S}}(E,t) = \Phi^{(n)}_{\substack{D\\S}}(E)\,e^{-\lambda_n t} \tag{11.308}$$

leads to the following equation for the eigenvalues

$$A_n = \lambda_n/\gamma \tag{11.309}$$

and the eigenfunctions $\Phi^{(n)}_{\substack{D\\S}}(E)$:

$E < E_0$

$$\frac{d}{dE} I(E)\left(1+\Theta\frac{d}{dE}\right)\Phi^{(n)}_S + A_n I'(E)\,\Phi^{(n)}_S = 0 \tag{11.310a}$$

$$\Phi_D = 0\,,$$

$E > E_0$

$$\frac{d}{dE}\left[I(E)\left(1+\Theta\frac{d}{dE}\right)\Phi^{(n)}_S - 2\pi F_0 \Phi^{(n)}_{\substack{S\\D}}\right] + A_n I'(E)\,\Phi^{(n)}_{\substack{D\\S}} = 0\,. \tag{11.310b}$$

Boundary Condition at $E = E_{min}$

Because $I(E)$ vanishes at $E = E_{min}$, we obtain from (11.310a) for $E = E_{min}$

$$I'(E_{min})\left(1+\Theta\frac{d}{dE}\right)\Phi^{(n)}_S(E)\Big|_{E=E_{min}} + A_n I'(E_{min})\,\Phi^{(n)}_S(E_{min}) = 0\,,$$

i.e.,

$$\Theta d\,\Phi^{(n)}_S(E)/dE\,|_{E=E_{min}} + (A_n+1)\,\Phi^{(n)}_S(E_{min}) = 0\,. \tag{11.311}$$

Thus the derivative of $\Phi^{(n)}_S$ can be expressed in terms of $\Phi^{(n)}_S$ and A_n at $E = E_{min}$.

Continuity Condition at $E = E_0$

Because the eigenfunctions (11.307) decay very slowly in time (the decay constant is of the order $\lambda_1 = A_1\gamma$) this time dependence need not be taken into account inside the very thin boundary layer of thickness $\sqrt{\gamma}$, where a quasistationary distribution is rapidly established. We therefore use the same continuity condition (11.121, 121b, 122) derived in Sects. 11.4.3, 4.4 for the stationary distribution, i.e.,

$$\Phi^{(n)}_S(E_0-0) = \Phi^{(n)}_S(E_0+0)\,, \tag{11.312}$$

$$d\,\Phi^{(n)}_S(E)/dE\,|_{E=E_0-0} = d\,\Phi^{(n)}_S(E)/dE\,|_{E=E_0+0} - \Phi_D(E_0+0)\,2\pi F_0/[I(E_0)\,\Theta] \tag{11.313}$$

$$\Phi^{(n)}_D(E_0+0) = \varkappa\, d\,\Phi^{(n)}_D(E)/dE\,|_{E=E_0+0}\sqrt{\gamma\Theta I(E_0)/\pi}\,. \tag{11.314}$$

The constant $\varkappa = 0.855(4)$ was determined in Sect. 11.4. Without the boundary layer, i.e., for $\gamma \to 0$, (11.313, 314) simplify to

$$\mathrm{d}\,\Phi_{\mathrm{S}}^{(n)}(E)/\mathrm{d}E\,|_{E=E_0-0} = \mathrm{d}\,\Phi_{\mathrm{S}}^{(n)}(E)/\mathrm{d}E\,|_{E=E_0+0}\,, \tag{11.313a}$$

$$\Phi_{\mathrm{D}}^{(n)}(E_0+0) = 0\,. \tag{11.314a}$$

The eigenvalues $A_n = \lambda_n/\gamma$ are thus determined by (11.310–314) and the requirement that $\Phi_{\mathrm{D}}^{(n)}$ and $\Phi_{\mathrm{S}}^{(n)}$ vanish for $E \to \infty$. The weak logarithmic singularity of $I'(E)$ at $E = E_0$ does not lead to serious difficulties.

$F_0 = 0$

Without the force F_0 the equations for $\Phi_{\mathrm{S}}^{(n)}$ and $\Phi_{\mathrm{D}}^{(n)}$ are decoupled. The equation for the sum function Φ_{S} does not contain any $\sqrt{\gamma}$ term and the eigenvalues $\lambda_n^{(\mathrm{S})}$ should therefore not depend very much on γ. The eigenvalues determined by a numerical integration (Sect. 5.9.2) agree fairly well with those obtained by the matrix continued-fraction method. The first non-zero eigenvalue for the cosine potential with $d/\Theta = 2$ is $\lambda_1^{(\mathrm{S})}/\gamma = 0.848$ whereas the matrix continued-fraction for $\gamma = 0.2$ leads to $\lambda_1^{(\mathrm{S})}/\gamma = 0.868$. Because for the difference function $\Phi_{\mathrm{D}}^{(n)}$, $\sqrt{\gamma}$ enters in the boundary condition, the eigenvalues $\lambda_n^{(\mathrm{D})}$ for the difference functions should be of the form

$$\lambda_n^{(\mathrm{D})}(\gamma)/\gamma = \lambda_n^{(\mathrm{D})}(0)/\gamma + B_n\sqrt{\gamma} \tag{11.315}$$

for small γ. For the cosine potential with $d/\Theta = 2$ the numerical integration leads to $\lambda_1^{(\mathrm{D})}(0)/\gamma = 3.19$ and $\lambda_1^{(\mathrm{D})}(0.2)/\gamma = 2.56$ for the first eigenvalue. The last expression agrees again fairly well with matrix continued-fraction result $\lambda_1^{(\mathrm{D})}(0.2)/\gamma = 2.64$.

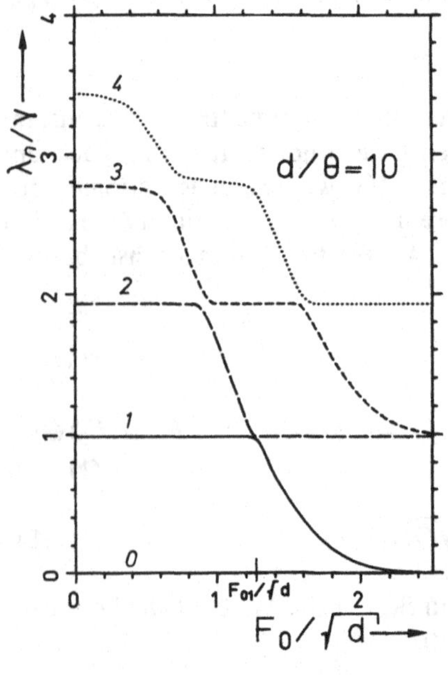

Fig. 11.46. The ratio λ_n/γ for $d/\Theta = 10$ as a function of $F_0/\sqrt{d}$ in the limit $\gamma \to 0$. The critical force $F_{01}/\sqrt{d} = 4/\pi$ is also indicated

$F_0 \neq 0$

For $F_0 \neq 0$ the differential equations are coupled. The eigenvalues can again be determined by numerical integration. (The eigenvalues A_n and values $\mathrm{d}\,\Phi_{\mathrm{D}}^{(n)}/\mathrm{d}E\,|_{E=E_0}$ are determined by the requirement that both $\Phi_{\mathrm{S}}^{(n)}$ and $\Phi_{\mathrm{D}}^{(n)}$ vanish for large E.) In Fig. 11.46 the dependence of some low eigenvalues obtained in [5.32] on the external force F_0 is shown for a low noise power in the limit $\gamma \rightarrow 0$, i.e., for the continuity condition (11.312, 313a, 314a). As seen the eigenvalues for small and large F_0 are essentially given by $\lambda_n/\gamma \approx n$ $(n = 0, 1, 2, \ldots)$ for low noise powers Θ; they are not degenerate for low F_0 but nearly twofold degenerate for large F_0. This can be interpreted as follows. For low F_0 the effective potential in Fig. 11.24 has only one well whereas for large F_0 two wells occur. For low noise powers Θ the wells are separated by a high barrier ($d \gg \Theta$). The eigenvalues in the left and right well nearly coincide and are given by $\lambda_n/\gamma \approx n$ for small Θ as may be derived by expanding $I(E)$ around the minima. Thus in the bistability region $F_0 > F_{01}$ a degeneracy occurs. In the plot the transition of the nondegenerate eigenvalues to the degenerate ones by changing F_0 from low to high values is clearly seen. This transition at F_{01} is similar to the transition at F_1 in Fig. 11.43a.

12. Statistical Properties of Laser Light

The Fokker-Planck equation has become a very useful tool for treating noise in quantum optics. In this chapter we investigate noise in a laser, which is the most important device in quantum optics. This subject together with other applications of the Fokker-Planck equation in quantum optics are already treated in a number of handbooks, books and review articles [12.1 – 13, 1.28, 4.8]. The main purpose of this chapter is to demonstrate how some of the methods of Chaps. 2 – 9 can be applied to a simple laser model (one mode, adiabatic elimination of all variables with the exception of the laser field, threshold region). The following two points make it difficult but also interesting to investigate the statistical properties of laser light.

The first difficulty arises because we have to deal with a nonlinear equation. The laser is a self-sustained oscillator in the optical frequency region. Such an oscillator, where for a small initial value the amplitude increases in an exponential way and finally oscillates with a finite fixed value, cannot be described by a linear equation. One of the simplest equations which shows the typical features of a self-excited oscillator is the *Van der Pol* equation [12.14] ($\beta > 0$)

$$\ddot{y} - 2\beta(d - y^2)\dot{y} + \omega^2 y = 0\,. \tag{12.1}$$

It was invented by Van der Pol to describe the amplitude of a self-sustained vacuum-tube oscillator. A similar equation was already used by *Rayleigh* [12.15] ($\beta > 0$)

$$\ddot{y} - 2\beta[d - \dot{y}^2/(3\,\omega^2)]\dot{y} + \omega^2 y = 0 \tag{12.2}$$

to deal with self-sustained oscillations in organ pipes. Both equations have an amplitude-dependent amplification or damping term. For small amplitudes and positive d the damping constant is negative and therefore the amplitude grows exponentially in time. For larger amplitudes this amplification decreases and at last the amplitude reaches a finite value.

Rotating Wave Approximation

If the amplification is small so that the increase of the amplitude is small in one period $1/\omega$, we can make the rotating wave approximation. Inserting

$$
\begin{aligned}
y(t) &= b(t)\,e^{-i\omega t} + \text{c.c.} \\
\dot{y}(t) &= \underline{\dot{b}(t)\,e^{-i\omega t}} - i\omega b(t)\,e^{-i\omega t} + \text{c.c.} \qquad (12.3)\\
\ddot{y}(t) &= \underline{\ddot{b}(t)\,e^{-i\omega t}} - 2i\omega \dot{b}(t)\,e^{-i\omega t} - \omega^2 b(t)\,e^{-i\omega t} + \text{c.c.}\,,
\end{aligned}
$$

(where +c.c. means that we have to add the complex conjugate) into (12.1 or 2) and neglecting the small underlined terms and higher harmonics, we obtain the rotating wave approximation to (12.1, 2)

$$\dot{b} - \beta(d - b^* b)\,b = 0\,. \qquad (12.4)$$

The explicit solution of this equation

$$b(t) = b(0)\,e^{\beta dt}[1 + (e^{2\beta dt} - 1)\,|b(0)|^2/d]^{-1/2} \qquad (12.5)$$

shows for $d > 0$ the behavior of a self-sustained oscillator described above, Fig. 12.1. For $d \leqq 0$ no final amplitude will build up.

The second difficulty arises because of the quantum nature of the spontaneous emission noise. There may of course be other noise sources in the laser, e.g., fluctuating pumping or fluctuating mirrors. Whereas the latter noise sources can be eliminated in principle, the spontaneous emission noise can never be eliminated, because it stems from the quantum nature of light. To include spontaneous noise in a laser one should therefore treat the laser field as well as the atoms or molecules, which drive the laser field, in a fully quantum-mechanical way. Because of the large number of photons in the laser cavity (even at threshold, this number is of the order 10^3), however, one may treat the electrical field classically, i.e., neglect its operator character, provided that a proper classical noise source is added. The strength of this classical noise force can be determined so that it leads to the correct spontaneous emission rate, as explained in Sect. 12.1.2. The atoms or molecules, which drive the laser field, must always be treated quantum mechanically. This procedure is called semi-classical treatment and will be used throughout this chapter. As also shown in Sect. 12.1.2, the main result is that one obtains the rotating wave Van der Pol equation (12.4) with a proper δ-correlated noise term,

$$\dot{b} - \beta(d - b^* b)\,b = \sqrt{q}\,\Gamma(t)\,, \qquad (12.6\text{a})$$

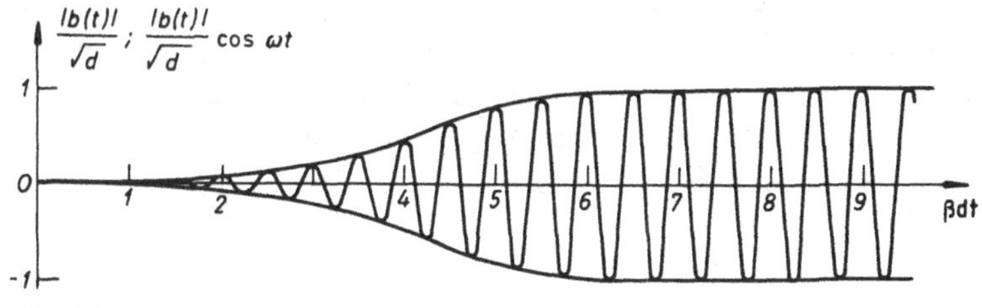

Fig. 12.1.
Solution (12.5) divided by $\sqrt{d}$ as a function of βdt. The initial value is given by $|b(0)|/\sqrt{d} = 10^{-2}$

$$\langle \Gamma(t) \rangle = 0\,, \quad \langle \Gamma(t)\Gamma^*(t') \rangle = 4\delta(t-t')\,, \quad \langle \Gamma(t)\Gamma(t') \rangle = 0\,. \tag{12.6b}$$

In a fully quantum-mechanical treatment, *Haken* [12.16] first derived (12.6), where b and b^* are the operators b and b^+ of the lasing field. If the operator character is neglected, one then obtains the nonlinear Langevin equation (12.6).

By a proper linearization procedure one may solve the nonlinear Langevin equation for large positive and large negative d [12.16]. For smaller d values the nonlinear Langevin equation (12.6a) must be used. As discussed in Chap. 4, the statistical properties of a process described by a nonlinear Langevin equation are best obtained from the corresponding Fokker-Planck equation. The Fokker-Planck equation corresponding to the laser Langevin equation (12.6) is set up in Sect. 12.1.3. It was derived and solved for various cases in [12.17 – 33, 9.18, 5.15]. In connection with the self-excited oscillator, the Fokker-Planck equation corresponding to Langevin equation (12.6) was also solved in [12.34] for the stationary state. A fully quantum mechanical derivation of this Fokker-Planck equation [12.35 – 37] (for solutions of similar equations see [12.38, 39]) has been obtained, whose main steps follow.

Starting with the equation of motion for the density operator for light field and atoms, one derives an equation of motion for continuous distribution functions. These distribution functions are defined in a way similar to that in App. A4, where the quantum-mechanical damped oscillator is treated. By eliminating the atom variables, one then obtains an equation for the distribution function of the light field alone. The leading term of this equation agrees with the Fokker-Planck equation corresponding to (12.6). A closer inspection shows that the correction terms to this Fokker-Planck equation contain derivatives which are of a higher order than two, i.e., the process is no longer described by an ordinary Fokker-Planck equation. (In the case of the damped quantum-mechanical linear oscillator in App. A4, only terms up to the second derivative occur.)

In Sect. 12.1 we derive the laser Langevin equation (12.6) and the corresponding Fokker-Planck equation. In Sect. 12.2 the stationary solution and stationary expectation values are obtained. In Sect. 12.3 the expansion of the instationary solution into eigenmodes is discussed and correlation functions are derived. By a proper expansion of the distribution function into Laguerre functions it is shown in Sect. 12.4 that the laser Fokker-Planck equation can be reduced to a system of ordinary differential equations, with only four nearest-neighbor coefficients coupled. By applying the method of Chap. 9, the system can be solved by matrix continued fractions. In Sect. 12.5 various methods for obtaining the transient of the laser are presented, and finally, in Sect. 12.6 the photon counting distribution, which follows from the laser Langevin equation (12.6), is investigated. These photon counting distributions, which have been measured in great detail [12.22, 40 – 46], confirm the theoretical predictions of the simple nonlinear one-mode laser Langevin equation (12.6). (For a generalization to two-mode and multi-mode lasers, see [12.1, 3, 47 – 49, 4.8].)

12.1 Semiclassical Laser Equations

12.1.1 Equations Without Noise

As discussed in the beginning of this chapter, semiclassical means that one neglects the operator character of the light field and treats it as a classical variable, but that the atoms are treated quantum mechanically. The laser equations comprise the wave equation for the electric field, where the polarization is the driving force, and the material equations, which express the polarization and inversion in terms of the electric field. To make the equations as simple as possible, we confine ourselves to a running wave single-mode ring-laser model with N two-level homogeneously broadened atoms. The wave equation for the electric field E polarized in x direction reads

$$\frac{\partial^2 E}{\partial t^2}+2\varkappa\frac{\partial E}{\partial t}-c^2\frac{\partial^2 E}{\partial z^2}=-\frac{1}{\varepsilon_0}\frac{\partial^2 P}{\partial t^2}, \tag{12.7}$$

where $\varkappa$ describes the losses of the electric field. The atoms near $z=z_\mu$ are described by a density operator $\rho^{(\mu)}$. Its equation of motion is given by [12.3]

$$\dot\rho^{(\mu)}=-(\mathrm{i}/\hbar)[H,\rho^{(\mu)}]\,. \tag{12.8}$$

The Hamilton operator

$$H=H_0-exE \tag{12.9}$$

is a sum of a free field part H_0 and an interaction part $-exE$. The electric field strength in (12.9) must be taken at $z=z_\mu$. Because we have assumed a two-level system, the density operator can be expanded in the two eigenstates $|1\rangle$ and $|2\rangle$ of the Hamilton operator H_0

$$\begin{aligned} H_0|1\rangle &= \varepsilon_1|1\rangle\,, & H_0|2\rangle &= \varepsilon_2|2\rangle \\ \langle i|j\rangle &= \delta_{ij}\,, & \hbar\omega_0 &= \varepsilon_2-\varepsilon_1\,. \end{aligned} \tag{12.10}$$

From (12.8) it follows that the equation of motion for the elements $\rho_{ij}^{(\mu)}=[\rho_{ji}^{(\mu)}]^*$ $=\langle i|\rho^{(\mu)}|j\rangle$ of the density operator $\rho^{(\mu)}$ has the form

$$\dot\rho_{12}^{(\mu)}=\mathrm{i}\,\omega_0\rho_{12}^{(\mu)}+\mathrm{i}(e/\hbar)x_{12}E[\rho_{22}^{(\mu)}-\rho_{11}^{(\mu)}]-\underline{\gamma_2\rho_{12}^{(\mu)}}\,, \tag{12.11}$$

$$\dot\rho_{22}^{(\mu)}-\dot\rho_{11}^{(\mu)}=\mathrm{i}(2e/\hbar)x_{12}E[\rho_{12}^{(\mu)}-\rho_{21}^{(\mu)}]+\underline{\gamma_1[\sigma_0/N-(\rho_{22}^{(\mu)}-\rho_{11}^{(\mu)})]}\,. \tag{12.12}$$

In deriving (12.11, 12) we assumed that there is no permanent dipole moment in the ground and excited states $\langle 1|x|1\rangle=\langle 2|x|2\rangle=0$. The phase factors of the ground and exited states were chosen so that the matrix element $x_{12} = \langle 1|x|2\rangle = x_{21}$ is real. The underlined terms in (12.11, 12) were added to describe damping of the off-diagonal elements $(-\gamma_2 \rho_{12}^{(\mu)})$ and diagonal elements $[-\gamma_1(\rho_{22}^{(\mu)} - \rho_{11}^{(\mu)})]$, and to include pumping $(\gamma_1 \sigma_0/N)$. Equations (12.11, 12) are the Bloch equations $(\gamma_2 = 1/T_2,\ \gamma_1 = 1/T_1)$ of the spin resonance theory [12.50]. The polarization in (12.7) is the expectation value of the dipole moment per volume

$$P(z,t) = \Delta^{-1} \sum_{(z_\mu - z)\in\Delta} e x_{12}[\rho_{12}^{(\mu)}(t) + \rho_{21}^{(\mu)}(t)] \,. \tag{12.13}$$

Here we sum up over a number of active atoms situated at z_μ, which in turn are placed in a volume element Δ around z. This volume element may be so small that in it the electric field $E(z,t)$ is practically constant, but so large that it contains a large number of active atoms. Since we are concerned with a one-dimensional ring laser with one allowed direction of propagation, the electric field has the form of a traveling wave. Because of the relatively weak interaction with the active atoms, the amplitudes will change slowly. Therefore we can make the ansatz (V is the cavity volume)

$$E(z,t) = \sqrt{\frac{\hbar\omega_0}{2\varepsilon_0 V}}\,\{b(t)\exp[-\mathrm{i}\,\omega_0(t - z/c)] + \text{c.c.}\}\,, \tag{12.14}$$

where $b(t)$ is a slowly varying complex function with respect to the period $2\pi/\omega_0$ (i.e., $|\dot{b}| \ll \omega_0|b|$) and where +c.c. means that one has to add the complex conjugate. That $b(t)$ does not depend on the space coordinate z restricts the field to one running mode only. Furthermore, it is implied in ansatz (12.14) that we are restricted to the tuned case $(\omega_0 L/c = n2\pi)$, because the field must be periodic with the cavity length L. The normalization in (12.14) was chosen in such a manner that b^*b gives the intensity of the electric light field in photon numbers. Introducing

$$s(t) = \sum_\mu \rho_{21}^{(\mu)} \exp[\mathrm{i}\,\omega_0(t - z_\mu/c)]\,, \tag{12.15}$$

$$\sigma(t) = \sum_\mu (\rho_{22}^{(\mu)} - \rho_{11}^{(\mu)})\,, \tag{12.16}$$

one obtains the following equations by inserting expressions (12.14 – 16) in (12.7, 11 – 13) and by neglecting antiresonant terms:

$$\begin{aligned} &\dot{b} + \varkappa b = \mathrm{i} g s \\ &\dot{s} + \gamma_2 s = -\mathrm{i} g b \sigma \\ &\dot{\sigma} + \gamma_1(\sigma - \sigma_0) = 2\mathrm{i} g(s^* b - s b^*)\,. \end{aligned} \tag{12.17}$$

Here the coupling constant is defined by

$$g = \omega_0 \sqrt{\alpha_S \frac{(x_{12})^2 \lambda}{V}}, \quad \alpha_S = \frac{e^2}{4\pi\varepsilon_0 \hbar c} = \frac{1}{137}, \tag{12.18}$$

where $\lambda = 2\pi c/\omega_0$ is the wavelength. The equations (12.17) have only a stationary nonzero solution if the pump parameter σ_0 is larger than the threshold value of the inversion

$$\sigma_0 > \sigma_{\mathrm{thr}} = \varkappa\gamma_2/g^2. \tag{12.19}$$

This relation is the *Schawlow-Townes* [12.51] formula specialized to our assumptions. The steady-state intensity is then given by

$$b^*b = \frac{\gamma_1}{4\varkappa}\left(\sigma_0 - \frac{\varkappa\gamma_2}{g^2}\right). \tag{12.20}$$

Near threshold ($\sigma_0 \approx \sigma_{\mathrm{thr}}$) one can simplify (12.17) further. If the time variation of b and s is slow in the times $1/\gamma_1$ and $1/\gamma_2$, the inversion is approximately given by

$$\sigma = \sigma_0 \Big/ \left(1 + \frac{4g^2}{\gamma_1\gamma_2} b^*b\right) \approx \sigma_0\left(1 - \frac{4g^2}{\gamma_1\gamma_2} b^*b\right). \tag{12.21}$$

Inserting this approximate inversion into (12.17) leads, after neglecting the (small) derivatives $\dot{s}$, to the rotating wave Van der Pol or Rayleigh equation (12.4). The parameters β and d in (12.4) are given by

$$\beta = 4g^2\varkappa/(\gamma_1\gamma_2), \quad d = (\sigma_0 - \sigma_{\mathrm{thr}})[\gamma_1/(4\varkappa)]. \tag{12.22}$$

12.1.2 Langevin Equation

To describe the spontaneous emission noise, we add a noise force to the rhs of (12.4), leading to (12.6a). Usually the time constant βd is much smaller than the decay rate γ_2 of the spontaneous emission process. In the slow time scale of the variable b we can therefore assume that the Langevin force $\Gamma(t)$ is δ correlated. If the spontaneous emission processes of the atoms are independent it is also reasonable to assume that $\Gamma(t)$ is a Gaussian variable. Because no phase can be induced by the spontaneous emission process, the correlation function of the Gaussian random variable is given by (12.6b), where the constant q must still be determined. In real notation

$$b = b_1 + \mathrm{i}b_2, \quad \Gamma = \Gamma_1 + \mathrm{i}\Gamma_2 \tag{12.23}$$

(12.6) takes the form

$$\dot{b}_i - \beta(d - b_1^2 - b_2^2)b_i = \sqrt{q}\,\Gamma_i, \quad i = 1, 2, \tag{12.24a}$$

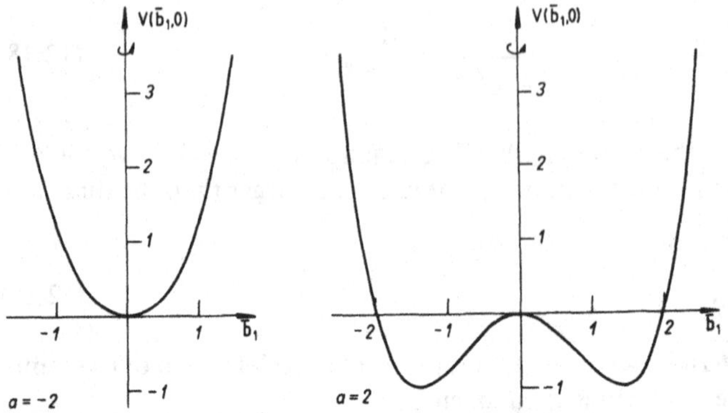

Fig. 12.2. The potential (12.25) with $\bar{b}_2 = 0$ for the normalization (12.37) as a function of $\bar{b}_1$ below ($a = -2$) and above ($a = 2$) threshold. The full potential, $V(\bar{b}_1, \bar{b}_2)$ with the $\bar{b}_2$ axis perpendicular to the plane, is the surface which is generated by rotating the potential curve in the figure as indicated by the *arrow*

$$\langle \Gamma_i(t)\Gamma_j(t')\rangle = 2\delta_{ij}\delta(t-t')\,. \tag{12.24b}$$

Equations (12.24) may be interpreted as overdamped Brownian motion in the potential

$$V(b_1, b_2) = -\tfrac{1}{2}\beta d(b_1^2+b_2^2) + \tfrac{1}{4}\beta(b_1^2+b_2^2)^2\,. \tag{12.25}$$

This potential is shown in Fig. 12.2 for the normalization (12.37).

To determine the strength q of the noise force and for later purposes we introduce the intensity and the phase of b defined by

$$I = b^*b = b_1^2+b_2^2\,, \qquad \varphi = \arctan(b_2/b_1)\,. \tag{12.26}$$

According to (3.126, 127) the Langevin equations (12.24a) in the variables I and φ now read

$$\begin{aligned} &\dot{I} - 2\beta(d-I)I = 2\sqrt{q}\sqrt{I}\cos\varphi\,\Gamma_1 + 2\sqrt{q}\sqrt{I}\sin\varphi\,\Gamma_2 \\ &\dot{\varphi} = -\sqrt{q}(\sin\varphi/\sqrt{I})\Gamma_1 + \sqrt{q}(\cos\varphi/\sqrt{I})\Gamma_2\,. \end{aligned} \tag{12.27}$$

These equations lead to the following drift and diffusion coefficients for the variables I and φ, see (3.118, 119);

$$\begin{aligned} D_I &= 2\beta(d-I)I + 4q \\ D_{II} &= 4qI \\ D_\varphi &= 0 \\ D_{\varphi\varphi} &= q/I\,. \end{aligned} \tag{12.28}$$

Instead of (12.27) we may equally well use the Langevin equations

$$\dot{I}-2\beta(d-I)I-2q=2\sqrt{qI}\,\Gamma_I$$
$$\dot{\varphi}=-\sqrt{q/I}\,\Gamma_\varphi \tag{12.29}$$

with

$$\langle\Gamma_I(t)\Gamma_I(t')\rangle=\langle\Gamma_\varphi(t)\Gamma_\varphi(t')\rangle=2\delta(t-t')$$
$$\langle\Gamma_I(t)\Gamma_\varphi(t')\rangle=0 \tag{12.30}$$

because they lead to the same drift and diffusion coefficients (12.28) in the Stratonovich definition.

Linearization of the Langevin Equation

In the limit far below and far above threshold we can linearize the laser Langevin equations in the following way.

Far below Threshold: ($d\ll-\sqrt{q/\beta}$)

Here we may neglect the nonlinearity in (12.24a) and thus obtain the Langevin equation of the Ornstein-Uhlenbeck process (Sect. 3.2)

$$\dot{b}_i+\beta|d|b_i=\sqrt{q}\,\Gamma_i\,. \tag{12.31}$$

Far above Threshold: ($d\gg\sqrt{q/\beta}$)

Here the amplitude of the intensity is stabilized at $I\approx d$. This follows from (12.49, 51) for $a\gg1$, if we use the normalized units (12.37) (see also Fig. 12.4). Writing

$$I=d+\Delta I\,, \tag{12.32}$$

we again obtain for ΔI an equation for an Ornstein-Uhlenbeck process

$$\Delta\dot{I}+2\beta d\,\Delta I=2\sqrt{qd}\,\Gamma_I\,, \tag{12.33 a}$$

whereas for φ we obtain the Wiener process (Sect. 3.2)

$$\dot{\varphi}=-\sqrt{q/d}\,\Gamma_\varphi \tag{12.33 b}$$

with Γ_I, Γ_φ given by (12.30).

Determination of the Constant q

As explained in Sect. 3.3.2, the drift coefficient D_I is the expectation value of the time derivative of I starting with a fixed value $I(t)$, i.e., ($\tau>0$)

$$\langle\dot{I}(t)\rangle=\lim_{\tau\to0}\langle I(t+\tau)-I(t)\rangle/\tau=D_I=2\beta(d-I)I+4q\,.$$

Using (12.19, 21, 22) it becomes

$$\langle \dot{I}(t) \rangle = 2(g^2/\gamma_2)(N_2 - N_1)I - 2\varkappa I + 4q\,, \tag{12.34}$$

where the inversion $N\sigma = N_2 - N_1$ is the difference of the number of atoms in the upper and lower states. The first term in (12.34) describes the induced emission rate, the second the loss rate due to the cavity losses. To obtain the correct quantum-mechanical spontaneous emission rate, we require

$$q = N_2 g^2/(2\gamma_2)\,, \tag{12.35}$$

according to the Einstein theory of radiation. (For simplicity we neglected the number of thermal quanta, which is very small for laser light.) The number N_2 of atoms in the upper state depends on b^*b, because the inversion depends on it, i.e., $N_2 = (N+\sigma)/2$; however, near threshold N_2 and therefore q may be regarded as constants.

For our simple model we thus now know the constants β, d and q of the basic Langevin equation (12.6a, b). If only one mode is involved, near threshold we get the same form as for other models, with only the constants β, d, q differing [12.1].

For a detuned laser β and d are complex [12.53].

Fluctuating Control Parameter

Whereas the spontaneous emission is described by an additive fluctuating force in the Langevin equation (12.6), we may also consider noise sources, where the pumping or control parameter d fluctuates. A theory of this process was given by *Graham* [12.52], see App. A6.

12.1.3 Laser Fokker-Planck Equation

With the help of (4.94a, 95, 99, 100), it is easy to write down the Fokker-Planck equation corresponding to (12.24):

$$\frac{\partial W}{\partial t} = \left[-\beta \sum_{i=1}^{2} \frac{\partial}{\partial b_i} (d - b_1^2 - b_2^2)\, b_i + q \sum_{i=1}^{2} \frac{\partial^2}{\partial b_i \partial b_i} \right] W\,, \tag{12.36}$$

or in vector notation [12.17a] $\boldsymbol{b} = (b_1, b_2)$

$$\frac{\partial W}{\partial t} + \beta \nabla [(d - |\boldsymbol{b}|^2)\boldsymbol{b} W] = q \Delta W\,. \tag{12.36a}$$

Here, the nabla operator ∇ and the Laplace operator Δ act with respect to $\boldsymbol{b}$. This final equation depends only on three parameters β, d, q. Whereas parameters β and q are constants for each laser, parameter d is variable and describes the strength of the pumping ($d<0$ below, $d=0$ at and $d>0$ above threshold).

Normalization

For numerical purposes it is convenient to introduce the normalized variables

$$\bar{b} = (\beta/q)^{1/4} b\,, \quad \bar{I} = \sqrt{\beta/q}\, I\,, \quad \bar{t} = \sqrt{\beta q}\, t\,, \quad a = \sqrt{\beta/q}\, d\,. \tag{12.37}$$

Equation (12.36a) is then transformed into

$$\partial W/\partial \bar{t} + \bar{\nabla}\, [(a - |\bar{b}|^2)\bar{b}\, W] = \bar{\Delta}\, W\,, \tag{12.38}$$

which depends only on the pump parameter a ($a<0$ below, $a=0$ at, and $a>0$ above threshold). The bar over ∇ and Δ indicates that one has to differentiate with respect to $\bar{b}$. Thus in the new variables (12.37) the constants β and q are normalized to 1, whereas d is replaced by a. In polar coordinates, $\bar{b} = \bar{r}\exp(\mathrm{i}\varphi)$, this Fokker-Planck equation has the form

$$\frac{\partial W}{\partial \bar{t}} + \frac{1}{\bar{r}}\frac{\partial}{\partial \bar{r}}[(a - \bar{r}^2)\bar{r}^2 W] = \frac{1}{\bar{r}}\frac{\partial}{\partial \bar{r}}\left(\bar{r}\frac{\partial W}{\partial \bar{r}}\right) + \frac{1}{\bar{r}^2}\frac{\partial^2 W}{\partial \varphi^2}\,. \tag{12.39}$$

The scaling parameters $\sqrt{(q/\beta)}$ and $1/\sqrt{\beta q}$ can be determined experimentally by measuring the photon number in the cavity and the linewidth of the intensity fluctuations near threshold [12.42a]. The Fokker-Planck equation corresponding to (12.27) or (12.29) is easily obtained. Using the same normalization (12.37) it reads

$$\frac{\partial \hat{W}}{\partial \bar{t}} = \left\{\frac{\partial}{\partial \bar{I}}[2(\bar{I} - a)\bar{I} - 4] + \frac{\partial^2}{\partial \bar{I}^2} 4\bar{I} + \frac{1}{\bar{I}}\frac{\partial^2}{\partial \varphi^2}\right\}\hat{W}\,. \tag{12.40}$$

This equation follows also from (12.39). Because the volume element transforms according to

$$\mathrm{d}\bar{b}_1\,\mathrm{d}\bar{b}_2 = \bar{r}\,\mathrm{d}\bar{r}\,\mathrm{d}\varphi = \tfrac{1}{2}\mathrm{d}\bar{I}\,\mathrm{d}\varphi\,,$$

the distributions W and $\hat{W}$, normalized according to

$$\int_{-\infty}^{\infty}\int_{-\infty}^{\infty} W\,\mathrm{d}\bar{b}_1\,\mathrm{d}\bar{b}_2 = \int_0^{\infty}\int_0^{2\pi} \hat{W}\,\mathrm{d}\bar{I}\,\mathrm{d}\varphi = 1\,,$$

differ by a factor of 2, i.e.,

$$W(\bar{b}_1, \bar{b}_2, \bar{t}) = 2\,\hat{W}(\bar{I}, \varphi, \bar{t})\,. \tag{12.41}$$

An equation for the averaged intensity

$$\langle \bar{I}^n(\bar{t})\rangle = \int_0^{\infty}\int_0^{2\pi} \bar{I}^n\,\hat{W}(\bar{I}, \varphi, \bar{t})\,\mathrm{d}\bar{I}\,\mathrm{d}\varphi \tag{12.42}$$

is obtained by multiplying (12.40) with $\bar{I}^n$ and by performing an integration by parts

$$\langle \dot{\bar{I}}^n(\bar{t})\rangle = 2na\langle \bar{I}^n(\bar{t})\rangle - 2n\langle \bar{I}^{n+1}(\bar{t})\rangle + 4n^2\langle \bar{I}^{n-1}(\bar{t})\rangle . \tag{12.43}$$

12.2 Stationary Solution and Its Expectation Values

The stationary solution of (12.40) cannot depend on the phase φ because no phase is preferred. The equation (12.40) for determining the stationary solution $\hat{W}_{st}(\bar{I})$ may be written as

$$\frac{\partial S_{\bar{I}}}{\partial \bar{I}} = 0\,, \quad S_{\bar{I}} = -2\bar{I}\left[(\bar{I}-a)\hat{W}_{st} + 2\frac{\partial \hat{W}_{st}}{\partial \bar{I}}\right]. \tag{12.44}$$

Here, $S_{\bar{I}}$ is the probability current in the $\bar{I}$ direction, which is a constant because of the first part of (12.44). This current must originate either from the origin or from infinity. In the present case, it has no physical meaning to assume a current $S_{\bar{I}}$ different from zero. Furthermore, it can be shown that for $S_{\bar{I}} \neq 0$ the distribution function $\hat{W}_{st}(\bar{I})$ does not go to zero sufficiently fast enough for $\bar{I} \to \infty$. When $S_{\bar{I}} = 0$, the stationary distribution function follows immediately from (12.44), viz.,

$$\begin{aligned} \hat{W}_{st}(\bar{I}) &= \frac{N}{2\pi}\exp(-\tfrac{1}{4}\bar{I}^2 + \tfrac{1}{2}a\bar{I}) = \frac{N}{2\pi}e^{a^2/4}\exp[-\tfrac{1}{4}(\bar{I}-a)^2]\,, \\ N^{-1} &= \int_0^\infty \exp(-\tfrac{1}{4}\bar{I}^2 + \tfrac{1}{2}a\bar{I})\,d\bar{I} = F_0(a)\,. \end{aligned} \tag{12.45}$$

The stationary distribution $W_{st} = 2\hat{W}_{st}$ is shown in Fig. 12.3 for different pump parameters near threshold. It is a Gaussian distribution in the intensity $\bar{I} = \bar{r}^2$ with the maximum value at $\bar{I} = a$ truncated at $\bar{I} = 0$.

Calculation of the Moments

The moments M_n and the generating function $M(u)$ of the moments of the stationary distribution function are given by

$$M_n = \langle \bar{I}^n\rangle = F_n(a)/F_0(a)\,, \tag{12.46}$$

$$M(u) = \langle e^{i\bar{I}u}\rangle = F_0(a+2iu)/F_0(a)\,, \tag{12.47}$$

where the integrals F_n are defined as

$$F_n(a) = \int_0^\infty \bar{I}^n \exp(-\tfrac{1}{4}\bar{I}^2 + \tfrac{1}{2}a\bar{I})\,d\bar{I}\,. \tag{12.48}$$

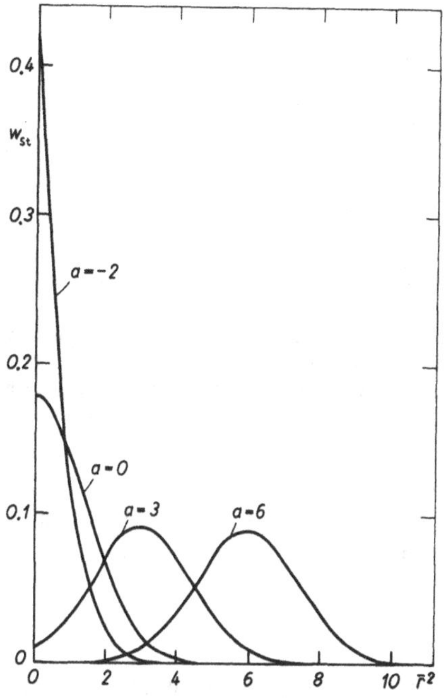

Fig. 12.3. The stationary distribution (12.45) as a function of the normalized intensity $\bar{I} = \bar{r}^2$ for various pump parameters a

The moments M_n can be reduced to the normalization integral $F_0(a)$ with the help of the recurrence relations

$$M_1 = a + 2/F_0(a)\,, \tag{12.49}$$

$$M_n = 2(n-1)M_{n-2} + aM_{n-1}\,, \quad (n \geqq 2)\,. \tag{12.50}$$

The recurrence relations may be derived by using partial integrations in (12.48). Equation (12.50) also follows from (12.43) for the stationary state. The normalization integral $F_0(a)$ can be expressed by the error functions erf and erfc (see [11.50] for a definition) according to

$$F_0(a) = \sqrt{\pi}\exp(\tfrac{1}{4}a^2)[1 + \mathrm{erf}(\tfrac{1}{2}a)]\,, \tag{12.51}$$

$$= \sqrt{\pi}\exp(\tfrac{1}{4}a^2)[2 - \mathrm{erfc}(\tfrac{1}{2}a)]\,, \tag{12.51 a}$$

$$= \sqrt{\pi}\exp(\tfrac{1}{4}a^2)\,\mathrm{erfc}(-\tfrac{1}{2}a)\,. \tag{12.51 b}$$

To calculate $F_0(a)$ one may use the series expansion of erf$(a/2)$ [Ref. 11.50, Eq. (7.1.5)] for $|a| \leqq 5$ and the continued fraction expansion of $\exp(a^2/4)\,\mathrm{erfc}(|a|/2)$ [Ref. 11.50, Eq. (7.1.14)] for $|a| > 5$, see also (9.43). The $F_n(a)$ may also be expressed by the parabolic cylinder function (9.49). If follows from the asymptotic expansion (9.50) that the moments $M_n(a)$ become infinitely large for $n \to \infty$.

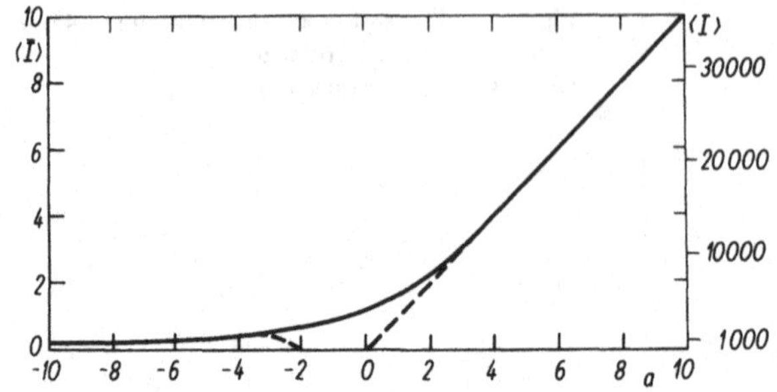

Fig. 12.4. The first moment $\langle \bar{I} \rangle = M_1(a)$ (*solid line*) and the asymptotic expansions $\langle \bar{I} \rangle = 2/|a| - 8/|a|^3$ for $a \ll -1$ and $\langle \bar{I} \rangle = a$ for $a \gg 1$ (*broken line*) as a function of the pump parameter a. The unnormalized intensity $\langle I \rangle$ (in photon numbers) is valid for threshold photon number ≈ 4000 from [12.42a]

The first moment $\langle \bar{I} \rangle = M_1(a)$ is shown in Fig. 12.4 as a function of the pump parameter a. In unnormalized units the intensity is

$$\langle I(a) \rangle = \sqrt{q/\beta}\langle \bar{I}(a) \rangle \,. \tag{12.52}$$

Thus $n_{\text{th}} = \sqrt{q/\beta}\langle \bar{I}(0) \rangle = \sqrt{q/\beta}\,2/\sqrt{\pi}$ is the number of photons at threshold, which, for example, *Arecchi* et al. [12.42a] have found to be about 4000 for their experiment.

Calculation of the Cumulants

Because the generating function $K(u)$ of the cumulants K_n is the logarithm of the generating function of the moments [see (2.25)], i.e.,

$$\begin{aligned} K(u) &= \sum_{n=1}^{\infty} [(\mathrm{i}u)^n/n!] K_n = \ln M(u) \\ &= \ln F_0(a+2\mathrm{i}u) - \ln F_0(a) \,, \end{aligned} \tag{12.53}$$

the cumulants $K_n(a)$ are derivatives of each other ($n \geqq 1$)

$$\begin{aligned} K_{n+1}(a) &= \frac{\mathrm{d}^n}{(\mathrm{d}\mathrm{i}u)^n} \frac{\mathrm{d}}{\mathrm{d}\mathrm{i}u} \ln F_0(a+2\mathrm{i}u)\bigg|_{u=0} = 2^n \frac{\mathrm{d}^n}{\mathrm{d}a^n} \frac{\mathrm{d}}{\mathrm{d}\mathrm{i}u} \ln F_0(a+2\mathrm{i}u)\bigg|_{u=0} \\ &= 2^n \frac{\mathrm{d}^n}{\mathrm{d}a^n} K_1(a) = 2 \frac{\mathrm{d}}{\mathrm{d}a} K_n(a) \,. \end{aligned} \tag{12.54}$$

In Fig. 12.5 the first seven cumulants are shown as a function of a. For relations between cumulants and the moments, see (2.26–29). It is worthwhile to note that the magnitude of the maximum of the cumulants $K_n(a)$ increases with increasing n in the threshold region for $n > 3$ and becomes infinite for $n \to \infty$. It is therefore not possible to solve the moment equations (12.43) in the threshold region by putting

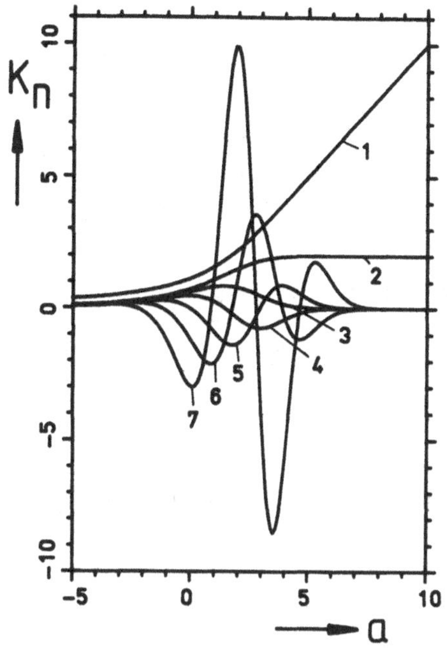

Fig. 12.5. The first seven cumulants (12.54) as functions of the pump parameter a

one of the high cumulants K_N to zero. (One would then obtain a finite closed system for the moments.) By putting some moment M_N equal to zero a solution of the tridiagonal moment equations (12.43) is not possible either above threshold, as shown in Sect. 2.2 [see the remark following (2.30)].

12.3 Expansion in Eigenmodes

As explained in Sect. 2.4.1, the complete information of a Markov process is given by the transition probability density. In the present case this transition probability is the Green's function of the Fokker-Planck equation (12.38, or 39 or 40). As discussed in Chaps. 5 and 6, the Green's function can be expanded into eigenmodes. Assuming periodic boundary conditions in φ, the expansion of the transition probability density may be put in the form [12.18–20]

$$P(\bar{r}, \varphi, \bar{t} | \bar{r}', \varphi', 0) = \frac{1}{2\pi\bar{r}} \frac{\psi_{00}(\bar{r})}{\psi_{00}(\bar{r}')} \times \sum_{n=0}^{\infty} \sum_{\nu=-\infty}^{\infty} \psi_{\nu n}(\bar{r}) \psi_{\nu n}(\bar{r}') \mathrm{e}^{\mathrm{i}\nu(\varphi-\varphi')} \mathrm{e}^{-\lambda_{\nu n}\bar{t}}, \tag{12.55}$$

where $\psi_{\nu n}$ are the eigenfunctions and $\lambda_{\nu n}$ are the eigenvalues of the one-dimensional Schrödinger equation

$$d^2\psi_{\nu n}/d\bar{r}^2+[\lambda_{\nu n}-V_\nu(\bar{r})]\,\psi_{\nu n}=0 \tag{12.56}$$

with the potential

$$V_\nu(\bar{r})=\frac{\nu^2}{\bar{r}^2}+\frac{\psi_{00}''(\bar{r})}{\psi_{00}(\bar{r})}=\left(\nu^2-\frac{1}{4}\right)\frac{1}{\bar{r}^2}+a+\left(\frac{1}{4}a^2-2\right)\bar{r}^2-\frac{1}{2}a\bar{r}^4+\frac{1}{4}\bar{r}^6. \tag{12.57}$$

Because of (12.41, 45, 55) the eigenfunction ψ_{00} belonging to the stationary eigenvalue $\lambda_{00}=0$ is

$$\psi_{00}(\bar{r})=\sqrt{2\bar{r}N}\exp\left(-\frac{1}{8}\bar{r}^4+\frac{1}{4}a\bar{r}^2\right). \tag{12.58}$$

The $\psi_{\nu n}(\bar{r})$ are assumed to be normalized. Thus, because of the orthogonality, we have

$$\int_0^\infty \psi_{\nu n}(\bar{r})\,\psi_{\nu n'}(\bar{r})\,d\bar{r}=\delta_{nn'}\,. \tag{12.59}$$

Expression (12.55) is a Green's function of the Fokker-Planck equation because, as one may verify by insertion, P is a solution of (12.39) and because of the completeness relations

$$\delta(\bar{r}-\bar{r}')=\sum_{n=0}^{\infty}\psi_{\nu n}(\bar{r})\,\psi_{\nu n}(\bar{r}')\,,\qquad \delta(\varphi-\varphi')=\frac{1}{2\pi}\sum_{\nu=-\infty}^{\infty}e^{i\nu(\varphi-\varphi')}\,, \tag{12.60}$$

the initial conditions are given by

$$P(\bar{r},\varphi,0\,|\,\bar{r}',\varphi',0)=\frac{1}{\bar{r}}\delta(\bar{r}-\bar{r}')\,\delta(\varphi-\varphi')=\delta(\bar{b}-\bar{b}')\,. \tag{12.61}$$

To obtain numerical results, one must calculate the eigenvalues and eigenfunctions of the Schrödinger equation (12.56). Only below ($a\ll-1$) and above ($a\gg1$) threshold can this Schrödinger equation be solved analytically (except of course for the stationary solution ψ_{00}). Thus, one may use either approximations (for instance, variational methods [12.17b, 5.15]) or a numerical integration of the Schrödinger equation [12.18–20]. For an alternative method, see Sect. 12.4.

The stationary joint probability distribution W_2, compare (2.75), now takes the form

$$W_2(\bar{b},\bar{t}+\bar{\tau};\bar{b}',\bar{t})=W_2(\bar{r},\varphi,\bar{t}+\bar{\tau};\bar{r}',\varphi',\bar{t})$$

$$=\frac{\psi_{00}(\bar{r})}{2\pi\bar{r}}\;\frac{\psi_{00}(\bar{r}')}{2\pi\bar{r}'}\sum_{n=0}^{\infty}\sum_{\nu=-\infty}^{\infty}\psi_{\nu n}(\bar{r})\,\psi_{\nu n}(\bar{r}')\,e^{i\nu(\varphi-\varphi')}\exp(-\lambda_{\nu n}|\bar{\tau}|)\,. \tag{12.62}$$

With the help of this stationary joint probability distribution we are able to obtain stationary two-time correlation functions.

Correlation Functions

The two most important correlation functions for the light field [12.54, 55] are that of the amplitude ($\mathrm{d}^2\bar{b} = \mathrm{d}(\mathrm{Re}\{\bar{b}\})\,\mathrm{d}(\mathrm{Im}\{\bar{b}\})$)

$$g(a,\bar{\tau}) = \langle \bar{b}^*(\bar{t}+\bar{\tau})\bar{b}(\bar{t})\rangle = \iint \bar{b}^*\bar{b}'\, W_2(\bar{b},\bar{t}+\bar{\tau};\bar{b}',\bar{t})\,\mathrm{d}^2\bar{b}\,\mathrm{d}^2\bar{b}' \;, \qquad (12.63)$$

and that of the intensity fluctuations

$$\begin{aligned} K(a,\bar{\tau}) &= \langle(|\bar{b}(\bar{t}+\bar{\tau})|^2 - \langle|\bar{b}|^2\rangle)(|\bar{b}(\bar{t})|^2 - \langle|\bar{b}|^2\rangle)\rangle \\ &= \iint (|\bar{b}|^2 - \langle|\bar{b}|^2\rangle)(|\bar{b}'|^2 - \langle|\bar{b}|^2\rangle)\, W_2(\bar{b},\bar{t}+\bar{\tau};\bar{b}',\bar{t})\,\mathrm{d}^2\bar{b}\,\mathrm{d}^2\bar{b}' \,. \end{aligned} \qquad (12.64)$$

Inserting here (12.62) and carrying out the integration leads to (in normalized units)

$$\begin{aligned} g(a,|\bar{\tau}|) &= g(a,0)\sum_{n=0}^{\infty} V_n^{(g)} \exp(-\lambda_{1n}|\bar{\tau}|) \\ V_n^{(g)} &= \left[\int_0^{\infty} \bar{r}\,\psi_{00}(\bar{r})\,\psi_{1n}(\bar{r})\,\mathrm{d}\bar{r}\right]^2 \Big/ g(a,0) \end{aligned} \qquad (12.65)$$

$$\begin{aligned} K(a,|\bar{\tau}|) &= K(a,0)\sum_{n=1}^{\infty} V_n^{(K)} \exp(-\lambda_{0n}|\bar{\tau}|) \\ V_n^{(K)} &= \left[\int_0^{\infty} \bar{r}^2\,\psi_{00}(\bar{r})\,\psi_{0n}(\bar{r})\,\mathrm{d}\bar{r}\right]^2 \Big/ K(a,0) \,. \end{aligned} \qquad (12.66)$$

The correlation functions for $\bar{\tau} = 0$ were calculated in Sect. 12.2:

$$\begin{aligned} g(a,0) &= \langle|\bar{b}|^2\rangle = K_1(a) = \langle\bar{I}\rangle \\ K(a,0) &= \langle(|\bar{b}|^2 - \langle|\bar{b}|^2\rangle)^2\rangle = K_2(a) \,. \end{aligned} \qquad (12.67)$$

Since all matrix elements V_n are positive and their sum is one

$$\sum_{n=0}^{\infty} V_n^{(g)} = \sum_{n=1}^{\infty} V_n^{(K)} = 1 \,, \qquad (12.68)$$

V_n give the relative influence of the nth order damping term. The Fourier transform of the correlation function gives the spectral profile. Because the correlation functions are sums of exponential functions, the spectral profile is a sum of Lorentzian functions.

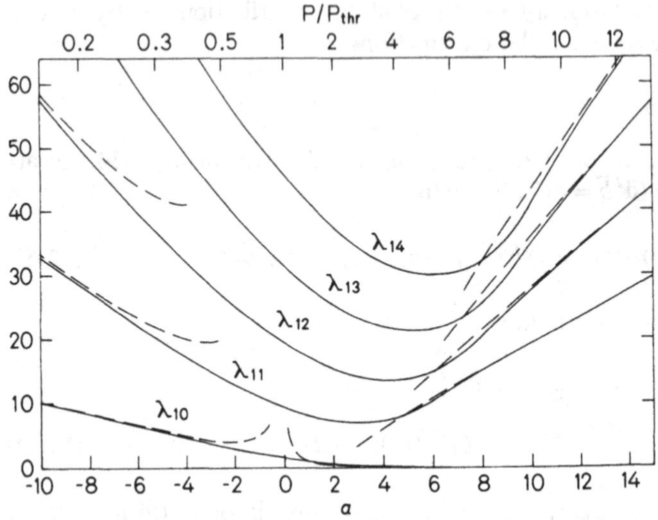

Fig. 12.6. The first five eigenvalues λ_{1n} as a function of the pump parameter a and as a function of the power output P/P_{thr}, respectively (*solid line*). The following asymptotic expansions for large positive and negative pump parameters, which can be obtained by a perturbation of the linearized equations, are also shown (*dashed line*):

$$a \gg 1: \quad \lambda_{10} = 1/a\,, \quad \lambda_{1n} = (n+1)a - n(3n+4)/a\,, \quad n \geqq 1\,,$$

$$a \ll -1: \quad \lambda_{1n} = (2n+1)|a| + 4(3n^2+4n+1)/|a|\,, \quad n \geqq 0\,.$$

Amplitude Correlation Functions

The first five eigenvalues λ_{1n}, determined in [12.53] by numerical integration of (12.56), are plotted in Fig. 12.6. Calculation of $V_0^{(g)}$ shows that $1 - V_0^{(g)} = \sum_{n=1}^{\infty} V_n^{(g)}$ is of the order of 2% near threshold and smaller outside. Therefore, the spectral profile is nearly a Lorentzian with a linewidth (in unnormalized units)

$$\Delta\nu = \sqrt{\beta q}\,\lambda_{10} = \alpha_{\text{L}} q/\langle I\rangle\,, \qquad \alpha_{\text{L}} = \lambda_{10}\langle \bar{I}\rangle\,. \tag{12.69}$$

The linewidth factor α_{L}, obtained in [12.17b, 20, 5.15], is plotted in Fig. 12.7. The factor α_{L} varies continuously from 2 to 1 by passing through the threshold

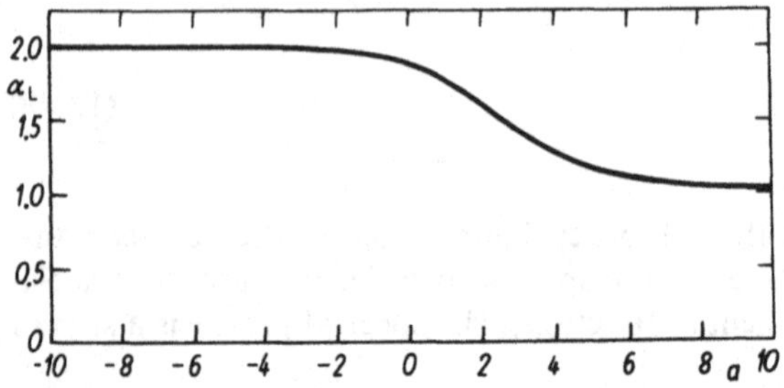

Fig. 12.7. The linewidth factor $\alpha_{\text{L}} = \lambda_{10}\langle\bar{I}\rangle$ as a function of the pump parameter a

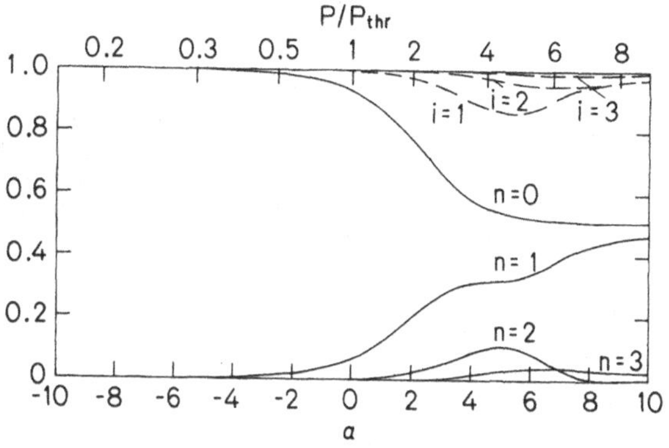

Fig. 12.8. The values $\lambda_{1n} V_n^{(g)} \times g(a,0)/2$ (*solid line*) for $n = 0, 1, 2, 3$ and the sum terms $\sum_{n=0}^{i} \lambda_{1n} \times V_n^{(g)} g(a,0)/2$ (*dashed line*) for $i = 1, 2, 3$ as a function of the pump parameter a and as a function of the power output P/P_{thr}, respectively

region. The ratio of 2 to 1 occurs because above threshold the laser amplitude is stabilized and therefore only half of the noise power (in φ direction) contributes to the linewidth. This linewidth factor was found experimentally by *Gerhardt* et al. [12.56]. For small $\bar{\tau}$ the slope of the correlation function (12.65) is, however, no longer given by the first term of the series at and above threshold. The slope at $\bar{\tau} = 0$ reads [12.53]

$$\frac{\mathrm{d}}{\mathrm{d}\bar{\tau}} g(a,\bar{\tau})\bigg|_{\bar{\tau}=0} = g(a,0) \sum_{n=0}^{\infty} V_n^{(g)} \lambda_{1n} = 2\,. \tag{12.70}$$

The last relation can be derived with the help of sum rules as explained in Sect. 7.2 (7.34). The first four terms of the series (12.70) are shown in Fig. 12.8. Evidently, at least two terms must be taken into account, since above threshold λ_{11} is much larger ($\lambda_{11} \sim 2a$ for $a \gg 1$) than λ_{10} ($\lambda_{10} \sim 1/a$ for $a \gg 1$) and $V_0^{(g)} \lambda_{10}$ becomes equal $V_1^{(g)} \lambda_{11}$ for $a \gg 1$. Thus the linewidth factor α_{L} can be determined by the slope of (12.65) only if the time $\bar{\tau}$ is large compared to $1/\lambda_{11}$, so that only the first term of the series will survive.

Intensity Correlation Functions

The first four nonzero eigenvalues λ_{0n} and the matrix elements $V_n^{(K)}$ obtained by a numerical integration of the Schrödinger equation (12.56) in [12.18] are shown in Figs. 12.9, 10. The potential (12.57), the first four eigenfunctions λ_{0n} and their eigenvalues above threshold are plotted in Fig. 12.11. As seen, the eigenvalues are nearly pairwise degenerate above threshold. Though the matrix elements $V_1^{(K)}$ and $V_2^{(K)}$ become nearly equal above threshold, only one decay constant of the series (12.66) prevails because of this degeneracy. In the whole threshold region we may approximate (12.66) by the single exponential function

$$K_{\text{eff}}(a,\bar{\tau}) = K_2(a) \exp(-\lambda_{\text{eff}} |\bar{\tau}|)\,, \qquad \frac{1}{\lambda_{\text{eff}}} = \sum_{n=1}^{\infty} \frac{V_n^{(K)}}{\lambda_{0n}}\,, \tag{12.71}$$

which has the same area and the same value at $\bar{\tau} = 0$ as the exact expression.

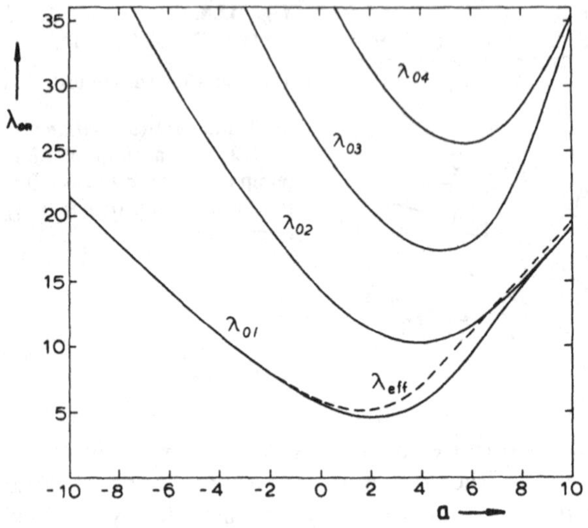

Fig. 12.9. The first four nonzero eigenvalues λ_{0n} and the effective eigenvalue λ_{eff} (12.71) as functions of the pump parameter a. Far below ($a \ll -1$) and high above ($a \gg 1$) threshold the effective eigenvalue can be approximated by $\lambda_{eff} = 2\,|a|$

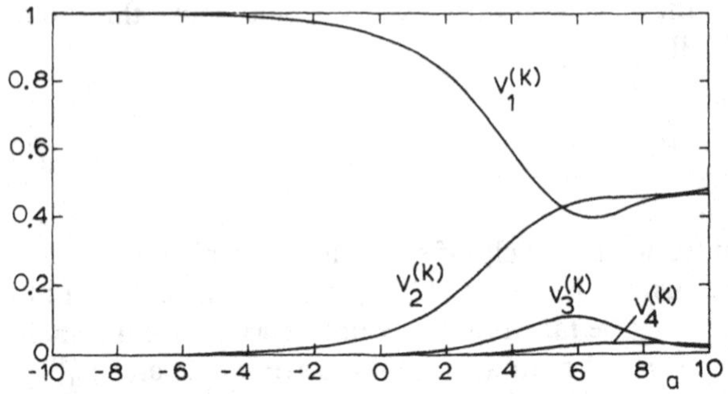

Fig. 12.10. The first four matrix elements $V_n^{(K)}$ as functions of the pump parameter a

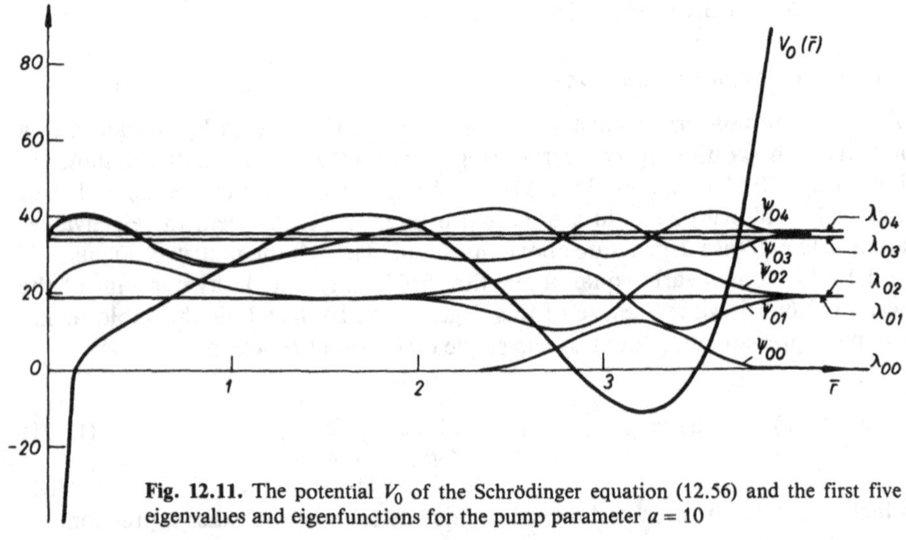

Fig. 12.11. The potential V_0 of the Schrödinger equation (12.56) and the first five eigenvalues and eigenfunctions for the pump parameter $a = 10$

The effective width λ_{eff} is 25% larger than the lowest decay constant for $a \approx 4.5$. This deviation of λ_{eff} from λ_{01} was found experimentally with measurements by *Arecchi* et al. [12.42b]. *Chopra* and *Mandel* [12.46] were able to obtain the lowest two matrix elements experimentally. (A table of eigenvalues, matrix elements and λ_{eff} is contained in [12.8].) For unnormalized units the linewidth of the spectrum of the intensity fluctuation is given by

$$\Delta\nu_I(a) = \sqrt{\beta q}\,\lambda_{\text{eff}}(a)\,.$$

Below threshold ($a \ll -1$) the correlation function of the intensity fluctuations is the square of the correlation function of the amplitude, viz.,

$$K(a,\bar{\tau}) = |g(a,\bar{\tau})|^2\,. \tag{12.72}$$

This can be shown by using the asymptotic values of the eigenvalues and of K_1 and K_2. It was proved by *Mandel* and *Wolf* [12.55] that (12.72) is valid provided the light field amplitude b is a Gaussian process. [It follows from the linearized Langevin equation (12.31) that the light field b is a Gaussian process far below threshold.]

Detuning

If the resonance frequency ω_c of the laser cavity and the atomic frequency ω_a are not exactly tuned, the normalized Langevin equation (12.6a) takes the form

$$\mathrm{d}\bar{b}/\mathrm{d}\bar{t} - (1+\mathrm{i}\delta)(a-|\bar{b}|^2)\bar{b} = \Gamma\,, \tag{12.73}$$

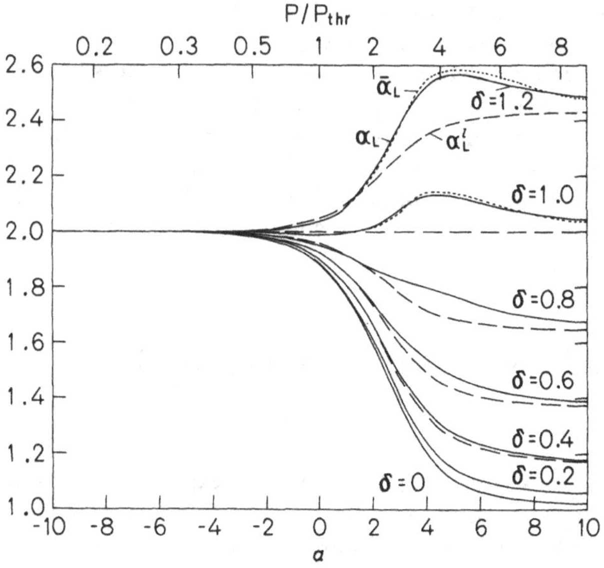

Fig. 12.12. For various detuning parameters δ the linewidth factor (12.75) is shown as a function of the pump parameter a and as a function of the normalized power output P/P_{thr}, respectively (*solid line*). The approximation $\alpha_{\text{L}}^{l}(a,\delta) = \alpha_{\text{L}}(a,0) + [2-\alpha_{\text{L}}(a,0)]\,\delta^2$ is shown as a *dashed line*, a perturbation expansion up to δ^2 [12.53] as a *dotted line*

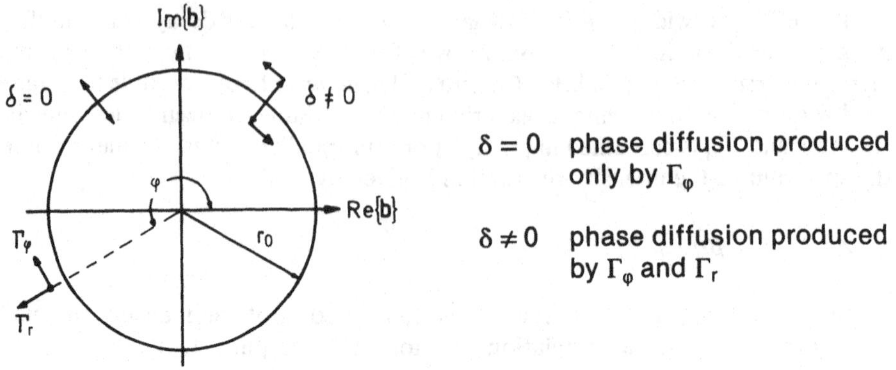

Fig. 12.13. Explanation of the enhancement of the linewidth factor due to detuning

where $\delta = (\omega_c - \omega_a)/\gamma_2$ is the detuning parameter [12.53]. The Schrödinger equation (12.56) then reads

$$d^2 \psi_{vn}/d\bar{r}^2 + [\lambda_{vn} - V_v(\bar{r}) + i\,v(\bar{r}^2 - a)\,\delta]\,\psi_{vn} = 0\,, \tag{12.74}$$

where the potential $V_v(\bar{r})$ is still given by (12.57). The eigenvalue λ_{10} which enters into the correlation function (12.63) now becomes complex. (The eigenfunction will, of course, become complex, too.) The real part of λ_{10} determines the linewidth. We may now define a linewidth factor α_L by

$$\alpha_L = \mathrm{Re}\{\lambda_{10}\} \cdot \langle \bar{I} \rangle\,. \tag{12.75}$$

The dependence on the detuning parameter, obtained in [12.53], is shown in Fig. 12.12. The linewidth factor increases with increasing detuning. This enhancement of the linewidth due to detuning may be explained qualitatively in the following way (Fig. 12.13): without detuning, the ratio of 2 to 1 occurs because above threshold the laser amplitude is stabilized and therefore only half of the noise power (only that in φ direction) contributes to the linewidth. Including detuning, the small fluctuations in $\bar{r}$ direction around $\bar{r}_0$ lead to a $(\bar{r} - \bar{r}_0)$-dependent additional motion of the phase φ, thus leading to additional phase diffusion above threshold. Below threshold, no such additional diffusion occurs.

12.4 Expansion into a Complete Set; Solution by Matrix Continued Fractions

To solve the laser Fokker-Planck equation (12.40), we may expand the distribution function $\hat{W}(\bar{I}, \varphi, \bar{t})$ into two complete sets, Sect. 6.6.5. Because we are looking for periodic functions in φ we use $(2\pi)^{-1/2}\exp(i\,v\varphi)$ with $v = 0, \pm 1, \pm 2, \ldots$

as one set. For the expansion with respect to the intensity we use generalized Laguerre functions as the other set. The expansion of $\hat{W}$ is written as

$$\hat{W}(\bar{I},\varphi,\bar{t}) = (2\pi)^{-1}\mathrm{e}^{-x}\sum_{n=0}^{\infty}\sum_{\nu=-\infty}^{\infty} c_n^{(\nu)}(\bar{t})x^{|\nu|/2}\mathrm{L}_n^{(|\nu|)}(x)\,\mathrm{e}^{\mathrm{i}\nu\varphi}, \tag{12.76}$$

where x is proportional to the intensity

$$\bar{I} = \alpha x. \tag{12.77}$$

The arbitrary scaling factor α is useful for a better adjustment of $\hat{W}$ to the Laguerre functions expansion. (By varying α one can change the speed of convergence.) We now insert the ansatz (12.76) into the Fokker-Planck equation (12.40), multiply the resulting equation by $(2\pi)^{-1/2}x^{|\nu'|/2}\mathrm{L}_n^{(|\nu'|)}(x)\exp(-\mathrm{i}\nu'\varphi)$ and integrate the expression over x and φ. By using the relations [9.26]

$$\begin{aligned} &x(\mathrm{d}/\mathrm{d}x)\mathrm{L}_n^{(\nu)} = n\mathrm{L}_n^{(\nu)} - (n+\nu)\mathrm{L}_{n-1}^{(\nu)},\\ &x(\mathrm{d}/\mathrm{d}x)^2\mathrm{L}_n^{(\nu)} + (\nu+1-x)(\mathrm{d}/\mathrm{d}x)\mathrm{L}_n^{(\nu)} + n\mathrm{L}_n^{(\nu)} = 0,\\ &(n+1)\mathrm{L}_{n+1}^{(\nu)} + (x-2n-\nu-1)\mathrm{L}_n^{(\nu)} + (n+\nu)\mathrm{L}_{n-1}^{(\nu)} = 0, \end{aligned} \tag{12.78}$$

$$\int_0^{\infty} x^{\nu}\mathrm{e}^{-x}\mathrm{L}_n^{(\nu)}(x)\mathrm{L}_m^{(\nu)}(x)\,\mathrm{d}x = \delta_{nm}(n+\nu)!/n! \tag{12.79}$$

for the generalized Laguerre polynomials, which are defined by

$$\mathrm{L}_n^{(\nu)}(x) = x^{-\nu}\mathrm{e}^{x}(\mathrm{d}/\mathrm{d}x)^n x^{n+\nu}\mathrm{e}^{-x}/n!, \tag{12.80}$$

one then obtains the following four-term recursion relation for the expansion coefficients [12.30] ($c_n^{(\nu)} = 0$ for $n < 0$):

$$\begin{aligned} \dot{c}_n^{(\nu)} = {}&(2n+\nu)(n+\nu+1)\,\alpha c_{n+1}^{(\nu)}\\ &+ [(2n+\nu)a - 2n(3n+3\nu+1)\,\alpha - \nu(\nu+1)\,\alpha]\,c_n^{(\nu)}\\ &+ n[(6n+3\nu-2)\,\alpha - 2a - 4/\alpha]\,c_{n-1}^{(\nu)}\\ &+ 2n(1-n)\,\alpha c_{n-2}^{(\nu)}. \end{aligned} \tag{12.81}$$

This infinite system of ordinary differential equations is equivalent to the Fokker-Planck equation (12.40). Notice that in system (12.81) coefficients with different upper indices are not coupled. The initial conditions for the expansion coefficients $c_n^{(\nu)}$ are determined by the initial conditions of $\hat{W}$. For example, the transition probability $\hat{P}$, i.e., the Green's function of the Fokker-Planck equation, takes the form

$$\hat{P}(\bar{I},\varphi,\bar{t}\,|\,\bar{I}',\varphi',0) = (2\pi)^{-1}\mathrm{e}^{-x}\sum_{n=0}^{\infty}\sum_{\nu=-\infty}^{\infty} c_n^{(\nu)}(x',\bar{t})x^{|\nu|/2}\mathrm{L}_n^{(|\nu|)}(x)\,\mathrm{e}^{\mathrm{i}\nu(\varphi-\varphi')}, \tag{12.82}$$

where the coefficients $c_n^{(\nu)}(x',\bar{t})$ have to satisfy (12.81) with the initial condition

$$c_n^{(\nu)}(x',0) = \alpha^{-1}[n!/(n+\nu)!]x'^{|\nu|/2}L_n^{(|\nu|)}(x') . \tag{12.83}$$

The infinite system (12.81) can be solved by numerical integration of the truncated system (12.81), Sect. 12.5.2. Because (12.81) is of the form (9.13), we may apply the methods of Chap. 9. The infinite system can be cast into the tridiagonal vector recurrence relation (9.10), where the matrix elements of the 2×2 matrices $\boldsymbol{Q}_n^{\pm}$, $\boldsymbol{Q}_n$ are given by [9.18]:

$$\begin{aligned} Q_n^{+11} &= Q_n^{+12} = Q_n^{+22} = 0 , \\ Q_n^{+21} &= (4n+2+\nu)(2n+2+\nu)\alpha , \end{aligned} \tag{12.84}$$

$$\begin{aligned} Q_n^{11} &= (4n+\nu)a - 4n(6n+3\nu+1)\alpha - \nu(\nu+1)\,\alpha \\ Q_n^{12} &= (4n+\nu)(2n+1+\nu)\alpha \\ Q_n^{21} &= (2n+1)(12n+4+3\nu)\alpha - (4n+2)a - (8n+4)/\alpha \\ Q_n^{22} &= (4n+2+\nu)a - (4n+2)(6n+4+3\nu)\alpha - \nu(\nu+1)\alpha , \end{aligned} \tag{12.85}$$

$$\begin{aligned} Q_n^{-11} &= 4n(1-2n)\alpha \\ Q_n^{-12} &= 2n(12n+3\nu-2)\alpha - 4na - 8n/\alpha \\ Q_n^{-21} &= 0 \\ Q_n^{-22} &= -4n(2n+1)\alpha . \end{aligned} \tag{12.86}$$

As discussed in Sect. 9.3, the stationary solution, the initial value problem as well as the eigenvalue problem can be obtained in terms of matrix continued fractions. Because here only 2×2 matrices occur, it is one of the simplest applications of the matrix continued-fraction method.

12.4.1 Determination of Eigenvalues

As an application we now discuss the determination of the eigenvalues. The determinant (9.119) for $m=0$ is plotted in Fig. 12.14 for the case $\nu=0$ as a function of λ. The matrix elements of the matrices $\boldsymbol{Q}_n^{\pm}$ and $\boldsymbol{Q}_n$ which enter in this determinant and in the continued fraction $\tilde{\boldsymbol{K}}_0$ (9.112) are now given by (12.84–86). The zeros of this determinant are the eigenvalues λ_{00}, λ_{01}, $\lambda_{02}, \dots$. To determine these eigenvalues, the scaling parameter may be chosen in the range $0.1, \dots, 10$, where the number of continued-fraction terms is then in the range $30, \dots, 1000$ [9.18]. For some α the determinant $D_0(\lambda)$ has poles. These poles can be shifted by changing the scaling factor α. The eigenvalues determined in this way agree very well with the eigenvalues obtained by numerical integration of the Schrödinger equation (12.56). (Some of the eigenvalues λ_{2n} obtained by the matrix continued fraction method are plotted in Fig. 12.15.) By comparing both

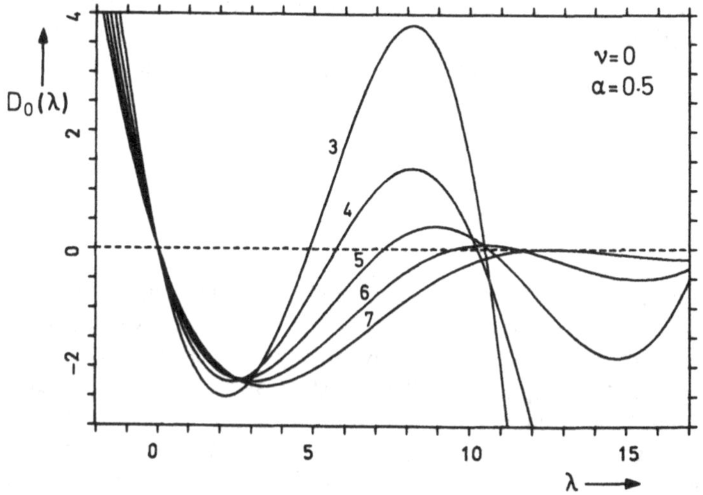

Fig. 12.14. The value of the determinant (9.119) for $m = 0$ as a function of λ for various pump parameters $a = 3, \ldots, 7$. The upper index ν is zero and the scaling parameter α is 0.5

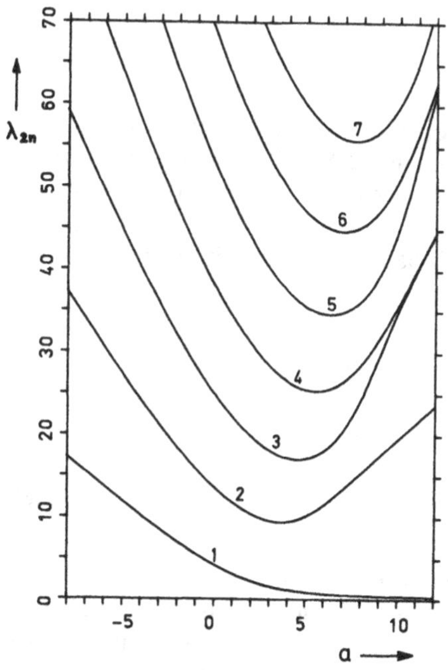

Fig. 12.15. The eigenvalues $\lambda_{21} \ldots \lambda_{27}$ as a function of the pump parameter a

methods, one may say that the matrix continued-fraction method is much simpler to put into program statements and it usually works faster than the numerical integration. Furthermore, the boundary conditions at $\bar{I} = 0$ and at $\bar{I} = \infty$ are automatically fulfilled by (12.76), whereas for the numerical integration method they have to be taken into account by a Taylor series expansion (for small $\bar{r}$) and by a truncation procedure (putting $\psi = 0$ for some large $\bar{r}$). The efficiency of the method is quite remarkable, because the expansion (12.76) is, especially well above threshold, in no way adapted to the eigenfunctions of the problem.

12.5 Transient Solution

If the pumping field of the laser is suddenly switched on, the laser amplitude $\bar{b}$ is initially zero. Neglecting noise, this solution $\bar{b} = 0$ is still a stationary solution of the laser equation (12.4), but it is unstable above threshold ($a > 0$). By the spontaneous emission noise, the amplitude will be pushed away from its unstable value $\bar{b} = 0$ and it then finally reaches its stable value $|\bar{b}| \approx \sqrt{a}$. This switching on process of the laser was treated in a number of papers [12.19, 21 – 25, 27, 30 – 33]. The theoretical predictions have been fully substantiated experimentally [12.22, 45]. (For other theoretical work on the decay of this and other unstable equilibrium states see [12.57 – 59] and references therein.) Here we present three methods by which this transient behavior of the laser amplitude can be handled.

12.5.1 Eigenfunction Method

The general expression for the transition probability density in terms of eigenfunctions and eigenvalues was already given in (12.55). To describe the switching on of the laser, we need only this expression for $\bar{r}' \to 0$. For $\bar{r}' \to 0$ the terms with $\nu \neq 0$ drop out in (12.55) (for small $\bar{r}$, $\psi_{\nu n}$ is proportional to $\bar{r}^{|\nu| + 1/2}$), i.e., the transient solution cannot depend on the phase φ. The special transition probability density then takes the form

$$W(\bar{r}, \varphi, \bar{t}) = (2\pi\bar{r})^{-1}\psi_{00}(\bar{r}) \sum_{n=0}^{\infty} A_n \psi_{0n}(\bar{r}) \exp(-\lambda_{0n}\bar{t}) \,, \tag{12.87}$$

where the coefficients A_n are given by

$$A_n = \lim_{\bar{r}\to 0} \psi_{0n}(\bar{r})/\psi_{00}(\bar{r}) = \lim_{\bar{r}\to 0} \psi_{0n}(\bar{r})/\sqrt{2\bar{r}N} \,. \tag{12.88}$$

It can be shown that $\psi_{0n}(\bar{r})$ is proportional to $\sqrt{\bar{r}}$ for small $\bar{r}$. Therefore the constants (12.88) have a well-defined value. For numerical calculations, the series (12.87) must be truncated. Because of this truncation the expansion (12.87) cannot be applied to very small times $\bar{t}$. For small times, however, the nonlinearity in the drift coefficient of the Fokker-Planck equation (12.38) can be neglected and the distribution function can thus be approximated by the Gaussian function

$$W(\bar{r}, \varphi, \bar{t}) = \frac{a}{2\pi(e^{2a\bar{t}} - 1)} \exp\left(-\frac{a}{2(e^{2a\bar{t}} - 1)}\bar{r}^2\right). \tag{12.89}$$

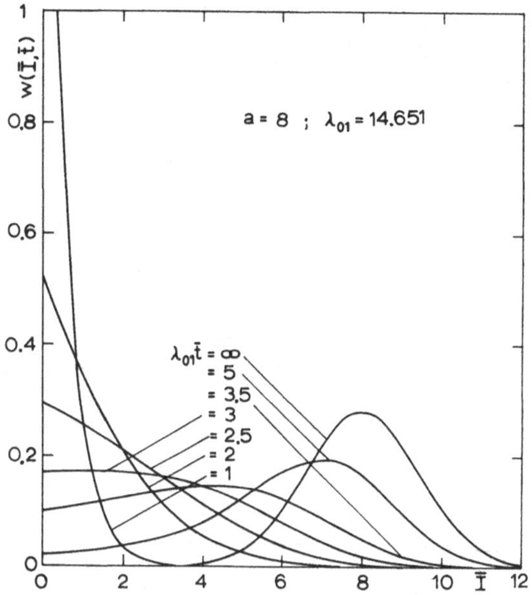

Fig. 12.16. The transient distribution function (12.90) for $a = 8$

The transient distribution

$$w(\bar{I},\bar{t}) = \frac{1}{2}\int_0^{2\pi} W(\sqrt{\bar{I}},\varphi,\bar{t})\,d\varphi\,, \tag{12.90}$$

which was obtained in [12.19] by taking into account 10 terms of expansion (12.87), is shown in Fig. 12.16. The transient mean intensity $\langle \bar{I}(\bar{t})\rangle$ or first cumulant $K_1(\bar{t})$ and the transient variance $\langle[\bar{I}(\bar{t}) - \langle \bar{I}(\bar{t})\rangle]^2\rangle$ or second cumulant $K_2(\bar{t})$ can be obtained from (12.90); the results, also obtained in [12.19], are shown in Figs. 12.17, 18. For small times one may derive the Taylor series expansion from the normalized equations (12.43) or from (12.89)

$$\langle \bar{I}(\bar{t})\rangle = 4\bar{t} + 4a\bar{t}^2 + \dots\,, \tag{12.91}$$

$$\langle[\bar{I}(\bar{t}) - \langle \bar{I}(\bar{t})\rangle]^2\rangle = 32\bar{t}^2 + 64a\bar{t}^3 + \dots\,. \tag{12.92}$$

These expansions were used in Figs. 12.17, 18 for small $\bar{t}$. In normalized units, all curves of the transient mean intensity start with the same finite slope. To find the physical significance of this effect, we rewrite the expression in unnormalized quantities, obtaining

$$\langle I(t)\rangle = AN_2t + B(N_2-N_1)_0N_2t^2 - B(N_2-N_1)_{\text{thr}}N_2t^2 + \dots\,, \tag{12.93}$$

where A, B, $(N_2-N_1)_{\text{thr}}$ and $(N_2-N_1)_0$ are given by

$$A = 2g^2/\gamma_2\,, \quad B = 2g^4/(\gamma_2)^2\,, \quad (N_2-N_1)_{\text{thr}} = N\sigma_{\text{thr}} = \varkappa\gamma_2/g^2\,,$$
$$(N_2-N_1)_0 = N\sigma_0\,. \tag{12.94}$$

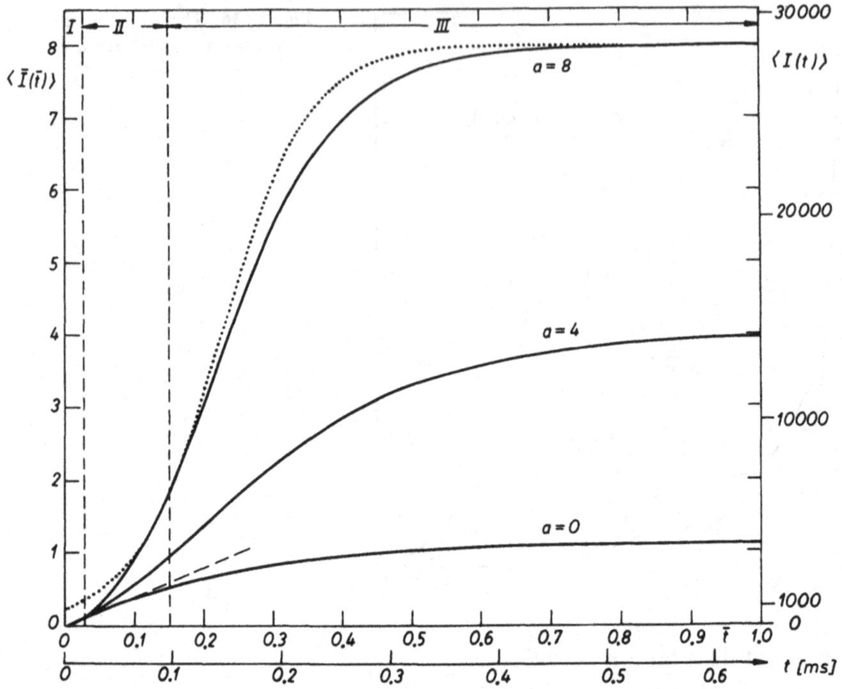

Fig. 12.17. The transient mean intensity of the distribution function (12.90) as a function of time for various pump parameters a. The unnormalized intensity $\langle I \rangle$ (in photon numbers) and t (in milliseconds) are valid for a threshold photon number 4000 and for a threshold linewidth of the intensity fluctuation 1400 c/s from [12.42c]. The three indicated regions are: I – spontaneous emission; II – spontaneous quanta are amplified by induced emission; III – saturation region. The solution $\bar{I}(\bar{t}) = a\bar{I}(0)\ \exp(2a\bar{t})/[a-\bar{I}(0)+\bar{I}(0)\ \exp(2a\bar{t})]$ of the normalized equation (12.4) ($\beta = 1,\ d = a$, $\bar{I} = \bar{b}^*\bar{b}$) without noise is dotted in for $a = 8$. The initial value $\bar{I}(0)$ was chosen so that the solutions with and without noise agreed at the end of Region II

We see that the term linear in t stems from the spontaneous emission rate. The second term in (12.93) stems from the spontaneous emission, which is amplified by induced emission. The third in (12.93) term describes the losses of the spontaneous quanta due to their finite lifetime in the cavity. Above threshold, the sum of both terms, which is quadratic in time, is positive. Higher expansion terms of (12.93) stem from induced emission and from the change of inversion. Thus, we can distinguish three main regions: spontaneous emission (Region I); amplification of spontaneous emission by induced emission (Region II); saturation effects of the inversion (Region III), Fig. 12.17. As shown in Fig. 12.18, the variance reaches a maximum at the beginning of the saturation region (III) above threshold. This means, of course, that for this time, the spread of the distribution function is largest. For larger times the variance becomes smaller and finally reaches its stationary value. This can be physically interpreted as follows. Since in the switching on process the spontaneous photons are amplified greatly, small fluctuations of the spontaneous photons lead to large fluctuations at the beginning of the saturated Region III. Because of the large stabilization of the nonlinearity,

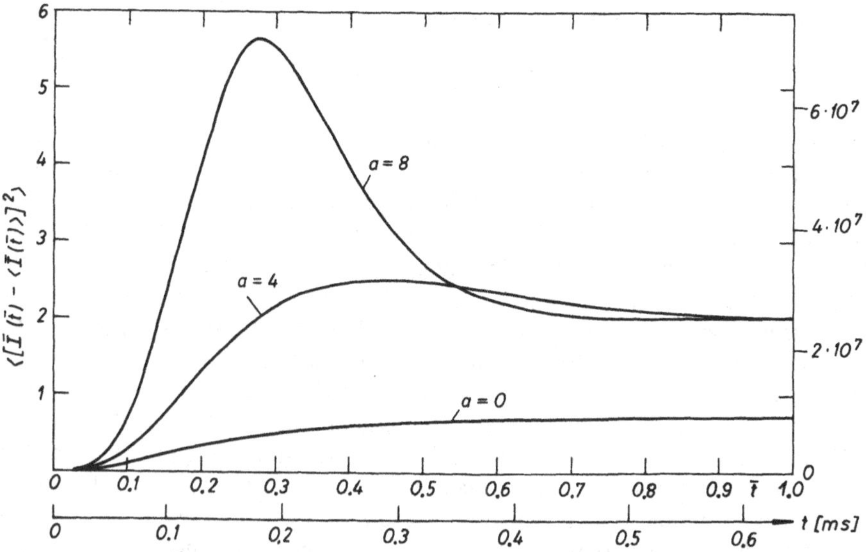

Fig. 12.18. The transient variance of distribution (12.90) as a function of time for various pump parameters. The unnormalized intensity is valid for the same threshold values as in Fig. 12.17

these fluctuations are then diminished and thus finally reach their relatively low values. At threshold ($a = 0$) no amplification occurs. Due to the low stabilization effect of the nonlinearity for $a = 0$, the variance reaches its largest value for $t \to \infty$. No maximum occurs for finite t and $a < 0$. These transient calculations have been substantiated experimentally [12.22, 45].

12.5.2 Expansion into a Complete Set

The transition probability density was expanded into two complete sets, see (12.82, 83). Because the Laguerre polynomials vanish at $x = 0$ for $\nu \neq 0$, the initial conditions (12.83) for the expansion coefficients $c_n^{(\nu)}$, which describe the switching on of the laser, now specialize to

$$c_n^{(\nu)}(0,0) = \delta_{\nu,0}/\alpha\,. \tag{12.95}$$

Thus, only the terms with $\nu = 0$ enter in the transition probability density (12.82) for $\bar{I}' = 0$. According to (2.26) the cumulants of the expansion can be expressed by the moments $\langle \bar{I}^n \rangle$. (The first moment is the first cumulant, the variance the second cumulant.) The moments can be expressed in terms of the expansion coefficients $c_n^{(0)}$. Using [Ref. 11.50, Table 22.10] and (12.79) one easily obtains

$$\begin{aligned} \langle \bar{I} \rangle &= \alpha^2 (c_0^{(0)} - c_1^{(0)}) \\ \langle \bar{I}^2 \rangle &= \alpha^3 (2c_0^{(0)} - 4c_1^{(0)} + 2c_2^{(0)}) \end{aligned} \tag{12.96}$$

$$\begin{aligned}\langle \bar{I}^3 \rangle &= \alpha^4(6c_0^{(0)} - 18c_1^{(0)} + 18c_2^{(0)} - 6c_3^{(0)}) \\ \langle \bar{I}^4 \rangle &= \alpha^5(24c_0^{(0)} - 96c_1^{(0)} + 144c_2^{(0)} - 96c_3^{(0)} + 24c_4^{(0)}) .\end{aligned} \tag{12.96}$$

Thus, the moments $\langle \bar{I}^n(\bar{t}) \rangle$ and the cumulants can be obtained from the infinite set of differential equations (12.81) for $\nu = 0$. To solve the infinite system numerically, one has to truncate it. Putting $c_n^{(0)} = 0$ for $n > N$, we obtain a finite system of differential equations, which was solved numerically in [12.30]. The truncation index N was determined in such a way that a further increase of N did not change the final result of K_1, K_2, K_3, K_4 beyond a given accuracy.

It turned out that the scaling factor α had a crucial influence on the truncation number N. To find the most suitable α the following observations were made in [12.30]. Generally, the absolute amount of the coefficients $c_n^{(\nu)}$ first increases with n and, after reaching a maximum, decreases. For large α the coefficients have a low or no maximum, but a slowly decreasing tail, whereas for small α the coefficients reach a high maximum, but then drop rapidly. Therefore, a large number of coefficients which only a few numbers of decimal digits has to be used for large α, whereas for small α less coefficients are necessary, but the number of digits has to be adjusted to the height of the maximum. The results of the numerical calculation obtained in [12.30] are shown in Fig. 12.19. In this figure the same normalization for the first two cumulants as in [12.23] was used. For large times the cumulants agree with the stationary cumulants in Fig. 12.5. Whereas for small pump parameters only a few equations need to be taken into account (typical values are $N = 10$ for $a = 5$ and $\alpha = 0.1$), the number N must be large for large pump parameters (typical values are $N = 220$ for $a = 20$ and $\alpha = 0.7$). Thus, the method ceases to be tractable for a appreciably larger than 20, but it is very accurate for $a \lesssim 20$. Notice that the only approximation made is the truncation of the expansion (12.76). By increasing N this approximation can easily be controlled.

Transient of the Amplitude

With the help of the coupled equations (12.81) for $\nu = 1$ it is easy to calculate the average amplitude. Here we assume that the amplitude starts with an initial sharp value $\bar{b}_0 = \sqrt{\bar{I}_0} = \sqrt{\alpha x_0}$ with zero phase $\varphi_0 = 0$. The average amplitude at a later time is then given by

$$\begin{aligned}\langle \bar{b}(\bar{t}) \rangle &= \langle \sqrt{\alpha x}\, \mathrm{e}^{\mathrm{i}\varphi} \rangle \\ &= \alpha^{3/2} \int_0^{2\pi} \int_0^{\infty} \sqrt{x}\, \mathrm{e}^{\mathrm{i}\varphi} \hat{P}(\bar{I}, \varphi, \bar{t} | \bar{I}_0, 0, 0)\, \mathrm{d}\varphi\, \mathrm{d}x \\ &= \alpha^{3/2} c_0^{(1)}(x_0, \bar{t}) .\end{aligned} \tag{12.97}$$

In [12.30] the truncated system (12.81) was solved for $\nu = 1$ with the initial condition, compare (12.83),

$$c_n^{(1)}(x_0, 0) = (b_0/\sqrt{\alpha})\, \mathrm{L}_n^{(1)}(\bar{b}_0^2/\alpha)\, [\alpha(n+1)]^{-1} , \tag{12.98}$$

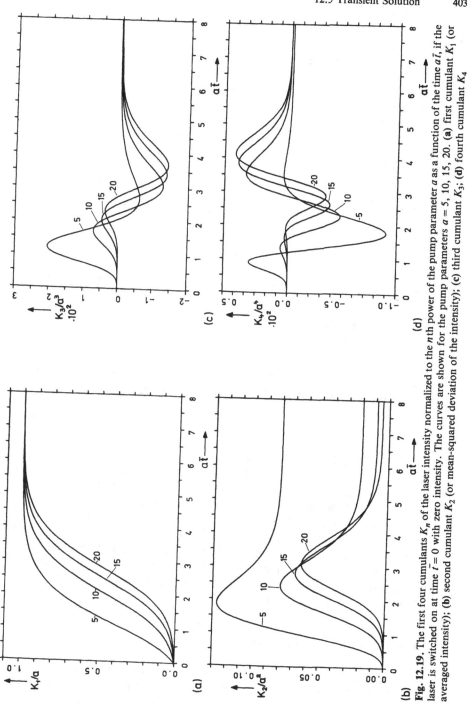

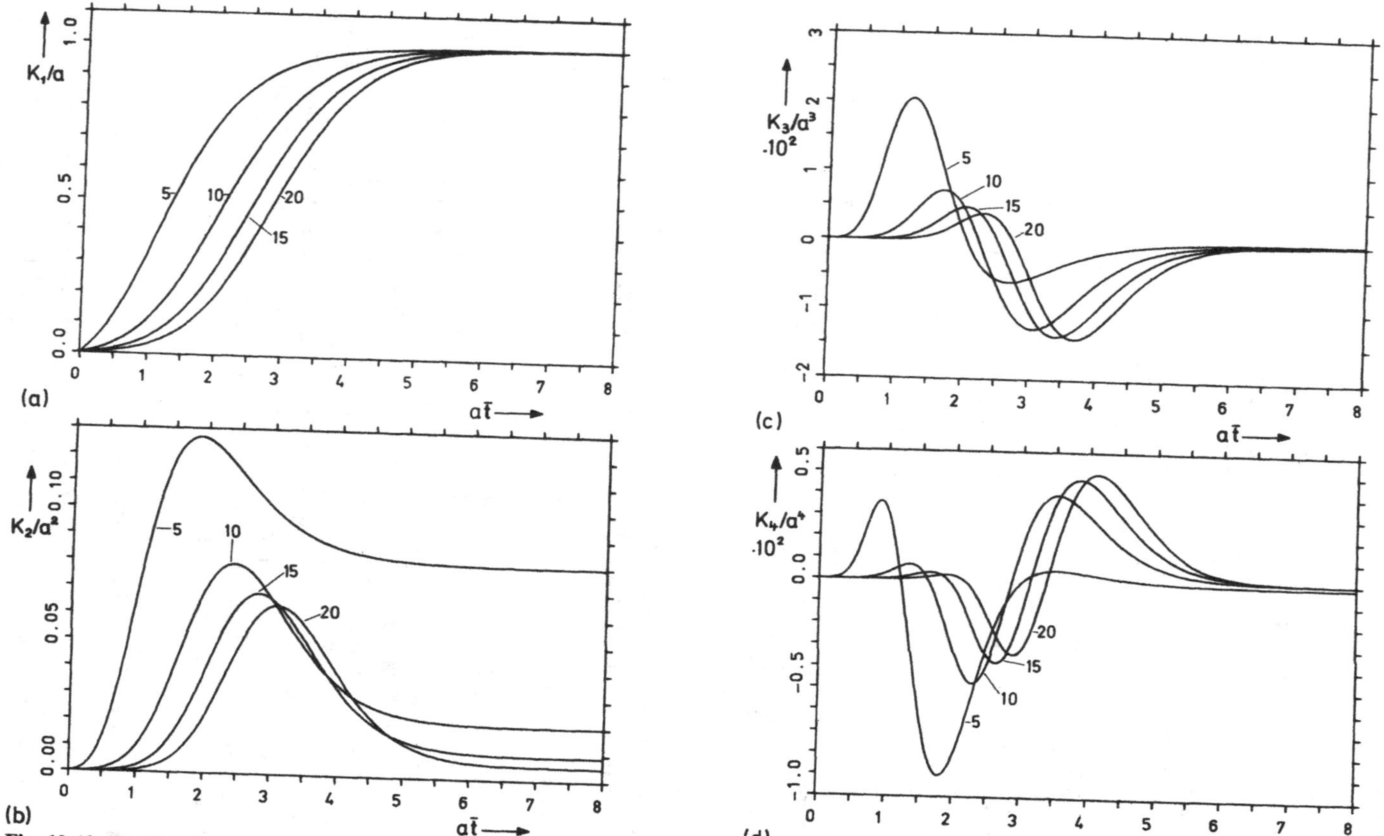

Fig. 12.19. The first four cumulants K_n of the laser intensity normalized to the nth power of the pump parameter a as a function of the time $a\bar{t}$, if the laser is switched on at time $\bar{t} = 0$ with zero intensity. The curves are shown for the pump parameters $a = 5, 10, 15, 20$. **(a)** first cumulant K_1 (or averaged intensity); **(b)** second cumulant K_2 (or mean-squared deviation of the intensity); **(c)** third cumulant K_3; **(d)** fourth cumulant K_4

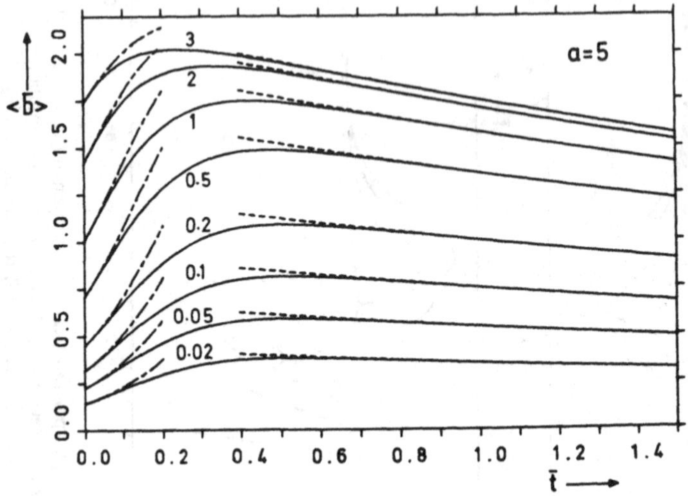

Fig. 12.20. The transient of the amplitude $\langle \bar{b} \rangle$ as a function of the time $\bar{t}$ for pump parameter $a = 5$. The curves are shown for zero initial phase and different initial intensities $\bar{b}(0)^2 = \bar{I}_0 = 0.02, \ldots, 3$. Curves – – – – – are the approximations (12.5) for small times without noise, curves --------- are the approximations (12.99) for large times

for various amplitudes $\bar{b}_0$, Fig. 12.20. Whereas the solution (12.5) without noise is a good approximation for small times, the exponential decay

$$\langle \bar{b}(\bar{t}) \rangle = \langle \bar{b}(\bar{t}_0) \rangle \exp[-\lambda_{10}(\bar{t}-\bar{t}_0)] \tag{12.99}$$

is a good approximation for large times. Here λ_{10} is the first eigenvalue of system (12.81) for $\nu = 1$ (Fig. 12.6) and $\bar{t}_0$ is a proper time at which both (12.97, 99) coincide. The $\bar{t}$ dependence of $\langle \bar{b}(\bar{t}) \rangle$ is easily understood. As already discussed, the motion of $\bar{b}(\bar{t})$ can be interpreted as the overdamped Brownian motion of particles in the potential (12.25), Fig. 12.2. Starting at the initial value $\bar{b}_0$ the particles first move down to the minimum of the potential at $|\bar{b}| = \sqrt{a}$. Then phase diffusion becomes important and the amplitude decays exponentially. For small initial values, however, the probability of the particle to diffuse over the top of the potential at $|\bar{b}| = 0$ is larger, therefore, the averaged amplitude $\langle \bar{b}(\bar{t}) \rangle$ has a lower maximum for smaller $\bar{b}_0$.

12.5.3 Solution for Large Pump Parameters

The eigenfunction method of Sect. 12.5.1 works very well for pump parameters up to $a = 8$ and the expansion method of Sect. 12.5.2 for pump parameters up to $a = 20$. We now discuss a method which works well for very large pump parameters ($a \geqq 20$). It was first developed by *Gordon* and *Aslaksen* [12.21] and further improved in [12.23 – 25, 27, 28, 32]. The main idea is the following. For large pump parameters the (spontaneous) noise is important only for small times, where it pushes the amplitude away from its unstable equilibrium state $\bar{b}=0$. Hence, saturation effects do not play any essential role for small times and we may thus use the linear but unstable laser equation, compare (12.31),

$$\dot{\bar{b}}_i - a\bar{b}_i = \bar{\Gamma}_i \tag{12.100}$$

for times $\bar{t}$ less than $\bar{t}_0$. Equation (12.100) corresponds to a Fokker-Planck equation without nonlinearity in the drift term. It leads to solution (12.89) or for the function (12.90) to

$$w(\bar{I},\bar{t}) = \frac{1}{2}\,\frac{a}{e^{2a\bar{t}}-1}\exp\left(-\frac{a}{2(e^{2a\bar{t}}-1)}\bar{I}\right), \quad (\bar{t} \leqq \bar{t}_0) \tag{12.101}$$

for the initial condition $w(\bar{I},0) = \lim_{\varepsilon\to 0}\delta(\bar{I}-\varepsilon)$. At time $\bar{t}=\bar{t}_0$ the distribution is thus given by

$$w_0(\bar{I}_0) \equiv w(\bar{I}_0,\bar{t}_0) = \frac{1}{2}\,\frac{a}{e^{2a\bar{t}_0}-1}\exp\left(-\frac{a}{2(e^{2a\bar{t}_0}-1)}\bar{I}_0\right). \tag{12.101 a}$$

For larger times $\bar{t} > \bar{t}_0$ we may neglect the noise. The noiseless solution (12.5), which reads for the intensity in normalized units

$$\bar{I}(\bar{t}) = \bar{I}_0 e^{2a(\bar{t}-\bar{t}_0)}/[1+(e^{2a(\bar{t}-\bar{t}_0)}-1)\bar{I}_0/a]\,, \tag{12.102}$$

takes into account the full nonlinearity. For different $\bar{I}_0$ (12.102) are the trajectories or paths, which connect the distribution at $\bar{t}=\bar{t}_0$ with the distribution $w(\bar{I},\bar{t})$ at a later time according to

$$\begin{aligned} w(\bar{I},\bar{t}) &= \int \delta(\bar{I}-\bar{I}(t))\,w_0(\bar{I}_0)\,d\bar{I}_0 \\ &= w_0(\bar{I}_0)\,d\bar{I}_0/d\bar{I}\,. \end{aligned} \tag{12.103}$$

Here, $\bar{I}_0$ has to be replaced by $\bar{I}$, i.e., one has to use the inverse relation of (12.102)

$$\bar{I}_0 = \bar{I}e^{-2a(\bar{t}-\bar{t}_0)}/[1+(e^{-2a(\bar{t}-\bar{t}_0)}-1)\bar{I}/a]\,. \tag{12.102a}$$

Performing the differentiation we finally obtain for $t \geqq t_0$

$$\begin{aligned} w(\bar{I},\bar{t}) &= \frac{a\,e^{-2a(\bar{t}-\bar{t}_0)}}{2(e^{2a\bar{t}_0}-1)[1+(e^{-2a(\bar{t}-\bar{t}_0)}-1)\bar{I}/a]^2} \\ &\quad\times\exp\left(-\frac{a\,e^{-2a(\bar{t}-\bar{t}_0)}\bar{I}}{2(e^{2a\bar{t}_0}-1)[1+(e^{-2a(\bar{t}-\bar{t}_0)}-1)\bar{I}/a]}\right). \end{aligned} \tag{12.104}$$

The moments of the distribution $w(\bar{I},\bar{t})$

$$\langle \bar{I}^n(\bar{t})\rangle = \int_0^\infty \bar{I}^n w(\bar{I},\bar{t})\,d\bar{I} \tag{12.105}$$

take the form for $\bar{t} \geqq \bar{t}_0$

$$\langle \bar{I}^n(\bar{t})\rangle = \int_0^\infty \left(\frac{\bar{I}_0 e^{2a(\bar{t}-\bar{t}_0)}}{1+(e^{2a(\bar{t}-\bar{t}_0)}-1)\bar{I}_0/a}\right)^n w_0(\bar{I}_0)\,d\bar{I}_0 .$$

Here we have substituted the variable $\bar{I}$ by the variable $\bar{I}_0$. Further substituting $x = a\bar{I}_0/\{2[\exp(2a\bar{t}_0)-1]\}$, we finally arrive at the following result for the moments:

$\underline{\bar{t} \leqq \bar{t}_0:}$

$$\langle \bar{I}^n(\bar{t})\rangle = [2(e^{2a\bar{t}}-1)/a]^n n! \;, \tag{12.106a}$$

$\underline{\bar{t} \geqq \bar{t}_0:}$

$$\langle \bar{I}^n(\bar{t})\rangle = a^n \int_0^\infty \{1-e^{-2a(\bar{t}-\bar{t}_0)}+a^2 e^{-2a\bar{t}}/[2x(1-e^{-2a\bar{t}_0})]\}^{-n} e^{-x}dx . \tag{12.106b}$$

The matching time $\bar{t}_0$ has to be chosen so that the following two conditions are fulfilled. First, $\bar{t}_0$ should be so large that the intensity $4\bar{t}_0$ due to the noise is small compared to the intensity $2[\exp(2a\bar{t}_0)-1]/a$ at that time, compare (12.106a). Second it should be so small that saturation plays no essential role. This is the case if the intensity is small compared to its final value a. This means one should require

$$4\bar{t}_0 \ll 2(e^{2a\bar{t}_0}-1)/a \ll a . \tag{12.107}$$

By equating the term in the middle with the geometric mean between the lower and the upper bound we obtain approximately for large a

$$a\bar{t}_0 \approx \tfrac{1}{2}\ln a + \tfrac{1}{4}\ln a\bar{t}_0 \approx \tfrac{1}{2}\ln a . \tag{12.108}$$

The first four cumulants obtained from (12.106) are shown in Fig. 12.21 for the choice $a\bar{t}_0 = \frac{1}{2}\ln a$ and $a\bar{t}_0 = \frac{1}{4}\ln a$ as a function of $a\bar{t}$. As seen, the difference between the two solutions for the above choices of $a\bar{t}_0$ gets smaller for increasing a. For $a = 20$ the first four cumulants agree fairly well with those obtained in Sect. 12.5.2, whereas for $a = 10$ we have, especially for K_2, K_3 and K_4, only a more qualitative agreement.

In the procedure described above, the noise is completely neglected for $\bar{t} > \bar{t}_0$. Therefore the distribution function (12.104) leads to the sharp distribution $\delta(\bar{I}-a)$ in the limit $\bar{t} \to \infty$ and consequently the second cumulant vanishes for large $\bar{t}$, in contradiction to the stationary distribution (12.45) and to the stationary value $K_2 = 2$ (Fig. 12.5). To remedy this drawback one may switch on

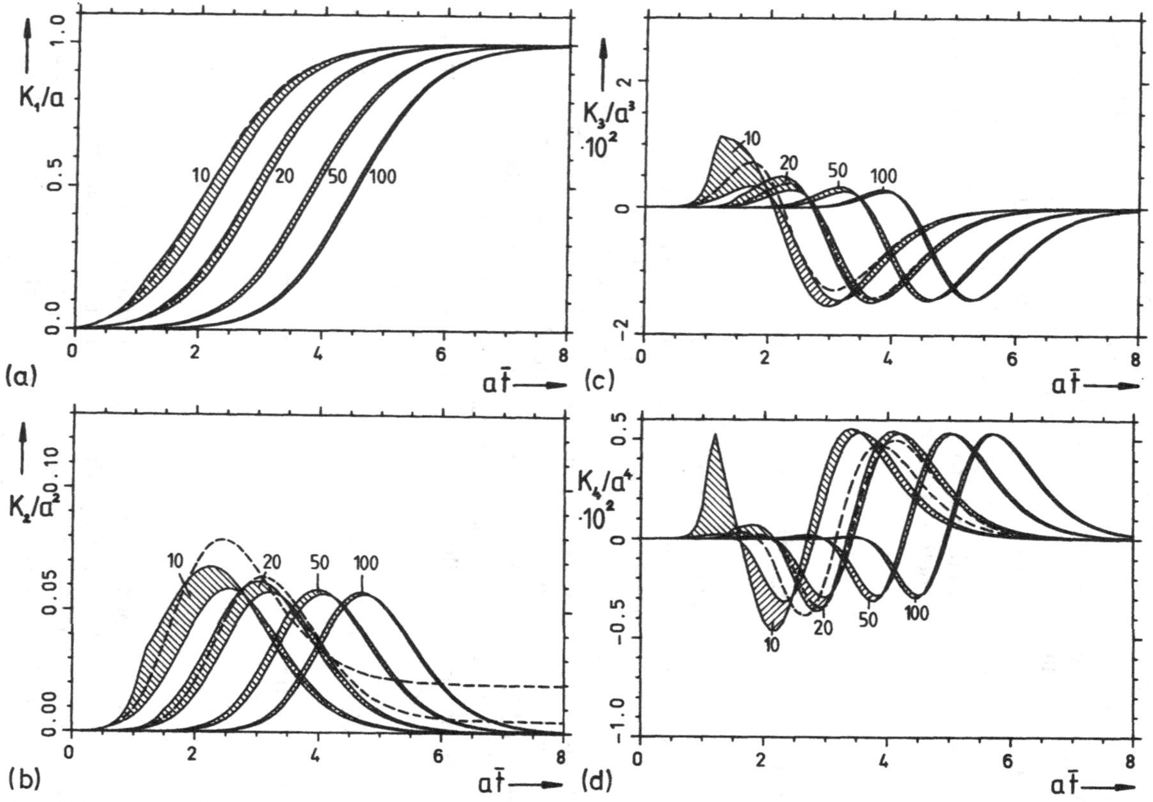

Fig. 12.21 a – d. The first four cumulants K_n/a^n as a function of $a\bar{t}$ obtained from (12.106) for the pump parameters $a = 10$, 20, 50, 100 and for $a\bar{t}_0 = 0.25 \ln a$ (*lower curve* for small times) and for $a\bar{t}_0 = 0.5 \ln a$ (*upper curve* for small times). The difference between these two curves (*lined region*) gives an estimate of the error made by changing $\bar{t}_0$. For $a = 10$, 20 the four cumulants of Fig. 12.19, which are accurate up to the linewidth in the drawing, are shown by the *broken lines*

the noise again at a time $\bar{t}_1 > \bar{t}_0$. This time must be so large that an appreciable amount of the intensity is in a region where the linearized equation (12.33a) is valid. Because the process is then again an Ornstein-Uhlenbeck process, the distribution for $\bar{t} > \bar{t}_1$ may be obtained by a convolution of the distribution $w(\bar{I}, \bar{t}_1)$ with the transition probability density (5.28) of the Ornstein-Uhlenbeck process [12.23, 24]. This distribution then has a finite width for $\bar{t} \to \infty$ and consequently the second cumulant is equal to 2 for $\bar{t} \to \infty$. The method described above resembles the path integral method [12.59].

12.6 Photoelectron Counting Distribution

The distribution function of the light field inside the laser cavity is not measured directly. Usually, one measures the intensity outside the laser cavity with a photon detector. Because only a small fraction of the light intensity is transmitted by the mirror and finally is absorbed by the detector, one usually counts only a few photoelectrons in a given time interval T. The connection between the continuous intensity distribution and the discrete photoelectron distribution was first derived by *Mandel* [12.60, 61]. (A fully quantum-mechanical derivation has been performed by *Kelly* and *Kleiner* [12.62]; see also [12.1, 3, 5].) Mandel's main result is that the probability $p(n, T)$ of finding n photoelectrons in the time interval $t, t+T$ is given by

$$p(n,T) = \langle (\Omega^n/n!)\mathrm{e}^{-\Omega}\rangle = \int (\Omega^n/n!)\mathrm{e}^{-\Omega} W(\Omega)\mathrm{d}\Omega \,, \tag{12.109}$$

where Ω is proportional to the time integral of the intensity

$$\Omega = \alpha \int_t^{t+T} I(t')\mathrm{d}t' \,. \tag{12.110}$$

The factor α is determined by the mirror transmittance, a geometrical factor giving the fraction of the intensity which falls on the photodetector, and the quantum efficiency of the photocounter. Because the intensity is a stochastic variable, Ω is a stochastic variable, too. In the Mandel expression (12.109) one has to average the Poisson distribution $\Omega^n \mathrm{e}^{-\Omega}/n!$ according to the distribution of the time-integrated intensity (12.110). In the stationary state, $W(\Omega)$ and therefore also $p(n, T)$ depend only on the length of the time interval T, but not on the time t itself. If one is interested in the expectation values

$$\langle n^r \rangle = \sum_{n=0}^{\infty} n^r p(n,T) \,, \quad r = 0,1,2,\dots \,, \tag{12.111}$$

the following relationship between the kth factorial moment $\langle n^{[k]}\rangle$ and the moments of $\langle \Omega^k \rangle$ is very useful:

$$\langle n^{[k]}\rangle = \sum_{n=0}^{\infty} n(n-1)\dots(n-k+1)\,p(n,T)$$

$$= \sum_{n=k}^{\infty} [n!/(n-k)!]\,p(n,T) = \langle \Omega^k\rangle\,. \tag{12.112}$$

The last relation follows immediately by insertion of (12.109).

The distribution $W(\Omega)$ can be obtained by the method in Sect. 8.2.1. The distribution $W(\Omega)$ for the laser Fokker-Planck equation (12.38) has been evaluated in this way by *Lax* and *Zwanziger* [12.63] for arbitrary time intervals T. Here we shall determine the full photoelectron counting distribution only for small time intervals T. The first moment of the variance of the counting distribution will be calculated, however, for arbitrary time intervals T.

12.6.1 Counting Distribution for Short Intervals

Here we assume that the time interval T is short compared to the time in which the intensity changes its value appreciably

$$T\sqrt{\beta q} = \bar{T} \ll 1/\lambda_{\text{eff}}\,. \tag{12.113}$$

Because the intensity does not change in the interval T, one may write

$$\Omega = \alpha \int_t^{t+T} I(t')\,\mathrm{d}t' = \alpha T I(t) = \alpha T\sqrt{q/\beta}\,\bar{I}(t)$$

$$= (\alpha\bar{T}/\beta)\,\bar{I}(t) = \nu\bar{I}(t) \tag{12.114}$$

and average according to the stationary distribution (12.45). We thus obtain

$$p(n,T) = N\int_0^{\infty} \frac{(\nu\bar{I})^n}{n!}\exp\left(-\nu\bar{I} - \frac{1}{4}\bar{I}^2 + \frac{1}{2}a\bar{I}\right)\mathrm{d}\bar{I}$$

$$= \frac{\nu^n}{n!}\,\frac{F_n(a-2\nu)}{F_0(a)}\,. \tag{12.115}$$

The parameter

$$\nu = \alpha T\sqrt{q/\beta} \tag{12.116}$$

relates the mean photon number with the mean intensity, see the first equation of (12.119), and F_n are the integrals (12.48). Typical counting distributions are shown in Fig. 12.22. For large average photon numbers (large ν), the photoelectron counting distribution is nearly identical to the continuous intensity distribution (Fig. 12.22c)

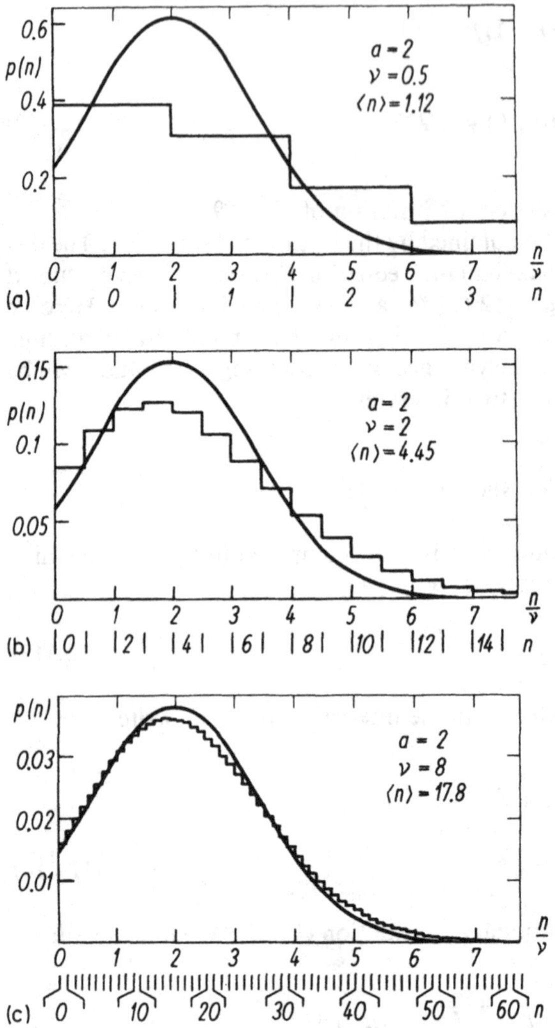

Fig. 12.22. The discrete photoelectron distribution (12.115) and the continuous distribution (12.117) for various ν parameters: **(a)** $\nu = 0.5$; **(b)** $\nu = 2$; **(c)** $\nu = 8$ but for the same pump parameter $a = 2$

$$p(n, T) \approx p_{\text{cont}}(n, T) = \frac{1}{\nu F_0(a)} \exp\left[-\frac{1}{4}(n/\nu)^2 + \frac{1}{2} a(n/\nu) \right]. \tag{12.117}$$

The relations between the first four cumulants of the photocounting distribution

$$k_i(a) = \left(\frac{\mathrm{d}}{\mathrm{d}s} \right)^i \ln \left[\sum_{n=0}^{\infty} \mathrm{e}^{ns} p(n, T) \right] \Bigg|_{s=0} \tag{12.118}$$

and the first four cumulants (12.54) of the stationary distribution (12.45) are given by [12.60, 61]

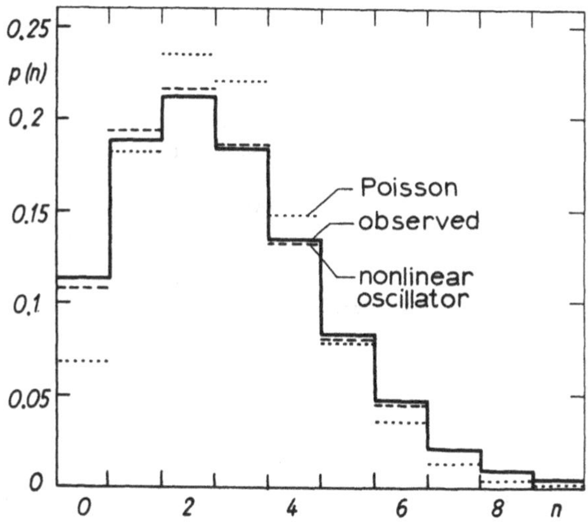

Fig. 12.23. Counting distribution just above threshold (*solid line*), the Poisson distribution for the same $\bar{n}$ (*dotted lines*), and the nonlinear oscillator distribution (12.115) giving the best fit (*dashed line*). For $n = 7$, 8 and 9 the nonlinear oscillator and observed distributions are coincident. (Redrawn after [12.40])

$$\begin{aligned} k_1(a) &= \langle n \rangle = \nu K_1(a) \\ k_2(a) &= \langle (n - \langle n \rangle)^2 \rangle = \nu K_1(a) + \nu^2 K_2(a) \\ k_3(a) &= \nu K_1(a) + 3\nu^2 K_2(a) + \nu^3 K_3(a) \\ k_4(a) &= \nu K_1(a) + 7\nu^2 K_2(a) + 6\nu^3 K_3(a) + \nu^4 K_4(a) \,. \end{aligned} \tag{12.119}$$

The cumulants $K_n(a)$ of the stationary intensity distribution are plotted in Fig. 12.5.

The photoelectron counting distribution $p(n, T)$ for short intervals and near threshold was first measured by *Smith* and *Armstrong* [12.40]. The excellent agreement between theory and measurements is shown in Fig. 12.23. One sees by inspection that the measured distribution is not a Bose-Einstein distribution

$$p_{\mathrm{B}}(n, T) = \frac{1}{1 + \langle n \rangle} \left(\frac{\langle n \rangle}{1 + \langle n \rangle} \right)^n \tag{12.120}$$

(a Bose-Einstein distribution has its largest probability always for $n = 0$), but that a Poisson distribution $\langle n \rangle^n \exp(-\langle n \rangle)/n!$ is a better approximation. The Bose-Einstein distribution is valid for pump parameters far below threshold ($a \ll -1$). This can be shown by using an asymptotic expansion of the integrals $F_n(a)$ for $a \ll -1$. (One can then neglect $\bar{I}^2/4$ in the exponential function.)

The transition from a Bose-Einstein to a Poisson distribution through the threshold region obtained by changing the pump parameter is best seen by looking at the variance

$$\langle (n - \langle n \rangle)^2 \rangle = \langle n \rangle + \eta(a) \langle n \rangle^2 \,, \tag{12.121}$$

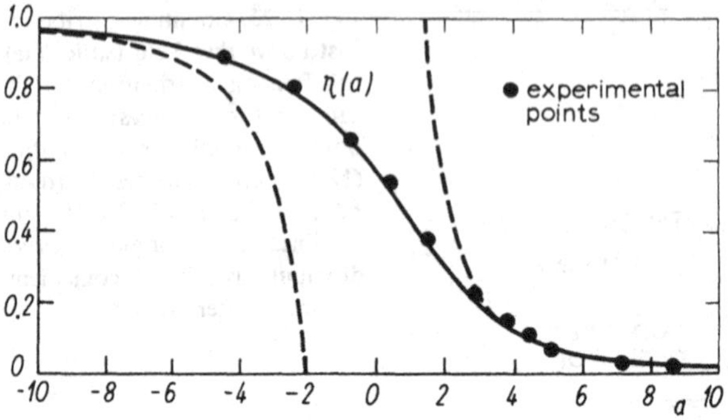

Fig. 12.24. The parameter $\eta(a)$ of (12.122) (*solid line*), and the asymptotic expansions $\eta = 1 - 4/a^2$ for $a \ll -1$ and $\eta = 2/a^2$ for $a \gg 1$ (*broken line*) as a function of the pump parameter a. The experimental points are taken from [12.42a]. (The abscissa $K_1(a)/K_1(0)$ from [12.42a] was rescaled and expressed as a function of a)

where we introduced the parameter $\eta(a)$ defined by

$$\eta(a) = K_2(a)/K_1^2(a) . \qquad (12.122)$$

As seen in Fig. 12.24 there is a smooth transition from the variance of the Bose-Einstein distribution $\eta = 1$ to the variance of the Poisson distribution $\eta = 0$. The variance (12.121) consists of two parts; the first part $\langle n \rangle$ stems from the discreteness of the photoelectrons, whereas the second part $\eta(a)\langle n \rangle^2$ stems from the fluctuation of the light beam. It should be noted that the last part $\eta(a)\langle n \rangle^2$ remains finite as the pump parameter a goes to infinity because $\langle n \rangle$ increases proportionally to a. However, if the light beam is attenuated so that $\langle n \rangle$ remains constant as a is increased, we finally obtain for $a \to \infty$ a pure Poisson distribution with variance $\langle n \rangle$.

12.6.2 Expectation Values for Arbitrary Intervals

We now drop the assumption (12.113) that the time interval T is short compared to the correlation time $(\beta q)^{-1/2}(\lambda_{\text{eff}})^{-1}$. In contrast to Sect. 12.6.1 we do not derive the counting distribution but merely discuss the expectation value $\langle n \rangle$ and the variance $\langle (n - \langle n \rangle)^2 \rangle$. Using (12.110, 112, 113, 116) we obtain

$$\langle n \rangle = \langle \Omega \rangle = \alpha \langle I \rangle = (\nu/\bar{T}) \int_{\bar{t}}^{\bar{t}+\bar{T}} \langle \bar{I}(\bar{t}') \rangle \mathrm{d}\bar{t}' = \nu K_1(a) , \qquad (12.123)$$

$$\begin{aligned}\langle (n - \langle n \rangle)^2 \rangle &= \langle n \rangle + \langle n(n-1) \rangle - \langle n \rangle^2 \\ &= \langle n \rangle + \langle \Omega^2 \rangle - \langle \Omega \rangle^2 \\ &= \langle n \rangle + (\nu/\bar{T})^2 \int_{\bar{t}}^{\bar{t}+\bar{T}} \int_{\bar{t}}^{\bar{t}+\bar{T}} [\langle \bar{I}(\bar{t}')\bar{I}(\bar{t}'') \rangle - \langle \bar{I}(\bar{t}') \rangle \langle \bar{I}(\bar{t}'') \rangle] \mathrm{d}t' \mathrm{d}t'' .\end{aligned}$$

The bracket under the double integral is the correlation function of the intensity (12.64) in normalized units. Inserting expansion (12.66), we thus obtain

$$\langle (n - \langle n \rangle)^2 \rangle = \nu K_1(a) + \nu^2 K_2(a) f(\bar{T})$$

$$f(\bar{T}) = \sum_{n=1}^{\infty} \frac{2\, V_n^{(K)}}{\lambda_{0n} \bar{T}} \left\{ 1 - \frac{1}{\lambda_{0n} \bar{T}} [1 - \exp(-\lambda_{0n} \bar{T})] \right\} \qquad (12.124)$$

$$\approx \frac{2}{\lambda_{\text{eff}} \bar{T}} \left\{ 1 - \frac{1}{\lambda_{\text{eff}} \bar{T}} [1 - \exp(-\lambda_{\text{eff}} \bar{T})] \right\} .$$

In the last line we have used the approximation (12.71).

For $\lambda_{\text{eff}} \bar{T} \ll 1$ we have $f(\bar{T}) = 1$ and thus obtain the result in the second equation of (12.119). For $\lambda_{\text{eff}} \bar{T} \gg 1$ we have $f(\bar{T}) = 2/(\lambda_{\text{eff}} \bar{T})$. The part of the variance $\nu^2 K_2(a) f(\bar{T})$ which stems from the fluctuation of the light beam is therefore diminished by using a long time interval $\bar{T} \gg \lambda_{\text{eff}}^{-1}$. Because λ_{eff} has a minimum near threshold, the part of the variance $\nu^2 K_2(a) f(\bar{T})$ is most important in the threshold region.

Appendices

A1. Stochastic Differential Equations with Colored Gaussian Noise

Here I want to show how the matrix continued-fraction method can be used to calculate expectation values for certain stochastic differential equations with colored Gaussian noise, i.e., the noise may have an arbitrary correlation time. This method was developed by *Zoller* [A1.1], *Dixit* et al. [9.22] and *Zoller* et al. [9.23] for treating the optical Bloch equations with multiplicative colored-noise terms.

The method is exemplified by the *Kubo* oscillator [A.1.2, 3, 3.1], whose frequency changes according to

$$\dot{u} = \mathrm{i}[\omega_0 + \varepsilon(t)]u\,, \tag{A1.1}$$

where $\varepsilon(t)$ is a Gaussian stochastic force with zero mean and an exponential correlation function

$$\langle\varepsilon(t)\rangle = 0\,, \qquad \langle\varepsilon(t)\varepsilon(t')\rangle = \gamma D \mathrm{e}^{-\gamma|t-t'|}. \tag{A1.2}$$

Formal integration of (A1.1) with the sharp initial value $u(0)$ leads to

$$u(t) = u(0)\exp\left[\mathrm{i}\omega_0 t + \mathrm{i}\int_0^t \varepsilon(t')\,dt'\right]. \tag{A1.3}$$

Using (3.75, 16) we obtain ($t \geqq 0$)

$$\langle u(t)\rangle = \langle u(0)\rangle \exp\{\mathrm{i}\omega_0 t - D[t-(1-\mathrm{e}^{-\gamma t})/\gamma]\}\,. \tag{A1.4}$$

For $\gamma \to \infty$ the stochastic force $\varepsilon(t)$ may be approximated by the δ-correlated Langevin force

$$\varepsilon(t) = \sqrt{D}\,\Gamma(t)\,, \qquad \langle\Gamma(t)\Gamma(t')\rangle = 2\delta(t-t')\,. \tag{A1.2a}$$

The result (A1.4) then reduces for $\omega_0 = 0$ to (3.76) with $a = \mathrm{i}\sqrt{D}$. For $\gamma \to 0$, $\langle u(t)\rangle = \langle u(0)\rangle \exp(\mathrm{i}\omega_0 t - \gamma D t^2/2)$, which could also be obtained by integrating (A1.1) for a fixed ε and then averaging it with the help of the stationary distribution $W_{\mathrm{st}}(\varepsilon) = (2\pi\gamma D)^{-1/2}\exp[-\varepsilon^2/(2\gamma D)]$.

The result (A1.4) can be obtained by the continued fraction method as follows. Equation (A1.1, 2) are equivalent to the Langevin equation (A1.1) and

$$\dot{\varepsilon} = -\gamma\varepsilon + \gamma\sqrt{D}\Gamma(t)\,, \quad \langle\Gamma(t)\Gamma(t')\rangle = 2\delta(t-t')\,, \quad \langle\Gamma(t)\rangle = 0 \qquad \text{(A1.5)}$$

because (A1.5) immediately leads to (A1.2) in the stationary state, Sect. 3.1. The Fokker-Planck equation corresponding to (A1.1, 5) reads $[W = W(u,\varepsilon,t)]$

$$\dot{W} = L_{\mathrm{FP}} W\,, \qquad \text{(A1.6)}$$

$$L_{\mathrm{FP}} = -\frac{\partial}{\partial u}\mathrm{i}(\omega_0+\varepsilon)u + L_\varepsilon\,, \qquad \text{(A1.7)}$$

$$L_\varepsilon = \gamma\frac{\partial}{\partial\varepsilon}\varepsilon + \gamma^2 D\frac{\partial^2}{\partial\varepsilon^2}\,. \qquad \text{(A1.8)}$$

Here, L_ε has the same form as the operator L_{ir} in Chap. 10. By multiplying (A1.6) by u and by integrating it with respect to u we obtain after integrating by parts

$$\dot{w} = \mathrm{i}(\omega_0+\varepsilon)w + L_\varepsilon w\,, \qquad \text{(A1.9)}$$

where the marginal distribution w [3.1] is given by

$$w(\varepsilon,t) = \int u\, W(u,\varepsilon,t)\,\mathrm{d}u\,. \qquad \text{(A1.10)}$$

By expanding this distribution into Hermite functions $\psi_n(\varepsilon)$ [see (10.38 – 40) with $v = \varepsilon$ and $v_{\mathrm{th}}^2 = \gamma D$]

$$w(\varepsilon,t) = \psi_0(\varepsilon)\sum_{n=0}^{\infty} c_n(t)\,\psi_n(\varepsilon)\,, \qquad \text{(A1.11)}$$

we obtain, similarly as in Sect. 10.1.4, the tridiagonal recurrence relation ($c_n = 0$ for $n \leqq -1$)

$$\dot{c}_n = (\mathrm{i}\,\omega_0 - n\gamma)c_n + \mathrm{i}\sqrt{\gamma D}\left(\sqrt{n+1}\,c_{n+1} + \sqrt{n}\,c_{n-1}\right). \qquad \text{(A1.12)}$$

The averaged value of u, see (A1.4), is then given by

$$\langle u(t)\rangle = \iint u\, W(u,\varepsilon,t)\,\mathrm{d}u\,\mathrm{d}\varepsilon = \int w(\varepsilon,t)\,\mathrm{d}\varepsilon = c_0(t)\,, \qquad \text{(A1.13)}$$

where $c_n(t)$ is a solution of the system (A1.12) with the initial condition (stationary distribution for ε)

$$c_n(0) = \delta_{n0}\,. \qquad \text{(A1.14)}$$

We may thus immediately apply the results of Sect. 9.2.2. The Laplace transform of $c_0(t)$ reads

$$\tilde{c}_0(s) = \tilde{G}_{0,0}(s)\langle u(0)\rangle = [s - \mathrm{i}\,\omega_0 - \tilde{K}_0(s)]^{-1}\langle u(0)\rangle\,, \tag{A1.15}$$

where $\tilde{K}_0(s)$ is the infinite ordinary continued fraction (9.72) with $m = 0$, i.e.,

$$\tilde{K}_0(s) = \frac{1\,\gamma D}{|\mathrm{i}\,\omega_0 - s - 1\,\gamma} + \frac{2\,\gamma D}{|\mathrm{i}\,\omega_0 - s - 2\,\gamma} + \dots\,. \tag{A1.16}$$

By using [Ref. 9.1, § 48, Eqs. (23, 26)], it may be shown that (A1.15, 16) are the Laplace transform of (A1.4), see also (10.148 – 152). By setting $s = \mathrm{i}\,\omega$ we have thus found a continued fraction for the half-sided Fourier transform of the solution of (A1.1, 2). This continued fraction is very convenient for numerical calculations.

A more general stochastic equation has the form [A1.1, 9.23]

$$\dot{u}_i = \sum_{j=1}^{N} [A^{ij} + B^{ij}\varepsilon(t)]\,u_j\,, \qquad i = 1, \dots, N\,, \tag{A1.17}$$

where $\varepsilon(t)$ is still given by (A1.2). By adding (A1.5) to (A1.17), we obtain Langevin equations for the variables $u_1, \dots, u_N, \varepsilon$. The corresponding Fokker-Planck equation is (A1.6), where L_{FP} is now given by

$$L_{\mathrm{FP}} = -\sum_{i=1}^{N}\sum_{j=1}^{N} \frac{\partial}{\partial u_i}(A^{ij} + B^{ij}\varepsilon)\,u_j + L_\varepsilon\,. \tag{A1.18}$$

By multiplying the Fokker-Planck equation with u_i and integrating the resulting equation over $u_1, \dots, u_N$, we obtain after performing a partial integration for the marginal averages

$$w_i(\varepsilon, t) = \int u_i W(u_1, \dots, u_N, \varepsilon, t)\,\mathrm{d}^N u \tag{A1.19}$$

the equations [3.1]

$$\dot{w}_i = \sum_{j=1}^{N} (A^{ij} + B^{ij}\varepsilon)\,w_j + L_\varepsilon w_i\,. \tag{A1.20}$$

Now we again expand the marginal averages in Hermite functions $\psi_n(\varepsilon)$

$$w_i(\varepsilon, t) = \psi_0(\varepsilon) \sum_{n=0}^{\infty} c_n^i(t)\,\psi_n(\varepsilon)\,. \tag{A1.21}$$

Using the vector and matrix notation

$$\boldsymbol{c}_n = \begin{pmatrix} c_n^1 \\ \vdots \\ c_n^N \end{pmatrix}, \qquad \boldsymbol{A} = (A^{ij})\,, \qquad \boldsymbol{B} = (B^{ij}) \tag{A1.22}$$

and inserting (A1.21) into (A1.20) we thus obtain the tridiagonal vector recurrence relation

$$\dot{c}_n = (A - n\gamma I)c_n + \sqrt{\gamma D}B(\sqrt{n+1}\,c_{n+1} + \sqrt{n}\,c_{n-1}) . \tag{A1.23}$$

The averaged value $\langle u_i(t)\rangle$ is given by

$$c_0(t) = \begin{bmatrix} \langle u_1(t)\rangle \\ \vdots \\ \langle u_N(t)\rangle \end{bmatrix} , \tag{A1.24}$$

where $c_n(t)$ is a solution of the system (A1.23) with the initial value

$$c_0(0) = \begin{bmatrix} \langle u_1(0)\rangle \\ \vdots \\ \langle u_N(0)\rangle \end{bmatrix} , \quad c_n(0) = \mathbf{0} \quad \text{for} \quad n \geqq 1 . \tag{A1.25}$$

As derived in Sect. 9.3.1, the Laplace transform of $c_0(t)$ reads

$$\tilde{c}_0(s) = \tilde{G}_{0,0}(s)c_0(0) = [sI - A - \tilde{K}_0(s)]^{-1}c_0(0) , \tag{A1.26}$$

where $\tilde{K}_0(s)$ is given by the infinite matrix continued fraction (9.112) (first derived in [A1.1, 9.22, 23])

$$\begin{aligned}\tilde{K}_0(s) = \gamma DB[(s+1\gamma)I - A - 2\gamma DB[(s+2\gamma)I - A \\ - 3\gamma DB[(s+3\gamma)I - A - \ldots]^{-1}B]^{-1}B]^{-1}B .\end{aligned} \tag{A1.27}$$

By setting $s = \mathrm{i}\omega$ we thus arrive at an expression for the half-sided Fourier transform of $\langle u_i(t)\rangle$, which is very convenient for numerical calculations.

$1/\gamma$-Expansion

For large γ we obtain the following $1/\gamma$ expansion for $\tilde{K}_0(s)$ and $K_0(t)$:

$$\begin{aligned}\tilde{K}_0(s) = &\frac{\gamma D}{s+\gamma}B^2 + \frac{\gamma D}{(s+\gamma)^2}BAB + \frac{2(\gamma D)^2}{(s+\gamma)^2(s+2\gamma)}B^4 \\ &+ \frac{\gamma D}{(s+\gamma)^3}BA^2B + \frac{2(\gamma D)^2}{(s+\gamma)^2(s+2\gamma)^2}B^2AB^2 \\ &+ \frac{2(\gamma D)^2}{(s+\gamma)^3(s+2\gamma)}(BAB^3 + B^3AB) \\ &+ \left[\frac{6(\gamma D)^3}{(s+\gamma)^2(s+2\gamma)^2(s+3\gamma)} + \frac{4(\gamma D)^3}{(s+\gamma)^3(s+2\gamma)^2}\right]B^6 + O(\gamma^{-3}) ,\end{aligned} \tag{A1.28}$$

$$
\begin{aligned}
\boldsymbol{K}_0(t) = {} & \gamma D e^{-\gamma t} \boldsymbol{B}^2 + D \gamma t e^{-\gamma t} \boldsymbol{BAB} \\
& + 2D^2(-e^{-\gamma t} + \gamma t e^{-\gamma t} + e^{-2\gamma t}) \boldsymbol{B}^4 \\
& + \frac{1}{\gamma} \left\{ D \frac{1}{2} (\gamma t)^2 e^{-\gamma t} \boldsymbol{BA}^2\boldsymbol{B} + 2D^2(-2e^{-\gamma t} + \gamma t e^{-\gamma t} + 2e^{-2\gamma t} \right. \\
& + \gamma t e^{-2\gamma t}) \boldsymbol{B}^2\boldsymbol{AB}^2 + 2D^2 \left[e^{-\gamma t} - \gamma t e^{-\gamma t} + \frac{1}{2}(\gamma t)^2 e^{-\gamma t} \right. \\
& \left. - e^{-2\gamma t} \right] (\boldsymbol{BAB}^3 + \boldsymbol{B}^3\boldsymbol{AB}) + D^3 \left[\frac{9}{2} e^{-\gamma t} - 5\gamma t e^{-\gamma t} \right. \\
& \left. \left. + 2(\gamma t)^2 e^{-\gamma t} - 6e^{-2\gamma t} + 2\gamma t e^{-2\gamma t} + \frac{3}{2} e^{-3\gamma t} \right] \boldsymbol{B}^6 \right\} + O(\gamma^{-2}) .
\end{aligned}
\tag{A1.29}
$$

It follows from (A1.26), compare (9.113), that $\boldsymbol{c}_0(t)$ obeys the integrodifferential equation

$$
\dot{\boldsymbol{c}}_0(t) = \boldsymbol{A}\boldsymbol{c}_0(t) + \int_0^t \boldsymbol{K}_0(t-\tau)\boldsymbol{c}_0(\tau)\,d\tau \tag{A1.30}
$$

with the initial condition (A1.25). Because the kernel $\boldsymbol{K}_0(t)$ falls off very rapidly in time for large γ we may use repeated partial integration similar to (10.183a). [The term $\boldsymbol{A}$ must now be added in (10.183).] After some lengthy calculations we thus obtain the following differential equation:

$$
\dot{\boldsymbol{c}}_0 = \boldsymbol{L}_0(t)\boldsymbol{c}_0 , \tag{A1.31}
$$

$$
\begin{aligned}
\boldsymbol{L}_0(t) = {} & \boldsymbol{A} + D(1 - e^{-\gamma t})\boldsymbol{B}^2 \\
& + \frac{1}{\gamma} D(1 - e^{-\gamma t} - \gamma t e^{-\gamma t}) \boldsymbol{B}[\boldsymbol{A},\boldsymbol{B}] \\
& + \frac{1}{\gamma^2} \left\{ D \left[1 - e^{-\gamma t} - \gamma t e^{-\gamma t} - \frac{1}{2}(\gamma t)^2 e^{-\gamma t} \right] \boldsymbol{B}[\boldsymbol{A},[\boldsymbol{A},\boldsymbol{B}]] \right. \\
& + D^2 \left[1 + e^{-\gamma t} - 2\gamma t e^{-\gamma t} - \frac{1}{2}(\gamma t)^2 e^{-\gamma t} - 2e^{-2\gamma t} - \gamma t e^{-2\gamma t} \right] \\
& \times \boldsymbol{B}[[\boldsymbol{B},\boldsymbol{A}],\boldsymbol{B}]\boldsymbol{B} + \frac{1}{2} D^2 [1 - 2e^{-\gamma t} - (\gamma t)^2 e^{-\gamma t} + e^{-2\gamma t}] \\
& \left. \times \boldsymbol{B}^2[[\boldsymbol{B},\boldsymbol{A}],\boldsymbol{B}] \right\} + O\left(\frac{1}{\gamma^3}\right) .
\end{aligned}
\tag{A1.32}
$$

For $\gamma t \gg 1$ (A1.32) reduces to

$$L_0(\infty) = A + DB^2 + \frac{1}{\gamma} DB[A,B] + \frac{1}{\gamma^2} DB[A,[A,B]] + \frac{1}{\gamma^2} D^2 \left\{ B[[B,A],B]B + \frac{1}{2} B^2[[B,A],B] \right\} + O\left(\frac{1}{\gamma^3}\right). \tag{A1.32a}$$

For commuting matrices $[A,B] = 0$ we can evaluate the inverse Laplace transform of the continued fraction $[sI - A - \tilde{K}_0(s)]^{-1}$ exactly (similar to the procedure at the end of Sect. 10.3.1) leading to the exact result

$$L_0(t) = A + D(1 - \mathrm{e}^{-\gamma t})B^2. \tag{A1.33}$$

For the Kubo oscillator (A1.1) we have $A = \mathrm{i}\,\omega_0$, $B = \mathrm{i}$ and thus obtain

$$\partial\langle u(t)\rangle/\partial t = [\mathrm{i}\,\omega_0 - D(1 - \mathrm{e}^{-\gamma t})]\,\langle u(t)\rangle \tag{A1.34}$$

in agreement with (A1.4).

Generalizations

Several generalizations of this method are possible.

(i) For the averages $\langle u_i(t)u_j(t)\rangle$, $\langle u_i(t)u_j(t)u_k(t)\rangle, \ldots$ the method is also applicable, leading to equations of motion for the marginal distribution functions w_{ij}, $w_{ijk}, \ldots$, which could also be solved by matrix continued-fraction methods.

(ii) If ε appears in some higher polynomial couplings of highest order M in (A1.17), the same expansion (A1.21) then leads to a form where $2M+1$ nearest coefficients c_n are coupled. As explained in Sect. 9.1, one can also cast this equation into a tridiagonal vector recurrence relation by using suitable vector notation.

(iii) If more stochastic forces $\varepsilon_1, \varepsilon_2, \varepsilon_3 \ldots$ appear in (A1.17), one has to use an expansion vector with more indices $c_{n_1,n_2,n_3}\ldots$. If ε_i appear linearly in (A1.17), one then generally gets a tridiagonal coupling in all the indices, which usually cannot be reduced to a tridiagonal coupling in one index and therefore the continued-fraction method cannot be used. One may, of course, still solve the coupled equations by a proper truncation.

It may, however, happen that for certain stochastic differential equations coupling may be reduced to tridiagonal coupling. This was the case in the problem treated in [A1, 9.22, 23], where a complex $\varepsilon(t)$ appeared and the phase of $\varepsilon(t)$ dropped out in the final expectation value.

(iv) If the variable u appears nonlinearly in (A1.1), one may still solve the problem by expanding $W(u,\varepsilon)$ into a complete set with respect to the u variable and into the set $\psi_n(\varepsilon)$. By truncating the expansion in u one may then derive a tridiagonal vector recurrence relation (A1.23).

(v) If $\varepsilon(t)$ is a random telegraph noise, *Wódkiewicz* [A1.4] has shown that the same method can still be used. The continued fractions will then, however, terminate.

(vi) The method may be applied to the partial differential equation

$$\partial\rho(x,t)/\partial t = [A + B\varepsilon(t)]\rho(x,t) \tag{A1.35}$$

where A and B are operators with respect to x. (The extension to N variables $\{x\} = x_1, \ldots, x_N$ is also possible.) If a proper expansion of $\rho(x)$ into a complete set is used, (A1.35) transforms to (A1.17). In x-representation the $1/\gamma$ expansion (A1.32) is now also useful where A and B are the operators A and B of (A1.35).

A2. Boltzmann Equation with BGK and SW Collision Operators

The one-dimensional Boltzmann equation with a BGK collision operator [1.23] or with the SW collision operator proposed by *Skinner* and *Wolynes* [A2.1] can also be treated by the matrix continued-fraction method. The SW collision operator is defined by

$$L_{\mathrm{SW}} W(x,v,t) = \int_{-\infty}^{\infty} [K(v',v)\, W(x,v',t) - K(v,v')\, W(x,v,t)]\, dv' , \tag{A2.1}$$

where the kernel K reads

$$K(v,v') = \gamma \frac{\gamma_s + 1}{2\sqrt{\gamma_s}} \sqrt{\frac{m}{2\pi kT}} \exp\left\{ - \frac{m}{8\gamma_s kT} [(\gamma_s - 1)v + (\gamma_s + 1)v']^2 \right\}. \tag{A2.2}$$

If the parameter γ_s is equal to 1, (A2.1) reduces to the BGK operator (1.32). As shown in [A2.1] the eigenfunctions of L_{SW},

$$L_{\mathrm{SW}}\psi_n(v) = -\lambda_n \psi_n(v) , \tag{A2.3}$$

are the Hermite functions $\psi_n(v)$ defined in (10.39) and the eigenvalues λ_n are given by ($n \geqq 1$)

$$\lambda_0 = 0, \quad \lambda_n = \gamma\left[1 - \left(\frac{1-\gamma_s}{1+\gamma_s}\right)^n\right]. \tag{A2.4}$$

For $\gamma_s \to 0$ we obtain the eigenvalues (10.37) of L_{ir} in Sect. 10.1.4 multiplied by $2\gamma_s$

$$\gamma_s \to 0: \quad \lambda_n = 2\gamma_s n \gamma, \quad n = 0, 1, 2, \ldots, \tag{A2.5}$$

whereas for $\gamma_s \to 1$ we obtain the eigenvalues of the BGK operator [A2.2]

$$\gamma_s \to 1: \quad \lambda_n = \begin{cases} 0 & \text{for} \quad n = 0 \\ \gamma & \phantom{\text{for}} \quad n \geqq 1 \end{cases}. \tag{A2.6}$$

Because the eigenvalues and the eigenfunctions are the same for both L_{ir} and $L_{\mathrm{SW}}/(2\gamma_s)$ in the limit $\gamma_s \to 0$, both operators must agree, i.e.,

$$\lim_{\gamma_s \to 0} \frac{1}{2\gamma_s} L_{\mathrm{SW}} = L_{\mathrm{ir}}(v) = \gamma \frac{\partial}{\partial v}\left(v + \frac{kT}{m}\frac{\partial}{\partial v}\right). \tag{A2.7}$$

This may also be derived explicitly as follows. Setting $\gamma_s = \gamma\tau/2$ we write

$$\frac{1}{2\gamma_s} K(v', v) = \frac{1+\gamma\tau/2}{\tau\sqrt{\pi\gamma\tau}} \sqrt{\frac{m}{kT}} \exp\left(-\frac{m}{kT}\frac{[v - v' + \gamma\tau(v+v')/2]^2}{4\gamma\tau}\right). \tag{A2.8}$$

In the limit $\gamma_s \to 0$, i.e., in the limit $\tau \to 0$, we may neglect $\gamma\tau$ in the first nominator on the right-hand side. Furthermore, in the limit $\tau \to 0$ we can replace $\tau(v+v')$ by $2\tau v'$ in the exponential. We thus obtain the transition probability (4.55) for small τ with $D^{(2)} = \gamma kT/m$ and $D^{(1)} = -\gamma v$, i.e.,

$$\begin{aligned} \frac{1}{2\gamma_s} K(v', v) &= \frac{1}{\tau} P(v, \tau | v', 0) = \frac{1}{\tau} \mathrm{e}^{L_{\mathrm{ir}}(v)\tau} \delta(v - v') \\ &= \left[\frac{1}{\tau} + L_{\mathrm{ir}}(v) + O(\tau)\right] \delta(v - v'). \end{aligned} \tag{A2.9}$$

Insertion of (A2.9) into (A2.1) leads to

$$\begin{aligned} \frac{1}{2\gamma_s} L_{\mathrm{SW}} W(x, v, t) = \tau^{-1} &\int [W(x, v', t) - W(x, v, t)]\, \delta(v - v')\, \mathrm{d}v' \\ &+ \int L_{\mathrm{ir}}(v)\, \delta(v - v')\, W(x, v', t)\, \mathrm{d}v' \\ &- \int L_{\mathrm{ir}}(v')\, \delta(v - v')\, W(x, v, t)\, \mathrm{d}v' + O(\tau). \end{aligned} \tag{A2.10}$$

Obviously, the first integral vanishes. The last integral also vanishes because the integration over the Fokker-Planck operator is zero. Therefore (A2.10) simplifies in the limit $\gamma_s \to 0$ to

$$\frac{1}{2\gamma_s} L_{\mathrm{SW}} W(x, v, t) = L_{\mathrm{ir}}(v)\, W(x, v, t),$$

which is equivalent to (A2.7).

The Boltzmann equation (1.31) with the collision operator (A2.1) can be expanded in the same way into Hermite functions as in Sect. 10.1.4 for the Kramers equation. The only difference now is that in the coupled equations (10.46, 46a) the diagonal damping terms $n\gamma$ have to be replaced by the eigenvalues λ_n given by (A2.4). Therefore, – with slight modifications – also the matrix continued-fraction method of Sect. 10.3 can be used for solving the hierarchy (10.46). The eigenvalues of the full Boltzmann equation (1.31) with a BGK collision operator were calculated in [9.16] for a cosine potential by this method.

A3. Evaluation of a Matrix Continued Fraction for the Harmonic Oscillator

In Sect. 10.3.1 we derived a general expression for the Green's function of the Kramers equation in terms of continued fractions (10.137 – 143). The Laplace transform for this Green's function in position only is given by

$$\tilde{G}_{0,0}(s) = [s\boldsymbol{I} - \tilde{\boldsymbol{K}}_0(s)]^{-1}, \tag{A3.1}$$

where $\tilde{\boldsymbol{K}}_0(s)$ is the infinite continued fraction

$$\begin{aligned}\tilde{\boldsymbol{K}}_0(s) = \boldsymbol{D}\,[(s+\gamma)\boldsymbol{I} - 2\boldsymbol{D}\,[(s+2\gamma)\boldsymbol{I} - 3\boldsymbol{D}\\ \times [(s+3\gamma)\boldsymbol{I} - \dots]^{-1}\hat{\boldsymbol{D}}]^{-1}\hat{\boldsymbol{D}}]^{-1}\hat{\boldsymbol{D}}\,.\end{aligned} \tag{A3.2}$$

On the other hand, the Green's function for a harmonic oscillator can be calculated exactly. In the x representation it simply follows from

$$G_{0,0}(x,x',t) = \iint P(x,v,t\,|\,x',v',0)\,(2\pi)^{-1/2} v_{\text{th}}^{-1} \exp[-\tfrac{1}{2}(v'/v_{\text{th}})^2]\,dv\,dv'\,, \tag{A3.3}$$

where P is the transition probability (10.55). By performing the integration and using (10.56 – 63), one thus obtains after some lengthy calculations

$$G_{0,0}(x,x',t) = \frac{\sqrt{m}\,\omega_0}{\sqrt{2\pi k_B T[1-y^2(t)]}} \exp\left(-\frac{m\omega_0^2[x-x'y(t)]^2}{2k_B T[1-y^2(t)]}\right). \tag{A3.4}$$

Here, $y(t)$ is given by

$$y(t) = (\lambda_1 e^{-\lambda_2 t} - \lambda_2 e^{-\lambda_1 t})/(\lambda_1 - \lambda_2)\,, \tag{A3.5}$$

where λ_1 and λ_2 are defined by (10.60).

The exact result (A3.4) should therefore agree with the exact result (A3.1, 2) for the harmonic oscillator taking $\varepsilon = 0$. To show the equivalence, we first have to evaluate the infinite matrix continued fraction (A3.2) for the harmonic oscillator.

For a harmonic oscillator the commutator of $\boldsymbol{D}$ and $\hat{\boldsymbol{D}}$ is proportional to the unit matrix, see (10.28, 52)

$$[\boldsymbol{D},\hat{\boldsymbol{D}}] = \omega_0^2 \boldsymbol{I} . \tag{A3.6}$$

Because of this relation we have the identity

$$\boldsymbol{D} F(\boldsymbol{D}\hat{\boldsymbol{D}}) = F(\boldsymbol{D}\hat{\boldsymbol{D}} + \omega_0^2 \boldsymbol{I})\boldsymbol{D} , \tag{A3.7}$$

where F is an arbitrary function. If we truncate the infinite continued fraction (A3.2), the last term only contains a $\boldsymbol{D}\hat{\boldsymbol{D}}$. Because of (A3.7) we then conclude that every denominator depends only on $\boldsymbol{D}\hat{\boldsymbol{D}}$. By shifting $\boldsymbol{D}$ in (A3.2) to the right we then have

$$\begin{aligned}\tilde{\boldsymbol{K}}_0(s) = [(s+\gamma)\boldsymbol{I} - 2(\boldsymbol{D}\hat{\boldsymbol{D}} + \omega_0^2 \boldsymbol{I})[(s+2\gamma)\boldsymbol{I} \\ - 3(\boldsymbol{D}\hat{\boldsymbol{D}} + 2\omega_0^2 \boldsymbol{I})[(s+3\gamma)\boldsymbol{I} - 4(\boldsymbol{D}\hat{\boldsymbol{D}} + 3\omega_0^2 \boldsymbol{I}) \\ \times [(s+4\gamma)\boldsymbol{I} - \dots]^{-1}]^{-1}]^{-1}]^{-1}\boldsymbol{D}\hat{\boldsymbol{D}} .\end{aligned} \tag{A3.8}$$

(The factors $1, 2, 3, \dots$ in front of $\omega_0^2 \boldsymbol{I}$ appear, because for each shift of $\boldsymbol{D}$ to the right a term $\omega_0^2 \boldsymbol{I}$ has to be added.) Because the operators in (A3.8) appear only in the combination $\boldsymbol{D}\hat{\boldsymbol{D}}$ and because the product $\boldsymbol{D}\hat{\boldsymbol{D}}$ commutes with itself we can now evaluate (A3.8) as an ordinary continued fraction. We therefore omit the matrix character and write

$$\boldsymbol{D}\hat{\boldsymbol{D}} = -\omega_0^2 \boldsymbol{\xi} \rightarrow -\omega_0^2 \xi . \tag{A3.9}$$

We thus have

$$\begin{aligned}[\tilde{G}_{0,0}(s)]^{-1} &= s - \tilde{K}_0(s) \\ &= s + \frac{\omega_0^2(\xi+1) - \omega_0^2|}{|\, s+\gamma} + \frac{2\omega_0^2(\xi+1) - 4\omega_0^2|}{|\, s+2\gamma} \\ &\quad + \frac{3\omega_0^2(\xi+1) - 9\omega_0^2|}{|\, s+3\gamma} + \dots .\end{aligned} \tag{A3.10}$$

This ordinary continued fraction fits the form of [Ref. 9.1, Vol. II, p. 288, Satz 6.5]. The result for $\tilde{G}_{0,0}(s)$ reads

$$\tilde{G}_{0,0}(s) = (\xi\lambda_2 + s)^{-1}{}_2F_1(-\xi, 1; (\xi\lambda_2 + s)/(\lambda_1 - \lambda_2) + 1; -\lambda_2/(\lambda_1 - \lambda_2)) , \tag{A3.11}$$

where ${}_2F_1$ is the hypergeometric function [9.26] and λ_1, λ_2 are defined in (10.60). If we use [9.26]

$$ {}_2F_1(-\xi,1;c;z) = (1-z)^{\xi}\,{}_2F_1(-\xi,c-1;c;z/(z-1)) \tag{A3.12} $$

and the integral representation [9.26]

$$ {}_2F_1(-\xi,c-1;c;\alpha) = (c-1)\int_0^1 u^{c-2}(1-\alpha u)^{\xi}\mathrm{d}u\,, \tag{A3.13} $$

we get

$$ \tilde{G}_{0,0}(s) = \frac{1}{\lambda_1-\lambda_2}\left(\frac{\lambda_1}{\lambda_1-\lambda_2}\right)^{\xi}\int_0^1 u^{(\xi\lambda_2+s)/(\lambda_1-\lambda_2)-1}(1-\lambda_2 u/\lambda_1)^{\xi}\mathrm{d}u\,. $$

The substitution

$$ u = \mathrm{e}^{-(\lambda_1-\lambda_2)t} $$

leads to

$$ \tilde{G}_{0,0}(s) = \int_0^{\infty}\mathrm{e}^{-st}[y(t)]^{\xi}\mathrm{d}t\,, \tag{A3.14} $$

where $y(t)$ is defined by (A3.5).

Hence, the Green's function $G_{0,0}(t)$ is given by $y(t)^{\xi}$. In x representation it thus takes the form

$$ G_{0,0}(x,x',t) = [y(t)]^{-D\hat{D}/\omega_0^2}\delta(x-x')\,, \tag{A3.15} $$

where D and $\hat{D}$ are the operators (10.27) with $\varepsilon = 0$. We may now expand the δ function in terms of eigenfunctions of $D\hat{D}$, Sect. 5.5.1. The remaining sum can then be evaluated by (5.65). Another method would be to obtain a solution of

$$ \dot{G}_{0,0} = -[\dot{y}(t)/y(t)](D\hat{D}/\omega_0^2)G_{0,0}\,, \tag{A3.16} $$

which follows from (A3.15) by differentiation. It is easily checked that the Fourier transform of $G_{0,0}$ with respect to x, i.e.,

$$ G_{0,0}(x,x',t) = (2\pi)^{-1}\int_{-\infty}^{\infty}\mathrm{e}^{\mathrm{i}kx}\tilde{G}(k,t|x')\mathrm{d}k\,, \tag{A3.17} $$

is given by

$$ \tilde{G}(k,t|x') = \exp\{-\mathrm{i}kx'y(t)-[k_{\mathrm{B}}T/(2m\omega_0^2)]k^2[1-y^2(t)]\}\,, \tag{A3.18} $$

compare (5.27) $[y(0) = 1]$. Insertion of (A3.18) into (A3.17) leads to (A3.4), which finally proves the equivalence of (A3.1, 2) with (A3.4) for the harmonic oscillator.

A4. Damped Quantum-Mechanical Harmonic Oscillator

To introduce damping in quantum mechanics, the system is coupled to a reservoir or heat bath. In the Schrödinger picture the equation of motion for the density operator ρ of an harmonic oscillator is then given by

$$\begin{aligned}\dot{\rho} &= -\mathrm{i}\omega_0[b^+b,\rho]+\varkappa(n_{\mathrm{th}}+1)\{[b,\rho b^+]+[b\rho,b^+]\}+\varkappa n_{\mathrm{th}}\{[b^+,\rho b]+[b^+\rho,b]\}\\ &= -\mathrm{i}\omega_0[b^+b,\rho]+\varkappa\{[b\rho,b^+]+[b,\rho b^+]+2n_{\mathrm{th}}[[b,\rho],b^+]\}\,. \qquad \text{(A4.1)}\end{aligned}$$

Here b^+ and b are the creation and annihilation operators of the harmonic oscillator obeying the Bose commutation relation

$$[b,b^+]=bb^+-b^+b=1\,, \qquad \text{(A4.2)}$$

ω_0 is the frequency and $\varkappa$ the damping constant. The number of the thermal quanta is denoted by

$$n_{\mathrm{th}}=\frac{1}{\mathrm{e}^{\hbar\omega_0/(kT)}-1}\,. \qquad \text{(A4.3)}$$

The damping constant $\varkappa$ is assumed to be small compared to the frequency ω_0. The first term on the right-hand side does not appear in the interaction picture. For a derivation of (A4.1) see [A4.1 – 3, 1.28, 12.1]. (In the Heisenberg picture one derives a Langevin type equation for the creation and annihilation operators $b^+(t)$ and $b(t)$, where the Langevin forces are operators [A4.4, 12.1].)

One way of handling the operator equation (A4.1) is to reduce it to a system of differential equations for the matrix elements

$$\rho_{n,m}=\langle n|\rho|m\rangle\,. \qquad \text{(A4.4)}$$

Here, $|n\rangle$ is the eigenstate of the number operator b^+b, i.e.,

$$b^+b|n\rangle=n|n\rangle\,,\qquad |n\rangle=(b^+)^n|0\rangle/\sqrt{n!}\,. \qquad \text{(A4.5)}$$

Because

$$b^+|n\rangle=\sqrt{n+1}\,|n+1\rangle\,,\qquad b|n\rangle=\sqrt{n}\,|n-1\rangle\,, \qquad \text{(A4.6)}$$

it is easy to obtain the following equation for the above density matrix elements

$$\begin{aligned}\dot{\rho}_{n,m} &= -\mathrm{i}\omega_0(n-m)\rho_{n,m}+2\varkappa(n_{\mathrm{th}}+1)\sqrt{n+1}\sqrt{m+1}\,\rho_{n+1,m+1}\\ &\quad -\varkappa[(1+2n_{\mathrm{th}})(n+m)+2n_{\mathrm{th}}]\rho_{n,m}+2\varkappa n_{\mathrm{th}}\sqrt{n}\sqrt{m}\,\rho_{n-1,m-1}\,. \qquad \text{(A4.7)}\end{aligned}$$

The diagonal elements $p_n=\rho_{n,n}$ obey the master equation (Sect. 4.5)

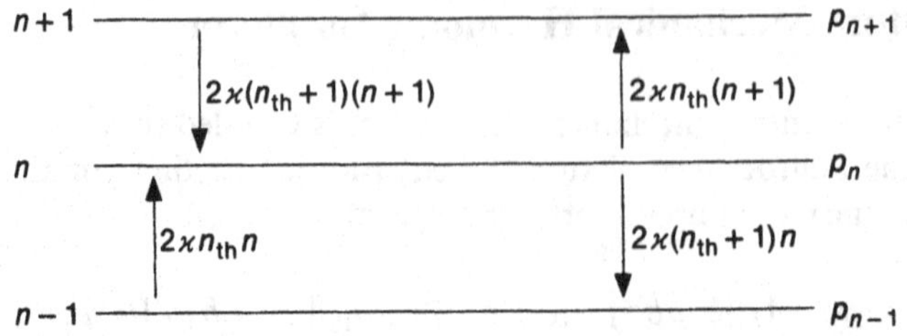

Fig. A4.1. Transition rates for the master equation (A4.8)

$$\dot{p}_n = 2\varkappa\{(n_{th}+1)(n+1)p_{n+1} - [(n_{th}+1)n + n_{th}(n+1)]p_n + n_{th} n p_{n-1}\}. \quad \text{(A4.8)}$$

The upward transition rate out of state n of the oscillator is the sum of

$2\varkappa n_{th} n$ (= induced emission rate due to the reservoir)

and

$2\varkappa n_{th} 1$ (= spontaneous emission rate due to the reservoir) ,

whereas the down transition rate out of state n is given by

$2\varkappa(n_{th}+1)n$ (= induced absorption rate due to the reservoir) ,

see Fig. A4.1. The stationary solution of (A4.8) is the Bose-Einstein distribution

$$p_n = \rho_{n,n} = \frac{1}{1+n_{th}}\left(\frac{n_{th}}{1+n_{th}}\right)^n = (1-\mathrm{e}^{-\hbar\omega_0/(kT)})\,\mathrm{e}^{-n\hbar\omega_0/(kT)}. \quad \text{(A4.9)}$$

Transformation to a Fokker-Planck Equation

We now want to show that the operator equation (A4.1) can be written in the form of a Fokker-Planck equation. To do this we first introduce a continous distribution function $W(u,u^*,t)$ defined by [A4.2, 3, 1.28, 12.1, 35]

$$\begin{aligned} W(u,u^*,t) &= \langle \delta(b^+ - u^*)\,\delta(b-u)\rangle \\ &= \pi^{-2}\iint_{-\infty}^{\infty} \langle \mathrm{e}^{\mathrm{i}(b^+ - u^*)\alpha^*}\mathrm{e}^{\mathrm{i}(b-u)\alpha}\rangle\,\mathrm{d}^2\alpha \\ &= \pi^{-2}\iint_{-\infty}^{\infty} \mathrm{tr}\{\mathrm{e}^{\mathrm{i}(b^+ - u^*)\alpha^*}\mathrm{e}^{\mathrm{i}(b-u)\alpha}\rho(t)\}\,\mathrm{d}^2\alpha . \end{aligned} \quad \text{(A4.10)}$$

Here, $\alpha = \alpha_r + \mathrm{i}\alpha_i$ is a complex quantity and $\mathrm{d}^2\alpha = \mathrm{d}\alpha_r\mathrm{d}\alpha_i$ is the two-dimensional volume element. Expression (A4.10) is formally the Fourier transform of the characteristic function (Sect. 2.2)

$$C(\alpha^*, \alpha, t) = \langle e^{ib^+\alpha^*} e^{ib\alpha} \rangle = \mathrm{tr}\{e^{ib^+\alpha^*} e^{ib\alpha} \rho(t)\} . \tag{A4.11}$$

The factor π^{-2} in front of the integral [instead of $(2\pi)^{-2}$] appears because we used complex notation. With the help of the distribution function every normally ordered product of b^+ and b (i.e., one, where all b^+ stand left of all b) can be calculated by using an integration over the distribution function

$$\langle [b^+(t)]^r [b(t)]^j \rangle = \mathrm{tr}\{(b^+)^r b^j \rho(t)\} = \int (u^*)^r u^j W(u, u^*, t) \, d^2u . \tag{A4.12}$$

The proof follows from the fact that the integration of (A4.12) in u space corresponds to a differentiation in α space

$$\int (u^*)^r u^j W d^2u = \left(\frac{\partial}{\partial i\alpha^*} \right)^r \left(\frac{\partial}{\partial i\alpha} \right)^j C(\alpha, \alpha^*) \bigg|_{\alpha=\alpha^*=0} . \tag{A4.13}$$

The distribution function $W(u, u^*)$ is the *Glauber-Sudarshan* [A4.5, A4.6] P representation of the density operator

$$\rho = \int |u\rangle W(u, u^*) \langle u| d^2u , \tag{A4.14}$$

where $|u\rangle$ are the eigenstates of the annihilation operator

$$b|u\rangle = u|u\rangle . \tag{A4.15}$$

[If a distribution function for antinormal ordering is needed, the exponentials in (A4.10) have to the interchanged.] To obtain an equation of motion for W we multiply (A4.1) by $\exp(ib^+\alpha^*) \exp(ib\alpha)$ and take the trace. By a proper cyclic permutation of the factors under the trace and by using

$$[b, e^{ib^+\alpha^*}] = i\alpha^* e^{ib^+\alpha^*}, \quad [e^{ib\alpha}, b^+] = i\alpha e^{ib\alpha} \tag{A4.16}$$

$$\begin{aligned} \mathrm{tr}\{e^{ib^+\alpha^*} e^{ib\alpha} b\rho\} &= \frac{\partial}{\partial i\alpha} C \\ \mathrm{tr}\{b^+ e^{ib^+\alpha^*} e^{ib\alpha} \rho\} &= \frac{\partial}{\partial i\alpha^*} C \\ \mathrm{tr}\{b^+ e^{ib^+\alpha^*} e^{ib\alpha} b\rho\} &= \frac{\partial^2}{\partial i\alpha \partial i\alpha^*} C , \end{aligned} \tag{A4.17}$$

we obtain from (A4.1) an equation for the characteristic function (A4.11)

$$\begin{aligned} \frac{\partial C}{\partial t} = &-i\omega_0 \left(\alpha \frac{\partial}{\partial \alpha} - \alpha^* \frac{\partial}{\partial \alpha^*} \right) C - \varkappa \left[i\alpha \frac{\partial}{\partial i\alpha} + i\alpha^* \frac{\partial}{\partial i\alpha^*} \right] C \\ &+ 2\varkappa n_{\mathrm{th}} (i\alpha^*)(i\alpha) C . \end{aligned} \tag{A4.18}$$

The distribution function (A4.10) is the Fourier transform of C. It therefore follows from W by the replacement

$$\mathrm{i}\alpha \to -\partial/\partial u\,, \quad \mathrm{i}\alpha^* \to -\partial/\partial u^*\,, \quad \partial/\partial(\mathrm{i}\alpha) \to u\,, \quad \partial/\partial(\mathrm{i}\alpha^*) \to u^*\,.$$

We thus obtain from (A4.18)

$$\partial W/\partial t = L_{\mathrm{FP}} W$$
$$L_{\mathrm{FP}} = \mathrm{i}\omega_0\left(\frac{\partial}{\partial u}u - \frac{\partial}{\partial u^*}u^*\right) + \varkappa\left[\frac{\partial}{\partial u}u + \frac{\partial}{\partial u^*}u^*\right] + 2\varkappa n_{\mathrm{th}}\frac{\partial^2}{\partial u\,\partial u^*}\,. \tag{A4.19}$$

If we use the real variables

$$u_1 = \mathrm{Re}\{u\}\,, \quad u_2 = \mathrm{Im}\{u\}\,,$$

(A4.19) takes the form

$$L_{\mathrm{FP}} = \omega_0\left(\frac{\partial}{\partial u_2}u_1 - \frac{\partial}{\partial u_1}u_2\right) + \varkappa\frac{\partial}{\partial u_1}\left(u_1 + \frac{1}{2}n_{\mathrm{th}}\frac{\partial}{\partial u_1}\right)$$
$$+ \varkappa\frac{\partial}{\partial u_2}\left(u_2 + \frac{1}{2}n_{\mathrm{th}}\frac{\partial}{\partial u_2}\right)\,. \tag{A4.20}$$

Obviously, the process described by the Fokker-Planck equation with the operator (A4.20) is an Ornstein-Uhlenbeck process, Sect. 6.5. An equation of motion for the averaged amplitude is easily obtained from the Fokker-Planck equation with (A4.20) by multiplying it with u_1 and u_2, respectively, and then integrating the expression over u_1 and u_2. We thus derive for $\langle u(t)\rangle = \langle u_1(t)\rangle + \mathrm{i}\langle u_2(t)\rangle$ the equation of motion

$$\langle \dot{u}(t)\rangle = -(\mathrm{i}\omega_0 + \varkappa)\langle u(t)\rangle\,, \tag{A4.21}$$

which clearly shows that the motion of the amplitude is damped.

Equation (A4.21) can also directly be derived from (A4.1) by multiplying (A4.1) with b, using the commutation relation (A4.2) and then taking the trace $[\langle b(t)\rangle = \langle u(t)\rangle = \mathrm{tr}\{b\rho(t)\}]$.

The stationary solution of the Fokker-Planck equation (A4.19) reads

$$W_{\mathrm{st}}(u, u^*) = (\pi n_{\mathrm{th}})^{-1}\exp(-uu^*/n_{\mathrm{th}})\,. \tag{A4.22}$$

By inserting this expression into (A4.14) we recover the result (A4.9) for the diagonal matrix elements. For the derivation

$$|\langle n|u\rangle|^2 = |u|^{2n}\mathrm{e}^{-|u|^2}/n! \tag{A4.23}$$

must be used [see [A4.5] for a derivation of (A4.23)].

A5. Alternative Derivation of the Fokker-Planck Equation

The nonlinear Langevin equation (3.67, 68), i.e.,

$$\dot{\xi} = h(\xi, t) + g(\xi, t)\Gamma(t)\,, \tag{A5.1}$$

$$\langle \Gamma(t) \rangle = 0\,, \qquad \langle \Gamma(t)\Gamma(t') \rangle = 2\delta(t-t')\,, \qquad \Gamma \text{ Gaussian} \tag{A5.2}$$

may be transformed into a linear partial differential equation. By introducing

$$\rho(t) = \delta(\xi(t) - x) \tag{A5.3}$$

it may be easily checked by insertion and by using (4.6) that $\rho(t)$ obeys the linear partial differential equation

$$\partial\rho(t)/\partial t \equiv \dot{\rho}(t) = [A(t) + B(t)\Gamma(t)]\,\rho(t)\,, \tag{A5.4}$$

where A and B are operators with respect to x and are given by

$$A(t) = -\frac{\partial}{\partial x} h(x,t)\,, \qquad B(t) = -\frac{\partial}{\partial x} g(x,t)\,. \tag{A5.5}$$

The distribution function $W(x,t)$ follows by averaging (A5.3) over the different realizations of $\Gamma(t)$ (2.7), i.e.,

$$W(x,t) = \langle \rho(t) \rangle = \langle \delta(\xi(t) - x) \rangle\,. \tag{A5.6}$$

An equation for the average $\langle \rho(t) \rangle$ can now be obtained using (A5.2). For a simple case [$A = 0$, $B(t)$ independent of time and no operator] this was already done in (3.76a). In the present case, A and B are noncommuting operators with respect to x and they may depend on time. By using a proper representation of the x dependence, relation (A5.4) can be cast into an equation for the vector ρ, where $\boldsymbol{A}(t)$ and $\boldsymbol{B}(t)$ are noncommuting matrices. For this case, the following result was derived by *Fox* [A5.1] for the Stratonovich rule (Sect. 3.3.3)

$$\langle \dot{\rho}(t) \rangle = [\boldsymbol{A}(t) + \boldsymbol{B}^2(t)]\,\langle \rho(t) \rangle\,. \tag{A5.7}$$

If we use the operators (A5.5) and the "x representation" for ρ, (A5.7) transforms to

$$\langle \dot{\rho}(t) \rangle = [A(t) + B^2(t)]\,\langle \rho(t) \rangle\,. \tag{A5.7a}$$

The derivation of (A5.7, 7a) is performed by formally integrating (A5.4) (using the time-ordered product) and then differentiating the averaged result [A5.1, 2]. Equation (A5.7) is a special case of (A1.32a) for $D = 1$ and $\gamma \to \infty$. The relation (A5.7a) was used by *Wódkiewicz* in a number of cases [A5.2, 3].

Because of (A5.6), (A5.7a) may be written as

$$\dot{W} = L_{\mathrm{FP}} W, \tag{A5.8}$$

$$\begin{aligned} L_{\mathrm{FP}} = A + B^2 &= -\frac{\partial}{\partial x} h(x,t) + \frac{\partial}{\partial x} g(x,t) \frac{\partial}{\partial x} g(x,t) \\ &= -\frac{\partial}{\partial x} D^{(1)}(x,t) + \frac{\partial^2}{\partial x^2} D^{(2)}(x,t), \end{aligned} \tag{A5.9}$$

where $D^{(1)}$ and $D^{(2)}$ are given by (3.95). Obviously, (A5.8, 9) is identical to the Fokker-Planck equation (4.44, 45).

For the multivariable Langevin equation (3.110, 111) we may proceed in the same way. By introducing

$$\rho(t) = \delta(\xi_1(t) - x_1) \ldots \delta(\xi_N(t) - x_N) \tag{A5.10}$$

the Langevin equations (3.110, 111) are transformed to the linear partial differential equation (summation convention)

$$\dot{\rho} = (A + B_k \Gamma_k(t)) \rho \tag{A5.11}$$

with

$$A = -\frac{\partial}{\partial x_i} h_i(\{x\}, t), \quad B_k = -\frac{\partial}{\partial x_i} g_{ik}(\{x\}, t), \tag{A5.12}$$

as may again be checked by insertion. Because of (3.111) we now have

$$\langle \dot{\rho} \rangle = (A + B_k B_k) \langle \rho \rangle, \tag{A5.13}$$

i.e., for the distribution function

$$W(\{x\}, t) = \langle \rho(t) \rangle = \langle \delta(\xi_1(t) - x_1) \ldots \delta(\xi_N(t) - x_N) \rangle \tag{A5.14}$$

we obtain (A5.8) with

$$\begin{aligned} L_{\mathrm{FP}} &= A + B_k B_k \\ &= -\frac{\partial}{\partial x_i} h_i(\{x\}, t) + \frac{\partial}{\partial x_i} g_{ik} \frac{\partial}{\partial x_j} g_{jk} \\ &= -\frac{\partial}{\partial x_i} D_i(\{x\}, t) + \frac{\partial^2}{\partial x_i \partial x_j} D_{ij}(\{x\}, t), \end{aligned} \tag{A5.15}$$

where D_i and D_{ij} are given by (3.118, 119). Obviously, (A5.15) is identical to (4.94, 95). A similar derivation of the Fokker-Planck equation was given by *Graham* [4.18].

A6. Fluctuating Control Parameter

Here we are interested in the laser Langevin equation (12.4), where the control parameter d fluctuates. Instead of (12.4) we use the normalized laser Langevin equation with detuning (12.73) but without an additive noise, i.e.

$$d\bar{b}/d\bar{t} - (1+i\delta)(\bar{a} - |\bar{b}|^2)\bar{b} = 0\,, \tag{A6.1}$$

where the control parameter $(1+i\delta)\,\bar{a}$ now fluctuates according to $(q \geqq s^2 \geqq 0)$

$$(1+i\delta)\bar{a} = (1+i\delta)a + \Gamma_1 + i(s\Gamma_1 + \sqrt{q-s^2}\,\Gamma_2) \tag{A6.2}$$

with

$$\langle \Gamma_i(\bar{t})\rangle = 0\,, \qquad \langle \Gamma_i(\bar{t})\Gamma_j(\bar{t}')\rangle = 2\delta_{ij}\delta(\bar{t}-\bar{t}')\,. \tag{A6.3}$$

Thus we consider the Langevin equation

$$d\bar{b}/d\bar{t} - (1+i\delta)(a - |\bar{b}|^2)\bar{b} = \bar{b}\Gamma_1 + i\bar{b}(s\Gamma_1 + \sqrt{q-s^2}\,\Gamma_2)\,. \tag{A6.4}$$

For the special case $\delta = q = s = 0$ we obtain the normalized laser Langevin equation (12.4) with fluctuating control parameter d.

In polar coordinates

$$\bar{b} = \bar{r}e^{i\bar{\varphi}} \tag{A6.5}$$

equation (A6.4) transforms to

$$d\bar{r}/d\bar{t} = (a - \bar{r}^2)\bar{r} + \bar{r}\Gamma_1\,, \tag{A6.6a}$$

$$d\bar{\varphi}/d\bar{t} = \delta\cdot(a - \bar{r}^2) + s\Gamma_1 + \sqrt{q-s^2}\,\Gamma_2\,. \tag{A6.6b}$$

These equations have been investigated by *Graham* [12.52]. He has shown that (A6.6a, b) are the normalized version of

$$db/dt = [(\alpha + i\beta) - (A + iB)|b|^2]b \tag{A6.7}$$

were α and β fluctuate according to

$$\alpha = \alpha_0 + \Gamma_\alpha(t)\,, \qquad \beta = \beta_0 + \Gamma_\beta(t) \tag{A6.8}$$

with $\Gamma_\alpha(t)$ and $\Gamma_\beta(t)$ given by

$$\begin{aligned} &\langle \Gamma_\alpha(t)\rangle = \langle \Gamma_\beta(t)\rangle = 0 \\ &\langle \Gamma_\alpha(t)\Gamma_\alpha(t')\rangle = Q_\alpha\delta(t-t') \\ &\langle \Gamma_\beta(t)\Gamma_\beta(t')\rangle = Q_\beta\delta(t-t') \\ &\langle \Gamma_\alpha(t)\Gamma_\beta(t')\rangle = Q_{\alpha\beta}\delta(t-t')\,. \end{aligned} \tag{A6.9}$$

The transformation of (A6.7 – 9) to (A6.6a, b) is achieved by

$$\begin{aligned}
&\bar{t} = (Q_\alpha/2)\,t\,, && \bar{r} = \sqrt{2A/Q_\alpha}\,r\,, && \bar{\varphi} = \varphi + \Omega t \\
&a = 2\alpha_0/Q_\alpha\,, && \delta = B/A\,, && \Omega = (\alpha_0 B - \beta_0 A)/A \\
&q = Q_\beta/Q_\alpha\,, && s = Q_{\alpha\beta}/Q_\alpha\,,
\end{aligned} \tag{A6.10}$$

where r and φ are the polar coordinates of b, i.e., $b = r\exp(\mathrm{i}\varphi)$.

The Fokker-Planck equation corresponding to (A6.3, 6a, b) reads

$$\begin{aligned}
\frac{\partial \bar{W}}{\partial \bar{t}} &= L_{\mathrm{FP}}\bar{W} \\
L_{\mathrm{FP}} &= -\frac{\partial}{\partial \bar{r}}\bar{r}(1 + a - \bar{r}^2) + \frac{\partial^2}{\partial \bar{r}^2}\bar{r}^2 \\
&\quad - \delta\cdot(a - \bar{r}^2)\frac{\partial}{\partial \bar{\varphi}} + q\frac{\partial^2}{\partial \bar{\varphi}^2} + 2s\frac{\partial^2}{\partial \bar{r}\,\partial \bar{\varphi}}\bar{r}\,.
\end{aligned} \tag{A6.11}$$

It should be noted that $\bar{W}$ is the distribution in $\bar{r}$ and $\bar{\varphi}$ space. It is connected to the distribution W used in Chap. 12 by $W = \bar{r}\bar{W}$.

Stationary Distribution

Because no phase is preferred the stationary distribution cannot depend on $\bar{\varphi}$. Because the probability current in $\bar{r}$ direction must be zero, we thus have

$$\left[\bar{r}(1 + a - \bar{r}^2) - \frac{\partial}{\partial \bar{r}}\bar{r}^2\right]\bar{W}_{\mathrm{st}} = 0\,. \tag{A6.12}$$

From this equation we obtain for $a > 0$ [8.5, 12.52]

$$\bar{W}_{\mathrm{st}}^{(1)}(\bar{r}) = \frac{\sqrt{2}}{\pi\Gamma(a/2)}\left(\frac{\bar{r}}{\sqrt{2}}\right)^{a-1}\exp\left(-\frac{\bar{r}^2}{2}\right) \tag{A6.13}$$

where Γ is the gamma function. Another stationary solution is given by

$$W_{\mathrm{st}}^{(2)}(\bar{r}) = \frac{1}{\pi}\delta(\bar{r})\,. \tag{A6.14}$$

If the system starts at $\bar{r} = 0$ the noise $b\Gamma_1$ cannot drive the amplitude away from $\bar{r} = 0$ and we then get (A6.14). For $a \leqq 0$ only the stationary solution (A6.14) is possible. For further considerations we assume $a > 0$ and that we start with an amplitude b being different from zero. Then we can omit the stationary solution (A6.14).

Transformation to a Fokker-Planck Equation with Additive Noise

First we assume that $\delta = s = q = 0$. Then we only need to consider (A6.6a) with the multiplicative noise force $\bar{r}\Gamma_1$. Similar as in Sect. 3.3.1 we may change the multiplicative noise to an additive noise by using the nonlinear transformation

$$y = \ln(\bar{r}/\sqrt{2+a}), \quad \bar{r} = \sqrt{2+a}\,e^y. \tag{A6.15}$$

Here the factor $\sqrt{2+a}$ is introduced because the potential of the Schrödinger equation then becomes the Morse potential (A6.20). The Langevin equation (A6.6a) for the variable y then takes the form

$$\dot{y} = a - (2+a)e^{2y} + \Gamma_1 \tag{A6.16}$$

and the Fokker-Planck equation corresponding to (A6.16, 3) reads

$$\begin{aligned} \frac{\partial \tilde{W}}{\partial t} &= \tilde{L}_{FP}\tilde{W} \\ \tilde{L}_{FP} &= -\frac{\partial}{\partial y}[a-(2+a)e^{2y}] + \frac{\partial^2}{\partial y^2}. \end{aligned} \tag{A6.17}$$

Here $\tilde{W}$ is the distribution in y, $\bar{\varphi}$ space, i.e., $\tilde{W}$ and $\bar{W}$ are connected by

$$\tilde{W} = \bar{W}\,d\bar{r}/dy = \sqrt{2+a}\,e^y\,\bar{W}. \tag{A6.18}$$

The force $F = a-(2+a)e^{2y} = -df/dy$ may be derived from the Toda potential [A6.3]

$$f(y) = (1+a/2)e^{2y} - ay. \tag{A6.19}$$

Transformation to a Schrödinger Equation

As discussed in Sect. 5.4 the Fokker-Planck equation (A6.17) can be transformed to a Schrödinger equation with the potential (5.55), i.e.,

$$V_S(y) = \tfrac{1}{4}[f'(y)]^2 - \tfrac{1}{2}f''(y) = (1+a/2)^2(e^{4y} - 2e^{2y}) + a^2/4. \tag{A6.20}$$

If we write $2y = -\alpha x$ the potential $V_S - a^2/4$ is exactly the Morse potential used in quantum mechanics to describe the binding of a molecule [A6.1]. The potentials $f(y)$ and $V_S(y)$ are plotted in Fig. (A6.1). It is clear from loocking at the figure that for $\lambda \geqq a^2/4$ a continuous spectrum of the eigenvalues occurs.

Discrete Eigenvalues

The discrete eigenvalues of the Morse potential or of the Fokker-Planck equation (A6.11) with $q = s = \delta = 0$ are given by [8.5, 12.52, A6.1]

$$\lambda_n(a) = 2na - 4n^2, \quad 0 \leqq n < a/4, \quad n = 0, 1, 2, \dots. \tag{A6.21}$$

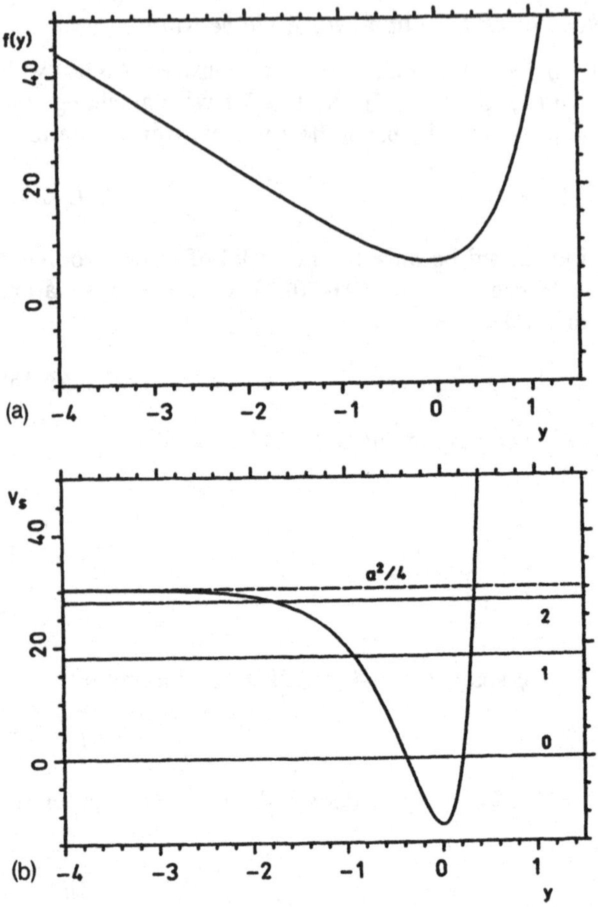

Fig. A6.1. The potential (A6.19) **(a)** and the Morse potential (A6.20) **(b)** for $a = 11$. The eigenvalues λ_n for $n = 0$, 1, 2 are also shown in **(b)**

The eigenfunctions of (A6.11) with $q = s = \delta = 0$ can be expressed in terms of Laguerre polynomials [8.5, 12.52]

$$\Phi_n(\bar{r}) = N_n \sqrt{\bar{W}_{\mathrm{st}}^{(1)}(\bar{r})}\,(\bar{r}^2/2)^{-1/4-n+a/4}\,\mathrm{L}_n^{(-2n+a/2)}(\bar{r}^2/2) \tag{A6.22}$$

Eigenvalues for the General Case $q \neq 0,\ s \neq 0,\ \delta \neq 0$

If we insert the separation ansatz

$$\bar{W}(\bar{r}, \bar{\varphi}, \bar{t}) = \Phi_{\nu n}(\bar{r})\,\mathrm{e}^{\mathrm{i}\nu\bar{\varphi}}\mathrm{e}^{-\lambda_{\nu n}\bar{t}} \tag{A6.23}$$

into the Fokker-Planck equation (A6.11) we can again transform the equation for $\Phi_{\nu n}$ to a Schrödinger equation with the Morse potential. Introducing

$$a_\nu = a + 2\mathrm{i}\,\nu(\delta - s)\,, \tag{A6.24}$$

$$y = \ln \bar{r}/\sqrt{2 + a_\nu}\,, \qquad \bar{r} = \sqrt{2 + a_\nu}\,\mathrm{e}^y\,, \tag{A6.25}$$

$$f(y) = (1 + a_\nu/2)\,e^{2y} - (a - 2is\nu)\,y \tag{A6.26}$$

we obtain for $\psi_{\nu n}(y)$ defined by

$$\Phi_\nu(\bar{r}) = \exp[-y - \tfrac{1}{2}f(y)]\,\psi_{\nu n}(y) \tag{A6.27}$$

the Schrödinger equation

$$\left[\frac{d^2}{dy^2} - V_S(y, a_\nu) + \lambda_{\nu n} - \nu^2(q + \delta^2 - 2\delta s)\right]\psi_{\nu n} = 0 \tag{A6.28}$$

with the Morse potential having the complex parameter a_ν

$$V_S(y, a_\nu) = (1 + a_\nu/2)^2(e^{4y} - 2e^{2y}) + a_\nu^2/4\,. \tag{A6.29}$$

By comparing (A6.28) with the eigenvalue equation for $q = \delta = s = 0$, i.e. with

$$\left[\frac{d^2}{dy^2} - V_S(y, a) + \lambda_n(a)\right]\psi_n = 0 \tag{A6.30}$$

we thus conclude from analytic continuation $a \to a_\nu$ that $\lambda_{\nu n}$ is given by [12.52]

$$\lambda_{\nu n} = \lambda_n(a_\nu) + (q + \delta^2 - 2\delta s)\,\nu^2 = 2na_\nu - 4n^2 + (q + \delta^2 - 2\delta s)\,\nu^2\,. \tag{A6.31}$$

The functions $\psi_{\nu n}$ can only be normalized if

$$0 \leqq n < a/4\,. \tag{A6.32}$$

(Because of $q \leqq s^2$, the factor in front of ν^2 is always positive and therefore the real part of the eigenvalues cannot become negative.) Because the constants a_ν are now complex the eigenvalue problem (A6.28) is not longer Hermitian. Therefore the adjoint functions must now be also obtained, which can be done by analytic continuation as explained at the end of Sect. 6.3. These normalized discrete eigenfunctions as well as the continuous eigenfunctions and their adjoints are given in [12.52]. (In this reference also correlation functions and in [A6.2] transient moments have been investigated.) Thus all eigenvalues and eigenfunctions can be obtained analytically for the Langevin equation (A6.4) with a multiplicative noise term, whereas the laser Langevin equation (12.6) with an additive noise term can be solved analytically only for the stationary case and well outside the threshold region.

S. Supplement to the Second Edition

In this supplement we give a short review of some material, which is closely related to the problems discussed in the first edition of this monograph. Most of these new results were developed after completion of the first edition. (For a list of some books and reviews, see [S.1 – 15]). In the following we mainly list some new references and, sometimes, explain the basic ideas of the reported work. The connection with the respective chapter or section of the main text is explicatly given.

S.1 Solutions of the Fokker-Planck Equation by Computer Simulation (Sect. 3.6)

Various methods have been developed to simulate Langevin equations by analog or digital computers, and to calculate stochastic integrals numerically [S1.1 – 7]. The main goal of these methods is, of course, to obtain good accuracy without requiring much computer time. Special methods have been developed for additive and multiplicative noise processes to obtain good accuracy for moments or for the distribution function. In [S1.7] the accuracies of some of these methods have been compared.

S.2 Kramers-Moyal Expansion (Sect. 4.6)

In Sect. 4.6 it was pointed out for a simple example that truncated Kramers-Moyal expansions of order higher than two may be quite useful. In Sect. 4.6 a δ-function initial condition was used. As was pointed out by *Pawula* [S2.1], a sum of δ-functions should finally appear for a very large truncation index N. This point and the problem of various other initial conditions were investigated in [S2.2]. As shown in this reference, the sum of δ-functions (each δ-function appears in an oscillating fashion) can be obtained for large N ($N \approx 67$). It was further demonstrated in this reference that for smooth initial conditions a better convergence with respect to the truncation index can be achieved. In particular, for the initial condition

$$W(x,0) = \sin \pi x/(\pi x) , \tag{S2.1}$$

which leads at the discrete points $x = m$ to the initial condition (4.66), the changes can not be seen in the plot for $N \geq 7$ for the parameters of Fig. 4.2. The continuous approximation of the random walk problem on a spatial lattice is related to the problem of approximating a discrete process by a continuous one, as discussed in Sect. 4.6. It was investigated by *Doering* et al. [S2.3].

S.3 Example for the Covariant Form of the Fokker-Planck Equation (Sect. 4.10)

A simple example for a non-vanishing curvature tensor is the Fokker-Planck equation on a sphere. Let us discuss the simplest case. This is a Fokker-Planck equation (FPE), without a drift-term and with a uniform diffusion term, i.e., it is the diffusion equation on a sphere of radius a. In polar coordinates the diffusion equation reads

$$\dot{W} = D\bar{\Delta}W = \frac{D}{a^2}\left(\frac{1}{\sin\Theta}\frac{\partial}{\partial\Theta}\sin\Theta\frac{\partial}{\partial\Theta} + \frac{1}{\sin^2\Theta}\frac{\partial^2}{\partial\phi^2}\right)W. \qquad (S3.1)$$

Here $\bar{\Delta}$ is the angle-dependent part of the Laplace operator, and W is the probability density on the surface of the sphere. For the probability density w of the Θ, ϕ coordinates we have

$$w = Wa^2\sin\Theta. \qquad (S3.2)$$

The FPE for w takes the form

$$\dot{w} = \frac{D}{a^2}\left(\frac{\partial}{\partial\Theta}\frac{\cos\Theta}{\sin\Theta} + \frac{\partial^2}{\partial\Theta^2} + \frac{\partial^2}{\partial\phi^2}\frac{1}{\sin^2\Theta}\right)w. \qquad (S3.3)$$

Thus we have a spurious-drift coefficient, and diffusion coefficients given by

$$D_\Theta = -\frac{D}{a^2}\frac{\cos\Theta}{\sin\Theta}, \quad D_\phi = 0 \qquad (S3.4)$$

$$D_{\Theta\Theta} = \frac{D}{a^2}, \quad \rho_{\Theta\phi} = 0, \quad D_{\phi\phi} = \frac{D}{a^2}\frac{1}{\sin^2\Theta}.$$

The invariant measure $\bar{w}$, see (4.139), is expressed by

$$\bar{w} = \sqrt{\operatorname{Det} D_{ij}}\, w = \frac{D}{a^2}\frac{1}{\sin\Theta}w = DW. \qquad (S3.5)$$

The component $R^{\Theta}_{\phi\Theta\phi}$ of the Riemann curvature tensor takes the form, see [S3.1] and (4.152),

$$R^{\Theta}_{\phi\Theta\phi} = \sin^2\Theta\,. \tag{S3.6}$$

It does not vanish everywhere and therefore one cannot find a global transformation so that the diffusion coefficient can be normalized to the unit matrix everywhere. For a discussion of Brownian motion in the presence of constraints, see [S3.2].

S.4 Connection to Supersymmetry and Exact Solutions of the One Variable Fokker-Planck Equation (Chap. 5)

In Sect. 5.8 we have investigated the solutions of the one-dimensional FPE with inverted potentials. As discussed by *Bernstein* and *Brown* [S4.1] there is a close connection between supersymmetric quantum mechanics [S4.2], see also [S4.3], and the inversion of the potential in the Fokker-Planck equation. In [S4.1] the lowest nonzero eigenvalue of a bistable potential was calculated by determining the lowest eigenvalue of the upside-down metastable potential by a variational method. *Jauslin* [S4.4] has used the supersymmetry property to construct potentials with arbitrarily prescribed eigenvalues. Exact solutions for bistable potentials have been obtained by *Hongler* and *Zheng* [S4.5] and recently by *Englefield* [S4.6]. In contrast to the model discussed in Sect. 5.7 these models have continuous potentials. It was shown by *Leiber* et al. [S10.14, 15] that the inverted potential is isospectral also for colored noise, i.e. the inverted problem has the same eigenvalues as the original problem.

S.5 Nondifferentiability of the Potential for the Weak Noise Expansion (Sects. 6.6 and 6.7)

In the stationary state w of (6.143, 144) is called the nonequilibrium potential. It must be minimal on the attractors of the deterministic motion [S5.1]. In the weak-noise limit it satisfies a Hamilton-Jacobi equation [(6.144) for $\dot{w} = 0$] [S5.2]

$$D_i \frac{\partial w}{\partial x_j} + D_{ij} \frac{\partial w}{\partial x_i}\, \frac{\partial w}{\partial x_j} = 0\,. \tag{S5.1}$$

It has been shown that the derivative of w may have discontinuities for certain surfaces if the Hamilton-Jacobi equation is not integrable [S5.2, 3]. It was further shown in [S5.4] that for coexisting attractors we may also have discontinuities even if the Hamiltonian belonging to (S5.1) is integrable. Discontinuities also occur for the Brownian motion in a periodic potential, as discussed in Chap. 11, in the limit of small friction (Fig. 11.24).

S.6 Further Applications of Matrix Continued-Fractions (Chap. 9)

The matrix continued-fraction (MCF) method for solving linear partial and ordinary differential equations has been applied to a variety of other problems. In [S6.1] this method has been applied to the one-dimensional Fokker-Planck equation for the quartic potential. For this simple example all the 2×2 matrices needed for determining the moments and the eigenvalues are given. In [S6.2] the method was applied to a Fokker-Planck equation describing the thermalization of neutrons in a heavy gas moderator. The problem of differential equations with a parametric excitation which is periodic in time was also solved by this method in [S6.3 – 5]. In [S6.6] we have applied the MCF algorithm for determining the eigenvalues of the Schrödinger equation with time-independent polynomial and nonpolynomial potentials. The method was applied to two-level atomic systems coupled to a vibrational mode [S6.7]. In [S6.8] the time-dependent expectation values for the Jaynes-Cummings model have been obtained by first deriving a tri-diagonal vector recurrence relation for the moments. After Fourier transformation the frequency-dependent recurrence relation was solved in terms of matrix continued fractions. The inverse Fourier-transform then finally leads to the time-dependent expectation values. Applications of the MCF method to the Kramers equation for the double-well potential, to the calculation of correlation times, to the escape problem in the presence of colored noise, and to Fokker-Planck equations with non-positive-definite diffusion matrices and to differential equations with derivatives up to third order have been discussed in supplements S7, S9, S10 and S11.

An important question concerns the complete sets, which are used for solving the differential equations. We have mainly used the classical polynomials and corresponding weight functions, like Hermite functions for coordinates defined in the interval $[-\infty, \infty]$ and Laguerre functions for coordinates defined in $[0, \infty]$. It is, of course, possible to employ other orthogonal polynomials. For the laser equation (12.40), for instance, it seems to be more appropriate to use orthogonal polynomials with the weight function being the stationary solution of (12.40). Shizgal and coworkers [S6.9, 10] have employed this type of polynomials. In this way an improvement of the convergence is expected because the orthogonal set is better adapted to the problem under consideration. The disadvantage, however, might be that the coefficients of the recurrence relations for these polynomials can only be generated by a numerically unstable algorithm. Therefore a large number of digits (e.g., 100) must be used to obtain these recurrence relation coefficients for large indices.

S.7 Brownian Motion in a Double-Well Potential (Chaps. 10 and 11)

In Chap. 11 the Brownian motion in an inclined periodic potential was investigated. Almost the same procedure can be applied to the Brownian motion in a double-well potential. In contrast to the periodic case, however, an expansion in

Hermite functions instead of a Fourier series for the position coordinate was used in [S7.1, 2]. Eigenvalues and the spectrum of correlation functions for the quartic potential have been obtained for normalized damping constants in the range 0.01 – 10 and for energy differences up to $\Delta E/(kT) = 10$. The smallest nonzero eigenvalue describes the transition rate from one potential well to the other. For this transition rate an analytic result (in terms of an integral) has been obtained by *Mel'nikov* and *Meshkov* [S7.3] for arbitrary damping constants in the limit of large $\Delta E/(kT)$. (For a metastable potential, see [S7.4]).

The spectrum of the position-correlation function has been measured [S7.5] and shown to be in good agreement with the numerical result of [S7.2]. For small damping constants the energy variable has to be introduced. Eigenvalues and eigenfunctions for this case have been obtained in [S7.6]. The zero-friction limit of the position-correlation function was investigated in [S7.5, 7, 8] and in the appendix of [S7.2]. A singular perturbation approach was also applied to the escape problem [S7.9]. More references on this subject may be found in the historical review by *Landauer* [S7.10] and in [S7.2, 11, 12].

S.8 Boundary Layer Theory (Sect. 11.4)

For very small damping the eigenfunction for the double-well Kramers problem describing the transition between the left and the right well (lowest nonzero eigenvalue) depends only on the energy inside the well. At the critical energy given by the barrier height separating the two minima, the eigenfunction must depend on x because above the critical trajectory particles leave the wells at the maximum of the potential. Thus particles of "opposite sign" travel to the other well, and therefore an x-dependence near the critical trajectory is observed. It turns out that the width of this boundary layer, in which an x-dependence must be taken into account, is proportional to the square root of the friction constant. This boundary layer theory leads to coupling of the eigenfunction $\Phi(E)$ and its derivative $\mathrm{d}\Phi/\mathrm{d}E$ according to

$$\Phi(E_c) = -\kappa \sqrt{\frac{\gamma I(E_c)\Theta}{2\pi}}\; \frac{\mathrm{d}\phi(E)}{\mathrm{d}E}\Bigg|_{E=E_c} \tag{S8.1}$$

where I is the action variable at the critical trajectory, γ the friction constant, Θ the normalized temperature (11.29) and κ a numerical constant. The connection of Φ and $\mathrm{d}\phi/\mathrm{d}E$ was proposed by *Büttiker* et al. [S8.1] and by *Büttiker* and *Landauer* [S8.2], see also [S8.3] for a review. The precise value of constant κ, however, was not obtained in these references. As explained in [S7.6] a similar boundary problem appears when calculating the stationary distribution for the Brownian motion in an inclined periodic potential. In [S7.6] it was shown that the constant κ is the same for the inclined periodic potential problem and the double-well potential. In [S7.1, 6], the value $\kappa_{\mathrm{DW}} = 0.8554$ was obtained, see also (11.123). It was shown by *Mel'nikov* and *Meshkov* [S7.3] that κ can be expressed analytically by Riemann's Zeta function according to

$$\kappa_{\rm DW} = -(2-\sqrt{2})\,\zeta(\tfrac{1}{2}) = 0.855455865\ldots\,. \tag{S8.2}$$

In [S7.6] we have also calculated the boundary layer distribution function. For an integral representation of the transition rate, see [S8.4].

For a metastable potential the constant κ, in [S8.1] is given by the Zeta function [S7.3]

$$\kappa_{\rm MS} = -\zeta(\tfrac{1}{2}) = 1.4603545088\ldots\,. \tag{S8.3}$$

The boundary layer distribution function was calculated for the metastable potential in [S8.5].

The Zeta function $\zeta(\frac{1}{2})$ also arises in the Kramers equation with an absorbing wall boundary condition as mentioned at the end of Sect. 8.1. This problem has been worked out in detail by *Marshall* and *Watson* [S8.6, 7].

The analytic expressions of the eigenfunctions and eigenvalues for small friction given in Sect. 11.9.1 have also been investigated by *Renz* [S8.8].

S.9 Calculation of Correlation Times (Sect. 7.12)

In Sect. 7.2 general expressions for correlation functions have been presented. For the one-variable case one may define a normalized correlation function by

$$\begin{aligned} \Phi(t) &= K(t)/K(0)\,, \\ K(t) &= \langle \Delta r(x(t))\,\Delta r(x(t))\rangle\,, \\ \Delta x(t) &= r(x(t)) - \langle r\rangle\,. \end{aligned} \tag{S9.1}$$

The subtraction of the average $\langle r\rangle$ guarantees that the normalized correlation function $\Phi(t)$ vanishes for large times. Obviously $\Phi(t)$ is normalized according to $\Phi(0) = 1$. A correlation time may be defined by

$$T = \int_0^\infty \Phi(t)\,dt\,. \tag{S9.2}$$

For an exponential dependence we then have $\Phi(t) = \exp(-t/T)$. For the one variable Langevin equation (3.67, 68) or the corresponding one-variable Fokker-Planck equation (4.44, 45) with $D^{(1)}$ and $D^{(2)}$ given by (3.95) an analytic expression for (S9.2) can be derived in the following way, see [S9.1, 2]. (In [S9.1, 2] only the special case $r(x) = x$ was considered. The generalization to $r(x)$ is straightforward.) The general expression (7.13) applied to $K(t)$ may be written in the form

$$K(t) = \int \Delta r(x)\,w(x,t)\,dx\,, \tag{S9.3}$$

where $w(x,t)$ obeys the one-dimensional Fokker-Planck equation

$$\frac{\partial w}{\partial t} = \left\{ -\frac{\partial}{\partial x}\,[h(x)+g'(x)g(x)] + \frac{\partial^2}{\partial x^2}\,g^2(x) \right\} w \tag{S9.4}$$

with the initial condition

$$w(x,0) = \Delta r(x)\, W_{\mathrm{St}}(x)\,. \tag{S9.5}$$

Introducing

$$\rho(x) = \int_0^\infty w(x,t)\,\mathrm{d}t \tag{S9.6}$$

(S9.2) takes the form (x is assumed to be defined in the range $-\infty < x < \infty$)

$$T = \frac{1}{K(0)} \int_{-\infty}^{\infty} \Delta r(x)\,\rho(x)\,\mathrm{d}x\,. \tag{S9.10}$$

Because of (S9.4 and 5) $\rho(x)$ must obey

$$-\Delta r(x)\, W_{\mathrm{St}}(x) = \left\{ -\frac{\mathrm{d}}{\mathrm{d}x}\,[h(x) + g'(x)\,g(x)] + \frac{\mathrm{d}^2}{\mathrm{d}x^2}\,g^2(x) \right\} \rho(x)\,. \tag{S9.11}$$

This equation can be integrated leading to

$$\rho(x) = W_{\mathrm{St}}(x) \int_{-\infty}^{x} \frac{f(x')}{g^2(x')\,W_{\mathrm{St}}(x')}\,\mathrm{d}x \tag{S9.12}$$

with $f(x)$ given by

$$f(x) = -\int_{-\infty}^{x} \Delta r(x')\,W_{\mathrm{St}}(x')\,\mathrm{d}x'\,. \tag{S9.13}$$

Inserting (S9.12) into (S9.10) we find, after integration by parts, the following analytical expression for the correlation time

$$T = \frac{1}{K(0)} \int_{-\infty}^{\infty} \frac{f^2(x)}{g^2(x)\,W_{\mathrm{St}}(x)}\,\mathrm{d}x\,. \tag{S9.14}$$

If x is not restricted to the interval $[-\infty, \infty]$ but defined as in [S9.1] in $[0, \infty]$ the lower limit of integration in (S9.13, 14) has to be replaced by 0. Using (S9.14) for the special case of [S9.1.2], *Nadler* and *Schulten* [S9.3] have obtained the following analytical expression for λ_{eff} defined in (12.71)

$$\lambda_{\mathrm{eff}} = K(a,0)/B\,,$$

$$B = \int_0^\infty \mathrm{d}\bar{I}\,F(\bar{I})^2/[4\bar{I}\,\hat{W}_{\mathrm{St}}(\bar{I})]\,,$$

$$F(\bar{I}) = -\frac{a}{\operatorname{erf}(a/2)}\{\operatorname{erf}[(\bar{I}-a)/2]-1\} + 2\hat{W}_{\mathrm{St}}(\bar{I}), \tag{S9.15}$$

where $\hat{W}_{\mathrm{St}}(\bar{I})$ is the stationary distribution (12.45). (For further investigations of λ_{eff}, see [S9.4] by *San Miguel* et al.). Correlation times for a laser model with white pump noise and quantum noise have also been obtained [S9.5] by solving (S9.11) numerically in terms of continued fractions.

S.10 Colored Noise (Appendix A1)

The simplest problem in colored noise may be stated as follows. Similarly to the stochastic differential equation (3.67) we have

$$\dot{x} = h(x) + g(x)\cdot\varepsilon(t), \tag{S10.1}$$

where the random Gaussian force $\varepsilon(t)$ has a finite correlation time τ and is determined through the correlations

$$\langle\varepsilon(t)\rangle = 0; \qquad \langle\varepsilon(t)\,\varepsilon(t')\rangle = (D/\tau)\,\mathrm{e}^{-|t-t'|/\tau}. \tag{S10.2}$$

For $\tau\to 0$ the correlation function in (S10.2) reduces to $2D\,\delta(t-t')$ and thus we recover the white-noise case. For $g(x)=1$ and $h(x)$ given by

$$h(x) = ax - bx^3 = -\mathrm{d}V(x)/\mathrm{d}x\ . \tag{S10.3}$$

Eq. (S10.1) describes the overdamped motion of a particle in the Landau potential

$$V(x) = -ax^2/2 + bx^4/4 \tag{S10.4}$$

driven by additive colored noise. By introducing the new variable ε, which obeys the Langevin equation

$$\dot{\varepsilon} = -\varepsilon/\tau + (\sqrt{D}/\tau)\,\Gamma(t), \tag{S10.5}$$

where $\Gamma(t)$ is the Gaussian white noise force (3.68), we obtain the correlation function (S10.2). Thus the two Langevin equations (S10.1, 5) describe the above colored noise problem. The Fokker-Planck equation for the two-dimensional distribution $W(x,\varepsilon)$ corresponding to (S10.1, 5) reads

$$\frac{\partial W}{\partial t} = \left\{-\frac{\partial}{\partial x}[h(x)+g(x)\,\varepsilon] + \frac{1}{\tau}\frac{\partial}{\partial\varepsilon}\,\varepsilon + \frac{D}{\tau^2}\frac{\partial^2}{\partial\varepsilon^2}\right\}W. \tag{S10.6}$$

A number of different methods have been developed by various schools to treat colored noise problems, see [S10.1–9].

Because (S10.6) is a two variable Fokker-Planck equation we may solve it by the matrix continued-fraction method. For the application of this algorithm we expand $W(x, \varepsilon)$ into two complete sets. One of these sets are the Hermite functions $\psi_n(\varepsilon)$ [see (10.38–40) with $v = \varepsilon$ and $v_{\text{th}}^2 = D/\tau$] the other one are suitable orthogonal functions $\phi_m(x)$. The explicit expansion reads

$$W(x, \varepsilon) = \rho_0(x)\, \psi_0(\varepsilon) \sum C_n^m \phi_m(x)\, \psi_n(\varepsilon)\,. \tag{S10.7}$$

Insertion of (S10.7) into (S10.6), multiplication by $\phi_{m'}(x)/\rho_0(x)$ and $\psi_{n'}(\varepsilon)/\psi_0(\varepsilon)$ leads after integration over ε to (A1.23) where the matrix elements of the matrices $\boldsymbol{A}$ and $\boldsymbol{B}$ are given by

$$\begin{aligned} A^{m'm} &= -\int \frac{\phi_{m'}(x)}{\rho_0(x)} \frac{\partial}{\partial x} [h(x)\, \rho_0(x)\, \phi_m(x)]\, \mathrm{d}x\,, \\ B^{m'm} &= -\int \frac{\phi_{m'}(x)}{\rho_0(x)} \frac{\partial}{\partial x} [g(x)\, \rho_0(x)\, \phi_m(x)]\, \mathrm{d}x\,. \end{aligned} \tag{S10.8}$$

The arbitrary function $\rho_0(x)$ may be chosen in such a way that integrals take simple forms. The integration boundaries in (S10.8) cover the whole accessible range of x. For short noise-correlation times τ we obtain the one variable Fokker-Planck equation (A1.31). For $t/\tau \gg 1$ (A1.31) reduces to (A1.32a). In x-representation explicit insertion of the operators

$$\boldsymbol{A} = -\frac{\partial}{\partial x} h(x)\,, \quad \boldsymbol{B} = -\frac{\partial}{\partial x} g(x) \tag{S10.9}$$

into (A1.32a) leads in first order in τ to

$$\begin{aligned} L_0(\infty) &= -\frac{\partial}{\partial x} h(x) + D\frac{\partial}{\partial x} g(x) \frac{\partial}{\partial x} g(x) + \tau D \left[\frac{\partial}{\partial x} g \frac{\partial}{\partial x} (gh' - g'h)\right] \\ &= -\frac{\partial}{\partial x} [h + Dg'g + D\tau g'(gh' - g'h)] + D\frac{\partial^2}{\partial x^2} [g^2 + \tau g(gh' - g'h)]\,. \end{aligned} \tag{S10.10}$$

This relation agrees for Gaussian noise with (4.180) of [1.10] Vol. I and in first order in τ with [Ref. S10.9, Eq. (2.36)].

The method described above was applied to a dye-laser model with colored noise by *Jung* and *Risken* [S10.10]. In this case we have $h(x) = 2(1-x)x$, $g(x) = 2x$ and, because the variable x is defined in the interval $[0, \infty]$, we have used Laguerre functions as the complete set in x. The stationary result for the intensity distribution $w(x) = \int W(x, \varepsilon)\, \mathrm{d}\varepsilon$ was found to be in good agreement with results obtained by digital simulation [S10.9].

The Landau potential (S10.4) with additive colored noise was treated by *Jung* and *Risken* [S10.11] with the MCF method. Here an expansion in Hermite functions was used for the x-variable. Stationary results for the one-variable distribution $w(x) = \int W(x, \varepsilon)\,d\varepsilon$ and for the two-variable distribution as well as results for the lowest nonzero eigenvalue have been obtained. Results obtained by analog simulation by *Moss* and *McClintock* [S10.12] are in good agreement with the two-variable distributions. In [S10.13] the small-τ equation with (S10.10) was used to calculate the small-τ dependence of the lowest nonzero eigenvalue. For the Landau-potential it was shown in that reference that the derivative of the eigenvalues with respect to τ can be obtained by the eigenvalues of the white-noise case in the limit $\tau \to 0$.

In [S10.14, 15] bistable periodic-potential models have been considered. Because of the good convergence of Fourier expansions eigenvalues could be obtained for much larger correlation times and lower noise intensities than for the Landau-potential in [S10.11]. Recently the MCF method [S10.11] was improved so that also larger correlation times could be handled for the Landau potential [S10.16] but not as large as in [S10.14, 15]. In [S10.17] a laser model with white quantum noise and colored pump noise was treated, see also [S10.18]. In [S10.19–21] (see [S10.22] for a review) the locking equation describing the phase dynamics of a ring-laser gyroscope with colored noise was investigated by the matrix continued-fraction method and by analog simulations.

Luciani and *Verga* [S10.23] have used a functional approach to obtain an expression for the mean first passage time valid for small and large correlation times. Furthermore an interpolating expression for the mean first passage time was given by *Tsironis* and *Grigolini* [S10.24] which approximates the numerical solution quite well. In [S20.25] an asymptotic expression for large τ was also obtained. Finally the paper by *Doering* et al. [S10.26] should be mentioned where the mean first passage time was calculated by assuming an absorbing wall at the top of the potential barrier which separates the two minima of the Landau potential. For this case a square root dependence of the correlation time was found for the mean first passage time.

S.11 Fokker-Planck Equation with a Non-Positive-Definite Diffusion Matrix and Fokker-Planck Equation with Additional Third-Order-Derivative Terms

In quantum optics one usually has to solve an equation of motion for the density operator, which describes the system under consideration. For simple model systems (one mode only, atomic coordinates are eliminated adiabatically) only the creation and annihilation operators of the light field enter in this equation. Continuous representations of the density operator such as the Glauber-Sudarshan P function [12.1, A4.5, 6], the Q functions or the Wigner function [S11.1] may be introduced, by which the equation of motion of the density operator is transformed into an equation of motion for these continuous quasi-distribution functions of a complex variable. (For a review on quasi-distribution functions see [S11.2].) For the model of *Drummond* and *Walls* [S11.3] describing dispersive optical bistability this equation is a Fokker-Planck equation with a non-positive definite diffusion matrix for the P and Q function or with additional third-order-derivative terms in it for the Wigner function. The question arises how to solve these equations.

For the P and Q function the Fokker-Planck equation has a diffusion matrix which is not positive definite or positive semidefinite. Such an equation has been termed pseudo-Fokker-Planck equation [S11.4] because it cannot be interpreted as describing the equation of motion for the probability of a Brownian particle under a suitable field of force. A simulation of this pseudo-Fokker-Planck equation is not possible. One may, however, obtain an equation with a positive definite diffusion matrix for the positive P function by doubling the phase space [S6]. A simulation is then possible as it was done, for instance, by *Dörfle* and *Schenzle* for a different problem [S11.5]. One may also use the complex P function [S6] and one can then obtain the stationary solution for this model [S11.3]. For direct numerical solutions, however, the doubling of the phase space, where one has to use two complex variables (four real ones) instead of one complex variable (two real ones) seems to be not very useful. *Vogel* and the author have solved these pseudo-Fokker-Planck equations by the matrix continued-fraction method [S11.4, 6 – 8]. For the application of this method one does not need a positive definite or a semidefinite diffusion matrix. It even works for the Fokker-Planck equation with third-order-derivative terms [S11.9] which arises in connection with the Wigner function for the dispersive optical bistability model.

For the model of *Drummond* and *Walls* squeezing of the light field [S11.10] occurs, see [S11.7]. In such a case the P function does not exist. As shown in [S11.4] the expansion coefficients of the P function do exist because they are connected with the moments. For determining eigenvalues, the matrix continued-fraction method can also be used for the P function as explained in [S11.4]. It was found that the eigenvalues agree within the numerical accuracy for the different quasi-distribution functions [S11.4, 9]. For optical bistability the lowest nonzero eigenvalue is very important because it determines the transition rate between the two nearly stable states. (For reviews on optical bistability, see [S11.11 – 14].) Without thermal fluctuations the transitions, which are caused by quantum fluctuations, and which lead to the quantum tunneling rates, determine the ultimate stability of the bistable states. It has been shown in [S11.4, 8, 9] that the tunneling rate (= lowest nonzero eigenvalue) shows an interesting oscillating variation as a function of a system parameter, by which the photon number inside the cavity scales. By using a completely different method applicable for small cavity damping [S11.15, 16] these oscillations have also been found [S11.16] in complete agreement with the results of the matrix continued-fraction method. It is interesting to note that the 'nonclassical' terms of the Fokker-Planck equation seem to be responsible for these oscillations. If the nonclassical terms are neglected (i.e., make the diffusion matrix positive definite, neglect third-order-derivative terms) these oscillations disappear [S11.9]. (Classical Fokker-Planck equations for dispersive optical bistability have been treated by *Graham* and *Schenzle* [S11.17] and by *Haug* et al. [6.20].)

It should be noted that in [S11.18] it was already mentioned that Fokker-Planck equations with a non-positive-definite diffusion matrix appear for absorptive optical bistability. In [S11.19, 20] a Fokker-Planck equation with linear drift and non-positive definite but constant diffusion matrix have been treated, but no distribution function was obtained. A simple model, where the diffusion matrix is not positive definite, was presented in [S11.4]. The Fokker-Planck equation for this model reads ($q > 0$)

$$\frac{\partial P}{\partial t} = \left[\frac{\partial}{\partial x}(x - \omega y) + \frac{\partial}{\partial y}(y + \omega x) + \frac{\partial^2}{\partial x^2} - q\frac{\partial^2}{\partial y^2}\right] P . \tag{S11.1}$$

The drift terms describe a damped rotation according to the deterministic equations

$$\dot{x} = -x + \omega y , \quad \dot{y} = -y - \omega x . \tag{S11.2}$$

The stationary solution exists if the conditions

$$q < 1 \quad \text{and} \quad 1 + \omega^2 > (1+q)^2/(1-q)^2 \tag{S11.3}$$

are fulfilled. Thus the negative diffusion coefficient should be smaller than the positive one and the rotation rate should be large enough. The stationary distribution is Gaussian with a positive definite variance matrix $\boldsymbol{\sigma}$ given by

$$\sigma_{11} = \frac{1}{2}\left(1 - q + \frac{1+q}{1+\omega^2}\right), \quad \sigma_{12} = -\frac{\omega}{2}\,\frac{1+q}{1+\omega^2},$$
$$\sigma_{22} = \frac{1}{2}\left(1 - q - \frac{1+q}{1+\omega^2}\right). \tag{S11.4}$$

For the time-dependent case, see the appendix of [S11.4].

References

Chapter 1

1.1 A. D. Fokker: Ann. Physik **43**, 810 (1914)
1.2 M. Planck: Sitzber. Preuß. Akad. Wiss. p. 324 (1917)
1.3 P. Langevin: Comptes rendus **146**, 530 (1908)
1.4 A. Einstein: Ann. Physik **17**, 549 (1905) and **19**, 371 (1906)
1.5 G. E. Uhlenbeck, L. S. Ornstein: Phys. Rev. **36**, 823 (1930)
1.6 S. Chandrasekhar: Rev. Mod. Phys.**15**, 1 (1943)
1.7 M. C. Wang, G. E. Uhlenbeck: Rev. Mod. Phys. **17**, 323 (1945)
1.8 References [1.5 – 7] and other articles are contained in: N. Wax (ed.): *Selected Papers on Noise and Stochastic Processes* (Dover, New York 1954)
1.9 A. T. Bharucha-Reid: *Elements of the Theory of Markov Processes and Their Applications* (McGraw-Hill, New York 1960)
1.10 R. L. Stratonovich: *Topics in the Theory of Random Noise,* Vols. I and II (Gordon & Breach, New York 1963 and 1967)
1.11 M. Lax: Rev. Mod. Phys. **32**, 25 (1960) (a), **38**, 359 (1966) (b) and **38**, 541 (1966) (c)
1.12 N. S. Goel, N. Richter-Dyn: *Stochastic Models in Biology* (Academic, New York 1974)
1.13 H. Haken: Rev. Mod. Phys. **47**, 67 (1975)
1.14 H. Haken: *Synergetics, An Introduction*, 3rd ed. Springer Ser. Synergetics, Vol. 1 (Springer, Berlin, Heidelberg, New York 1983)
1.15 Z. Schuss: *Theory and Applications of Stochastic Differential Equations* (Wiley, New York 1980)
1.16 O. Klein: Arkiv for Mathematik, Astronomi, och Fysik **16**, No 5 (1921)
1.17 H. A. Kramers: Physica **7**, 284 (1940)
1.18 M. v. Smoluchowski: Ann. Physik **48**, 1103 (1915)
1.19 J. E. Moyal: J. Roy. Stat. Soc. (London) B **11**, 150 (1949)
1.20 L. Boltzmann: *Lectures on Gas Theory*, transl. by S. Brush (University of California Press, Berkley 1964)
1.21 R. Balescu: *Equilibrium and Nonequilibrium Statistical Mechanics* (Wiley, New York 1975) Chap. 11
1.22 P. Résibois, M. De Leener: *Classical Kinetic Theory of Fluids* (Wiley, New York 1977) Chap. 9
1.23 P. L. Bhatnagar, E. P. Gross, M. Krook: Phys. Rev. **94**, 511 (1954)
1.24 N. G. van Kampen: Adv. Chem. Phys. **34**, 245 (1976)
1.25 I. Prigogine, P. Résibois: Physica **24**, 795 (1958)
1.26 S. Nakajima: Prog. Theor. Phys. **20**, 948 (1958)
1.27 R. W. Zwanzig: J. Chem. Phys. **33**, 1338 (1960)
1.28 F. Haake: *Springer Tracts Mod. Phys. 66*, 98 (Springer, Berlin, Heidelberg, New York 1973)
1.29 V. M. Kenkre: In *Statistical Mechanics and Statistical Methods in Theory and Applications*, ed. by U. Landmann (Plenum, New York 1977) p. 441
1.30 H. Grabert: *Springer Tracts Mod. Phys. 95* (Springer, Berlin, Heidelberg, New York 1982)

Chapter 2

2.1 R. von Mises: *Mathematical Theory of Probability and Statistics* (Academic, New York 1964)
2.2 B. W. Gnedenko: *Lehrbuch der Wahrscheinlichkeitsrechnung* (Akademie, Berlin 1957)
2.3 W. Feller: *An Introduction to Probability Theory and its Applications*, Vols. 1 and 2 (Wiley, New York 1968 and 1971)
2.4 Yu. V. Prohorov, Yu. A. Rozanov: *Probability Theory*, Grundlehren der mathematischen Wissenschaften in Einzeldarstellungen, Bd. 157 (Springer, Berlin, Heidelberg, New York 1968)
2.5 M. Loève: *Probability Theory.* Vols. 1 and 2, Graduate Texts Math. (Springer, New York, Heidelberg, Berlin 1977 and 1978)
2.6 J. L. Doob: *Stochastic Processes* (Wiley, New York 1953)
2.7 J. D. Jackson: *Classical Electrodynamics* (Wiley, New York 1962) p. 4
2.8 T. Muir: *A Treatise on the Theory of Determinants* (Dover, New York 1960) p. 719
2.9 P. Hänggi, P. Talkner: J. Stat. Phys. **22**, 65 (1980)
2.10 S. O. Rice: Mathematical Analysis of Random Noise, in *Selected Papers on Noise and Stochastic Processes*, ed. by N. Wax (Dover, New York 1954) p. 133

Chapter 3

3.1 N. G. van Kampen: Phys. Rep. **24**, 171 (1976)
3.2 K. Itô: Proc. Imp. Acad. **20**, 519 (1944)
3.3 R. L. Stratonovich: *Conditional Markov Processes and Their Application to the Theory of Optimal Control* (Elsevier, New York 1968)
3.4 R. E. Mortensen: J. Stat. Physics **1**, 271 (1969)
3.5 L. Arnold: *Stochastische Differentialgleichungen* (Oldenbourg, München 1973)
3.6 D. Ryter, U. Deker: J. Math. Phys. **21**, 2662 (1980), U. Deker, D. Ryter: J. Math. Phys. **21**, 2666 (1980)
3.7 N. G. van Kampen: Phys. Lett. **76A**, 104 (1980)
3.8 H. Haken, H. D. Vollmer: Z. Physik **242**, 416 (1971)
3.9 H. Risken, C. Schmid, W. Weidlich: Z. Physik **193**, 37 (1966)
3.10 G. N. Mil'shtein: Theory Probab. Appl. **XIX**, 557 (1974)
3.11 R. H. Morf, E. P. Stoll: In *Numerical Analysis*, ed. by J. Descloux and J. Marti (Birkhäuser, Basel 1977) p. 139

Chapter 4

4.1 P. Hänggi: Helv. Phys. Acta **51**, 183 (1978)
4.2 F. J. Dyson: Phys. Rev. **75**, 486 (1949)
4.3 R. F. Pawula: Phys. Rev. **162**, 186 (1967)
4.4 I. S. Gradshteyn, I. M. Ryzhik: *Tables of Integrals, Series and Products* (Academic, New York 1965) p. 338
4.5 H. Haken: Z. Physik **B24**, 321 (1976)
4.6 C. Wissel: Z. Physik **B35**, 185 (1979)
4.7 L. Onsager, S. Machlup: Phys. Rev. **91**, 1505, 1512 (1953)
4.8 R. Graham: *Springer Tracts in Mod. Phys. 66*, 1, (Springer, Berlin, Heidelberg, New York 1973)
4.9 R. Graham: Z. Physik **B26**, 281 (1977)
4.10 R. Graham: In *Stochastic Processes in Nonequilibrium Systems*, Proc., Sitges (1978), Lecture Notes.Phys. Vol. 84 (Springer, Berlin, Heidelberg, New York 1978) p. 83
4.11 H. Leschke, M. Schmutz: Z. Physik **B27**, 85 (1977)
4.12 U. Weiss: Z. Physik **B30**, 429 (1978)
4.13 H. Risken, H. D. Vollmer: Z. Physik **B35**, 313 (1979)
4.14 H. D. Vollmer: Z. Physik **B33**, 103 (1979)
4.15 H. Margenau, G. M. Murphy: *The Mathematics of Physics and Chemistry* (Van Nostrand, Princeton, NJ 1964)

4.16 J. Mathews, R. Walker: *Mathematical Methods of Physics* (Benjamin, Menlo Park, CA 1973) Chap. 15
4.17 A. Duschek, A. Hochrainer: *Grundzüge der Tensorrechnung in analytischer Darstellung*, Teil II Tensoranalysis (Springer, Wien 1950)
4.18 R. Graham: Z. Physik **B26**, 397 (1977)
4.19 H. Grabert, R. Graham, M. S. Green: Phys. Rev. **A21**, 2136 (1980)

Chapter 5

5.1 J. Mathews, R. L. Walker: *Mathematical Methods of Physics* (Benjamin, Menlo Park, CA 1973)
5.2 R. Courant, D. Hilbert: *Methoden der Mathematischen Physik*, Vol. I (Springer, Berlin 1931)
5.3 C. Cohen-Tannoudji, B. Diu, F.Laloë: *Quantum Mechanics* I (Wiley, New York 1977)
5.4 P. M. Morse, H. Feshbach: *Methods of Theoretical Physics* (McGraw-Hill, New York 1953)
5.5 E. Nelson: Phys. Rev. **150**, 1079 (1966)
5.6 N. G. van Kampen: J. Stat. Phys. **17**, 71 (1977)
5.7 K. Yasue: Phys. Rev. Lett. **40**, 665 (1978)
5.8 D. L. Weaver: Phys. Rev. Lett. **40**, 1473 (1978)
5.9 H. Brand, A. Schenzle: Phys. Lett. **68A**, 427 (1978)
5.10 E. R. Hansen: *A Table of Series and Products* (Prentice-Hall, Englewood Cliffs, NJ 1975) p. 329, eq. (49.6.1)
5.11 C. Cohen-Tannoudji, B. Diu, F. Laloë: *Quantum Mechanics* II (Wiley, New York 1977) p. 1360
5.12 M. Mörsch, H. Risken, H. D. Vollmer: Z. Physik **B32**, 245 (1979)
5.13 D. H. Weinstein: Proc. Nat. Acad. Sci. (USA) **20**, 529 (1934)
5.14 E. Kamke: *Differentialgleichungen*, Vol. I (Geest & Portig, Leipzig 1961) p. 232
5.15 H. Brand, A. Schenzle, G. Schröder: Phys. Rev. **A25**, 2324 (1982)
5.16 F. G. Tricomi: *Vorlesungen über Orthogonalreihen* (Springer, Berlin 1955)
5.17 K. Voigtlaender: *Der durch äußere Strahlung angetriebene Josephson-Effekt; Methoden zur Lösung der zugehörigen Fokker-Planck-Gleichung,* Diplomthesis, Ulm (1982)
5.18 R. Landauer, J. A. Swanson: Phys. Rev. **121**, 1668 (1961)
5.19 R. Landauer: J. Appl. Phys. **33**, 2209 (1962)
5.20 J. S. Langer: Ann. Phys. **54**, 258 (1969)
5.21 N. G. van Kampen: Suppl. Progr. Theor. Phys. **64**, 389 (1978)
5.22 K. Matsuo, K. Lindenberg, K. E. Shuler: J. Stat. Phys. **19**, 65 (1978)
5.23 R. S. Larson, M. D. Kostin, J. Chem. Phys. **69**, 4821 (1978)
5.24 B. Caroli, C. Caroli, B. Roulet: J. Stat. Phys. **21**, 415 (1979)
5.25 O. Edholm, O. Leimar: Physica **98A**, 313 (1979)
5.26 R. Gilmore: Phys. Rev. **A20**, 2510 (1979)
5.27 H. Dekker: *Critical Dynamics,* Proefschrift, Utrecht (1980); J. Chem. Phys. **72**, 189 (1980)
5.28 U. Weis, W. Häffner: In *Functional Integration, Theory and Application*, ed. by J. P. Antoine and E. Tirapeyii (Plenum, New York 1980)
5.29 W. Weidlich, H. Grabert: Z. Physik **B36**, 283 (1980)
5.30 W. Bez, P. Talkner: Phys. Lett. **82A**, 313 (1981)
5.31 J. Kevorkian, J. D. Cole: *Perturbation Methods in Applied Mathematics* (Springer, New York 1981)
5.32 P. Jung: *Brownsche Bewegung im periodischen Potential; Untersuchung für kleine Dämpfungen* (Diplom Thesis, Ulm 1983); P. Jung, H. Risken: Z. Physik **B54**, 537 (1984)

Chapter 6

6.1 R. L. Stratonovich: *Topics on the Theory of Random Noise*, Vol. I (Gordon and Breach, New York 1963) p. 77
6.2 N. G. van Kampen: Physica **23**, 707, 816 (1957)
6.3 R. Graham, H. Haken: Z. Physik **243**, 289 (1971); **245**, 141 (1971)

6.4 U. Ulhorn: Arkiv Fysik **17**, 361 (1960)
6.5 J. L. Lebowitz, P. G. Bergmann: Ann. Physik **1**, 1 (1957)
6.6 F. Schlögl: Z. Physik **243**, 303 (1971); **244**, 199 (1971)
6.7 J. P. La Salle, S. Lefschetz: *Stability by Ljapunov's Direct Method* (Academic, New York 1961)
6.8 A. Rényi: *Wahrscheinlichkeitsrechnung* (VEB Verlag der Wissenschaften, Berlin 1966)
6.9 J. H. Wilkinson: *The Algebraic Eigenvalue Problem* (Clarendon, Oxford 1965) Chap. 1
6.10 H. Risken: Z. Physik **251**, 231 (1972)
6.11 R. Graham: Z. Physik **B40**, 149 (1980)
6.12 S. R. De Groot, P. Mazur: *Non-Equilibrium Thermodynamics* (North Holland, Amsterdam 1962)
6.13 M. Lax: In *Symmetries in Sciences*, ed. by B. Gruber, R. S. Millman (Plenum, New York 1980) p. 189, Eqs. (2.16, 26)
6.14 R. Courant, D. Hilbert: *Methoden der Mathematischen Physik*, Bd. II (Springer, Berlin 1937)
6.15 K. Seybold: *Die Fokker-Planck-Gleichung in der Nichtgleichgewichtsstatistik; Lösungsmethoden und Lösungen,* Dissertation, Ulm (1978)
6.16 L. Collatz: *Numerische Behandlung von Differentialgleichungen* (Springer, Berlin 1951)
6.17 G. E. Forsythe, W. R. Wasow: *Finite-Difference Methods for Partial Differential Equations* (Wiley, New York 1967)
6.18 G. D. Smith: *Numerical Solution of Partial Differential Equations* (Oxford University Press, London 1965)
6.19 M. Mörsch: *Lösung einer Fokker-Planck-Gleichung des Lasers mit Matrizenkettenbrüchen* (Dissertation, Ulm 1982); see also M. Mörsch, H. Risken, H. D. Vollmer: Z. Phys. **B49**, 47 (1982)
6.20 H. Haug, S. W. Koch, R. Neumann, H. E. Schmidt: Z. Phys. **49**, 79 (1982)
6.21 B. Caroli, C. Caroli, B. Roulet, J. F. Gouyet: J. Stat. Phys. **22**, 515 (1980)
6.22 J. K. Cohen, R. M. Lewis: J. Inst. Maths. Applics. **3**, 266 (1967)
6.23 A. Messiah: *Quantum Mechanics*, Vol. I (North-Holland, Amsterdam 1966) p. 234ff.

Chapter 7

7.1 M. S. Green: J. Chem. Phys. **19**, 1036 (1951)
7.2 H. B. Callen, T. A. Welton: Phys. Rev. **83**, 34 (1951)
7.3 R. Kubo: J. Phys. Soc. Japan **12**, 570 (1957); Rep. Prog. Phys. **29**, 255 (1966)
7.4 K. M. Case: Transp. Th. Stat. Phys. **2**, 129 (1972)
7.5 G. S. Agarwal: Z. Physik **252**, 25 (1972)
7.6 B. K. P. Scaife: *Complex Permittivity* (English University Press, London 1971)
7.7 J. McConnel: *Rotational Brownian Motion and Dielectric Theory* (Academic, London 1980)
7.8 L. Landau, E. M. Lifschitz: *Statistical Physics* (Pergamon, London 1958)

Chapter 8

8.1 R. L. Stratonovich: *Topics in the Theory of Random Noise*, Vol. I (Gordon and Breach, New York 1963) p. 79ff.
8.2 G. H. Weiss: Ad. Chem. Phys. **13**, 1 (1966)
8.3 M. A. Burschka, U. M. Titulaer: J. Stat. Phys. **25**, 569 and **26**, 59 (1981)
8.4 M. A. Burschka, U. M. Titulaer: Physica **112A**, 315 (1982)
8.5 A. Schenzle, H. Brand: Phys. Rev. **A20**, 1628 (1979)
8.6 H. Brand, A. Schenzle: Phys. Lett. **81A**, 321 (1981)
8.7 K. Kaneko: Progr. Theor. Phys. **66**, 129 (1981)
8.8 O. Madelung: *Introduction to Solid-State Theory*, Springer Ser. Solid-State Sci., Vol. 2 (Springer, Berlin, Heidelberg, New York 1978) p. 9
8.9 R. W. Zwanzig: *Lectures in Theoretical Physics*, Vol. 3 (Wiley-Interscience, New York 1961)

Chapter 9

9.1 O. Perron: *Die Lehre von den Kettenbrüchen*, Vols. I, II (Teubner, Stuttgart 1977)
9.2 H. S. Wall: *Analytic Theory of Continued Fractions* (Chelsea, Bronx, NY 1973)
9.3 W. B. Jones, W. J. Thron: *Continued Fractions*, Encyclopedia of Mathematics and its Applications, Vol. 11 (Addison-Wesley, Reading, MA 1980)
9.4 G. A. Baker, Jr.: *Essentials of Padé Approximants* (Academic, New York 1975)
9.5 P. Hänggi, F. Rösel, D. Trautmann: Z. Naturforsch. **33a**, 402 (1978)
9.6 W. Götze: Lett. Nuovo Cimento **7**, 187 (1973)
9.7 J. Killingbeck: J. Phys. **A10**, L 99 (1977)
9.8 G. Haag, P. Hänggi: Z. Physik **B34**, 411 (1979) and **B39**, 269 (1980)
9.9 S. H. Autler, C. H. Townes: Phys. Rev. **100**, 703 (1955)
9.10 S. Stenholm, W. E. Lamb: Phys. Rev. **181**, 618 (1969)
9.11 S. Stenholm: J. Phys. **B5**, 878 (1972)
9.12 S. Graffi, V. Grecchi: Lett. Nuovo Cimento **12**, 425 (1975)
9.13 M. Allegrini, E. Arimondo, A. Bambini: Phys. Rev. **A15**, 718 (1977)
9.14 H. Risken, H. D. Vollmer: Z. Physik **B33**, 297 (1979)
9.15 H. D. Vollmer, H. Risken: Z. Physik **B34**, 313 (1979)
9.16 H. D. Vollmer, H. Risken: Physica **110A**, 106 (1982)
9.17 H. Risken, H. D. Vollmer: Mol. Phys. **46**, 555 (1982)
9.18 H. Risken, H. D. Vollmer: Z. Physik **B39**, 339 (1980)
9.19 H. Risken, H. D. Vollmer, M. Mörsch: Z. Physik **B40**, 343 (1981)
9.20 W. Dieterich, T. Geisel, I. Peschel: Z. Physik **B29**, 5 (1978)
9.21 H. J. Breymayer, H. Risken, H. D. Vollmer, W. Wonneberger: Appl. Phys. **B28**, 335 (1982)
9.22 S. N. Dixit, P. Zoller, P. Lambropoulos: Phys. Rev. **A21**, 1289 (1980)
9.23 P. Zoller, G. Alber, R. Salvador: Phys. Rev. **A24**, 398 (1981)
9.24 H. Denk, M. Riederle: J. Appr. Theory **35**, 355 (1982)
9.25 H. Meschkowski: *Differenzengleichungen*, Studia Mathematica Vol. XIV (Vanderhoeck & Ruprecht, Göttingen 1959) Chap. X
9.26 W. Magnus, F. Oberhettinger, R. P. Soni: *Formulas and Theorems for the Special Functions of Mathematical Physics* (Springer, New York 1966)

Chapter 10

10.1 H. C. Brinkman: Physica **22**, 29 (1956)
10.2 U. M. Titulaer: Physica **91A**, 321 (1978)
10.3 P. Resibois: *Electrolyte Theory* (Harper & Row, New York 1968) pp. 78, 150
10.4 J. W. Dufty: Phys. Fluids **17**, 328 (1974)
10.5 R. M. Mazo: *Lecture Notes in Physics 84*, 58 (Springer, Berlin, Heidelberg, New York 1978)
10.6 R. A. Sack: Physica **22**, 917 (1956)
10.7 P. C. Hemmer: Physica **27**, 79 (1961)
10.8 G. H. Weiss, A. A. Maradudin: J. Math. Phys. **3**, 771 (1962)
10.9 J. L. Skinner, P. G. Wolynes: Physica **96A**, 561 (1979)
10.10 R. I. Stratonovich: *Topics in the Theory of Random Noise*, Vol. I (Gordon and Breach, New York 1963) p. 115, Eq. (4.245)
10.11 G. Wilemski: J. Stat. Phys. **14**, 153 (1976)
10.12 M. San Miguel, J. M. Sancho: J. Stat. Phys. **22**, 605 (1980)
10.13 S. Chaturvedi, F. Shibata: Z. Physik **B35**, 297 (1979)
10.14 N. G. Van Kampen: Physica **74**, 215 and 239 (1974)
10.15 P. Hänggi, H. Thomas, H. Grabert, P. Talkner: J. Stat. Phys. **18**, 155 (1978)
10.16 U. Geigenmüller, U. M. Titulaer, B. U. Felderhof: Physica **119A**, 41 (1983)
10.17 F. Haake, M. Lewenstein: Phys. Rev. **A28**, 3606 (1983)

Chapter 11

11.1 R. L. Stratonovich: Radiotekhnika; elektronika **3**, No 4, 497 (1958)
11.2 R. L. Stratonovich: *Topics in the Theory of Random Noise*, Vol. II (Gordon and Breach, New York 1967) Chap. 9

11.3 A. J. Viterbi: Proc. IEEE **51**, 1737 (1963)
11.4 A. J. Viterbi: *Principles of Coherent Communication* (McGraw-Hill, New York 1966)
11.5 W. C. Lindsey: *Synchronization Systems in Communication and Control* (Prentice Hall, Englewood Cliffs, NJ 1972)
11.6 H. Haken, H. Sauermann, Ch. Schmid, H. D. Vollmer: Z. Physik **206**, 369 (1967)
11.7 Y. M. Ivanchenko, L. A. Zil'berman: Sov. Phys. JETP **28**, 1272 (1969)
11.8 V. Ambegaokar, B. I. Halperin: Phys. Rev. Lett. **22**, 1364 (1969)
11.9 J. D. Cresser, W. H. Louisell, P. Meystre, W. Schleich, M. O. Scully: Phys. Rev. **A25**, 2214 (1982)
11.10 J. D. Cresser, D. Hammonds, W. H. Louisell, P. Meystre, H. Risken: Phys. Rev. **A25**, 2226 (1982)
11.11 P. Fulde, L. Pietronero, W. R. Schneider, S. Strässler: Phys. Rev. Lett. **35**, 1776 (1975)
11.12 W. Dieterich, I. Peschel, W. R. Schneider: Z. Physik **B27**, 177 (1977)
11.13 H. Risken, H. D. Vollmer: Z. Physik **B31**, 209 (1978)
11.14 T. Geisel: In *Physics of Superionic Conductors*, ed. by M. B. Saloman, Topic Current Phys., Vol. 15 (Springer, Berlin, Heidelberg, New York 1979) p. 201
11.15 W. Dieterich, P. Fulde, I. Peschel: Adv. Phys. **29**, 527 (1980)
11.16 A. K. Das, P. Schwendimann: Physica **89A**, 605 (1977)
11.17 W. T. Coffey: Adv. Molecular Relaxation and Interaction Processes **17**, 169 (1980)
11.18 G. Wyllie: Phys. Reps. **61**, 329 (1980)
11.19 R. W. Gerling: Z. Physik **B45**, 39 (1981)
11.20 E. Praestgaard, N. G. van Kampen: Molec. Phys. **43**, 33 (1981)
11.21 V. I. Tikhonov: Avtomatika i Telemekhanika **21**, 301 (1960)
11.22 P. A. Lee: J. Appl. Phys. **42**, 325 (1971)
11.23 K. Kurkijärvi, V. Ambegaokar: Phys. Lett. **A31**, 314 (1970)
11.24 T. Schneider, E. P. Stoll, R. Morf: Phys. Rev. **B18**, 1417 (1978)
11.25 P. Nozières, G. Iche: J. Physique **40**, 225 (1979)
11.26 E. Ben-Jacob, D. J. Bergman, B. J. Matkowsky, Z. Schuss: Phys. Rev. **A26**, 2805 (1982)
11.27 H. D. Vollmer, H. Risken: Z. Physik **B37**, 343 (1980)
11.28 H. D. Vollmer, H. Risken: Z. Physik **52**, 259 (1983)
11.29 H. Risken, H. D. Vollmer: Phys. Lett. **69A**, 387 (1979)
11.30 H. Risken, H. D. Vollmer: Z. Physik **B35**, 177 (1979)
11.31 B. D. Josephson: Phys. Lett. **1**, 251 (1962)
11.32 L. Solymar: *Superconductive Tunneling and Applications* (Chapman and Hall, London 1972)
11.33 A. Barone, G. Paterno: *Physics and Applications of the Josephson Effect* (Wiley, New York 1982)
11.34 P. Debye: Ber. dt. phys. Ges. **15**, 777 (1913); translated in *The Collected Papers of Peter J. W. Debye* (Interscience, New York 1954)
11.35 A. Seeger: In *Continuum Models of Discrete Systems*, ed. by E. Kröner and K. H. Anthony (University of Waterloo Press, Waterloo 1980) p. 253
11.36 R. D. Parmentier: In *Solitons in Action*, ed. by K. Longren, A. Scott (Academic, New York 1978) p. 173
11.37 G. L. Lamb: *Elements of Soliton Theory* (Wiley, New York 1980)
11.38 R. K. Bullough, P. J. Caudrey (eds.): *Solitons*, Topics Current Phys., Vol. 17 (Springer, Berlin, Heidelberg, New York 1980)
11.39 G. Eilenberger: *Solitons*, Springer Ser. Solid-State Sci., Vol. 19 (Springer, Berlin, Heidelberg, New York 1981)
11.40 M. Büttiker, R. Landauer: In *Nonlinear Phenomena at Phase Transitions and Instabilities*, ed. by T. Riste (Plenum, New York 1982)
11.41 R. A. Guyer, M. D. Miller: Phys. Rev. **A17**, 1774 (1978)
11.42 H. Jorke: *Modellrechnungen zur Brownschen Bewegung im periodischen Potential*, Diplom-thesis, Ulm (1981)
11.43 J. Mathews, R. L. Walker: *Mathematical Methods of Physics* (Benjamin, Menlo Park, CA 1973) p. 198ff.
11.44 L. Brillouin: *Wave Propagation in Periodic Structures* (McGraw-Hill, New York 1946)
11.45 A. H. Wilson: *The Theory of Metals* (University Press, Cambridge 1958)

11.46 R. Festa, E. G. d'Agliano: Physica **90A**, 229 (1978)
11.47 D. L. Weaver: Physica **98A**, 359 (1979)
11.48 R. A. Guyer: Phys. Rev. **B21**, 4484 (1980)
11.49 H. Risken, K. Voigtlaender: unpublished
11.50 M. Abramowitz, I. A. Stegun: *Handbook of Mathematical Functions* (Dover, New York 1965)
11.51 D. E. McCumber: J. Appl. Phys. **39**, 3113 (1968)
11.52 C. M. Falco: Am. J. Phys. **44**, 733 (1976)
11.53 W. Dieterich, T. Geisel, I. Peschel: Z. Physik **B29**, 5 (1978)
11.54 T. Springer: *Quasielastic Neutron Scattering for the Investigation of Diffusive Motion in Solids and Liquids*. Springer Tracts Mod. Phys. 64 (Springer, Berlin, Heidelberg, New York 1972)
11.55 J. L. Synge, B. A. Griffith: *Principles of Mechanics* (McGraw-Hill, New York 1959)
11.56 A. H. Nayfeh, D. T. Mook: *Nonlinear Oscillations* (Wiley, New York 1979) p. 55
11.57 H. Goldstein: *Classical Mechanics* (Addison-Wesley, Reading, Mass. 1950)

Chapter 12

12.1 H. Haken: *Laser Theory*, Encyclopedia of Physics, Vol. XXV/2c (Springer, Berlin, Heidelberg, New York 1970)
12.2 F. T. Arecchi, E. O. Schulz-Dubois (eds.): *Laser Handbook* (North-Holland, Amsterdam 1972)
12.3 M. Sargent III, M. O. Scully, W. E. Lamb: *Laser Physics* (Addison-Wesley, Reading, MA 1974)
12.4 A. Yariv: *Quantum Electronics* (Wiley, New York 1967)
12.5 B. Saleh: *Photoelectron Statistics*, Springer Ser. Opt. Sci., Vol. 6 (Springer, Berlin, Heidelberg, New York 1978)
12.6 H. Haken: *Licht und Materie* II (Bibliographisches Institut, Mannheim 1981)
12.7 M. Lax: *1966 Brandeis University Summer Institute in Theoretical Physics* (Gordon and Breach, New York 1968)
12.8 H. Risken: Fortschr. Physik **16**, 261 (1968)
12.9 H. Risken: *Progress in Optics 8* (North-Holland, Amsterdam 1970) p. 239
12.10 S. M. Kay, A. Maitland (eds.): *Quantum Optics* (Academic, London 1970)
12.11 J. Perina: *Coherence of Light* (Van Nostrand Reinhold, London 1972)
12.12 R. Graham: In *Fluctuations, Instabilities, and Phase Transitions*, ed. by T. Riste (Plenum, New York 1975)
12.13 V. Dohm: *Phasenübergänge und Chaos im Laser,* Ferienkurs über nichtlineare Dynamik in kondensierter Materie, Kernforschungsanlage, Jülich (1983)
12.14 B. Van der Pol: Phil. Mag **3**, 65 (1927)
12.15 J. W. S. Rayleigh: *Theory of Sound*, Vol. 1 (1894), reprint (Dover, New York 1945)
12.16 H. Haken: Z. Physik **181**, 96 (1964)
12.17 H. Risken: Z. Physik **186**, 85 (1965) (a) and **191**, 302 (1966) (b)
12.18 H. Risken, H. D. Vollmer: Z. Physik **201**, 323 (1967)
12.19 H. Risken, H. D. Vollmer: Z. Physik **204**, 240 (1967)
12.20 R. D. Hempstead, M. Lax: Phys. Rev. **161**, 350 (1967)
12.21 J. P. Gordon, E. W. Aslaksen: IEEE J. **QE-6**, 428 (1970)
12.22 F. T. Arecchi, V. Degiorgio: Phys. Rev. **A3**, 1108 (1971)
12.23 M. Suzuki: Prog. Theor. Phys. **56**, 77 (1976) and **56**, 477 (1976)
12.24 M. Suzuki: Physica **A86**, 622 (1977) and T. Arimitsu, M. Suzuki: Physica **A90**, 303 (1978)
12.25 F. Haake: Phys. Lett. **41**, 1685 (1978)
12.26 S. Grossmann: Phys. Rev. **A17**, 1123 (1978)
12.27 F. De Pasquale, P. Tartaglia, P. Tombesi: Physica **A99**, 581 (1979)
12.28 H. King, U. Deker, F. Haake: Z. Physik **B36**, 205 (1979)
12.29 K. Ziegler, H. Horner: Z. Physik **B37**, 339 (1980)
12.30 H. Risken, H. D. Vollmer: Z. Physik **B39**, 89 (1980)
12.31 J. Fiutak, J. Mizerski: Z. Physik **B39**, 347 (1980)
12.32 T. Arimitsu: Physica **107A**, 71 (1981)

12.33 J. Mizerski: Z. Physik **B49**, 173 (1982)
12.34 R. L. Stratonovich: *Topics in the Theory of Random Noise*, Vol. II (Gordon and Breach, New York 1967) Chap. 5
12.35 H. Haken, H. Risken, W. Weidlich: Z. Physik **206**, 355 (1967)
12.36 M. Lax: Phys. Rev. **157**, 213 (1967)
12.37 M. Lax, W. H. Louisell: IEEE J. **QE-3**, 47 (1967)
12.38 M. O. Scully, W. E. Lamb: Phys. Rev. **159**, 208 (1967)
12.39 J. P. Gordon: Phys. Rev. **161**, 367 (1967)
12.40 A. W. Smith, J. A. Armstrong: Phys. Rev. Lett. **16**, 1169 (1966)
12.41 J. A. Armstrong, A. W. Smith: *Progress in Optics 6*, 211 (North-Holland, Amsterdam 1967)
12.42 F. T. Arecchi, G. S. Rodari, A. Sona: Phys. Lett. **25A**, 59 (1967) (a); F. T. Arecchi, M. Giglio, A. Sona: Phys. Lett. **25A**, 341 (1967) (b); F. T. Arecchi, V. Degiorgio, B. Querzola: Phys. Rev. Lett. **19**, 1168 (1967) (c)
12.43 F. Davidson, L. Mandel: Phys. Lett. **25A**, 700 (1967)
12.44 R. F. Chang, V. Korenman, C. O. Alley, R. W. Detenbeck: Phys. Rev. **178**, 612 (1969)
12.45 D. Meltzer, L. Mandel: Phys. Rev. **A3**, 1763 (1971)
12.46 S. Chopra, L. Mandel: IEEE J. **QE-8**, 324 (1972)
12.47 S. Grossmann, P. H. Richter: Z. Physik **249**, 43 (1971)
12.48 F. T. Hioe: J. Math. Phys. **19**, 1307 (1978)
12.49 F. T. Hioe, S. Singh: Phys. Rev. **A24**, 2050 (1981)
12.50 R. K. Wangsness, F. Bloch: Phys. Rev. **89**, 728 (1953)
12.51 A. L. Schawlow, C. H. Townes: Phys. Rev. **112**, 1940 (1958)
12.52 R. Graham: Phys. Rev. **25A**, 3234 (1982)
12.53 K. Seybold, H. Risken: Z. Physik **267**, 323 (1974)
12.54 M. Born, E. Wolf: *Principles of Optics* (Pergamon, London 1964)
12.55 L. Mandel, E. Wolf: Phys. Rev. **124**, 1696 (1961)
12.56 H. Gerhardt, H. Welling, A. Güttner: Z. Physik **253**, 113 (1972)
12.57 F. Haake, J. W. Haus, R. Glauber: Phys. Rev. **A23**, 3255 (1981)
12.58 M. Mangel: Phys. Rev. **A24**, 3226 (1981)
12.59 U. Weiss: In *Chaos and Order in Nature*, ed. by H. Haken, Springer Ser. Synergetics, Vol. 11, (Springer, Berlin, Heidelberg, New York 1981) p. 177
12.60 L. Mandel: Proc. Phys. Soc. **72**, 1037 (1958)
12.61 L. Mandel: *Progress in Optics 2*, 181 (North-Holland, Amsterdam 1963)
12.62 P. L. Kelley, W. H. Kleiner: Phys. Rev. **136A**, 316 (1964)
12.63 M. Lax, M. Zwanziger: Phys. Rev. Lett. **24**, 937 (1970)

Appendices

A1.1 P. Zoller: *Laser Temporal Coherence Effects in Resonant Multiphoton-Processes,* Habilitationsschrift, Innsbruck, Austria (1980)
A1.2 R. Kubo: *A Stochastic Theory of Line-Shape and Relaxation,* in: *Fluctuation, Relaxation, and Resonance in Magnetic Systems*; D. ter Haar (ed.) (Oliver and Boyd, Edinburgh-London 1962)
A1.3 R. Fox: Phys. Reps. **48**, 179 (1978)
A1.4 K. Wódkiewicz: Z. Phys. **B42**, 95 (1981)

A2.1 J.L. Skinner, P. G. Wolynes: J. Chem. Phys. **72**, 4913 (1980)
A2.2 J.L. Skinner, P. G. Wolynes: J. Chem. Phys. **69**, 2143 (1978)

A4.1 W. Weidlich, F. Haake: Z. Physik **185**, 30 (1965)
A4.2 W.H. Louisell, J.H. Marburger: IEEE J. **QE-3**, 348 (1967)
A4.3 G.S. Agarwal: *Progress in Optics 11*, 27 (North-Holland, Amsterdam 1973)
A4.4 I.R. Senitzky: Phys. Rev. **119**, 670 (1960) and **124**, 642 (1961)
A4.5 R.J. Glauber: Phys. Rev. **130**, 2529 and **131**, 2766 (1963)
A4.6 E.C.G. Sudarshan: Phys. Rev. Lett. **10**, 277 (1963)

A5.1 R.F. Fox: J. Math. Phys. **13**, 1196 (1972)
A5.2 K. Wódkiewicz: J. Math. Phys. **20**, 45 (1979)
A5.3 K. Wódkiewicz: Z. Phys. **B47**, 239 (1982)

A6.1 S. Flügge: *Practical Quantum Mechanics* I (Springer, Berlin, Heidelberg, New York 1971) problem 70
A6.2 C.W. Gardiner, R. Graham: Phys. Rev. **A25**, 1851 (1982)
A6.3 M. Toda: Phys. Rep. **18**, 1 (1975)

Supplement

General References (Books and Reports)

S1 W. Ebeling, R. Feistel: *Physik der Selbstorganisation und Evolution* (Akademie-Verlag, Berlin 1982)
S2 P. Hänggi, H. Thomas: Stochastic processes: time evolution, symmetries and linear response. Phys. Rep. **88**, 207 – 319 (1982)
S3 H. Haken: *Advanced Synergetics*, Springer Ser. Syn., Vol. 20 (Springer, Berlin, Heidelberg 1983)
S4 N.G. van Kampen: *Stochastic Processes in Physics and Chemistry*, first reprint (North-Holland, Amsterdam 1983)
S5 W. Horsthemke, R.L. Lefever: *Noise-Induced Transitions*, Springer Ser. Syn., Vol. 15 (Springer, Berlin, Heidelberg 1984)
S6 C.W. Gardiner: *Handbook of Stochastic Methods*, 2nd. ed., Springer Ser. Syn., Vol. 13 (Springer, Berlin, Heidelberg 1985)
S7 W.T. Coffey, M.W. Evans, P. Grigolini: *Molecular Diffusion and Spectra* (Wiley, New York 1984)
S8 M.I. Dykman, M.A. Krivoglaz: Theory of nonlinear oscillator interacting with a medium, in *Physics Reviews* (Soviet Scientific Reviews) (Harwood Academic, New York 1984) pp. 265 – 441
S9 H. Malchow, L. Schimansky-Geier: *Noise and Diffusion in Bistable Nonequilibrium Systems* (Teubner, Leipzig 1985)
S10 K.-H. Li: Physics of open systems. Phys. Rep. **134**, 1 – 85 (1986)
S11 G. Röpke: Statistische Mechanik für das Nichtgleichgewicht (Physik-Verlag, Weinheim 1987)
S12 S. Dattagupta: Relaxation Phenomena in Condensed Matter Physics (Academic Press, New York 1987)
S13 C. Cercignani: *The Boltzmann Equation and Its Applications* (Springer, Berlin, Heidelberg 1989)
S14 F. Moss, P.V.E. McClintock (eds.): *Noise in Nonlinear Dynamical Systems; Theory, Experiment, Simulation* (3 Volumes) (Cambridge Univ. Press, Cambridge 1989)
S15 H. Grabert, P. Schramm, G.L. Ingold: Quantum brownian motion: the functional integral approach. Phys. Rep. **168**, 115 – 207 (1988)

Supplement 1

S1.1 N.J. Rao, J.D. Borwankar, D. Ramkrishna: Numerical solution of Ito integral equation. SIAM J. Control **12**, 124 – 139 (1974)
S1.2 G.N. Milshtein: A method of second-order accuracy integration of stochastic differential equations. Theory Prob. Appl. **23**, 396 – 401 (1978)
S1.3 G.L. Blankenship, J.S. Baras: Accurate evaluation of stochastic Wiener integrals with applications to scattering in random media and to nonlinear filtering. SIAM J. Appl. Math. **41**, 518 – 552 (1981)
S1.4 W. Rümelin: Numerical treatment of stochastic differential equations. SIAM J. Num. Anal. **19**, 604 – 613 (1982)
S1.5 J.R. Klauder, W.P. Peterson: Numerical integrations of multiplicative-noise stochastic differential equations. SIAM Num. Anal. **22**, 1153 – 1166 (1985)

S1.6 D.W. Heerman: *Computer Simulation Methods in Theoretical Physics* (Springer, Berlin, Heidelberg 1986)
S1.7 A. Greiner, W. Strittmatter, J. Hohnerkamp: Numerical integration of stochastic differential equations. J. Stat. Phys. **51**, 95 – 108 (1988)

Supplement 2

S2.1 R.F. Pawula: Private communication (July 1985)
S2.2 H. Risken, H.D. Vollmer: On solutions of truncated Kramers Moyal-expansions; continuum approximations to the Poisson process. Z. Physik **B66**, 257 – 262 (1987)
S2.3 C.R. Doering, P.S. Hagan, P. Rosenau: Random walk in a quasi-continuum. Phys. Rev. **A36**, 985 – 988 (1987)

Supplement 3

S3.1 C.W. Misner, K.S. Thorne, J.A. Wheeler: *Gravitation* (Freeman, San Francisco 1973) p. 341
S3.2 N.G. van Kampen: Brownian motion on a manifold. J. Stat. Phys. **44**, 1 – 24 (1986)

Supplement 4

S4.1 M. Bernstein, L.S. Brown: Supersymmetry and the bistable Fokker-Planck equation. Phys. Rev. Lett. **52**, 1933 – 1935 (1984)
S4.2 E. Witten: Dynamical breaking of supersymmetry. Nucl. Phys. **B185**, 513 – 554 (1981)
S4.3 C.V. Sukumar: Supersymmetry, factorisation of the Schrödinger equation and a Hamilton hierarchy. J. Phys. **A18**, L57 – L61 (1984)
S4.4 H.R. Jauslin: Exact propagator and eigenfunctions for multistable models with arbitrarily prescribed N lowest eigenvalues. J. Phys. **A21**, 2337 – 2350 (1988)
S4.5 M.O. Hongler, W.M. Zheng: Exact solution for the diffusion in bistable potentials. J. Stat. Phys. **29**, 317 – 327 (1982)
S4.6 M.J. Englefield: Explicit solution of the Fokker-Planck equation. J. Stat. Phys. **52**, 369 – 381 (1988)

Supplement 5

S5.1 M.I. Freidlin, A.D. Ventsel: *Random Perturbations of Dynamical Systems* (Springer, New York 1984)
S5.2 R. Graham, T. Tél: On the weak-noise limit of Fokker-Planck models. J. Stat. Phys. **35**, 729 – 748 (1984)
S5.3 H.R. Jauslin: Nondifferentiable potentials for nonequilibrium steady states. Physica **144A**, 179 – 191 (1987)
S5.4 R. Graham, T. Tél: Nonequilibrium potential for coexisting attractors. Phys. Rev. **A33**, 1322 – 1337 (1986)

Supplement 6

S6.1 H. Risken: Solutions of Fokker-Planck equations in terms of matrix continued fractions, in *Lasers and Synergetics*, ed. by R. Graham and A. Wunderlin, Springer Proc. Phys. **19** (Springer, Berlin, Heidelberg 1987) pp. 148 – 164
S6.2 H. Risken, K. Voigtlaender: Solutions of the Fokker-Planck equation describing the thermalization of neutrons in a heavy gas moderator. Z. Physik **B54**, 253 – 262 (1984)
S6.3 W. Schleich, C.-S. Cha, J.D. Cresser: Quantum noise in a dithered ring-laser gyroscope. Phys. Rev. **A29**, 230 – 238 (1984)
S6.4 A. Bambini, S. Stenholm: Theory of a dithered-ring laser gyroscope. Phys. Rev. **A31**, 329 – 337 (1985)
S6.5 T. Leiber, H. Risken: Stability of parametrically excited dissipative systems. Phys. Lett. **A129**, 214 – 218 (1988)
S6.6 H. Scherrer, H. Risken, T. Leiber: Eigenvalues of the Schrödinger equation with rational potentials. Phys. Rev. **A38**, 3949 – 3959 (1988)
S6.7 C. Durst, E. Sigmund, P. Reineker, A. Scheuing: Treatment of nonadiabatic Hamiltonians by matrix continued fractions: I. Electronic two level system coupled to a single vibrational mode. J. Phys. **C19**, 2701 – 2720 (1986)

S6.8 N. Nayak, R.K. Bullough, B.V. Thompson, G.S. Agarval: Quantum collapse and revival of Rydberg atoms in cavities of arbitrary Q at finite temperature. IEEE J. Q.E. **24**, 1331 – 1337 (1988)

S6.9 B. Shizgal, R. Blackmore: Discrete ordinate method of solution of linear boundary value and eigenvalue problems. J. Comput. Phys. **55**, 313 – 327 (1984)

S6.10 R. Blackmore, B. Shizgal: Discrete-ordinate method of Fokker-Planck equations with nonlinear coefficients. Phys. Rev. **A31**, 1855 – 1868 (1985)

Supplement 7

S7.1 K. Voigtlaender, H. Risken: Eigenvalues of the Fokker-Planck and BGK operators for a double-well potential. Chem. Phys. Lett. **105**, 506 – 510 (1984)

S7.2 K. Voigtlaender, H. Risken: Solutions of the Fokker-Planck equation for a double-well potential in terms of matrix continued fractions. J. Stat. Phys. **40**, 397 – 429 (1985)

S7.3 V.I. Mel'nikov, S.V. Meshkov: Theory of activated rate processes: Exact solution of the Kramers problem. J. Chem. Phys. **85**, 1018 – 1027 (1986)

S7.4 V.I. Mel'nikov: Activated tunneling decay of metastable states. Sov. Phys. JETP **60**, 380 – 385 (1984)

S7.5 M.I. Dykman, R. Mannella, P.V. McClintock, F. Moss, S.M. Soskin: Spectral density of fluctuations of a double-well duffing oscillator driven by white noise. Phys. Rev. **A37**, 1303 – 1313 (1988)

S7.6 H. Risken, K. Voigtlaender: Eigenvalues and eigenfunctions of the Fokker-Planck equation for the extremly underdamped Brownian motion in a double-well potential. J. Stat. Phys. **41**, 825 – 863 (1985)

S7.7 Y. Onodera: Dynamic susceptibility of classical anharmonic oscillator. Progr. Theor. Phys. **44**, 1477 – 1499 (1970)

S7.8 M.I. Dykman, S.M. Soskin, M.A. Krivoglaz: Spectral distribution of a nonlinear oscillator performing Brownian motion in a double-well potential. Physica **133A**, 53 – 73 (1985)

S7.9 M.M. Dygas, B.J. Matkowsky, Z. Schuss: A singular perturbation approach to non-Markovian escape rate problems. SIAM J. Appl. Math. **46**, 265 – 298 (1986)

S7.10 R. Landauer: Noise-activated escape from metastable states: An historical review, in *Noise in Nonlinear Dynamical Systems; Theory, Experiment, Simulation*, Vol. I, ed. by F. Moss and P.V.E. McClintock (Cambridge Univ. Press, Cambridge 1989) pp. 1 – 15

S7.11 P. Hänggi: Escape from a metastable state. J. Stat. Phys. **42**, 105 – 148 (1986)

S7.12 H. Risken, H.D. Vollmer: Methods for solving Fokker-Planck equations with applications to bistable and periodic potentials, in *Noise in Nonlinear Dynamical Systems; Theory, Experiment, Simulation*, Vol. I, ed. by F. Moss and P.V.E. McClintock (Cambridge Univ. Press, Cambridge 1989) pp. 191 – 224

Supplement 8

S8.1 M. Büttiker, E. Harris, R. Landauer: Thermal activation in extremly underdamped Josephson junction circuits. Phys. Rev. **B28**, 1268 – 1275 (1983)

S8.2 M. Büttiker, R. Landauer: Escape energy distribution for particles in an extremly underdamped potential well. Phys. Rev. **B30**, 1551 – 1553 (1984)

S8.3 M. Büttiker: Escape from the underdamped potential well, in *Noise in Nonlinear Dynamical Systems: Theory, Experiment, Simulation*, Vol. II, ed. by F. Moss, P.V.E. McClintock (Cambridge Univ. Press, Cambridge 1989) pp. 45 – 64

S8.4 D. Ryter: Noise-induced transitions in a double-well potential at low friction. J. Stat. Phys. **49**, 751 – 765 (1987)

S8.5 H. Risken, K. Vogel, H.D. Vollmer: Boundary-layer theory for the extremely underdamped Brownian motion in a metastable potential. IBM J. Res. Develop. **32**, 112 – 118 (1988)

S8.6 T.W. Marshall, E.J. Watson: A drop of ink falls from my pen ... it comes to earth, I know not when. J. Phys. **A18**, 3531 – 3559 (1985)

S8.7 T.W. Marshall, E.J. Watson: The analytic solutions of some boundary layer problems in the theory of Brownian motion. J. Phys. **A20**, 1345 – 1354 (1987)

S8.8 W. Renz: Derivation and solution of a low-friction Fokker-Planck equation for a bound Brownian particle. Z. Physik **B59**, 91 – 102 (1985)

Supplement 9

S9.1 P. Jung, H. Risken: Correlations functions and correlations times for models with multiplicative white noise. Z. Physik **B59**, 469–481 (1985)
S9.2 W. Nadler, K. Schulten: Mean relaxation time approximation for dynamical correlation functions in stochastic systems near instabilities. Z. Physik **B59**, 53–61 (1985)
S9.3 W. Nadler, K. Schulten: Mean relaxation time approximation for dynamical correlation functions in stochastic systems near instabilities: II. The single mode laser. Z. Physik **B72**, 535–543 (1988)
S9.4 M. San Miguel, L. Pesquera, M. A. Rodriguez, A. Hernández-Machado: Effective eigenvalue for the intensity correlations of single-mode and two-mode lasers. Phys. Rev. **A35**, 208–217 (1987)
S9.5 P. Jung, Th. Leiber, H. Risken: Dye laser model with pump and quantum fluctuations; white noise. Z. Physik **B66**, 397–407 (1987)

Supplement 10

S10.1 M. San Miguel, J. M. Sancho: Langevin equations with colored noise, in *Noise in Nonlinear Dynamical Systems: Theory, Experiment, Simulation*, Vol. I, ed. by F. Moss, P. V. E. McClintock (Cambridge Univ. Press, Cambridge 1989) pp. 110–160
S10.2 K. Lindenberg, B. J. West, J. Masoliver: First passage time problems for non-Markovian processes, in *Noise in Nonlinear Dynamical Systems: Theory, Experiment, Simulation*, Vol. I, ed. by F. Moss, P. V. E. McClintock (Cambridge Univ. Press, Cambridge 1989) pp. 110–160
S10.3 P. Grigolini: The projection approach to the Fokker-Planck equation: Applications to phenomenological stochastic equations with colored noises, in *Noise in Nonlinear Dynamical Systems: Theory, Experiment, Simulation*, Vol. I, ed. by F. Moss, P. V. E. McClintock (Cambridge Univ. Press, Cambridge 1989) pp. 161–190
S10.4 P. Hänggi: Noise in continuous dynamical systems: A functional calculus approach, in *Noise in Nonlinear Dynamical Systems: Theory, Experiment, Simulation*, Vol. I, ed. by F. Moss, P. V. E. McClintock (Cambridge Univ. Press, Cambridge 1989) pp. 307–328
P. Jung, P. Hänggi: Dynamical systems: A unified colored-noise approximation. Phys. Rev. **A35**, 4464–4466 (1987)
S10.5 R. F. Fox: Functional-calculus approach to stochastic differential equations. Phys. Rev. **A33**, 467–476 (1986)
R. F. Fox: Uniform convergence to an effective Fokker-Planck equation for weakly colored noise. Phys. Rev. **A34**, 4525–4527 (1986)
R. F. Fox: Mean first-passage times and colored noise. Phys. Rev. **A37**, 911–917 (1988)
S10.6 M. M. Klosek-Dygas, B. J. Matkowsky, Z. Schuss: Colored noise in dynamical systems. SIAM J. Appl. Math. **48**, 425–441 (1988)
M. M. Klosek-Dygas, B. J. Matkowsky, Z. Schuss: Uniform asymptotic expansions in dynamical systems driven by colored noise. Phys. Rev. **A38**, 2605–2613 (1988)
S10.7 K. Wódkiewicz: Matrix-continued fraction solutions of some stochastic equations with random telegraph noise. Z. Physik **B42**, 95–98 (1981)
K. Wódkiewicz, B. W. Shore, J. H. Eberly: Noise in strong laser-atom interactions: Frequency fluctuations and nonexponential correlations. Phys. Rev. **A30**, 2390–2398 (1984)
S10.8 F. Moss, P. V. E. McClintock: Analoge techniques for the study of problems in stochastic nonlinear dynamics, in *Noise in Nonlinear Dynamical Systems: Theory, Experiment, Simulation*, Vol. III, ed. by F. Moss, P. V. E. McClintock (Cambridge Univ. Press, Cambridge 1989) pp. 243–274
S10.9 J. M. Sancho, M. San Miguel, S. L. Katz, J. D. Gunton: Analytical and numerical studies of multiplicative noise. Phys. Rev. **A26**, 1589–1609 (1982)
S10.10 P. Jung, H. Risken: Distribution function for a nonlinear laser model with multiplicative colored noise. Phys. Lett. **103A**, 38–40 (1984)
S10.11 P. Jung, H. Risken: Motion in a double-well potential with additive colored Gaussian noise. Z. Physik **B61**, 367–379 (1985)
S10.12 F. Moss, P. V. E. McClintock: Measurements of two-dimensional densities for a bistable device driven by colored noise. Z. Physik **B61**, 381–386 (1985)
S10.13 T. Leiber, H. Risken: Decay rates in bistable Landau potentials driven by weakly colored Gaussian noise. Phys. Rev. **A38**, 3789–3791 (1988)

S10.14 T. Leiber, F. Marchesoni, H. Risken: Colored noise and bistable Fokker-Planck equations. Phys. Rev. Lett. **59**, 1381 – 1384 (1987)
ERRATUM: Phys. Rev. Lett. **60**, 659 (1988)

S10.15 T. Leiber, F. Marchesoni, H. Risken: Numerical analysis of stochastic relaxation in bistable systems driven by colored noise. Phys. Rev. **A38**, 983 – 993 (1988)

S10.16 P. Jung, P. Hänggi: Bistability and colored noise in nonequilibrium systems: Theory versus exact results. Phys. Rev. Lett. **61**, 11 – 14 (1988)

S10.17 T. Leiber, P. Jung, H. Risken: Dye laser model with pump and quantum fluctuations; colored noise. Z. Physik **B68**, 123 – 133 (1987)

S10.18 R. Roy, A.W. Yu, S. Zhu: Colored noise in dye laser fluctuations, in *Noise in Nonlinear Dynamical Systems: Theory, Experiment, Simulation*, Vol. III, ed. by F. Moss, P.V.E. McClintock (Cambridge Univ. Press, Cambridge 1989) pp. 90 – 158

S10.19 K. Vogel, H. Risken, W. Schleich, M. James, F. Moss, P.V.E. McClintock: Skewed probability densities in the ring laser gyroscope: A colored noise effect. Phys. Rev. **A35**, 463 – 465 (1987) Rapid Communications

S10.20 K. Vogel, H. Risken, W. Schleich, M. James, F. Moss, R. Mannella, P.V.E. McClintock: Colored noise in the ring laser gyroscope: Theory and simulation. J. Appl. Phys. **62**, 721 – 723 (1987)

S10.21 K. Vogel, H. Risken, T. Leiber, P. Hänggi, W. Schleich: Locking equation with colored noise: Continued fraction solution versus decoupling theory. Phys. Rev. **A35**, 4882 – 4885 (1987) Rapid Communications

S10.22 K. Vogel, H. Risken, W. Schleich: Noise in a ring-laser gyroscope, in *Noise in Nonlinear Dynamical Systems: Theory, Experiment, Simulation*, Vol. II, ed. by F. Moss, P.V.E. McClintock (Cambridge Univ. Press, Cambridge 1989) pp. 271 – 292

S10.23 J.F. Luciani, A.D. Verga: Functional integral approach to bistability in the presence of correlated noise. Europhys. Lett. **4**, 255 – 261 (1987)
J.F. Luciani, A.D. Verga: Bistability driven by correlated noise: Functional integral treatment. J. Stat. Phys. **50**, 567 – 597 (1988)

S10.24 G.P. Tsironis, P. Grigolini: Color induced transition to a non-conventional diffusion regime. Phys. Rev. Lett. **61**, 7 – 10 (1988)
G.P. Tsironis, P. Grigolini: Escape over a potential barrier in the presence of colored noise: Predictions of a local linearization theory. Phys. Rev. **A38**, 3749 – 3757 (1988)

S10.25 P. Hänggi, P. Jung, F. Marchesoni: Escape driven by strongly correlated noise. J. Stat. Phys. March (1989)

S10.26 C.R. Doering, P.S. Hagan, C.D. Levermore: Bistability driven by weakly colored Gaussian noise: Mean first-passage times. Phys. Rev. Lett. **59**, 2129 – 2132 (1987)

Supplement 11

S11.1 E.P. Wigner: On the quantum correction for thermodynamic equilibrium. Phys. Rev. **40**, 749 – 759 (1932)

S11.2 M. Hillery, R.F. O'Connell, M.O. Scully, E.P. Wigner: Distribution functions in physics: Fundamentals. Phys. Rep. **106**, 121 – 167 (1984)

S11.3 P. Drummond, D. Walls: Quantum theory of optical bistability. I: Nonlinear polarisability model. J. Phys. **A13**, 725 – 741 (1980)

S11.4 K. Vogel, H. Risken: Quantum-tunneling rates and stationary solutions in dispersive optical bistability. Phys. Rev. **A38**, 2409 – 2422 (1988)

S11.5 M. Dörfle, A. Schenzle: Bifurcations and the positive *P*-representation. Z. Physik **B65**, 113 – 131 (1986)

S11.6 K. Vogel, H. Risken: Solution of the quantum-Fokker-Planck equation for dispersive optical bistability in terms of matrix continued fractions. Opt. Commun. **62**, 45 – 48 (1987)

S11.7 H. Risken, K. Vogel: Expectation values, *Q*-functions and eigenvalues for dispersive optical bistability, in *Fundamentals of Quantum Optics*, Vol. II, ed. by F. Ehlotzky, Springer Lecture Notes Phys. **282** (Springer, Berlin, Heidelberg 1987) pp. 225 – 239

S11.8 H. Risken, K. Vogel: Eigenvalues of the Fokker-Planck equation for dispersive optical bistability, in *Optical Bistability, Instability and Optical Computing*, ed. by H.-Y. Zhang and K.K. Lee (World Scientific, Singapore 1988) pp. 136 – 147

S11.9 H. Risken, K. Vogel: Quantum treatment of dispersive optical bistability. Lecture Notes of the X Sitges Summerschool on "Far from Equilibrium Phase Transitions." Springer, to be published

S11.10 D.F. Walls: Squeezed states of light. Nature **306**, 141 – 146 (1983)

S11.11 R. Bonifacio (ed.): *Dissipative systems in quantum optics*. Topics Curr. Phys., Vol. 27 (Springer, Berlin, Heidelberg 1982)

S11.12 L.A. Lugiato: Theory of optical bistability, in *Progress in Optics XXI*, 69 – 216 (North-Holland, Amsterdam 1984)

S11.13 J.C. Englund, R.R. Snapp, W.C. Schieve: Fluctuations, instabilities and chaos in the laser-driven nonlinear ring cavity, in: *Progress in Optics XXI*, 355 – 428 (North-Holland, Amsterdam 1984)

S11.14 H.M. Gibbs, P. Mandel, N. Peyghambarian, S.D. Smith (eds.): *Optical Bistability* III, Springer Proc. Phys., Vol. 8 (Springer, Berlin, Heidelberg 1986)

S11.15 H. Risken, C. Savage, F. Haake, D.F. Walls: Quantum tunneling in dispersive optical bistability. Phys. Rev. **A35**, 1729 – 1739 (1987)

S11.16 H. Risken, K. Vogel: Quantum tunneling rates in dispersive optical bistability for low cavity damping. Phys. Rev. **A38**, 1349 – 1357 (1988)

S11.17 R. Graham, A. Schenzle: Dispersive optical bistability with fluctuations. Phys. Rev. **A23**, 1302 – 1321 (1981)

S11.18 M. Gronchi, L.A. Lugiato: Fokker-Planck equation for optical bistability. Lett. Nuovo Cim. **23**, 593 – 598 (1978)

S11.19 H.P. Yuen, P. Tombesi: Langevin equations with negative diffusion coefficients. A new approach to quantum optics. Opt. Commun. **59**, 155 – 159 (1986)

S11.20 W. Tan, Y. Li, W. Zhang: The solution of the Fokker-Planck equation with zero or negative diffusion coefficients in quantum optics. Opt. Commun. **64**, 195 – 199 (1987)

Subject Index